Lecture Notes in Mathematics

Volume 2273

T0171996

More information about this series at http://www.springer.com/series/304

Chaire Jean-Morlet

The CIRM Jean-Morlet Series is a collection of scientific publications centering on the themes developed by successive holders of the Jean Morlet Chair.

This chair has been hosted by the *Centre International de Rencontres Mathématiques* (CIRM, Luminy, France) since its creation in 2013. The Chair is named in honour of Jean Morlet (1931–2007). He was an engineer at the French oil company Elf (now Total) and, together with the physicist Alex Grossman, conducted pioneering work in wavelet analysis. This theory has since become a building block of modern mathematics. It was at CIRM that they met on several occasions, and the center then played host to some of the key conferences in this field.

Appointments to the *Jean-Morlet* Chair are made to world-class researchers based outside France and who work in collaboration with local project leaders in order to conduct original and ambitious scientific programs. The Chair is supported financially by CIRM, Aix-Marseille Université and the City of Marseille.

A key feature of the Chair is that it does not focus solely on the research themes developed by Jean Morlet. The idea is to support the freedom of pioneers in mathematical sciences and to nurture the enthusiasm that comes from opening new avenues of research.

CIRM: a beacon for international cooperation
Situated at the heart of the *Parc des Calanques*, an area of outstanding natural beauty, CIRM is one of the largest conference centers dedicated to mathematical and related sciences in the world, with close to 3500 visitors per year. Jointly supervised by SMF (the French Mathematical Society) and CNRS (French National Center for Scientific Research), CIRM has been a hub for international research in mathematics since 1981. CIRM's *raison d'être* is to be a venue that fosters exchanges, pioneering research in mathematics in interaction with other sciences and the dissemination of knowledge to the younger scientific community.

www.chairejeanmorlet.com
www.cirm-math.fr

Shigeki Akiyama • Pierre Arnoux

Editors

Substitution and Tiling Dynamics: Introduction to Self-inducing Structures

CIRM Jean-Morlet Chair, Fall 2017

 Springer

Editors
Shigeki Akiyama (iD)
Institute of Mathematics
University of Tsukuba
Tsukuba, Japan

Pierre Arnoux
Institut de Mathématiques de
Marseille (I2M)
Aix-Marseille University
Marseille Cedex 09, France

ISSN 0075-8434   ISSN 1617-9692   (electronic)
Lecture Notes in Mathematics
ISBN 978-3-030-57665-3   ISBN 978-3-030-57666-0   (eBook)
https://doi.org/10.1007/978-3-030-57666-0

Jointly published with Société Mathématique de France (SMF), Paris, France

Mathematics Subject Classification: 37B10, 05B45, 37B50, 68R15, 52C23

This Springer imprint is published by the registered company Springer Nature Switzerland AG.
The registered company address is: Gewerbestrasse 11, 6330 Cham, Switzerland

Preface

Tilings have been drawn and studied for centuries, in art and science, from the Sumerian patterns, Roman mosaics, Alhambra wall tilings (see Fig. 1), and the attempts of Johannes Kepler to tile the plane with fivefold symmetric patterns (see Fig. 2), a goal which cannot be realized by a periodic tiling. These attempts were one of the inspirations for the now-classic fivefold Penrose tiling, see the Foreword of Ref. [7] in Chap. 7.

The study of periodic tilings and their symmetry groups, the so-called crystallographic group, was developed to understand the structure of physical crystals; detailed account on the classification of periodic tilings and their symmetries can be found in the book by Grünbaum–Shephard (See Ref. [58] in Chap. 2), which still contains a lot of interesting open questions and attracts many researchers. The book

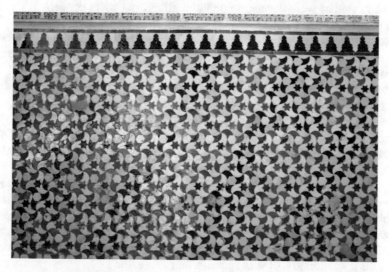

Fig. 1 One of the many Alhambra wall tilings. Source: https://commons.wikimedia.org/wiki/File: Tassellatura_alhambra.jpg

Fig. 2 Kepler Aa tiling: an
attempt to tile the plane with
fivefold symmetric patterns.
Source: https://gallica.bnf.fr/
ark:/12148/btv1b26001687/
f4.item

also contains many examples of aperiodic tilings: however, they wrote at the time
*"Unlike the material in the earlier sections of the book, many of our assertions here
are **not** supported by published proofs"* because the subject was in its infancy when
[58] was published.

If early studies were devoted to periodic tilings, our main concern in this book
is the presentation of several recent developments in the study of aperiodic tilings.
One could say that the relation of aperiodic tilings to periodic tilings is similar to the
relation of irrational numbers to rational numbers and similarly opens a vast range
of new phenomena.

The modern theory of aperiodic tilings started in the beginning of the 1960s,
with the proof by Berger of the undecidability of the Domino problem, using
aperiodic tilings, followed by the invention of the Penrose tiling, the Heighway
dragon, and the Knuth–Davis twindragon, soon followed by the discovery of quasicrystals and the study of tiling spaces which are one of the basic examples of
non-commutative geometry. Self-similar tilings, like the Fibonacci and Penrose
tilings, were a fundamental element of this development. They have known many
generalizations, and the recent years have seen a number of advances in the domain.

This book presents lecture notes delivered at the research school *Tiling dynamical
systems*, one of the Morlet Chair events organized in the second semester of 2017 as
part of the program *Tilings and Discrete Geometry* organized by Shigeki Akiyma
during his invitation to CIRM as Morlet Chair Professor.

The simplest aperiodic tilings are defined by substitutions; there are various types
of substitutions (these various types, and their generalizations, are a central theme

in this book), but in the simplest form, a substitution consists of replacing each letter of a finite alphabet by a word on this alphabet; the first one is probably the Fibonacci substitution, on the alphabet $\{a, b\}$, which replaces a by ab and b by a. To this substitution, one can associate infinite words which can be infinitely recoded by first decomposing them in the words a and ab (which implies that bb does not appear) and then replacing ab by a and a by b. The set of all these words is invariant by the shift, and this defines an interesting discrete dynamical system associated with the substitution, a *substitutive dynamical system* (see Fig. 3 for the associated self-similar tiling). Tiling dynamical systems are continuous-time analogues of the substitutive dynamical systems; these are tilings of the line (or the plane or space) by a finite number of shapes (the tiles) that can be either subdivided (*inflation*) in smaller similar tiles or regrouped (*deflation*) in larger similar tiles (see Fig. 4). This leads to remarkable structures, like the celebrated Penrose tiling, with "self-inducing" properties: they contain smaller and larger copies of themselves.

The main object of the research school, as well as of the whole semester, was to study these self-inducing systems and tiling dynamical systems, and their

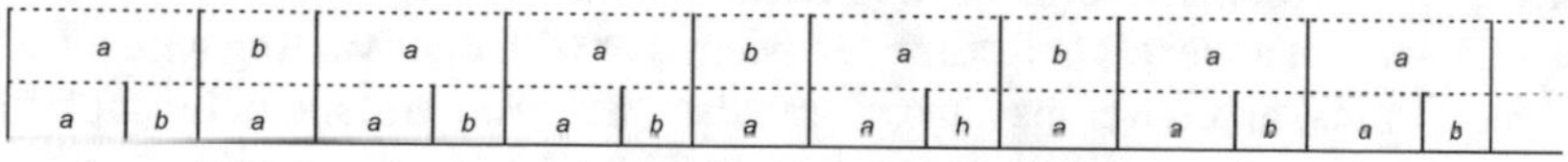

a		b	a		a	b	a		b	a		a	
a	b	a	a	b	a	b	a	a	h	a	a	b	a	b

Fig. 3 The Fibonacci tiling, associated with the substitution $a \mapsto ab, b \mapsto a$

Fig. 4 Rauzy dragon: an example of a pseudo-self-similar structure, generated by a substitution rule, or alternatively by a fusion rule. Source: courtesy of T. Fernique

generalizations, from many different points of views and to increase research interactions among related people. The semester provided excellent opportunities for research discussion among people working on discrete geometry, number theory, fractal geometry, theoretical computer science, and dynamical systems.

The book consists of eight chapters, the first six of which are expanded lecture notes, and the last two of which are selected contributions.

Chapter 1, by B. Solomyak, is an introduction to the domain of tilings. It starts by defining the fundamental notions of Delone sets and Meyer sets, the Delone sets with inflation symmetry and their number-theoretic properties (in relation with Pisot and Perron numbers). It then studies the related substitution tilings and their associated dynamical systems, clarifying the notions of self-affine and pseudo-self-affine tilings and their properties. The last section presents some developments (infinite local complexity, pure discrete spectrum, and Fourier quasi-crystals) and some open problems.

Chapter 2, by N. P. Frank, deals with the various models of tilings, from the simplest, discrete tilings in one dimension (symbolic systems), to continuous tilings of the line, discrete systems in higher dimension ($\mathbb{Z}^d$ systems), and tilings in higher dimension. It presents several ways to build systems with a hierarchical structure, either the same at all levels (substitutive systems) or with rules varying with the level (S-adic systems) and develops a new formalism, the fusion rule to build supertiles, which regroups several different ways to build recurrent tilings with interesting hierarchical properties; the properties of these tilings are studied from the dynamical and the diffraction (Fourier analysis) perspectives, and this chapter also contains a historical introduction of the spectral study of tiling generated by substitution.

Chapter 3, by J.Thuswaldner, discusses in depth S-adic systems and their geometric realizations. S-adic symbolic systems are a natural generalization of substitutive systems. They are systems which are generated by an infinite sequence of substitutions belonging to a finite set S. The most elementary case is that of a rotation on a circle of unit length; a rotation by an irrational quadratic number with periodic continued fraction has very particular diophantine properties (this is a simple example of a self-induced dynamical system) and is the geometric realization of a substitutive system determined by this periodic continued fraction. A circle rotation by an irrational number, with an infinite nonperiodic continued fraction is infinitely renormalizable, but not self-induced, and it is a geometric realization of a corresponding S-adic system. Is the same thing true in higher dimension, for toral rotations? In fact, it is almost always true in higher dimension. This is shown, in a precise sense, by using properties of some generalized continued fractions.

The study of tiling dynamics is historically motivated by quasi-crystals: a real material having long-range order but no translational periods. The recurrence properties of quasi-crystals have been studied for a long time, in relation to the diffraction spectrum and the spectral properties of the corresponding Schrödinger operators. In pursuing these directions, topological properties of tiling dynamics play an essential role. They determine, for instance, the labelling of the gaps in the spectrum. The study of the topological invariants involves elements of non-commutative geometry as developed by Connes. One of the first examples of non-commutative spaces he proposed, the space of all Penrose tilings foliated by the translation action, stands at the beginning of the developments described in Chap. 4. In this chapter, J. Kellendonk gives an account on the construction of spaces, dynamical systems and algebras for tilings, and how their topological invariants help to understand the topological properties of the underlying material.

Chapter 5, by M. Rigo, deals with an apparently very different domain, that of combinatorial games. These games provide an original introduction to the historical relations between tilings and language theory, logics, and computer science. The best-known combinatorial game is certainly the Nim game, where each player takes out tokens from one or several piles of tokens, and the winner is the player who takes the last token. Since the game is deterministic, there are winning and losing positions; it turns out that this set of winning positions has a remarkable structure, which can be determined by a finite automaton. It is related to substitutive sequences, numeration systems, Pisot numbers, and to higher dimensional extensions associated with tilings of $\mathbb{N}^d$ having interesting properties.

The birth of the modern theory of tilings is associated with the famous proof by Berger of the undecidability of the Domino problem. In Chap. 6, E. Jeandel and P. Vanier revisit this proof and give us a modern view of this problem, including the necessary foundations on automata, Turing machines, and shift spaces. They carefully expose four different proofs of the undecidability result. A crucial point of any proof of the undecidability of the Domino problem is to exhibit a set of tiles which can tile the plane, but only in an aperiodic way, and the authors emphasize on the various constructions of these sets of tiles.

To a tiling space, one can associate a dynamical spectrum, coming from the $\mathbb{R}^d$-action by translation, which allows to define a spectral measure; one can also define a diffraction measure, coming initially from physics, with close connections to the spectral measure; these two types of spectra already appear in the first two chapters. Chapter 7 of the book, contributed by M. Baake and U. Grimm, deals with the properties of these measures, and in particular with the question of the existence of a nontrivial continuous or singular component. An essential element to answer this question is the study of autocorrelation and the pair correlation function. Using this technique, conditions are given to ensure that the diffraction measure is singular.

The last chapter, by P. Mercat and S. Akiyama, deals with the Pisot substitution conjecture, one of the main open problems in the field of substitutive tilings. This conjecture, which has several forms and generalizations, says that any substitution dynamical system of irreducible Pisot type has pure discrete spectrum, see Sect. 2.6.3.1 for more details (by contrast, when the inflation factor is not a Pisot number, complicated behaviour such as infinite local complexity become possible, see Fig. 5, and also Fig. 2.6 of Chap. 2); it occurs in several chapters of this book. Chapter 8 contains a noteworthy simple new characterization of subshifts with a discrete spectrum, which is basically checkable by automata computation; this gives an algorithm to verify that a given subshift of Pisot type satisfies the Pisot conjecture. This characterization is shown to be an equivalence for subshifts generated by irreducible Pisot substitutions and applies as well to S-adic systems.

Fig. 5 The Frank–Robinson tiling: an example of tiling with a non-Pisot inflation factor and infinite local complexity

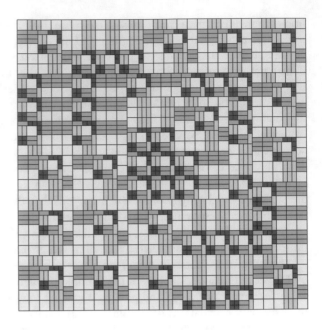

We thank the lecturers for giving a series of introductory talks at CIRM as well as providing us detailed expositions of this developing area. We believe that this book will give a nice access to these subjects and plenty of research directions to the related researchers.

Tsukuba, Japan   Shigeki Akiyama
Marseille, France   Pierre Arnoux
January–July 2020

Contents

4 Operators, Algebras and Their Invariants for Aperiodic Tilings 193
Johannes Kellendonk

Chapter 1
Delone Sets and Dynamical Systems

Boris Solomyak

Abstract In these notes we focus on selected topics around the themes: Delone sets as models for quasicrystals, inflation symmetries and expansion constants, substitution Delone sets and tilings, and associated dynamical systems.

1.1 Introduction

This is an expository article, based on a 1.5-hour talk delivered at the School on Tilings and Dynamics, held in CIRM, Luminy (November 2017). There are no new results here; most of the content is at least 10 years old. The goal of the talk was to supplement the lectures of N. P. Frank [16], and the notes are written in the same spirit. There is, of course, too much material here for a single lecture, but we tried to present at least a somewhat complete picture of the subject, from a personal viewpoint. Two themes that we tried to emphasize are (a) the characterization of expansion symmetries, and (b) the duality of substitution Delone sets and substitution tilings.

In Sects. 1.2–1.4 we develop the notions of Delone sets and their classification, as well as characterization of (scalar) inflation symmetries. We largely follow the paper by J. C. Lagarias [37], although many original ideas are due to Y. Meyer [49, 50]. Some relatively easy statements are proved completely; other proofs are sketched, and still others are referred to the original papers. Several examples are worked out in detail to illustrate the results.

Section 1.5 is central for us, although it contains almost no proofs. Here we develop the notions of substitution Delone sets and substitution tilings and discuss the relation between them. In particular, *representable* substitution Delone sets (or rather, "multi-color" Delone sets, or m-sets) are introduced. Much of the section is based on the paper by J. C. Lagarias and Y. Wang [39], with a few important

B. Solomyak (✉)
Bar-Ilan University, Ramat-Gan, Israel

© The Editor(s) (if applicable) and The Author(s), under exclusive license
to Springer Nature Switzerland AG 2020
S. Akiyama, P. Arnoux (eds.), *Substitution and Tiling Dynamics: Introduction to Self-inducing Structures*, Lecture Notes in Mathematics 2273,
https://doi.org/10.1007/978-3-030-57666-0_1

additions from [40]. We include a discussion of a theorem about pseudo-self-affine tilings being mutually locally derivable with self-affine tilings, following [60], since this is a nice application of the duality between substitution Delone sets and tilings.

Section 1.6 is devoted to dynamical systems arising from tilings and Delone sets. Since this is one of the main themes of [16], we have limited ourselves to a very brief exposition, highlighting the aspects related to the topics developed in earlier sections, specifically, the number-theoretic issues arising from the description of eigenvalues and the Meyer property.

The last Sect. 1.7 contains some general remarks and references on several topics not included in the main text: infinite local complexity, pure discrete spectrum, and Fourier quasicrystals. We conclude with a short list of open problems on the characterization of expansion maps for tilings with inflation symmetry.

1.2   Delone Sets of Finite Type

Here we borrow much from the paper [37].

Definition 1.2.1 A discrete set X in $\mathbb{R}^d$ is a *Delone set* if it is

(a) *Uniformly discrete:* there exists $r > 0$ such that every open ball of radius r contains at most one point of X; equivalently, the distance between distinct points in X is at least $2r$.
(b) *Relatively dense:* there exists $R > 0$ such that every closed ball of radius R contains at least one point of X.

Delone sets are also sometimes called *separated nets*. The notion of Delone sets as fundamental objects of study in crystallography was introduced by the Russian school in the 1930s; in particular, by Boris Delone (or Delaunay). One can think about a Delone set as an idealized model of an atomic structure of a material without "holes". This is obviously too general to be considered an "ordered" structure, so we impose some conditions on the set.

For a set $X \subset \mathbb{R}^d$ denote by $[X]$ the abelian group (equivalently, $\mathbb{Z}$-module) generated by X. Recall that $[X] := \left\{\sum_{i=1}^{k} n_i x_i : k \in \mathbb{N}, \ n_i \in \mathbb{Z}\right\}$. The following classes of Delone sets have been studied:

Definition 1.2.2 Let X be a Delone set in $\mathbb{R}^d$.

(a) X is *finitely generated* if $[X - X]$ (equivalently, $[X]$) is a finitely generated abelian group.
(b) X is of *finite type* if $X - X$ is a discrete closed subset of $\mathbb{R}^d$; i.e., the intersection of $X - X$ with any ball is a finite set.

In the next section we will add to these the notion of *Meyer set*.

Exercise 1.2.3 (Easy) Give an example of a finitely generated Delone set which is not of finite type.

We will see many such examples in Sect. 1.3.

Exercise 1.2.4 (Easy) Show that a Delone set X is of finite type if and only if it has *finite local complexity* (FLC), that is, for any $\rho > 0$ there are finitely many "local clusters" $X \cap B(\mathbf{y}, \rho)$, up to translation.

We will prove that every Delone set of finite type is finitely generated, but first we introduce the important notion of *address map*. Recall that a finitely generated subgroup of $\mathbb{R}^d$ is free. We can choose a free basis, its cardinality is independent of the choice of basis; it is called the *rank* of the free group.

Definition 1.2.5 Let X be a finitely generated Delone set. Choose a basis of $[X]$, say,

$$[X] = [\mathbf{v}_1, \mathbf{v}_2, \ldots, \mathbf{v}_s].$$

The address map $\phi : [X] \rightarrow \mathbb{Z}^s$ associated to this basis is

$$\phi\left(\sum_{i=1}^{s} n_i \mathbf{v}_i\right) = (n_1, \ldots, n_s).$$

The address map for the Delone set of control points of a self-similar tiling has been used by Thurston [64] (we describe it in Sect. 1.4). The address map is not unique, it depends on the choice of a basis, so it is determined up to left-multiplication by an element of $GL(s, \mathbb{Z})$. Observe that $s \geq d$ since the linear span of a Delone set is obviously the entire $\mathbb{R}^d$. If $s = d$, then X is a subset of a *lattice*; this is a special case—usually, $s > d$. Quoting J. Lagarias [37]:

> The structure of a finitely generated Delone set is to some extent analyzable by studying its image in $\mathbb{R}^s$ under the address map. The address map describes X using s "internal dimensions."

Theorem 1.2.6 (See [37]) *For a Delone set X in $\mathbb{R}^d$, the following properties are equivalent:*

(i) *X is a Delone set of finite type.*
(ii) *X is finitely generated and any address map $\phi : [X] \rightarrow \mathbb{Z}^s$ is **globally Lipshitz on X**:*

$$\|\phi(\mathbf{x}) - \phi(\mathbf{x}')\| \leq C_0 \|\mathbf{x} - \mathbf{x}'\|, \quad \textit{for all } \mathbf{x}, \mathbf{x}' \in X,$$

for some constant C_0 depending on ϕ.

***Proof** (Sketch)* (ii) $\Rightarrow$ (i). If $\mathbf{x}, \mathbf{x}' \in X$ are such that $\|\mathbf{x} - \mathbf{x}'\| \leq T$, then by hypothesis,

$$\|\phi(\mathbf{x} - \mathbf{x}')\| = \|\phi(\mathbf{x}) - \phi(\mathbf{x}')\| \leq C_0 T.$$

Addresses all lie in $\mathbb{Z}^s$, hence there are only finitely many choices for $\phi(\mathbf{x} - \mathbf{x}')$, hence for $\mathbf{x} - \mathbf{x}'$, since $\phi : X \to \mathbb{Z}^s$ is one-to-one. It follows that $(X - X) \cap B(\mathbf{0}, T)$ is finite, so X is of finite type. $\qquad\square$

Lemma 1.2.7 (See [37]) *Let X be a Delone set in $\mathbb{R}^d$ with parameters (r, R). Then there exist constants $k > 1$ and $C = C(R, r)$, such that given any two points $\mathbf{x}, \mathbf{x}' \in X$ there is a chain of points*

$$\mathbf{x} = \mathbf{x}_0, \mathbf{x}_1, \mathbf{x}_2, \ldots, \mathbf{x}_m = \mathbf{x}'$$

in X such that

(a) $\|\mathbf{x}_i - \mathbf{x}_{i-1}\| \le kR$ *for all i;*
(b) $m \le C\|\mathbf{x} - \mathbf{x}'\|$.

Lagarias [37] proved that one can take $k = 2$ and $C(R, r) = 4R/r^2$. We give a simple proof for $k = 4$ and $C(R, r) = (2R)^{-1} + r^{-1}$.

***Proof** (Lemma 1.2.7)* Let $m = \lfloor \frac{\|\mathbf{x} - \mathbf{x}'\|}{2R} \rfloor + 1$ and consider the points $\mathbf{x} = \mathbf{x}'_0, \mathbf{x}'_1, \ldots, \mathbf{x}'_{m-1}, \mathbf{x} = \mathbf{x}'_m$ at equal distance $\frac{\|\mathbf{x} - \mathbf{x}'\|}{m} \le 2R$ from each other. By the (r, R)-property, we can choose $\mathbf{x}_i \in X$, for $i = 1, \ldots, m - 1$, such that $\|\mathbf{x}_i - \mathbf{x}'_i\| \le R$. Then $\|\mathbf{x}_i - \mathbf{x}_{i+1}\| \le R + 2R + R = 4R$ for all $i = 0, \ldots, m - 1$. It remains to note that

$$m \le \frac{\|\mathbf{x} - \mathbf{x}'\|}{2R} + 1 \le \frac{\|\mathbf{x} - \mathbf{x}'\|}{2R} + \frac{\|\mathbf{x} - \mathbf{x}'\|}{r},$$

and the proof is complete. $\qquad\square$

Now we can deduce the implication (i) $\Rightarrow$ (ii). By Lemma 1.2.7, $[X - X]$ is generated by $\{\mathbf{x} - \mathbf{x}' : \|\mathbf{x} - \mathbf{x}'\| \le kR\}$, which is finite; thus X is finitely generated. Let $\phi : [X] \to \mathbb{Z}^s$ be an address map. Let

$$C_1 := \max\{\|\phi(y)\| : y \in (X - X) \cap B(\mathbf{0}, kR)\}.$$

Given $\mathbf{x}, \mathbf{x}'$, by Lemma 1.2.7 we have a chain in X connecting $\mathbf{x}$ to $\mathbf{x}'$ which satisfies $\|\mathbf{x}_i - \mathbf{x}_{i+1}\| \le kR$, and the number of points is at most $C\|\mathbf{x} - \mathbf{x}'\|$. Using $\mathbb{Z}$-linearity of ϕ on $[X]$, we can write

$$\|\phi(\mathbf{x}) - \phi(\mathbf{x}')\| \le \sum_{i=1}^{m} \|\phi(\mathbf{x}_i) - \phi(\mathbf{x}_{i-1})\|$$

$$= \sum_{i=1}^{m} \|\phi(\mathbf{x}_i - \mathbf{x}_{i-1})\|$$

$$\le C_1 m \le C_1 C \|\mathbf{x} - \mathbf{x}'\|,$$

which proves (ii). $\qquad\square$

1.3 Meyer Sets

Meyer sets were introduced by Meyer [49] in 1972 in the form of "relatively dense harmonious sets," but they remained in obscurity for a long time (see the fascinating interview [14] with Yves Meyer on the occasion of the 2017 Abel Prize awarded to him). The paper [50] introduced them to the quasicrystal community under the name "quasicrystal". Now they are generally called Meyer sets, see [36, 51].

Definition 1.3.1 Let X be a Delone set in $\mathbb{R}^d$. It is called a *Meyer set* if the self-difference set $X - X$ is uniformly discrete, equivalently, a Delone set.

Actually, this was not the original definition. An interesting feature of Meyer sets is that they can be characterized in seemingly very different terms: using discrete geometry, almost linear mappings, and cut-and-project sets.

Definition 1.3.2 Let $\mathbb{R}^n = E^d \oplus H$, where $E^d \approx \mathbb{R}^d$ (the "physical space") and $H \approx \mathbb{R}^m$ (the "internal space") are linear subspaces of $\mathbb{R}^n$, not necessarily orthogonal. Let π and π_{int} be the projection onto E^d parallel to H, and the projection to H parallel to E^d, respectively. A *window* Ω is a bounded open subset of H. Let Γ be a full rank lattice in $\mathbb{R}^n$ (that is, Γ is a discrete subgroup of rank n; equivalently, it is a subgroup which forms a Delone set). The *cut-and-project set* $X(\Gamma, \Omega)$ associated to the data (Γ, Ω) is

$$X(\Gamma, \Omega) = \pi\big(\{\mathbf{w} \in \Gamma : \pi_{\mathrm{int}}(\mathbf{w}) \in \Omega\}\big).$$

A cut-and-project set is called *nondegenerate* if π is one-to-one on Γ. It is *irreducible* if $\pi_{\mathrm{int}}(\Gamma)$ is dense in H. Cut-and-project sets (sometimes with different requirements for the window) are also called *model sets*.

Theorem 1.3.3 (Y. Meyer, J. Lagarias, see [37]) *For a Delone set X in $\mathbb{R}^d$, the following properties are equivalent:*

(i) *X is a Meyer set, that is, $X - X$ is a Delone set.*
(ii) *there is a finite set F such that $X - X \subseteq X + F$.*
(iii) *X is a finitely generated Delone set and the address map $\phi : [X] \to \mathbb{Z}^s$ is an almost linear mapping, i.e., there is a linear map $L : \mathbb{R}^d \to \mathbb{R}^s$ and $C > 0$ such that*

$$\|\phi(\mathbf{x}) - L\mathbf{x}\| \le C \ \text{ for all } \mathbf{x} \in X.$$

(iv) *X is a subset of a non-degenerate cut-and-project set.*

The theorem is mostly due to Meyer, except that the implication (i) $\Rightarrow$ (ii) was proved by Lagarias [36]. R. V. Moody [51] did much to popularize the concept; his article [51] contains other interesting characterizations of Meyer sets, e.g., in terms of Harmonic Analysis and a certain kind of "duals". For a more complete characterization of Meyer sets, see [63] by N. Strungaru.

Proof *(Partial Sketch)* (ii) $\Rightarrow$ (iii) First note that X is of finite type. Indeed, $(X - X) \cap B(\mathbf{0}, N) \subset (X + F) \cap B(\mathbf{0}, N)$ is finite for any $N > 0$. Thus it is finitely generated, and we can consider the address map ϕ. We construct $L : \mathbb{R}^d \to \mathbb{R}^s$ as an "ideal address map," as follows. For each $\mathbf{y} \in \mathbb{R}^d$ define

$$L(\mathbf{y}) := \lim_{k \to \infty} \frac{\phi(\mathbf{x}_k)}{2^k},$$

where $\mathbf{x}_k \in X$ satisfies $\|\mathbf{x}_k - 2^k \mathbf{y}\| \le R$. Using (ii) one proves that this limit exists and is unique (independent of the choice of $\mathbf{x}_k$), and is linear, roughly along these lines: By the definition of $\mathbf{x}_k, \mathbf{x}_{k+1}$ and the triangle inequality we have $\|2\mathbf{x}_k - \mathbf{x}_{k+1}\| \le 3R$. But

$$2\mathbf{x}_k - \mathbf{x}_{k+1} = \mathbf{x}_k - (\mathbf{x}_{k+1} - \mathbf{x}_k) = \mathbf{x}_k - (\mathbf{z} + \mathbf{w}),$$

for some $\mathbf{z} \in X$ and $\mathbf{w} \in F$ (the finite set from (ii)). Thus,

$$\|\mathbf{x}_k - \mathbf{z}\| \le 3R + C_1, \quad \text{where } C_1 = \max_{\mathbf{u} \in F} \|\mathbf{u}\|.$$

Since ϕ is $\mathbb{Z}$-linear on X, we have

$$2\phi(\mathbf{x}_k) - \phi(\mathbf{x}_{k+1}) = [\phi(\mathbf{x}_k) - \phi(\mathbf{z})] - \phi(\mathbf{w}),$$

hence

$$\|2\phi(\mathbf{x}_k) - \phi(\mathbf{x}_{k+1})\| \le \|\phi(\mathbf{x}_k) - \phi(\mathbf{z})\| + \|\phi(\mathbf{w})\|$$

$$\le C_0(3R + C_1) + C_2 =: C'.$$

where $C_2 = \max_{\mathbf{u} \in F} \|\phi(\mathbf{u})\|$ and C_2 is from Theorem 1.2.6. Therefore,

$$\left\| \frac{\phi(\mathbf{x}_k)}{2^k} - \frac{\phi(\mathbf{x}_{k+1})}{2^{k+1}} \right\| \le C'/2^{k+1},$$

and convergence follows. Now it is not difficult to check linearity of L. For $\mathbf{x} = \mathbf{y} \in X$ we can take $\mathbf{x}_0 = \mathbf{x}$ and, summing the geometric series, obtain

$$\|\phi(\mathbf{x}) - L(\mathbf{x})\| \le C'.$$

This shows almost linearity of the address map.

(iii) $\Rightarrow$ (i) It is clear that $X - X$ is relatively dense, so we only need to show that it is uniformly discrete. Equivalently, that there is a lower bound on the norm of $\mathbf{z} \in (X - X) - (X - X)$, whenever $\mathbf{z} \ne \mathbf{0}$. Suppose that $\|\mathbf{z}\| \le R$. By the hypothesis (iii), we have $\|\phi(\mathbf{z}) - L\mathbf{z}\| \le 4C$, since ϕ is $\mathbb{Z}$-linear on $[X]$ and L is linear on $\mathbb{R}^d$.

Therefore,

$$\|\phi(\mathbf{z})\| \leq 4C + \|L\|R.$$

Since ϕ is one-to-one on $[X]$ and $\phi(\mathbf{z}) \in \mathbb{Z}^s$, we see that there are only finitely many possibilities for $\mathbf{z}$. The desired claim follows.

For the equivalence of (iv) and (i)–(iii), see [37] or [51]. The direction (iv) $\Rightarrow$ (i) is simple: one needs to verify that a relatively dense subset of a cut-and-project set is Delone, then note that for a lattice Λ and a window Ω we have $X(\Lambda, \Omega) - X(\Lambda, \Omega) \subseteq X(\Lambda, \Omega - \Omega)$.

The direction (iii) $\Rightarrow$ (iv) proceeds by taking the physical space $E^d = L(\mathbb{R}^d) \subset \mathbb{Z}^s$ and a complementary subspace H as the internal subspace. We then "lift" X to $\mathbb{Z}^s$ via the address map (which is, of course, a lattice in $\mathbb{R}^s$), project it onto H and note that the projection is bounded by the property (iii). We can then choose any bounded open set containing the projection as a window. See [37] for details. $\square$

1.4 Inflation Symmetries

This section is largely based on [37].

Definition 1.4.1 A Delone set X in $\mathbb{R}^d$ has an inflation symmetry by the real number $\eta > 1$ if $\eta X \subseteq X$.

Recall that a (complex) number η is an algebraic integer if $p(\eta) = 0$ for some monic polynomial $p \in \mathbb{Z}[x]$, that is, $p(x) = x^s + \sum_{j=0}^{s-1} c_j x^j$, with $c_j \in \mathbb{Z}$. The degree of η is the minimal degree of $p(x) \in \mathbb{Z}[x]$ such that $p(\eta) = 0$. The algebraic (or Galois) conjugates of η are the roots of the minimal polynomial for η, other then η. (Although in Galois theory, η itself is sometimes included in the list of conjugates, for us it is more convenient to exclude it.) Several classes of algebraic integers appear:

Definition 1.4.2 Let η be a real algebraic integer greater than one.

(a) η is a *Pisot number* or *Pisot-Vijayaraghavan (PV)*-number if all algebraic conjugates satisfy $|\eta'| < 1$.
(b) η is a *Salem number* if for all conjugates $|\eta'| \leq 1$ and at least one satisfies $|\eta'| = 1$.
(c) η is a *Perron number* if for all conjugates $|\eta'| < \eta$.
(d) η is a *Lind number* if for all conjugates $|\eta'| \leq \eta$ and at least one satisfies $|\eta'| = \eta$.

"Lind numbers" were introduced by Lagarias [37], but this apparently did not become standard terminology.

Definition 1.4.3 Let X be a Delone set in $\mathbb{R}^d$. Finite subsets of X are called *X-clusters*. The Delone set X in $\mathbb{R}^d$ is called *repetitive* if for any $T > 0$ there exists

$M_X(T) > 0$ such that every ball of diameter $M_X(T)$ contains a translated copy of every X-cluster of diameter T.

Theorem 1.4.4 (J. Lagarias [37], Y. Meyer [49]) *Let X be a Delone set in $\mathbb{R}^d$ such that $\eta X \subseteq X$ for a real number $\eta > 1$.*

(i) *If X is finitely generated, then η is an algebraic integer.*
(ii) *If X is a Delone set of finite type, then η is a Perron number or a Lind number.*
(ii′) *If X is repetitive Delone set of finite type, then η is a Perron number.*
(iii) *If X is a Meyer set, then η is a Pisot number or a Salem number.*

Proof *(Partial Sketch)*

(i) We will prove a more general statement:

$\square$

Lemma 1.4.5 *Let X be a finitely generated Delone set in $\mathbb{R}^d$ such that $QX \subseteq X$ for some expanding linear map Q. Then all eigenvalues of Q are algebraic integers.*

Proof of Lemma Since X is finitely generated, it can be written $[X] = [\mathbf{v}_1, \ldots, \mathbf{v}_s]$ for some free generators and the address map $\phi : [X] \to \mathbb{Z}^s$. Since $QX \subseteq X$, we also have $Q([X]) \subseteq [X]$. It follows that $Q\mathbf{v}_j$ is an integer linear combination of the vectors $\mathbf{v}_j$ (recall that these are the free generators of $[X]$). Define the matrix $V = [\mathbf{v}_1, \ldots, \mathbf{v}_s]$ of size $d \times s$. We thus obtain an integer matrix M of size $s \times s$ such that

$$QV = VM. \qquad (1.1)$$

It is clear that $\{\mathbf{v}_j\}_{j \leq s}$ spans $\mathbb{R}^d$, because X does, hence $\mathrm{rank}(V) = d$. Let $\mathbf{e}$ be a left eigenvector of Q corresponding to an eigenvalue λ. Then $\lambda \mathbf{e} V = \mathbf{e} QV = \mathbf{e} VM$. Notice that $\mathbf{e} V$ is not zero, because the rows of V are linearly independent. Thus $\mathbf{e} V$ is an eigenvector for M corresponding to λ. But M is an integer matrix, so all its eigenvalues are algebraic integers. $\qquad \square$

(ii) Let γ be a conjugate of η. We continue the argument of the lemma. Since η is an eigenvalue of the integer matrix M, so is γ. Thus there is an eigenvector $\mathbf{e}_\gamma \in \mathbb{R}^s$ corresponding to γ. We want to prove that $|\gamma| \leq \eta$. Let ϕ be the address map as in the proof of the lemma. Then $\{\phi(\mathbf{v}_j)\}_{j \leq s}$ is the canonical basis of $\mathbb{R}^s$, by definition. Since $\mathbf{v}_j$'s are the generators of $[X]$, we must have that $\phi(X)$ spans $\mathbb{R}^s$. It follows that we can find $\mathbf{x}_0, \mathbf{x} \in X$ such that $\phi(\mathbf{x} - \mathbf{x}_0) = \phi(\mathbf{x}) - \phi(\mathbf{x}_0)$ has a non-zero coefficient corresponding to $\mathbf{e}_\gamma$ in the canonical eigen(root)vector expansion corresponding to M. We have $\eta^n \mathbf{x} \in X$ for all $n \in \mathbb{N}$, and

$$\phi(\eta^n \mathbf{x}) = M^n \phi(\mathbf{x}), \qquad (1.2)$$

by the definition of M. Now, by Theorem 1.2.6, we have,

$$\|\phi(\eta^n \mathbf{x}) - \phi(\eta^n \mathbf{x}_0)\| \leq C_0 \|\eta^n \mathbf{x} - \eta^n \mathbf{x}_0\| = C_0 \eta^n \|\mathbf{x} - \mathbf{x}_0\|.$$

On the other hand, by (1.2) and the choice of $\mathbf{x}, \mathbf{x}_0$,

$$\|\phi(\eta^n \mathbf{x}) - \phi(\eta^n \mathbf{x}_0)\| = \|M^n(\phi(\mathbf{x} - \mathbf{x}_0))\| \geq c|\gamma|^n,$$

for some $c > 0$, and we can conclude that $|\gamma| \leq |\eta|$.   □

(ii′) This can derived be similarly to [64, §10], see also [61]; we omit the proof.

(iii) Consider the address map ϕ and the matrix M as above. Recall that for a Meyer set the address map is almost linear, that is, there exists a linear map $L : \mathbb{R}^d \to \mathbb{R}^s$ such that $\|\phi(\mathbf{x}) - L\mathbf{x}\| \leq C$ for all $\mathbf{x} \in X$. Consider the range $H := L(\mathbb{R}^d)$, a d-dimensional subspace of $\mathbb{R}^s$.

Claim *The subspace H is invariant under M; in fact, H is contained in the eigenspace of M corresponding to η.*

Proof of Claim Choose any unit vector $\mathbf{w} \in H$, then $\mathbf{w} = L\mathbf{z}$ for some $\mathbf{z} \in \mathbb{R}^d$. For any $k > 1$ we can find $\mathbf{x} \in X$ such that $\|k\mathbf{z} - \mathbf{x}\| \leq R$. Then $\|k\mathbf{w} - L\mathbf{x}\| \leq \|L\|R$, hence

$$\|k\mathbf{w} - \phi(\mathbf{x})\| \leq C + \|L\|R. \tag{1.3}$$

Since $\eta\mathbf{x} \in X$, we also have $\|\phi(\eta\mathbf{x}) - L(\eta\mathbf{x})\| \leq C$, therefore,

$$\|\phi(\eta\mathbf{x}) - \eta\phi(\mathbf{x})\| \leq \|\phi(\eta\mathbf{x}) - L(\eta\mathbf{x})\| + \eta\|\phi(\mathbf{x}) - L\mathbf{x}\| \leq C(1 + \eta).$$

Recall from (1.2) that $\phi(\eta\mathbf{x}) = M\phi(\mathbf{x})$, thus $\|(M - \eta I)\phi(\mathbf{x})\| \leq C(1 + \eta)$. Combining this with (1.3) yields

$$\|(M - \eta I)k\mathbf{w}\| \leq C(1 + \eta) + \|M - \eta I\| \cdot (C + \|L\|R) =: \tilde{C}.$$

Therefore, $\|(M - \eta I)\mathbf{w}\| \leq \tilde{C}/k$ for all $k > 1$, and the claim follows.

Now we repeat part of the argument from (ii): given a conjugate γ of η and the corresponding eigenvector $\mathbf{e}_\gamma$ for M, choose $\mathbf{x}$ with $\phi(\mathbf{x})$ having non-zero coefficient corresponding to $\mathbf{e}_\gamma$. By the Claim, $\mathbf{e}_\gamma \notin H$. As above, we have $\phi(\eta^n \mathbf{x}) = M^n\phi(\mathbf{x})$, hence

$$c|\gamma|^n \leq \|M^n\phi(\mathbf{x}) - \eta^n L\mathbf{x}\| = \|\phi(\eta^n \mathbf{x}) - L(\eta^n \mathbf{x})\| \leq C,$$

for some $c > 0$, since $L\mathbf{x}$ has a zero coefficient corresponding to $\mathbf{e}_\gamma$. It follows that $|\gamma| \leq 1$, as desired.   □

Example 1.4.6

(i) Let $\eta > 1$ be an irrational algebraic integer. One can construct a finitely generated Delone set $X \subseteq \mathbb{R}$, with $\eta X \subseteq X$ as follows. We will have $X \subseteq \mathbb{Z}[\eta]$ (the ring generated by $\mathbb{Z}$ and η) and $X = -X$. Start with $X \subset [0, \eta) = \{0, 1\}$ and proceed by induction, adding points to X from $[\eta^k, \eta^{k+1})$, once we did this in $[\eta^{k-1}, \eta^k)$. First make sure that $X \cap [\eta^k, \eta^{k+1}) \supseteq \eta(X \cap [\eta^{k-1}, \eta^k))$, and then add more points from $\mathbb{Z}[\eta]$, if necessary, to maintain the relative denseness, but also preserve uniform discreteness. The latter is easy, since $\mathbb{Z}[\eta]$ is dense in $\mathbb{R}$. Finally, observe that $\mathbb{Z}[\eta] = [1, \eta, \ldots, \eta^{s-1}]$ is finitely generated as a $\mathbb{Z}$-module, where s is the degree of η.

(ii) Let $\eta > 1$ be a Perron number. One can construct a Delone set $X \subseteq \mathbb{R}$ with inflation symmetry η. In fact, it will even be a *substitution Delone set*, discussed in the next section. It is obtained as a set of endpoints of a self-similar tiling of $\mathbb{R}$ corresponding to a primitive substitution. By a theorem of D. Lind [47], for any Perron number η there exists a primitive integer matrix M with η as a dominant eigenvalue. Moreover, by a minor modification of the construction, one can make sure that the entry $(1, 1)$ of the matrix is positive and the first column sum is at least three, see [61]. Then simply choose any substitution with substitution matrix M.

It is known that such an X is Meyer if and only if η is a Pisot number.

(iii) β-**integers.** Fix $\beta > 1$, with $\beta \notin \mathbb{N}$. Let $X_\beta = X_\beta^+ \bigcup (-X_\beta^+)$, where

$$X_\beta^+ = \left\{ \sum_{j=0}^{N} a_j \beta^j, \ a_j \in \{0, 1, \ldots, \lfloor \beta \rfloor\}, \text{ "greedy" expansion} \right\}.$$

Then X_β is relatively dense in $\mathbb{R}$ and $\beta X \subset X$.

- X_β is Delone if and only if the orbit of 1 under $T_\beta(x) = \beta x \pmod 1$ does not accumulate to 0.
- Delone X_β is finitely generated iff β is an algebraic integer.
- Delone X_β is of finite type iff β is a Parry β-number (see [53]), i.e., the orbit $\{T_\beta^n(1)\}_{n \geq 0}$ is finite.
- If β is Pisot (or Salem of degree four [12]), then X_β is Meyer.

We will now explain the last claim, that Pisot β implies the Meyer property, at the same time illustrating some of the concepts in the proofs of the theorem above.

Fix $\beta > 1$ algebraic integer, such that $\{T_\beta^n 1\}_{n \geq 0}$ does not accumulate to zero. Then $X_\beta \subset \mathbb{R}$ is Delone and $[X_\beta] = \mathbb{Z}[\beta]$. Free generators for $[X_\beta]$ can be chosen $v_j = \beta^{j-1}$, $j \leq s$, where s is the degree of β. Let $c_0 + c_1 x + \cdots + c_{s-1} x^{s-1} + x^s$ be the minimal integer polynomial for β.

We have $Qx = \beta x$ on $\mathbb{R}$, and $QX_\beta \subset X_\beta$. Then $QV = VM$, where $V = [v_1, \ldots, v_s]$ is a row and

$$
M = \begin{pmatrix}
0 & \cdots \cdots & 0 & -c_0 \\
1 & 0 & \cdots & 0 & -c_1 \\
0 & 1 & \cdots & 0 & -c_2 \\
& \cdots \cdots \cdots & & \cdots \\
& \cdots \cdots \cdots & 0 & -c_{s-2} \\
0 & 0 & \cdots & 1 & -c_{s-1}
\end{pmatrix}
\tag{1.4}
$$

Let ϕ be the associated address map, $\phi : [X_\beta] = \mathbb{Z}[\beta] \to \mathbb{R}^s$. We have

$$
\phi(\beta^n) = M^n \phi(1) = M^n \begin{pmatrix} 1 \\ 0 \\ \vdots \\ 0 \end{pmatrix}.
$$

Now suppose that β is Pisot. Then we have

$$
\phi(\beta^n) = \beta^n e_\beta + O(\varrho^n),
\tag{1.5}
$$

where e_β is the eigenvector of M corresponding to β and $\varrho \in (0, 1)$ is the maximal absolute value of the Galois conjugates of β. Define $L : \mathbb{R} \to \mathbb{R}^s$ by $L(x) = x e_\beta$, a linear map. We want to show that $\|\phi(x) - Lx\| \le C$ on X_β, whence X_β is a Meyer set by Theorem 1.3.3. In view of (1.5), we have for $x = \sum_{j=0}^{N} a_j \beta^j \in X_\beta^+$:

$$
\|\phi(x) - Lx\| = \left\| \phi\left(\sum_{j=0}^{N} a_j \beta^j \right) - L\left(\sum_{j=0}^{N} a_j \beta^j \right) \right\|
$$

$$
= O\left(\max_j |a_j| \cdot \sum_{j=0}^{N} \varrho^j \right) = O(1),
$$

as desired.

The same proof works, e.g., for the set of endpoints of a self-similar tiling on $\mathbb{R}$ with a Pisot inflation factor.

(iv) **Salem inflation factors.** For every Salem number β there exists a Meyer set in $\mathbb{R}$ with inflation β; see [49] for Meyer's original construction. I am grateful to Shigeki Akiyama who showed me the following example, which is, apparently, "folklore".

Let β be a Salem number of degree $s \ge 4$, and let $p(x) = c_0 + c_1 x + \cdots + c_{s-1} x^{s-1} + x^s$ be the minimal polynomial for β. Let $\beta_2, \ldots, \beta_s$ be the

Galois conjugates of β. It is well-known that $p(x)$ is a reciprocal polynomial, i.e., $c_0 = 1$ and $c_{s-j} = c_j$ for $j = 1, \ldots, s-1$, and the conjugates satisfy $|\beta_2| = \ldots = |\beta_{s-1}| = 1$ and $|\beta_s| < 1$, see [56]. Consider

$$X_\beta := \left\{ x = \sum_{n=0}^{N} a_n \beta^n : a_n \in \mathbb{Z}, \ \max_{2 \leq j \leq s} \left| \sum_{n=0}^{N} a_n \beta_j^n \right| < 1 \right\}.$$

It is immediate that $\beta X_\beta \subset X_\beta$. Moreover, X_β is a non-degenerate model set, hence it is Meyer.

In order to prove the last claim, we consider the matrix M from (1.4) above and identify X_β with $X_\beta e_\beta$, where e_β is the right eigenvector of M corresponding to β. The subspace spanned by e_β is our "physical space". The "internal space" H is the linear span of the other eigenvectors. One can check that (with appropriate normalization) the coordinate a_k of a vector $\mathbf{y} = (y_j)_1^s \in \mathbb{R}^s$ with respect to the eigenvector of M corresponding to β_k is given by $\sum_{j=0}^{s-1} y_j \beta_k^j$. Thus, taking $\mathbb{Z}^s$ as a lattice and the window in H given by the condition

$$\|\mathbf{y}\| := \max_{2 \leq k \leq s} |a_k| < 1 \quad \text{for} \quad \mathbf{y} = \sum_{k=2}^{s} a_k e_{\beta_k},$$

we get the desired representation as a cut-and-project set. The details are left to the reader. (Instead of the unit ball in the ℓ^∞ norm as a window we can choose a different radius and a different norm to get other examples of Meyer sets with the same inflation symmetry.)

1.5   Substitution Delone Sets and Substitution Tilings

This section is based on [39] and [41]; see also [10].

If we think about Delone sets as being models of atomic structures, it is natural to add the feature of "color" or "type" of a point/atom. Thus we are going to talk about "m-sets". (Sometimes, the term "multiset" is used, but it usually refers to a set with multiplicities, and we want to avoid this.)

1.5.1   Substitution Delone m-Sets

Definition 1.5.1 An m-multiset in $\mathbb{R}^d$ is a subset $\boldsymbol{\Lambda} = \Lambda_1 \times \cdots \times \Lambda_m \subset \mathbb{R}^d \times \cdots \times \mathbb{R}^d$ (m copies) where $\Lambda_i \subset \mathbb{R}^d$. We also write $\boldsymbol{\Lambda} = (\Lambda_1, \ldots, \Lambda_m) = (\Lambda_i)_{i \leq m}$. We say that $\Lambda = (\Lambda_i)_{i \leq m}$ is a *Delone m-set* in $\mathbb{R}^d$ if each Λ_i is Delone and $\text{supp}(\Lambda) := \bigcup_{i=1}^{m} \Lambda_i \subset \mathbb{R}^d$ is Delone.

Although Λ is a product of sets, it is convenient to think of it as a set with types or colors, i being the color of points in Λ_i. (However, we do not assume that Λ_i are pairwise disjoint!) A *cluster* of Λ is, by definition, a family $\mathbf{P} = (P_i)_{i \leq m}$ where $P_i \subset \Lambda_i$ is finite for all $i \leq m$. For a bounded set $A \subset \mathbb{R}^d$, let $A \cap \Lambda := (A \cap \Lambda_i)_{i \leq m}$. There is a natural translation $\mathbb{R}^d$-action on the set of Delone m-sets and their clusters in $\mathbb{R}^d$. The translate of a cluster $\mathbf{P}$ by $x \in \mathbb{R}^d$ is $x + \mathbf{P} = (x + P_i)_{i \leq m}$. We say that two clusters $\mathbf{P}$ and $\mathbf{P}'$ are translationally equivalent if $\mathbf{P} = x + \mathbf{P}'$, i.e. $P_i = x + P_i'$ for all $i \leq m$, for some $x \in \mathbb{R}^d$.

Recall that a linear map $Q : \mathbb{R}^d \to \mathbb{R}^d$ is *expanding* if its every eigenvalue lies outside the unit circle.

Definition 1.5.2 $\Lambda = (\Lambda_i)_{i \leq m}$ is called a *substitution Delone m-set* if Λ is a Delone m-set and there exist an expanding map $Q : \mathbb{R}^d \to \mathbb{R}^d$ and finite sets $\mathcal{D}_{ij}$ for $i, j \leq m$ (possibly empty) such that

$$\Lambda_i = \biguplus_{j=1}^{m}(Q\Lambda_j + \mathcal{D}_{ij}), \quad i \leq m, \tag{1.6}$$

where $\biguplus$ denotes disjoint union. The *substitution matrix* S is defined by $\mathsf{S}_{ij} = \sharp(\mathcal{D}_{ij})$. The substitution m-set is *primitive* if S is primitive, i.e., some power of S has only strictly positive entries.

With an abuse of terminology, we say that $\text{supp}(\Lambda)$, or sometimes even Λ, is simply a substitution Delone set.

There is a connection between substitution Delone sets and Delone sets with inflation symmetries, discussed in Sect. 1.3. Of course, here Q is more general, whereas in Sect. 1.3 it was a homothety $\mathbf{x} \mapsto \eta\mathbf{x}$. It is not always true that $\text{supp}(\Lambda) \supset Q(\text{supp}(\Lambda))$; a sufficient condition is that for every $j \leq m$ there exists $i \leq m$ such that $\mathcal{D}_{ij} \ni 0$. This may be achieved by passing from Λ_i to $\Lambda_i + x_i$, which satisfy a system of equations as in (1.6), with modified $\mathcal{D}_{ij}$; the only issue is whether the new m-set is still a Delone m-set. In any case, the following holds:

Lemma 1.5.3 *Suppose that Λ is a substitution Delone m-set with expansion map Q. If $\text{supp}(\Lambda)$ is finitely generated, then all eigenvalues of Q are algebraic integers.*

For the proof, it is convenient to consider the set of "inter-atomic vectors" (more precisely, translation vectors between points of the same color):

$$\Xi(\Lambda) := \bigcup_{i=1}^{m}(\Lambda_i - \Lambda_i). \tag{1.7}$$

It is immediate from (1.6) that $Q(\Xi(\Lambda)) \subset \Xi(\Lambda)$. Even though it need not be a Delone set (it is Delone when supp(Λ) is a Meyer set), the proof proceeds similarly to the proof of Lemma 1.4.5.

There is another important necessary condition for Q to be an expansion map.

Theorem 1.5.4 ([39, Thm. 2.3]) *If Λ is a primitive substitution Delone m-set with expansion map Q, then the Perron-Frobenius (PF) eigenvalue $\lambda(S)$ of the substitution matrix S equals $|\det(Q)|$.*

In fact, it is not difficult to see that

$$\text{supp}(\Lambda) \text{ is relatively dense} \implies \lambda(S) \geq |\det(Q)|;$$

$$\text{supp}(\Lambda) \text{ is uniformly discrete} \implies \lambda(S) \leq |\det(Q)|.$$

For each primitive substitution Delone m-set Λ (1.6) one can set up an *adjoint system* of equations

$$QA_j = \bigcup_{i=1}^{m}(\mathcal{D}_{ij} + A_i), \quad j \leq m. \tag{1.8}$$

From the theory of graph-directed iterated function systems, it follows that (1.8) always has a unique solution for which $\mathcal{A} = \{A_1, \ldots, A_m\}$ is a family of non-empty compact sets of $\mathbb{R}^d$. It is proved in [39, Thm. 2.4 and Thm. 5.5] that if Λ is a primitive substitution Delone m-set, then all the sets A_i from (1.8) have non-empty interiors and, moreover, each A_i is the closure of its interior. From Theorem 1.5.4 it follows that the interiors of the sets in the right-hand side of (1.8) are disjoint, hence we have a natural candidate for a tiling. We next review briefly the relevant tiling definitions.

1.5.2 Tilings and Substitution Tilings

This section has a significant overlap with [16].

We begin with a set of types (or colors) $\{1, \ldots, m\}$, which we fix once and for all. A *tile* in $\mathbb{R}^d$ is defined as a pair $T = (A, i)$ where $A = \text{supp}(T)$ (the support of T) is a compact set in $\mathbb{R}^d$ which is the closure of its interior, and $i = \ell(T) \in \{1, \ldots, m\}$ is the type of T. We let $g + T = (g + A, i)$ for $g \in \mathbb{R}^d$. Given a tile T and a set $X \subseteq \mathbb{R}^d$ we use the notation:

$$T + X = \{T + x : x \in X\}.$$

A finite set P of tiles is called a *patch* if the tiles of P have mutually disjoint interiors (strictly speaking, we have to say "supports of tiles," but this abuse of language should not lead to confusion). A *tiling* of $\mathbb{R}^d$ is a set $\mathcal{T}$ of tiles such that $\mathbb{R}^d = \bigcup\{\text{supp}(T) : T \in \mathcal{T}\}$ and distinct tiles have disjoint interiors. Given a tiling $\mathcal{T}$, finite sets of tiles of $\mathcal{T}$ are called $\mathcal{T}$-patches.

We always assume that any two $\mathcal{T}$-tiles with the same color are translationally equivalent. (Hence there are finitely many $\mathcal{T}$-tiles up to translation.)

Definition 1.5.5 Let $\mathcal{A} = \{T_1, \ldots, T_m\}$ be a finite set of tiles in $\mathbb{R}^d$ such that $T_i = (A_i, i)$; we will call them *prototiles*. Denote by $\mathcal{P}_\mathcal{A}$ the set of patches made of tiles each of which is a translate of one of T_i's. We say that $\omega : \mathcal{A} \to \mathcal{P}_\mathcal{A}$ is a *tile-substitution* (or simply *substitution*) with expanding map Q if there exist finite sets $\mathcal{D}_{ij} \subset \mathbb{R}^d$ for $i, j \leq m$, such that

$$\omega(T_j) = \bigcup_{i=1}^{m} (T_i + \mathcal{D}_{ij}). \tag{1.9}$$

Since $\omega(T_j)$ is a patch, it follows that for all $j \leq m$,

$$QA_j = \bigcup_{i=1}^{m} (A_i + \mathcal{D}_{ij}),$$

and the sets in the right-hand side have disjoint interiors.

The substitution (1.9) is extended to all translates of prototiles by $\omega(x + T_j) = Qx + \omega(T_j)$, and to patches and tilings by $\omega(P) = \bigcup\{\omega(T) : T \in P\}$. The substitution ω can be iterated, producing larger and larger patches $\omega^k(T_j)$. As for a substitution Delone m-set, we associate to ω its $m \times m$ substitution matrix S, with $\mathsf{S}_{ij} := \sharp(\mathcal{D}_{ij})$. The substitution ω is said to be primitive if S is primitive. The tiling $\mathcal{T}$ is called a fixed point of a substitution if $\omega(\mathcal{T}) = \mathcal{T}$.

Definition 1.5.6 A tiling is called *self-affine* if it is a fixed point of a primitive tile-substitution. Usually it is also assumed that the tiling has finite local complexity (FLC). If the expansion map Q is a similitude, that is, $Q = \eta\mathcal{O}$ for some $\eta > 1$ and an orthogonal linear transformation $\mathcal{O}$, then we say that the tiling is *self-similar*. For a self-similar tiling in $\mathbb{R}^2$ one also considers the *complex expansion factor* $\lambda \in \mathbb{C}$, $|\lambda| > 1$, by identifying the plane with $\mathbb{C}$ and the map Q with $z \mapsto \lambda z$.

Example 1.5.7 Figures 1.1 and 1.2 show a self-affine tiling found by G. Gelbrich: the tile substitution is

$$\omega(T) = T \cup (-T) \cup (-T + (1, 0)) \cup (-T + (0, 1)),$$

Fig. 1.1 Tile substitution

Fig. 1.2 Patch of the
self-affine tiling

where $\mathrm{supp}(\omega(T)) = Q(\mathrm{supp}(T)) + (1/2, -1/2)$, with

$$Q = \begin{pmatrix} -2 & 1 \\ 2 & 1 \end{pmatrix}.$$

It is actually a p2 crytallographic tiling, and therefore periodic, see [25].

Below, in Example 1.5.18, one can see a figure showing a self-similar tiling.

I am grateful to S. Akiyama and to an anonymous volunteer who helped with the examples and the figures.

Notice that a fixed point of a substitution naturally defines a substitution Delone m-set, as follows: By definition, we can write $\mathcal{T} = \bigcup_{j=1}^{m} (T_j + \Lambda_j)$ for some Delone sets Λ_j. Then we have

$$
\bigcup_{i=1}^{m}(T_i + \Lambda_i) = \mathcal{T} = \omega(\mathcal{T}) = \bigcup_{j=1}^{m}\big(\omega(T_j) + Q\Lambda_j\big)
$$

$$
= \bigcup_{j=1}^{m}\bigg(\bigcup_{i=1}^{m}(T_i + \mathcal{D}_{ij}) + Q\Lambda_j\bigg)
$$

$$
= \bigcup_{i=1}^{m}\bigg(T_i + \bigcup_{j=1}^{m}(Q\Lambda_j + \mathcal{D}_{ij})\bigg).
$$

It follows that $\Lambda = (\Lambda_i)_{i=1}^{m}$ satisfies the system of equations (1.6). In general, it is not necessarily true that Λ_i are disjoint, but we can ensure this, e.g., by taking $T_j := \widetilde{T}_j - c(\widetilde{T}_j)$, where $\widetilde{T}_j$ is a $\mathcal{T}$-tile of type j and $c(\widetilde{T}_j)$ is a point chosen in its interior. Then we obtain that Λ is a substitution Delone m-set.

A natural question is when this procedure can be reversed.

1.5.3 Representable Delone m-Sets

Definition 1.5.8 A Delone m-set $\Lambda = (\Lambda_i)_{i \leq m}$ is called *representable* (by tiles) for a tiling if there exists a set of prototiles $\mathcal{A} = \{T_i : i \leq m\}$ so that

$$
\Lambda + \mathcal{A} := \{x + T_i : x \in \Lambda_i,\ i \leq m\} \quad \text{is a tiling of } \mathbb{R}^d, \tag{1.10}
$$

that is, $\mathbb{R}^d = \bigcup_{i \leq m}\bigcup_{x \in \Lambda_i}(x + A_i)$ where $T_i = (A_i, i)$ for $i \leq m$, and the sets in this union have disjoint interiors. In the case that Λ is a primitive substitution Delone m-set we will understand the term representable to mean relative to the tiles $T_i = (A_i, i)$, for $i \leq m$, arising from the solution to the adjoint system (1.8). We call $\Lambda + \mathcal{A}$ the associated tiling of Λ.

Definition 1.5.9 For a substitution Delone m-set $\Lambda = (\Lambda_i)_{i \leq m}$ satisfying (1.6), define a matrix $\Phi = (\Phi_{ij})_{i,j=1}^{m}$ whose entries are finite (possibly empty) families of linear affine transformations on $\mathbb{R}^d$ given by

$$
\Phi_{ij} = \{f : x \mapsto Qx + a :\ a \in \mathcal{D}_{ij}\}.
$$

We define $\Phi_{ij}(\varXi) := \bigcup_{f \in \Phi_{ij}} f(\varXi)$ for a set $\varXi \subset \mathbb{R}^d$. For an m-set $(\varXi_i)_{i \leq m}$ let

$$
\Phi\big((\varXi_i)_{i \leq m}\big) = \bigg(\bigcup_{j=1}^{m}\Phi_{ij}(\varXi_j)\bigg)_{i \leq m}.
$$

Thus $\Phi(\Lambda) = \Lambda$ by definition. We say that Φ is an *m-set substitution*.

Let Λ be a substitution Delone m-set and Φ the associated m-set substitution.

Definition 1.5.10 Let Λ be a primitive substitution Delone m-set and let $\mathbf{P}$ be a cluster of Λ. The cluster $\mathbf{P}$ will be called *legal* if it is a translate of a subcluster of $\Phi^k(\{x_j\})$ for some $x_j \in \Lambda_j$, $j \leq m$ and $k \in \mathbb{N}$. (Here $\{x_j\}$ is an m-set which is empty in all coordinates other than j, for which it is a singleton.)

Lemma 1.5.11 ([41]) *Let Λ be a primitive substitution Delone m-set such that every Λ-cluster is legal. Then Λ is repetitive.*

Not every substitution Delone m-set is representable (see [41, Ex. 3.12]), but the following theorem provides the sufficient condition for it.

Theorem 1.5.12 ([41]) *Let Λ be a repetitive primitive substitution Delone m-set. Then every Λ-cluster is legal if and only if Λ is representable.*

Remark 1.5.13 In [39, Lemma 3.2] it is shown that if Λ is a substitution Delone m-set, then there is a finite m-set (cluster) $\mathbf{P} \subset \Lambda$ for which $\Phi^{n-1}(\mathbf{P}) \subset \Phi^n(\mathbf{P})$ for $n \geq 1$ and $\Lambda = \lim_{n \to \infty} \Phi^n(\mathbf{P})$. We call such a m-set $\mathbf{P}$ a *generating m-set*. Note that, in order to check that every Λ-cluster is legal, we only need to see if some cluster that contains a finite generating m-set for Λ is legal.

1.5.4 Characterization of Expansion Maps

An important question, first raised by Thurston [64], is to characterize which expanding linear maps may occur as expansion maps for self-affine (self-similar) tilings. It is pointed out in [64] that in one dimension, $\eta > 1$ is an expansion factor if and only if it is a Perron number (necessity follows from the Perron-Frobenius theorem and sufficiency follows from a result of Lind [47] as in Example 1.4.6(ii)). In two dimensions, Thurston [64] proved that if λ is a complex expansion factor of a self-similar tiling, then λ is a *complex Perron number*, that is, an algebraic integer whose Galois conjugates, other than $\bar{\lambda}$, are all less than $|\lambda|$ in modulus.

The following theorem was stated in [30], but complete proof was not available until much later.

Theorem 1.5.14 ([30, 33]) *Let ϕ be a diagonalizable (over $\mathbb{C}$) expansion map on $\mathbb{R}^d$, and let $\mathcal{T}$ be a self-affine tiling of $\mathbb{R}^d$ with expansion ϕ. Then*

(i) *every eigenvalue of ϕ is an algebraic integer;*
(ii) *if λ is an eigenvalue of ϕ of multiplicity k and γ is an algebraic conjugate of λ, then either $|\gamma| < |\lambda|$, or γ is also an eigenvalue of ϕ of multiplicity greater or equal to k.*

Here part (i) is included for completeness; it is a folklore result, proved similarly to Lemma 1.4.5. Recently Theorem 1.5.14 was finally extended to the general, not necessarily diagonalizable, case by J. Kwapisz [35]; we don't state precise "generalized Perron" conditions here, but refer the reader to his paper. Basically, one has to take into account the multiplicity of Jordan blocks as well. It is conjectured that the necessary condition is also sufficient, at least, in the weaker form: one should be able to construct a self-affine tiling with expansion map Q^n for some $n \in \mathbb{N}$. Sufficiency (in the stronger form) in dimension one follows from [47], as discussed in Sect. 1.3, Example (ii), and the construction in the two-dimensional self-similar case is found in [32]. A natural approach to the conjecture would be to construct first a substitution Delone m-set with the desired inflation symmetry, and then apply Theorem 1.5.12. This way, the geometric shape of tiles comes from the adjoint equation, and one has more freedom.

The starting point in the proofs of necessity of the Perron condition is defining the *control points* for the tiles [64].

Definition 1.5.15 Let $\mathcal{T}$ be a fixed point of a primitive tile-substitution with expanding map Q. For each $\mathcal{T}$-tile T, fix a tile γT in the patch $\omega(T)$; choose γT with the same relative position for all tiles of the same type. This defines a map $\gamma : \mathcal{T} \to \mathcal{T}$ called the *tile map*. Then define the *control point* for a tile $T \in \mathcal{T}$ by

$$\{c(T)\} = \bigcap_{n=0}^{\infty} Q^{-n}(\gamma^n T).$$

The control points have the following properties:

(a) $T' = T + c(T') - c(T)$, for any tiles T, T' of the same type;
(b) $Q(c(T)) = c(\gamma T)$, for $T \in \mathcal{T}$.

It immediately follows from these properties that $\Lambda = (\Lambda_i)_{i \leq m}$, where $\Lambda_i = \{c(T) : T \in \mathcal{T} \text{ of type } i\}$, is a substitution Delone m-set. Moreover, $X := \text{supp}(\Lambda) = \{c(T) : T \in \mathcal{T}\}$ is a Delone set satisfying $QX \subset X$. Thurston [64] defined the address map $\phi : [X] \to \mathbb{Z}^s$, as in Sect. 1.1, and considered the induced action of the linear expanding map Q. See the references for the rest.

1.5.5 Pseudo-Self-Affine Tilings

We mention briefly another instance where the duality between substitution Delone m-sets and substitution tilings was useful. The reader is referred to [16, 4.4.2] and the original papers for more details.

Definition 1.5.16 ([5]) To a subset $F \subset \mathbb{R}^d$ and a tiling $\mathcal{T}$ of $\mathbb{R}^d$ we associate a $\mathcal{T}$-patch by $[F]^{\mathcal{T}} = \{T \in \mathcal{T} : \operatorname{supp}(T) \cap F \neq \emptyset\}$. Let $\mathcal{T}_1$ and $\mathcal{T}_2$ be two tilings. Say that $\mathcal{T}_2$ is *locally derivable* (LD) from $\mathcal{T}_1$ with radius $R > 0$ if for all $x, y \in \mathbb{R}^d$,

$$[B_R(x)]^{\mathcal{T}_1} = [B_R(y)]^{\mathcal{T}_2} + (x - y) \;\Rightarrow\; [\{x\}]^{\mathcal{T}_2} = [\{y\}]^{\mathcal{T}_2} + (x - y).$$

If $\mathcal{T}_2$ is LD from $\mathcal{T}_1$ and $\mathcal{T}_1$ is LD from $\mathcal{T}_2$, the tilings are *mutually locally derivable* (MLD).

Definition 1.5.17 Let $Q : \mathbb{R}^d \to \mathbb{R}^d$ be an expanding linear map. A repetitive FLC tiling of $\mathbb{R}^d$ is called a *pseudo-self-affine tiling* with expansion Q if $\mathcal{T}$ is LD from $Q\mathcal{T}$. If Q is a similitude, the pseudo-self-affine tiling is called *pseudo-self-similar*.

E. A. Robinson, Jr. [55] conjectured that every pseudo-self-affine tiling is MLD with a self-affine tiling. This was settled in the affirmative: for pseudo-self-similar tilings in $\mathbb{R}^2$ in [21], and in [60] in full generality. A few comments:

- In [21] the method of "redrawing the boundary" was used (following [32] to some extent); as a result we obtained an MLD self-similar tiling where each tile is a topological disk bounded by a Jordan curve.
- In contrast, in [60] we first constructed an MLD substitution Delone m-set in $\mathbb{R}^d$, and then applied Theorem 1.5.12. This way we only know that each tile is a compact set, which is a closure of its interior; the tiles need not even be connected.
- In both papers [21, 60] we needed to pass from the expansion Q for the original tiling to the expansion Q^k for k sufficiently large for the resulting self-similar (or self-affine) tiling, in order for the construction to work, similarly to the weaker form of the conjecture, discussed after Theorem 1.5.14.

Example 1.5.18 Here is an example of a pseudo-self-similar tiling, due to J. Socolar (the help of S. Akiyama and an anonymous volunteer is gratefully acknowledged). See the online Tiling Encyclopedia:
https://tilings.math.uni-bielefeld.de/substitution/limhex/
Note that there is one prototile $\widetilde{T}$, up to translation and rotation, which appears in 6 orientations, so that up to translation there are 6 prototiles. It is pseudo-self-similar, with the expansion map given by $z \mapsto e^{2\pi i/3} z$ in the complex plane. The subdivision rule leads to an iterated function system and the set equation

$$2T = T \cup (T - 1) \cup (w^5 T + w^2) \cup (w^2 T + w), \quad \text{where } w = e^{2\pi i/3}.$$

The solution looks rather complicated: see T and the subdvision of $2T$ in Fig. 1.3 and the patch of the resulting self-similar tiling in Figs. 1.4, 1.5, and 1.6.

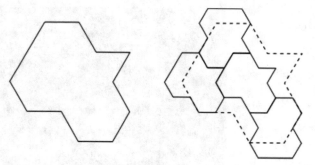

Fig. 1.3 Subdivision rule $2\widetilde{T} \mapsto \widetilde{T} \cup (\widetilde{T} - 1) \cup (w^5\widetilde{T} + w^2) \cup (w^2\widetilde{T} + w)$ where $w = e^{2\pi i/3}$

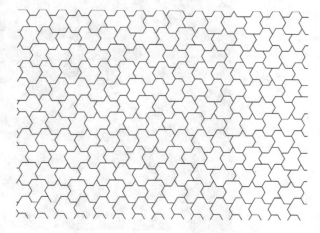

Fig. 1.4 Patch of the pseudo-self-similar tiling

Fig. 1.5 The self-similar tile and the tile-substitution

Fig. 1.6 Patch of the
self-similar tiling

1.6 Dynamical Systems from Delone Sets in $\mathbb{R}^d$

This section is not intended to be a comprehensive survey (even in the "local" sense);
we only touch on some aspects of this topic. For more on the background and other
topics see [16] and references therein.

Let Λ be an FLC Delone m-set and let X_Λ be the collection of all Delone m-sets
each of whose clusters is a translate of a Λ-cluster. We introduce a "big ball" metric
on X_Λ in the standard way: two Delone m-sets are close if they agree exactly in a
large neighborhood of the origin, possibly after a small translation. Precisely:

$$\rho(\Lambda_1, \Lambda_2) := \min\{\widetilde{\rho}(\Lambda_1, \Lambda_2), 2^{-1/2}\}, \tag{1.11}$$

where

$$\widetilde{\rho}(\Lambda_1, \Lambda_2) = \inf\{\epsilon > 0 : \exists\, x, y \in B_\epsilon(0),$$
$$B_{1/\epsilon}(0) \cap (-x + \Lambda_1) = B_{1/\epsilon}(0) \cap (-y + \Lambda_2)\}.$$

For the proof that ρ is a metric, see [40]. It is well-known that (X_Λ, ρ) is compact (here the FLC assumption is crucial). Here we restrict ourselves to the FLC case; the general case is discussed briefly in Sect. 1.7.

Observe that $X_\Lambda = \overline{\{-h + \Lambda : h \in \mathbb{R}^d\}}$ where the closure is taken in the topology induced by the metric ρ. The group $\mathbb{R}^d$ acts on X_Λ by translations which are homeomorphisms, and we get a topological dynamical system $(X_\Lambda, \mathbb{R}^d)$. We should point out that most of the definitions and statements in this section have a parallel version in the tiling setting. It is usually not a problem to pass from the tiling framework to the Delone set framework and back, and we will do so freely.

Certain discrete-geometric and statistical properties of the Delone m-set Λ correspond to properties of the associated dynamical system. For instance, Λ is repetitive if and only if $(X_\Lambda, \mathbb{R}^d)$ is minimal (i.e., every orbit is dense), see [55].

We next discuss cluster (patch) frequencies and invariant measures. For a cluster $\mathbf{P}$ and a bounded set $A \subset \mathbb{R}^d$ denote

$$L_\mathbf{P}(A) = \sharp\{x \in \mathbb{R}^d : x + \mathbf{P} \subset A \cap \Lambda\},$$

where $\sharp$ denotes the cardinality. In plain language, $L_\mathbf{P}(A)$ is the number of translates of $\mathbf{P}$ contained in A, which is clearly finite. For a bounded set $F \subset \mathbb{R}^d$ and $r > 0$, let $(F)^{+r} := \{x \in \mathbb{R}^d : \text{dist}(x, F) \leq r\}$ denote the r-neighborhood of F. A *van Hove sequence* for $\mathbb{R}^d$ is a sequence $\mathcal{F} = \{F_n\}_{n \geq 1}$ of bounded measurable subsets of $\mathbb{R}^d$ satisfying

$$\lim_{n \to \infty} \text{Vol}((\partial F_n)^{+r}) / \text{Vol}(F_n) = 0, \text{ for all } r > 0. \tag{1.12}$$

Definition 1.6.1 Let $\{F_n\}_{n \geq 1}$ be a van Hove sequence. The Delone m-set Λ has *uniform cluster frequencies* (UCF) (relative to $\{F_n\}_{n \geq 1}$) if for any non-empty cluster $\mathbf{P}$, the limit

$$\text{freq}(\mathbf{P}, \Lambda) = \lim_{n \to \infty} \frac{L_\mathbf{P}(x + F_n)}{\text{Vol}(F_n)} \geq 0$$

exists uniformly in $x \in \mathbb{R}^d$.

Recall that a topological dynamical system is *uniquely ergodic* if there is a unique invariant probability measure (which is then automatically ergodic). It is known (see e.g. [40, Thm. 2.7]) that for a Delone m-set Λ with FLC, the dynamical system $(X_\Lambda, \mathbb{R}^d)$ is uniquely ergodic if and only if Λ has UCF. A primitive FLC substitution Delone m-set is known to have UCF (see [41]), hence we get a uniquely ergodic $\mathbb{R}^d$-action $(X_\Lambda, \mathbb{R}^d, \mu)$.

1.6.1 Eigenvalues of Delone Set Dynamical Systems

Let μ be an ergodic invariant Borel probability measure for the dynamical system $(X_\Lambda, \mathbb{R}^d)$. A point $\alpha = (\alpha_1, \ldots, \alpha_d) \in \mathbb{R}^d$ is called an eigenvalue (or dynamical eigenvalue) for the $\mathbb{R}^d$-action if there exists an eigenfunction $f \in L^2(X_\Lambda, \mu)$, that is, $f \not\equiv 0$ and

$$f(-g + \mathcal{S}) = e^{2\pi i \langle g, \alpha \rangle} f(\mathcal{S}), \quad \text{for all } g \in \mathbb{R}^d. \tag{1.13}$$

Here $\langle g, \alpha \rangle$ is the usual scalar product in $\mathbb{R}^d$ and the equality is understood in L^2, that is, for μ-a.e. $\mathcal{S}$.

An eigenvalue α is a *continuous* or *topological* eigenvalue if (1.13) has a continuous solution. A characterization of topological eigenvalues was obtained in [59].

Definition 1.6.2 Let Λ be a Delone m-set of finite type. We say that $\mathbf{y} \in \mathbb{R}^d$ is a *topological ε-almost-period* for Λ if

$$\Lambda \cap B_{1/\varepsilon}(0) = (\Lambda - y) \cap B_{1/\varepsilon}(0).$$

Denote by $\Psi_\varepsilon(\Lambda)$ the set of topological ε-almost-periods.

Theorem 1.6.3 ([59]) *Let Λ be a repetitive Delone m-set of finite type. Then α is a topological eigenvalue for $(X_\Lambda, \mathbb{R}^d)$ if and only if*

$$\lim_{\varepsilon \to 0} \sup_{y \in \Psi_\varepsilon(\Lambda)} |e^{2\pi i \langle y, \alpha \rangle} - 1| = 0. \tag{1.14}$$

In [59] the theorem is proved for a single Delone set dynamical system, but the proof transfers to the case of Delone m-sets without any changes.

There is a connection between the *diffraction spectrum* of a Delone m-set and the dynamical spectrum, going back to Dworkin [15], but this topic is beyond the scope of these Notes. On this matter, the reader should consult e.g., [3, 7] and references therein.

N. Strungaru [62] proved that any Meyer set has a relatively dense set of Bragg peaks, which implies, via the link with the dynamical spectrum, that the set of eigenvalues for the associated dynamical system is relatively dense as well. More recently, J. Kellendonk and L. Sadun [28] proved that the latter property holds for topological eigenvalues too. In fact, they established the following

Theorem 1.6.4 ([28, Thm. 1.1]) *A repetitive FLC Delone set dynamical system in $\mathbb{R}^d$ has d linearly independent topological eigenvalues if and only if it is topologically conjugate to a Meyer set dynamical system.*

In the case of substitution systems, we obtained the following result earlier, jointly with J.-Y. Lee:

Theorem 1.6.5 ([42, Thm. 4.14]) *Let* $\Lambda = (\Lambda_j)_{j=1}^m$ *be a representable primitive FLC substitution Delone m-set. The set of eigenvalues for the* $\mathbb{R}^d$*-action* $(X_\Lambda, \mathbb{R}^d, \mu)$ *is relatively dense in* $\mathbb{R}^d$ *if and only if* $\mathrm{supp}(\Lambda) = \bigcup_{j=1}^m \Lambda_j$ *is a Meyer set.*

As a corollary, we showed in [42] that if the $\mathbb{R}^d$-action $(X_\Lambda, \mathbb{R}^d, \mu)$ has purely discrete spectrum, then $\mathrm{supp}(\Lambda)$ is a Meyer set. This was an answer to a question of J. Lagarias from [38].

For the proof of sufficiency of the Meyer property we relied on the result of Strungaru [62] quoted above. The proof of necessity in Theorem 1.6.4 proceeds via the notion of *Pisot families*. Let Q be the expansion map for the substitution Delone set. By Lemma 1.5.3, the set of eigenvalues of Q consists of algebraic integers. Following Mauduit [48], we say that a set $\mathfrak{P}$ of algebraic integers forms a *Pisot family* if for every $\lambda \in \mathfrak{P}$ and every Galois conjugate λ' of λ, if $\lambda' \notin \mathfrak{P}$, then $|\lambda'| < 1$.

The link from eigenvalues to Number Theory comes from the following, which we restate here in different terms:

Theorem 1.6.6 ([58, Thm. 4.3]) *Let* Λ *be a repetitive substitution Delone m-set with expansion map* Q, *which has FLC. Let* $\Xi(\Lambda)$ *be the set of "inter-atomic" vectors defined in (1.7). If* $\alpha \in \mathbb{R}^d$ *is an eigenvalue for* $(X_\Lambda, \mathbb{R}^d, \mu)$, *then for any* $x \in \Xi(\Lambda)$ *we have* $\|\langle Q^n x, \alpha \rangle\| \to 0$ *as* $n \to \infty$.

Here $\|t\|$ denotes the distance from t to the nearest integer. Then we apply the following result, a generalization of the classical Pisot theorem, which we only partially state:

Theorem 1.6.7 ([34, 48]) *Let* $\lambda_1, \ldots, \lambda_r$ *be distinct algebraic numbers such that* $|\lambda_i| \geq 1$, $i = 1, \ldots, r$, *and let* $P_1, \ldots, P_r$ *be nonzero polynomials with complex coefficients. If* $\sum_{i=1}^r P_i(n)\lambda_i^n$ *is real for all n and*

$$\lim_{n \to \infty} \left\| \sum_{i=1}^r P_i(n)\lambda_i^n \right\| = 0,$$

Then $\{\lambda_1, \ldots, \lambda_r\}$ *is a Pisot family.*

1.7 Concluding Remarks and Open Problems

This material was not presented in the talk; we hope it provides a useful supplement to the Notes, without any claims of being comprehensive.

1.7.1 Infinite Local Complexity

The FLC—finite type assumption—is often too restrictive. In the general case the
metric (1.11) is replaced by

$$\varrho(\Lambda_1, \Lambda_2) := \min\{\widetilde{\varrho}(\Lambda_1, \Lambda_2), 2^{-1/2}\},$$

where we define $\widetilde{\varrho}(\Lambda_1, \Lambda_2)$ as

$$\inf\{\varepsilon > 0 : B_{1/\varepsilon}(0) \cap \Lambda_1 \subset \Lambda_2 + B_\varepsilon(0), \; B_{1/\varepsilon}(0) \cap \Lambda_2 \subset \Lambda_1 + B_\varepsilon(0)\}.$$

The induced topology is called the *local rubber topology* in [6], see also [11]; it
agrees with the topology induced by the vague topology on measures, if we identify
a Delone set Λ with the measure $\delta_\Lambda = \sum_{\lambda \in \Lambda}$, called the "Dirac comb". The "hull"
X_Λ of Λ is then defined as the closure of the translation orbit of Λ in this metric.
The space X_Λ is compact, and we get a topological dynamical system $(X_\Lambda, \mathbb{R}^d)$,
where the action is by translations. A similar construction works for tilings as well.
Dynamical properties of general non-FLC Delone dynamical systems have been
studied in [23, 52], in particular, questions of minimality and unique ergodicity.

There is a large literature on non-FLC tiling substitutions. In particular, two large
classes have been considered:

(I) The tiling has finite local complexity with respect to the larger group of
    Euclidean isometries, but the prototiles appear in infinitely many (dense)
    orientations. This is caused by a presence of irrational rotation in the expansive
    linear map associated with the tile substitution. The best known example in this
    class is the pinwheel tiling of Conway and Radin [54]; see also the recent [24]
    and references therein.

(II) There are finitely many prototiles up to translation, but the FLC breaks down.
    Substitution tilings of this kind were constructed by Danzer [13] and Kenyon
    [31], and investigated more systematically by Frank and Robinson [17]. This
    phenomenon is usually associated with a non-Pisot expansion factor. In this
    setting we have also recently studied the question of unique ergodicity and
    spectral properties, in a joint work with J.-Y. Lee [43].

A very general class of hierarchical tilings, which includes all of the above
and much more ("fusion tilings") was considered by Frank and Sadun [18–20];
see [16] in this volume for details and additional references. See also [22] by
D. Frettlöh for a comprehensive review of classes of hierarchical tilings.

1.7.2 Pure Discrete Spectrum

We would be amiss without any mention of the problem: *when is the dynamical
system $(X_\Lambda, \mathbb{R}^d, \mu)$ pure discrete?* There is a huge literature devoted to it, so we only

make a few remarks, referring the reader to [7] and references there for additional information. Here we do not give precise definitions, and also do not necessarily distinguish between Delone sets, m-sets, and multisets.

First of all, pure discrete dynamical spectrum is equivalent to pure discrete diffraction spectrum, under very general conditions: [6, 26, 40, 44] are some of the papers devoted to this question.

Lagarias [38] calls a Delone set Λ *purely diffractive*, or *Patterson set*, if it has a unique autocorrelation measure γ_Λ, for which the associated diffraction measure $\widehat{\gamma}_\Lambda$ is pure point. Lagarias [38, Problem 4.10] asked whether for a repetitive Delone set of finite type, being purely diffractive implies being a Meyer set. In this generality the answer is "no", a counter-example, called the scrambled Fibonacci tiling, was given in [19] (see also [28]). As already mentioned, if we additionally assume that Λ is a primitive substitution Delone m-set, then the answer is "yes". This was [38, Problem 4.11].

Pure discrete spectrum has been linked with various notions of *almost periodicity*. Let us say that $\mathbf{y} \in \mathbb{R}^d$ is a *statistical ε-period* of a Delone set Λ if $\overline{\mathrm{dens}}(\Lambda \triangle (\Lambda - \mathbf{y})) \leq \varepsilon$, where $\overline{\mathrm{dens}}$ is the upper density of a set defined by

$$\overline{\mathrm{dens}} = \limsup_{R \to \infty} \frac{1}{\mathrm{Vol}(B_R)} \sharp(\Lambda \cap B_R(0)).$$

From our result in [59] it follows that if the set of statistical ε-periods is relatively dense in $\mathbb{R}^d$ for all $\varepsilon > 0$, then the dynamical spectrum is pure discrete (for any ergodic invariant measure). Baake and Moody [9] proved, in particular, that if Λ is Meyer, then this condition is also necessary (their result is much more general; this is just one of the consequences of the theory developed in [9]). Gouéré extended this further and proved that a Delone set is a Patterson set if and only if it is almost periodic in the Besicovitch pseudo-metric defined by

$$\overline{\varrho}(\Lambda, \Lambda') = \limsup_{R \to \infty} \frac{1}{\mathrm{Vol}(B_R)} \int_{B_R(0)} \varrho(\Lambda - t, \Lambda' - t)\, dt.$$

For the most up-to-date results on this topic see the chapter by N. Strungaru [63] in [4].

Given a Delone set Λ, the corresponding Dirac comb δ_Λ is called *strongly almost periodic* if $\delta_\Lambda * \phi$ is a Bohr almost periodic function, for any continuous compactly supported function ϕ. Lagarias [38, Problem 4.4] asked with this property implies that Λ is completely periodic. Kellendonk and Lenz [29] demonstrate that the answer is "yes" in the FLC case, but "no" in the general case. They also show that the strong almost periodicity is equivalent to the dynamical system $(X_\Lambda, \mathbb{R}^d)$ being equicontinuous.

Regular model sets form an important class of pure diffractive Delone sets. We defined cut-and-project sets in Definition 1.3.2, with a Euclidean internal space H. More generally, the internal space may be taken to be a locally compact Abelian group G, and the cut-and-project set corresponding to a window $\Omega \subset H$ and a lattice $\Gamma < \mathbb{R}^d \times G$ is defined similarly. It is called a regular model set if the boundary $\partial\Omega$ has zero Haar measure in G. Repetitive regular modular sets are known to be pure point diffractive, and they have only topological eigenvalues [27, 57]. Baake et al. [8] characterized the dynamical systems associated to repetitive regular model sets as those which satisfy the following four conditions: (a) all elements of the space $X = X_\Lambda$ are Meyer sets; (b) the translation $\mathbb{R}^d$-action is minimal and uniquely ergodic; (c) the measure-preserving $\mathbb{R}^d$-action has pure point dynamical spectrum and all eigenvalues are topological, and (d) the continuous eigenfunctions separate almost all points of the space X.

Many papers are devoted to the question of pure point spectrum for substitutions, in the symbolic, tiling, and Delone m-set frameworks, but we only mention a few. On the *Pisot discrete spectrum conjecture*, the reader should see [1]. For self-affine tilings and substitution Delone m-sets criteria for pure pointedness based on "coincidence conditions" were obtained in [41, 58], and an efficient algorithm for checking them was developed in [2].

1.7.3 Fourier Quasicrystals

Following Lagarias [38], say that a uniformly discrete set $\Lambda \subset \mathbb{R}^d$ is a *Fourier quasicrystal*, if there exists a translation bounded measure μ supported on Λ, such that the Fourier transform $\widehat{\mu}$ (in the distribution sense) is a discrete measure. (Sometimes the measure μ is called a Fourier quasicrystal.) The set of point masses of $\widehat{\mu}$ is called the *spectrum of* μ. A Dirac comb δ_Λ for a lattice Λ is a classical example.

A Fourier quasicrystal is necessarily a Patterson set (see [38]), but the converse does not hold. Meyer's model sets are Fourier quasicrystals [49], with a dense spectrum of the corresponding measure.

Roughly speaking, in building a Fourier quasicrystal on a Patterson set, the problem is how to choose the "weights" $\psi(y)$ for the measure $\mu = \sum_{y \in \Lambda} \psi(y)\delta_y$, in order for the Fourier transform $\widehat{\mu}$ to be a discrete measure. This requires careful "accounting" of the "phase information" coming from $\widehat{\mu}$, which is lost when considering the diffraction measure (see [38] for details).

[38, Problem 4.1(a)] asked whether a Fourier quasicrystal Λ, such that the spectrum of the corresponding measure $\mu = \sum_{y \in \Lambda} \psi(y)\delta_y$ is also a uniformly discrete set, is necessarily contained in $L + F$, where L is a lattice in $\mathbb{R}^d$ and F is a finite set. Lev and Olevskii [45] proved that the answer is "yes" in $\mathbb{R}^d$ for any $d \geq 1$, if the weights $\psi(y)$ are positive, and also for $d = 1$ for arbitrary weights. The case of $\mathbb{R}^d$, $d \geq 2$, and arbitrary weights is apparently still open. [38, Problem 4.1(b)] asked whether the conclusion still holds if the support of μ and its spectrum are only assumed to be closed discrete sets. Here the answer is "no," as shown by the same authors in [46].

1.7.4 Open Problems (Expansion Maps)

In Sect. 1.5.4 we discussed results on the characterization of expansion maps for self-similar and self-affine tilings, assuming FLC. Even there, not everything is understood:

Problem 1 *Suppose that Q is a "generalized Perron expansion map" in $\mathbb{R}^d$ (as described in Sect. 1.5.4). Does there exists a self-affine tiling with expansion Q? Perhaps, with expansion Q^n for n sufficiently large?*

The problem seems to be completely open without the FLC assumption.

Problem 2 *Which expansive linear maps Q can arise as expansions for a substitution tiling in $\mathbb{R}^d$, without FLC? Assume that there are finitely many prototiles, either up to translation, or up to Euclidean isometries. Maybe assume that Q is a pure dilation by $\lambda > 1$ for the beginning.*

The only thing we know about this is that $\det(Q)$ must be a Perron number (assuming the substitution is primitive), and of course, the examples.

Acknowledgments Thanks to CIRM, Luminy, for hospitality and for providing a perfect work environment. The author is grateful to Shigeki Akiyama, the 2017 Morlet Chair, and to Pierre Arnoux for the invitation and for running a very successful and stimulating program. Valuable comments and suggestions from M. Baake and anonymous referees are gratefully acknowledged. Additional thanks are due to S. Akiyama and an anonymous volunteer for their help with the figures. The research of B.S. was supported by the Israel Science Foundation (Grant 396/15).

References

1. S. Akiyama, M. Barge, V. Berthé, J.-Y. Lee, A. Siegel, On the Pisot substitution conjecture, in *Mathematics of Aperiodic Order*, ed. by J. Kellendonk, D. Lenz, J. Savinien, Progr. Math., vol. 309 (Birkhäuser/Springer, Basel, 2015), pp. 33–72
2. S. Akiyama, J.-Y. Lee, Algorithm for determining pure pointedness of self-affine tilings. Adv. Math. **226**, 2855–2883 (2011)
3. M. Baake, U. Grimm, *Aperiodic Order. Vol. 1: A Mathematical Invitation* (Cambridge University Press, Cambridge, 2013)
4. M. Baake, U. Grimm (eds.), *Aperiodic Order, Vol. 2. Crystallography and Almost Periodicity*. Encyclopedia of Mathematics and Its Applications, vol. 166 (Cambridge University Press, Cambridge, 2017)
5. M. Baake, M. Schlottman, P.D. Jarvis, Quasiperiodic tilings with tenfold symmetry and equivalence with respect to local derivability. J. Phys. A **24**, 4637–4654 (1991)
6. M. Baake, D. Lenz, Dynamical systems on translation bounded measures: pure point dynamical and diffraction spectra. Ergodic Theory Dyn. Syst. **24**, 1867–1893 (2004)
7. M. Baake, D. Lenz, Spectral notions of aperiodic order. Discrete Contin. Dyn. Syst. **10**, 161–190 (2017)
8. M. Baake, D. Lenz, R.V. Moody, Characterization of model sets by dynamical systems. Ergodic Theory Dyn. Syst. **26**, 1–42 (2006)

9. M. Baake, R.V. Moody, Weighed Dirac combs with pure point diffraction. J. Reine Angew. Math. (Crelle) **573**, 61–94 (2004)
10. C. Bandt, Self-similar tilings and patterns described by mappings, in *The Mathematics of Long-Range Aperiodic Order (Waterloo, ON, 1995)*, ed. by R.V. Moody. NATO Adv. Sci. Inst. Ser. C Math. Phys. Sci., vol. 489 (Kluwer Acad. Publ., Dordrecht, 1997), pp. 45–83
11. J. Bellissard, D.J.L. Herrmann, M. Zarrouati, Hulls of aperiodic solids and gap labeling theorems, in *Directions in Mathematical Quasicrystals*, ed. by M. Baake, R.V. Moody. CRM Monograph Series, vol. 13 (AMS, Providence, RI, 2000), pp. 207–258
12. D.W. Boyd, Salem numbers of degree four have periodic expansions, *Théorie des nombres (Quebec, PQ, 1987)* (de Gruyter, Berlin, 1989), pp. 57–64
13. L. Danzer, Inflation species of planar tilings which are not of locally finite complexity. Proc. Steklov Inst. Math. **230**, 118–126 (2002)
14. B. Dundas, C. Skau, Interview with Abel laureate Yves Meyer. Eur. Math. Soc. Newsl. **105**, 14–22 (2017)
15. S. Dworkin, Spectral theory and X-ray diffraction. J. Math. Phys. **34**, 2964–2967 (1993)
16. N.P. Frank, Introduction to hierarchical tiling dynamical systems, in *Substitution and Tiling Dynamics: Introduction to Self-inducing Structures*, ed. by S. Akiyama, P. Arnoux. Lecture Notes in Mathematics (Springer, Cham, 2020)
17. N.P. Frank, E.A. Robinson, Jr., Generalized β-expansions, substitution tilings, and local finiteness. Trans. Am. Math. Soc. **360**, 1163–1177 (2008)
18. N.P. Frank, L. Sadun, Topology of (some) tiling spaces without finite local complexity. Discrete Contin. Dyn. Syst. **23**, 847–865 (2009)
19. N.P. Frank, L. Sadun, Fusion: a general framework for hierarchical tilings of $\mathbb{R}^d$. Geom. Dedicata **171**(1), 149–186 (2014)
20. N.P. Frank, L. Sadun, Fusion tilings with infinite local complexity. Topology Proc. **43**, 235–276 (2014)
21. N.P. Frank, B. Solomyak, A characterization of planar pseudo-self-similar tilings. Discrete Comput. Geom. **26**, 289–306 (2001)
22. D. Frettlöh, More inflation tilings, in *Aperiodic Order*, vol. 2, Encyclopedia Math. Appl., 166, (Cambridge University Press, Cambridge, 2017), pp. 1–37
23. D. Frettlöh, C. Richard, Dynamical properties of almost repetitive Delone sets. Discrete Contin. Dyn. Syst. **34**(2), 531–556 (2014)
24. D. Frettlöh, A.L.D. Say-awen, M.L.A.N. De Las Peñas, Substitution tilings with dense tile orientations and n-fold rotational symmetry. Indag. Math. (N.S.) **28**(1), 120–131 (2017)
25. G. Gelbrich, Crystallographic reptiles. Geom. Dedicata **51**, 235–256 (1994)
26. J.-B. Gouéré, Quasicrystals and almost periodicity. Commun. Math. Phys. **255**, 655–681 (2005)
27. A. Hof, On diffraction by aperiodic structures. Commun. Math. Phys. **169**, 25–43 (1995)
28. J. Kellendonk, L. Sadun, Meyer sets, topological eigenvalues, and Cantor fiber bundles. J. Lond. Math. Soc. **89**, 114–130 (2014)
29. J. Kellendonk, D. Lenz, Equicontinuous Delone dynamical systems. Canad. J. Math. **65**, 149–170 (2013)
30. R. Kenyon, Self-Similar Tilings. Ph.D Thesis (Princeton University, NJ, 1990)
31. R. Kenyon, Self-replicating tilings, in *Symbolic Dynamics and Its Application (New Haven, CT, 1991)*, ed. by P. Walters. Contemp. Math., vol. 135 (American Mathematical Society, Providence, RI, 1992), pp. 239–263
32. R. Kenyon, The construction of self-similar tilings. Geom. Funct. Anal. **6**, 471–488 (1996)
33. R. Kenyon, B. Solomyak, On the characterization of expansion maps for self-affine tilings. Discrete Comput. Geom. **43**, 577–593 (2010)
34. I. Környei, On a theorem of Pisot. Publ. Math. Debrecen **34**(3–4), 169–179 (1987)
35. J. Kwapisz, Inflations of self-affine tilings are integral algebraic Perron. Invent. Math. **205**, 173–220 (2016)
36. J.C. Lagarias, Meyer's concept of quasicrystal and quasiregular sets. Commun. Math. Phys. **179**, 365–376 (1996)

37. J.C. Lagarias, Geometric models for quasicrystals I. Delone sets of finite type. Discrete Comput. Geom. **21**, 161–191 (1999)
38. J.C. Lagarias, Mathematical quasicrystals and the problem of diffraction, in *Directions in Mathematical Quasicrystals*, ed. by M. Baake, R.V. Moody. CRM Monograph Series, vol. 13 (AMS, Providence, RI, 2000), pp. 61–93
39. J.C. Lagarias, Y. Wang, Substitution Delone sets. Discrete Comput. Geom. **29**, 175–209 (2003)
40. J.-Y. Lee, R.V. Moody, B. Solomyak, Pure point dynamical and diffraction spectra. Ann. Henri Poincaré **3**, 1003–1018 (2002)
41. J.-Y. Lee, R.V. Moody, B. Solomyak, Consequences of pure point diffraction spectra for multiset substitution systems. Discrete Comput. Geom. **29**, 525–560 (2003)
42. J.-Y Lee, B. Solomyak, Pure point diffractive substitution Delone sets have the Meyer property. Discrete Comput. Geom. **39**, 319–338 (2008)
43. J.-Y. Lee, B. Solomyak, On substitution tilings and Delone sets without finite local complexity. Discrete Contin. Dyn. Syst. **39**(6), 3149–3177 (2019)
44. D. Lenz, N. Strungaru, Pure point spectrum for measure dynamical systems on locally compact Abelian groups. J. Math. Pures Appl. **92**, 323–341 (2009)
45. N. Lev, A. Olevskii, Quasicrystals and Poisson's summation formula. Invent. Math. **200**, 585–606 (2015)
46. N. Lev, A. Olevskii, Quasicrystals with discrete support and spectrum. Rev. Mat. Iberoam. **32**(4), 1341–1352 (2016)
47. D. Lind, The entropies of topological Markov shifts and a related class of algebraic integers. Ergodic Theory Dyn. Syst. **4**, 283–300 (1984)
48. C. Mauduit, Caractérisation des ensembles normaux substitutifs. Invent. Math. **95**, 133–147 (1989)
49. Y. Meyer, *Algebraic Numbers and Harmonic Analysis* (North-Holland, Amsterdam, 1972)
50. Y. Meyer, Quasicrystals, Diophantine approximation and algebraic numbers, in *Beyond quasicrystals (Les Houches, 1994)*, ed. by F. Axel, D. Gratias (Springer, Berlin, 1995), pp. 3–16
51. R.V. Moody, Meyer sets and their duals, in *The Mathematics of Long-Range Aperiodic Order (Waterloo, ON, 1995)*, ed. by R.V. Moody. NATO Adv. Sci. Inst. Ser. C Math. Phys. Sci. vol. 489 (Kluwer, Dordrecht, 1997), pp. 403–441
52. P. Müller, C. Richard, Ergodic properties of randomly coloured point sets. Canad. J. Math. **65**, 349–402 (2013)
53. W. Parry, On the β-expansions of real numbers. Acta Math. Acad. Sci. Hungar. **11**, 401–416 (1960)
54. C. Radin, The pinwheel tilings of the plane. Ann. Math. **139**, 661–702 (1994)
55. E.A. Robinson, Symbolic dynamics and tilings of $\mathbb{R}^d$, in *Symbolic Dynamics and Its Applications*, ed. by S.G. Williams. Proc. Sympos. Appl. Math., vol. 60 (American Mathematical Society, Providence, RI, 2004), pp. 81–119
56. R. Salem, *Algebraic Numbers and Fourier Analysis* (D. C. Heath and Co., Boston, MA, 1963)
57. M. Schlottman, Generalized model sets and dynamical systems, in *Directions in Mathematical Quasicrystals*, ed. by M. Baake, R.V. Moody. CRM Monograph Series, vol. 13 (American Mathematical Society, Providence, RI, 2000), pp. 143–159
58. B. Solomyak, Dynamics of self-similar tilings. Ergodic Theory Dyn. Syst. **17**, 695–738 (1997) [Corrections: *ibid.* **19** (1999), 1685.]
59. B. Solomyak, Spectrum of dynamical systems arising from Delone sets, in *Quasicrystals and Discrete Geometry*, ed. by J. Patera. The Fields Institute for Research in Mathematical Sciences Monograph Series, vol. 10 (American Mathematical Society, Providence, RI, 1998), pp. 265–275
60. B. Solomyak, Pseudo-self-affine tilings in $\mathbb{R}^d$. Zap. Nauchn. Semin. POMI **326**, 198–213 (2005); translation in *J. Math. Sci. (N.Y.)* **140**, 452–460 (2007)
61. B. Solomyak, *Tilings and Dynamics*. EMS Summer School on Combinatorics, Automata and Number Theory, 8–19 May 2006, Liege, available at address http://u.math.biu.ac.il/~solomyb/PREPRINTS/liege.pdf

62. N. Strungaru, Almost periodic measures and long-range order in Meyer sets. Discrete Comput. Geom. **33**, 483–505 (2005)
63. N. Strungaru, Almost periodic pure point measures, in *Aperiodic Order*, vol. 2. Encyclopedia Math. Appl., vol. 166 (Cambridge University Press, Cambridge, 2017), pp. 271–342
64. W. Thurston, Groups, Tilings, and Finite State Automata. AMS Lecture Notes (1989)

Chapter 2
Introduction to Hierarchical Tiling Dynamical Systems

Natalie Priebe Frank

Abstract This chapter is about the tiling dynamical systems approach to the study of aperiodic order. We compare and contrast four related types of systems: ordinary (one-dimensional) symbolic systems, one-dimensional tiling systems, multidimensional $\mathbb{Z}^d$-systems, and multidimensional tiling systems. Aperiodically ordered structures are often hierarchical in nature, and there are a number of different yet related ways to define them. We will focus on what we are calling "supertile construction methods": symbolic substitution in one and many dimensions, S-adic sequences, self-similar and pseudo-self-similar tilings, and fusion rules. The techniques of dynamical analysis of these systems are discussed and a number of results are surveyed. We conclude with a discussion of the spectral theory of supertile systems from both the dynamical and diffraction perspectives.

2.1 Introduction

The central objects in these lecture notes are tilings constructed via a variety of methods that together we call *supertile methods*. These tilings display hierarchical structure that is highly ordered yet not periodic. Their study is truly multidisciplinary, having originated in fields as disparate as logic, chemistry and geometry. To motivate the topic we offer three examples from the history of the field that are relevant to these lectures.

First, imagine square tiles whose edges come in given combinations of colors, and you are only allowed to put two tiles next to each other if the edge colors match. Can you make an infinite tiling of the plane with these tiles? This is the question logician Hao Wang was considering in 1961 [105]. In particular he was thinking about the decidability of what is now known as the domino problem: "Given a finite

N. P. Frank (✉)
Department of Mathematics and Statistics, Vassar College, Poughkeepsie, NY, USA
e-mail: nafrank@vassar.edu

© The Editor(s) (if applicable) and The Author(s), under exclusive license to Springer Nature Switzerland AG 2020
S. Akiyama, P. Arnoux (eds.), *Substitution and Tiling Dynamics: Introduction to Self-inducing Structures*, Lecture Notes in Mathematics 2273, https://doi.org/10.1007/978-3-030-57666-0_2

33

set of tiles in the plane, can it be made to form an infinite tiling?" The answer depended on whether an *aperiodic prototile set* exists, i.e. a set of tiles that are able to form an infinite tiling of the plane, but every tiling they make must be nonperiodic. The question was proved to be undecidable by Robert Berger [22] with the discovery of an aperiodic set of prototiles. That prototile set had over 20,000 tiles in it, but in 1971 Raphael Robinson published an aperiodic set with only 6 tiles [94]. In Robinson's version the hierarchical structure is clearly evident and in fact drives the proof of aperiodicity. In this volume in [63] we find four proofs of undecidability, including how to construct the aperiodic tile set(s).

A second development, which entered the public consciousness through a *Scientific American* article by Martin Gardner [54], was Penrose's 1974 discovery of an aperiodic set of two tiles. In the middle of the twentieth century, Roger Penrose began to develop an interest in tiling questions in part because of Hilbert's Problem 18. The interest intensified as Penrose and his father developed a collaboration with M. C. Esher (see the foreword to [11]). Penrose was trying to create a hierarchical tiling and found his original tiling (which in that foreword he tells us is [82, Fig. 4]) by experimentation.

There are a number of versions of Penrose tilings, all of which can be generated by a supertile method. In Figs. 2.1 and 2.2 we show a *pseudo-self-similar* version (see Sect. 2.4.4.2), for which the tiles also form an aperiodic tile set. Figure 2.1 shows the rule for inflating and replacing the tiles, and Fig. 2.2 shows the result of inflating and replacing a central patch twice.

The third and possibly most invigorating development we mention here is the discovery of quasicrystals in 1982 [97]. This earned Dan Shechtman the Wolf Prize in Physics in 1999 and the Nobel Prize in Chemistry in 2011 [73]. In his laboratory in what is now the U. S. National Institute of Standards and Technology, Shechtman analyzed an aluminum-magnesium alloy and found that its diffraction image revealed contradictory properties: it had bright spots indicative of a periodic atomic structure, but had symmetries impossible for such a structure. The discovery went against all conventional wisdom at the time, but eventually the scientific community accepted that there was no mistake, this alloy did indeed display 'quasi'-crystalline structure. In some of the images in Fig. 2.3 one can see the 'forbidden' tenfold rotational symmetries.

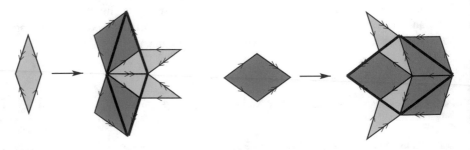

Fig. 2.1 The Penrose rhombuses and their inflate-and-subdivide rules

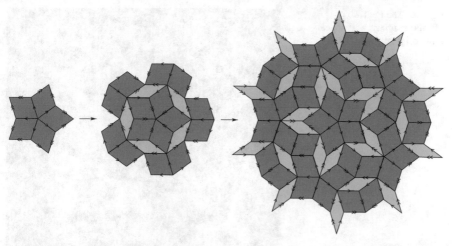

Fig. 2.2 1-, 2-, and 3-supertiles for the Penrose rhombus tiling

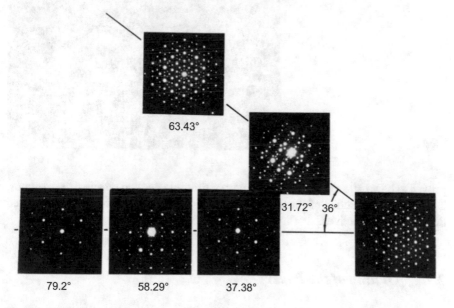

Fig. 2.3 The quasicrystal diffraction images as they appear in the original paper [97]

Coincidentally, in 1982 Alan Mackay [77] published the diffraction pattern of a Penrose tiling, shown in its original form in Fig. 2.4. Once Shechtman's diffraction pattern was published, it did not take long for similarities between it and Mackay's to be noticed. This established Penrose tilings and highly structured tilings like them (including some generated by supertile methods), as mathematical models of quasicrystals. It is apparent that spectral methods, then, are an interesting way to study aperiodic tilings. Spectral analysis, including mathematical diffraction, has

Fig. 2.4 The Penrose
diffraction image as it appears
in the original paper [77]

proved to be an effective tool for the study of tilings generated by supertile methods.
The last three sections of these notes discuss spectral theory from dynamical and
diffraction perspectives.

2.1.1   Outline of the Paper

In Sect. 2.2 we begin by giving general definitions of the four types of structures
of interest and the basic relationships between them. Specifics of why and how the
dynamical systems viewpoint is used appears in Sect. 2.3. In this section we compare
and contrast how the metrics are related, show how standard dynamical properties
like minimality can be interepreted, and talk about invariant measures and their
connection to the idea of frequency. In Sect. 2.4 we learn about the various supertile
construction methods and give examples of many of them. In Sect. 2.5 we introduce
the idea of transition matrices and how their properties allow us to extend dynamical
results to supertile systems. Section 2.6 is devoted to the dynamical spectrum of

supertile systems, while Sect. 2.7 is devoted to their diffraction spectra. Section 2.8 presents results on the connection between the two types of spectrum. We conclude in Sect. 2.9 with a selection of references.

2.1.2   Not Covered

The field of aperiodic order and tiling dynamical systems spans a broad range of topics and we have not attempted to give a complete survey. Topics we do not discuss include tiling cohomology, matching rules, the projection method, tilings with infinite local complexity, K-theory of C^*-algebras for tilings, spectral triples, decidability and tiling problem questions. Moreover we do not consider tilings of hyperbolic space or other spaces, or anything about the spectrum of Schrodinger operators modeled on tilings.

2.2   The Fundamental Objects

2.2.1   Motivation: Shift Spaces

The way that we study tiling spaces is a generalization of symbolic dynamics, a large branch of dynamical systems theory. Thus we begin by describing the basic setup in this situation.

Let $\mathcal{A}$ be a finite set we will call our *alphabet*. A *sequence* is a function $\mathbf{x}$: $\mathbb{Z} \to \mathcal{A}$ and we denote the set of all sequences to be $\mathcal{A}^{\mathbb{Z}}$. We equip the space with a metric that defines the product topology, as follows. Let $N(\mathbf{x}, \mathbf{y}) = \min\{n \geq 0$ such that $\mathbf{x}(j) \neq \mathbf{y}(j)$ for some $|j| = n\}$, and define $d(\mathbf{x}, \mathbf{y})$ to be $\exp(-N(\mathbf{x}, \mathbf{y}))$. That is, $\mathbf{x}$ and $\mathbf{y}$ are very close if they agree on a large ball centered at the origin.

For each $j \in \mathbb{Z}$, we can shift a sequence $\mathbf{x}$ by j, yielding the sequence $\mathbf{x} - j$ defined by $(\mathbf{x} - j)(k) = \mathbf{x}(k + j)$. This is known as the shift map, and is a $\mathbb{Z}$-action on sequences. (Notice that $(\mathbf{x} - j)(0) = \mathbf{x}(j)$, meaning that $\mathbf{x}$ has been shifted so that what was at j is now at the origin.) The space $\mathcal{A}^{\mathbb{Z}}$ along with the shift map is known as the *full shift* on $|\mathcal{A}|$ symbols.[1] The shift map allows us to investigate the long-range structure of sequences by moving distant parts 'into view' of the origin. This perspective is consistent with our choice of metric topology.

There is already some dynamics to study for the full shift, but things get much more interesting when we restrict our attention to closed, nonempty, shift-invariant subsets $\Omega \subset \mathcal{A}^{\mathbb{Z}}$. We call such an Ω a *shift space* or *subshift of* $\mathcal{A}^{\mathbb{Z}}$ and use the terminology *shift dynamical system* for $(\Omega, \mathbb{Z})$. Subshifts are handy tools for

---

[1] Ordinarily in the literature the shift map is given with notation like $\sigma(\mathbf{x})$, so that $\mathbf{x} - j = \sigma^j(\mathbf{x})$. We use the notation "$\mathbf{x} - j$" instead to be consistent with the more general case.

encoding the dynamics of many types of systems, and they also arise in natural processes. Readers interested in diving into the vast literature on this subject might find [6, 42, 67, 74, 76] in their libraries.

Example 1 Let Ω consist of the periodic sequences $\ldots 0101.0101 \ldots$ and its shift $\ldots 1010.1010 \ldots$, where the decimal point is there to denote where the origin is. Shifting by $j \in \mathbb{Z}$ just moves the decimal point j units to the right (or left, if j is negative). One sees quickly that Ω is shift-invariant and that the dynamical system is periodic with period 2.

Because periodic systems like these are completely understood we will tend to assume that the sequences in our sequence spaces are not periodic. Instead, supertile construction methods generate sequences with just the right amount of long-range order to be tractable for analysis.

2.2.2 Straightforward Generalization: Sequences in $\mathbb{Z}^d$

Let $\mathcal{A}$ be a finite alphabet and consider $\mathcal{A}^{\mathbb{Z}^d}$ to be the set of all sequences in $\mathbb{Z}^d$, that is, functions from $\mathbb{Z}^d$ to $\mathcal{A}$. Given $\mathbf{x}, \mathbf{y} \in \mathcal{A}^{\mathbb{Z}^d}$, let $N(\mathbf{x}, \mathbf{y}) = \min\{n \geq 0$ such that $\mathbf{x}(\jmath) \neq \mathbf{y}(\jmath)$ for some $|\jmath| = n\}$, where $|\jmath|$ is the largest absolute value of the components of $\jmath$. Then $d(\mathbf{x}, \mathbf{y}) = \exp(-N(\mathbf{x}, \mathbf{y}))$ provides an origin-centric metric as before.

Translation by elements of $\mathbb{Z}^d$ is defined as before and provides a way to analyze the structure of multidimensional sequences. There are complications and considerations due to the extra dimensions that we will discuss as we encounter them.

2.2.3 Straightforward Generalization: Tilings of $\mathbb{R}$

We choose a closed interval for each symbol in $\mathcal{A}$. For any element $\mathbf{x} \in \mathcal{A}^{\mathbb{Z}}$ make a tiling by placing the interval corresponding to $\mathbf{x}(0)$ with its left endpoint at 0, and placing copies of all the other symbols of $\mathbf{x}$ in the corresponding order, with overlap at the interval endpoints. In this perspective a tile is a closed interval labelled by an element of $\mathcal{A}$. Tiles and tilings can be translated by elements of $\mathbb{R}$ and there is an origin-centric tiling metric that we will describe in the general situation in the next section.

Figure 2.5 depicts a tiling of $\mathbb{R}$ with two tile types, a longer interval pictured in dark blue and a shorter interval pictured in light blue. (The colors represent the labels). The patch shown corresponds to the symbolic sequence $\ldots abbbaaaabbbabbbabbba \ldots$

Fig. 2.5 A patch of a one-dimensional tiling with two tile lengths

2.2.4 Geometric Generalization: Tilings of $\mathbb{R}^d$

The finite alphabet $\mathcal{A}$ is replaced by a finite *prototile set* $\mathcal{P}$. A prototile $p \in \mathcal{P}$ is a closed topological disk in $\mathbb{R}^d$ carrying a label (for instance, a color). The closed set is known as the *support* of p (denoted supp(p)) and the label is there to distinguish any tiles that may have congruent shapes. We can apply any self-map of $\mathbb{R}^d$ to a prototile by applying it to the support and carrying the label along. Although it is common to use some or all elements of the Euclidean group to move tiles around, we restrict our attention to translations only. A $\mathcal{P}$-*tile* or just *tile* is any translate of a prototile from $\mathcal{P}$. Two tiles are *equivalent* if their supports are translates of each other and they carry the same label.

Consider some fixed set of prototiles $\mathcal{P}$. A $\mathcal{P}$-*patch* (or *patch* when the prototile set is understood) is a set of tiles that intersect at most on their boundaries that is supported on a connected set in $\mathbb{R}^d$. For technical reasons it is often assumed that the supports form a topological disk. A $\mathcal{P}$-*tiling* or just *tiling* of $\mathbb{R}^d$ is a collection of tiles that 'cover' $\mathbb{R}^d$ in the sense that the union of the tile supports is $\mathbb{R}^d$, but also 'pack' $\mathbb{R}^d$ in the sense that any two supports intersect only on their boundaries. Let $\Omega_{\mathcal{P}}$ denote the space of all $\mathcal{P}$-tilings. As with the full shift $\mathcal{A}^{\mathbb{Z}}$, in which an element is an infinite sequence, elements of $\Omega_{\mathcal{P}}$ are infinite tilings of $\mathbb{R}^d$.

Like tiles, patches and tilings can be translated by elements of $\mathbb{R}^d$. We write $\mathcal{T} - \mathbf{v}$ to denote the tiling obtained by translating the support of every tile of $\mathcal{T}$ by $\mathbf{v}$. Note that the origin in $\mathcal{T} - \mathbf{v}$ corresponds to $\mathbf{v}$ in $\mathcal{T}$, so this translation brings the neighborhood of $\mathbf{v}$ into view of the origin.

Analogous to the simpler cases, we say a tiling $\mathcal{T}$ is *nonperiodic* if there is no $\mathbf{v}$ for which $\mathcal{T} - \mathbf{v} = \mathcal{T}$. In higher dimensions it is possible to be periodic in some directions but not *fully periodic*: the directions of periodicity must form a basis for $\mathbb{R}^d$ for full periodicity.

Now geometry plays a fundamental role, and there is the possibility that tiles in a tiling can be adjacent in many different ways. Consider the tiling in Fig. 2.6, which is constructed from a set of four rectangular tiles, with side lengths given by 1 and $\frac{1+\sqrt{17}}{2}$. There are many offsets where vertices meet edges, and the number of those offsets will go to infinity as we consider larger and larger patches.

Definition 2.2.1 We say a tiling $\mathcal{T} \in \Omega_{\mathcal{P}}$ has *finite local complexity (FLC)* if it contains only finitely many two-tile patches up to translation. A subset of tilings $\Omega \subset \Omega_{\mathcal{P}}$ is said to have finite local complexity if there are only finitely many two-tile patches up to translation in Ω.

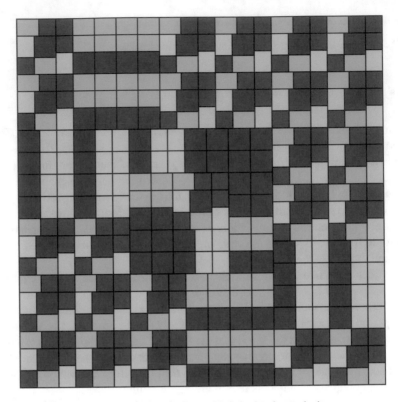

Fig. 2.6 A patch of a tiling with four prototiles and infinite local complexity

For the purposes of this work we assume finite local complexity in all tilings and tiling spaces unless otherwise stated.[2]

It is convenient to have notation for the patch of a tiling that intersects a subset of $\mathbb{R}^d$. Let $\mathcal{T}$ be a tiling of $\mathbb{R}^d$ and let $B \subset \mathbb{R}^d$. The patch of tiles in $\mathcal{T}$ whose supports intersect B is denoted $\mathcal{T} \cap B$. One could say that $\mathcal{T}$ has finite local complexity if the set of patches

$$\left\{ \mathcal{T} \cap \{x\} \text{ such that } x \in \mathbb{R}^d \right\}$$

is finite up to translation.

---

[2]FLC is a common restriction, but if you want to learn about the infinite local complexity case see [47] and references within.

2.3 The Dynamical Systems Viewpoint

2.3.1 Tiling Spaces

A tiling space is a translation-invariant subset Ω of the full tiling space $\Omega_{\mathcal{P}}$ that is closed in the metric we describe next. We form a dynamical system by letting an additive subgroup G of $\mathbb{R}^d$ act on it by translation. Ordinarily G is just $\mathbb{R}^d$ itself, but occasionally it might be $\mathbb{Z}^d$ or some other full rank subgroup. We use the notation (Ω, G) to denote the tiling dynamical system.

2.3.1.1 The "big ball" Metric

The metric used in dynamical systems theory for tilings is modeled on the metric for shift spaces, and therefore is also origin-centric. The definition of the metric becomes technical because translation is a continuous action and because the prototiles can have interesting geometry. Still the basic idea is that two tilings are close if they very nearly agree on a ball around the origin.

We give the definition of metric for tilings of finite local complexity. Let $\mathcal{T}$ and $\mathcal{T}'$ be tilings of $\mathbb{R}^d$ from a prototile set $\mathcal{P}$. Informally, we say $\mathcal{T}$ and $\mathcal{T}'$ are within ϵ of one another if they agree on a ball of radius $1/\epsilon$, except for a small translation. Here is a formal definition.

Definition 2.3.1 Let $R(\mathcal{T}, \mathcal{T}')$ be the supremum of all $r \geq 0$ such that there exists $\mathbf{x}, \mathbf{y} \in \mathbb{R}^d$ with

1. $|\mathbf{x}| < 1/2r$ and $|\mathbf{y}| < 1/2r$, and
2. On the ball of radius r around the origin, $(\mathcal{T} - \mathbf{x}) \cap B_r(0) = (\mathcal{T}' - \mathbf{y}) \cap B_r(0)$.

We define

$$d(\mathcal{T}, \mathcal{T}') := \min\left\{\frac{1}{R(\mathcal{T}, \mathcal{T}')}, 1\right\}.$$

There are various versions in the literature; this version parallels [100].

2.3.1.2 Two Common Ways to Construct Tiling Spaces

There are two main ways that tiling spaces are constructed. One is to make a closed, translation-invariant space around a given tiling; in this situation the space is called the 'hull' of the tiling. The other is to specify an 'atlas' of allowed patches and include tilings that contain these patches only.

For the first method, suppose there is some tiling $\mathcal{T} \in \Omega_{\mathcal{P}}$ that is of particular interest. We can construct the *hull of the tiling* $\mathcal{T}$ as the orbit closure of $\mathcal{T}$:

$$\Omega_{\mathcal{T}} = \overline{\{\mathcal{T} - \mathbf{v} \text{ for all } \mathbf{v}\}}.$$

By definition it is closed and it is not difficult to show that it is translation-invariant.

The second method is akin to making a shift space from a language of allowed words. Let $\mathcal{R}$ be a set of $\mathcal{P}$-patches. We say that $\mathcal{T} \in \Omega_{\mathcal{P}}$ is *allowed* by $\mathcal{R}$ if every patch in $\mathcal{T}$ is translation-equivalent to a subpatch of an element of $\mathcal{R}$. The tiling space $\Omega_{\mathcal{R}}$ is the set of all allowed tilings.

We will ignore all questions of which types of rules $\mathcal{R}$ admit non-trivial tiling spaces, referring them to theoretical computer scientists and/or logicians. But it should be clear that if nontrivial, $\Omega_{\mathcal{R}}$ should be translation-invariant and closed in the big ball metric.

2.3.1.3 Cylinder Sets and the Metric Topology

Because both are important to our further analysis we discuss these topics for both the motivating symbolic case and for the tiling situation.

In symbolic dynamics the fundamental sets are *cylinder sets*. Consider a shift space Ω and suppose w is a finite word in $\mathcal{A}^*$, where $\mathcal{A}^*$ is the set of non-empty words on $\mathcal{A}$. The *cylinder set* Ω_w *generated by* w is given by $C_w = \{\mathbf{x} \in \Omega$ such that $\mathbf{x}(U) = w\}$; it is the set of all sequences that contain the word w in the location given by U, with no other restrictions. One can check that cylinder sets are both closed and open in the metric topology. One can also check that for any $\epsilon > 0$ and any $\mathbf{x} \in \Omega$, the ball of radius ϵ around x is a cylinder set for a word around the origin in $\mathbf{x}$. Thus cylinder sets form a basis for the topology in shift spaces. When we are considering sequences in $\mathbb{Z}^d$ the cylinder sets are completely analogous.

The situation becomes somewhat more complicated for tilings of $\mathbb{R}^d$ when the translation group G is uncountable. Let P be a $\mathcal{P}$-patch, let $U \subset \mathbb{R}^d$, and let $\Omega \subset \Omega_{\mathcal{P}}$ be a tiling space. The cylinder set $\Omega_{P,U}$ is the set of all tilings in Ω that contain a copy of P translated by an element of U. That is,

$$\Omega_{P,U} = \{\mathcal{T} \in \Omega \text{ such that } P - u \text{ is a } \mathcal{T} - \text{patch for some } u \in U\}.$$

The reader can check that if ϵ is sufficiently small and $U = B_\epsilon(0)$ (the open ball of radius ϵ around the origin), the cylinder set is open. In [93, p. 13] we see how to get a countable basis for the topology by discretizing $\epsilon_n \to 0$, since there are only a countable number of patches of any size up to translation.

The basic fact of compactness is proved in several works, see for example [92]. We include a short argument here for the tiling situation.

Lemma 2.3.2 *If $\Omega \subset \Omega_{\mathcal{P}}$ is closed and of finite local complexity, then Ω is complete and compact.*

Proof Let $\{\mathcal{T}_n\}$ be a Cauchy sequence in Ω and fix some $K \in \mathbb{Z}$. Consider $\epsilon > 0$ for which $1/\epsilon > K$. There is some M such that for $n, m \geq M$, $d(\mathcal{T}_n, \mathcal{T}_m) < \epsilon$. This means that the patches $\mathcal{T}_n \cap B_{1/\epsilon}(0)$ and $\mathcal{T}_m \cap B_{1/\epsilon}(0)$ agree up to translation by at most ϵ. Thus the patches $\mathcal{T}_n \cap B_K(0)$ and $\mathcal{T}_m \cap B_K(0)$ agree up to translation $< \epsilon$. As $\epsilon \to 0$ there is a patch P_K covering $B_K(0)$ that is the limit of the patches $\mathcal{T}_n \cap B_K(0)$. We obtain a nested sequence of patches P_K and therefore there is a tiling $\mathcal{T}$ such that $P_K \subset \mathcal{T}$ for all $K \in \mathbb{Z}$, and this tiling must be the limit of the Cauchy sequence. All of the tiles in $\mathcal{T}$ belong in a P_K for some K and so are $\mathcal{P}$-tiles, thus $\mathcal{T} \in \Omega_{\mathcal{P}}$. Since Ω is closed we know $\mathcal{T} \in \Omega$, proving sequential compactness.

Under many conditions, for instance topological transitivity, Ω is connected. Each tiling in Ω defines a path component that is a continuous embedding of $\mathbb{R}^d$. In general there are uncountably many distinct $\mathcal{P}$-tilings up to translation, and therefore there are uncountably many path components.

2.3.2 Notions of Equivalence for Symbolic and Tiling Dynamical Systems

Suppose we have a sequence in $\mathcal{A} = \{0, 1\}$. In what way does it change if we make every 0 into an a and every 1 into a b? What about if we had a checkerboard tiling with black and white squares, and split each black square horizontally into two rectangles? In the symbolic case there are local maps called "sliding block codes" which determine factor maps and topological conjugacies between shift spaces. The tiling equivalent is local derivability through local mappings.

2.3.2.1 Sliding Block Codes

We follow [74]. Let $\mathcal{A}$ and $\mathcal{A}'$ be finite alphabets and suppose Ω is a shift space in $\mathcal{A}^{\mathbb{Z}}$. Choose nonnegative integers m and n and let $B_{m,n}$ denote the set of all words of length $m + n + 1$ that appear in Ω. Let $\Phi : B_{m,n} \to \mathcal{A}'$ be any map. Then the *sliding block code* $\phi : \Omega \to (\mathcal{A}')^{\mathbb{Z}}$ is defined by this map via

$$y_i = \Phi(x_{i-m} x_{i-m+1} \cdots x_{i+n-1} x_{i+n}) = (\phi(\mathbf{x}))_i.$$

Thus we see that a sliding block code will convert every sequence $\mathbf{x}$ to a sequence $\mathbf{y}$ entry by entry, examining the block in $\mathbf{x}$ around x_i and using it to determine the value of y_i. It is not difficult to check that sliding block codes are continuous. This powerful theorem tells us that sliding block codes are the only maps on shift spaces that are both continuous and shift-commuting:

Theorem 2.3.1 (Curtis-Lyndon-Hedlund, See [74]) *Suppose Ω and Ω' are shift spaces, not necessarily on the same alphabet, and let $\theta : \Omega \to \Omega'$. Then θ is a sliding block code if and only if it is shift-commuting and continuous.*

In particular this means that topological conjugacies between shift dynamical systems are invertible sliding block codes and vice versa.

2.3.2.2 Local Derivability

Local mappings are the analogue to sliding block codes for tilings of $\mathbb{R}^d$. We give a brief definition here; a full exposition appears in section 5.2 of [11].

Definition 2.3.3 A continuous surjective mapping between tiling spaces $Q : \Omega \to \Omega'$ is a *local mapping* if there is an $r > 0$ such that for any $x \in \mathbb{R}^d$ and $\mathcal{T}_1, \mathcal{T}_2 \in \Omega$, if $\mathcal{T}_1 \cap B_r(x) = \mathcal{T}_2 \cap B_r(x)$, then $Q(\mathcal{T}_1) \cap \{x\} = Q(\mathcal{T}_2) \cap \{x\}$.

That is to say, the patch in $\mathcal{T}$ containing the ball $B_r(x)$ completely determines the tile at the center of the ball in $Q(\mathcal{T})$. If such a local mapping exists we say $Q(\mathcal{T})$ is *locally derivable* from $\mathcal{T}$. If Q is invertible we say $\mathcal{T}$ and $Q(\mathcal{T})$ are *mutually locally derivable*, and we also use this terminology for their tiling spaces. It is not difficult to show that any local mapping is continuous in the big ball topology.

Lemma 2.3.4 *If Ω and Ω' are mutually locally derivable tiling spaces, then their dynamical systems are topologically conjugate.*

If there were to be a tiling analogue of the Curtis-Lyndon-Hedlund theorem, it would mean that the only continuous translation-commuting maps between tiling spaces are local mappings. That is, the above lemma would be an "if and only if". The fact that it is not was first shown in [85] and [91].

Nonlocal homeomorphisms for tilings tend to require information from far distances in $\mathcal{T}$ to settle the precise location of the origin in $Q(\mathcal{T})$. In Example 13 of Sect. 2.4.7 we describe how to make a nonlocal homeomorphism between two tiling spaces generated by a related pair of supertile methods.

2.3.3 Repetitivity and Minimality

Recall that a dynamical system is called *transitive* if there is a dense orbit and *minimal* if every orbit is dense.

Definition 2.3.5 A tiling $\mathcal{T}$ is said to be *repetitive*[3] iff for every finite patch P in $\mathcal{T}$ there is an $R = R(P) > 0$ such that $\mathcal{T} \cap B_R(x)$ contains a translate of P for every

---

[3] Also known in the literature as $\mathcal{T}$ being *uniformly recurrent*, *almost periodic*, and having the *local isomorphism property*.

$x \in \mathbb{R}^d$. It is *linearly repetitive* iff there is a $C > 0$ such that for any $\mathcal{T}$-patch P there is a translate of P in any ball of radius $C\,diam(P)$ in $\mathcal{T}$.

In other words, a tiling is repetitive if for every patch P there is some radius R such that every ball of that radius contains a copy of P. Moreover, it is linearly repetitive if R can be taken to be $C\,diam(P)$, that is, the radius depends only linearly on the size of P. In [35, 36] it is shown that a symbolic system is linearly repetitive if and only if it is "primitive and proper" S-adic (a supertile method discussed in Sect. 2.4.5.2).

Standard arguments show the following, stated here using tiling terminology but applicable to symbolic spaces as well.

Lemma 2.3.6 (See E.g. [92, 93, 100]) *Let $\mathcal{T} \in \Omega_{\mathcal{P}}$ and let $\Omega_{\mathcal{T}}$ denote its hull. The tiling dynamical system $(\Omega_{\mathcal{T}}, \mathbb{R}^d)$ is minimal if and only if $\mathcal{T}$ is repetitive.*

Large classes of supertile methods produce sequences or tilings that are repetitive, and therefore their dynamical systems are minimal.

2.3.4   Invariant and Ergodic Measures

Let Ω be a shift or tiling space with topology given by the appropriate metric, and let G be the group of translations defining its dynamical system. A Borel probability measure μ on Ω is said to be *invariant* with respect to translation if $\mu(A - g) = \mu(A)$ for all Borel measurable sets A and all $g \in G$. We say μ is *ergodic* with respect to translation if whenever A is a translation-invariant set, then $\mu(A)$ equals 0 or 1. The set of invariant Borel probability measures is convex and its extremal elements are ergodic (see for example [84] or [104] for the general theory of ergodic measures).

A dynamical system is said to be *uniquely ergodic* if it possesses only one ergodic measure. Because the set of all invariant measures is convex and the ergodic measures are the extremal measures from that set, this implies that the ergodic measure is also the only invariant measure.

Let P be a $\mathcal{P}$-patch and let U be a fixed and very small ball so that if $\mathcal{T} \in \Omega_{P,U}$, then $P - g \in \mathcal{T}$ for at most one $g \in U$. Denote by $\mathbb{I}_{P,U}$ the indicator function for $\Omega_{P,U}$ and suppose μ is some ergodic measure for translation. Then from elementary measure theory along with the ergodic theorem[4] we know that for μ-almost every $\mathcal{T}_0 \in \Omega$,

$$\mu(\Omega_{P,U}) = \int_{\Omega} \mathbb{I}_{P,U}(\mathcal{T})d\mu(\mathcal{T}) = \lim_{r \to \infty} \frac{1}{Vol(B_r(0))} \int_{B_r(0)} \mathbb{I}_{P,U}(\mathcal{T}_0 - x)dx.$$

---

[4]For a more indepth discussion of the meaning of the word 'frequency' and the appropriate ergodic theorem for this setting, see [50, Section 3.3].

Consider the integral on the right. For every copy of P in $\mathcal{T}_0 \cap B_r(0)$ that isn't too close to the boundary of $B_r(0)$ the indicator function will be 1 over a set of size $Vol(U)$. Any copy of P that is too close to the boundary will only yield a portion of that, but as $r \to \infty$ this effect is negligible. Letting the notation $\#(P \in \mathcal{T}_0 \cap B_r(0))$ mean the number of copies of P in $\mathcal{T}_0 \cap B_r(0)$, it is straightforward to show that the term on the right therefore becomes $\lim_{r \to \infty} \frac{\#(P \in \mathcal{T}_0 \cap B_r(0))}{Vol(B_r(0))} Vol(U)$. For μ-almost every $\mathcal{T}_0$ we get the same answer and so we can say that μ defines a frequency measure on the set of $\mathcal{P}$-patches as:

$$\mu(\Omega_{P,U}) = Vol(U) freq_\mu(P).$$

2.4 Supertile Construction Techniques

When we consider a sequence or tiling space Ω under the action of translation it is not particularly interesting if the elements of Ω are periodic. Considering the dynamics on the full shift $\mathcal{A}^{\mathbb{Z}^d}$ or full tiling space $\Omega_{\mathcal{P}}$ is more interesting, since the spaces have many properties including carrying many different measures, having many possible letter/tile frequencies, and having nontrivial positive topological and measure-theoretic entropies, for instance. But in this study we wish to apply the theory to spaces whose elements all have common properties that arise from given construction techniques. In particular we look at sequences and tilings constructed via *substitution* or *fusion*, which we are generically terming "supertile constructions".

2.4.1 Motivation: Symbolic Substitutions

Introduced as examples of symbolic dynamical systems by Gottschalk in [57], these are the fundamental (and simplest) objects on which our other supertile methods are based. Much is known about substitution sequences and the books [42, 89] are devoted to results on the subject. We will expose many of these results and see when they have generalizations or fail to generalize to higher-dimensional structures.

Given a finite alphabet $\mathcal{A}$, a *substitution* is a map $\sigma : \mathcal{A} \to \mathcal{A}^*$, where $\mathcal{A}^*$ is the set of non-empty words on $\mathcal{A}$. The substitution can be applied to words by concatenating the substitution of the letters in the word. We use the terminology *n-superword* to mean a word of the form $\sigma^n(a)$ for some $a \in \mathcal{A}$.

A sequence $\mathbf{x} \in \mathcal{A}^{\mathbb{Z}}$ is said to be *admitted* by σ if every subword of $\mathbf{x}$ is a subword of a superword of some size. We define $\Omega_\sigma \subset \mathcal{A}^{\mathbb{Z}}$ to be the set of all sequences admitted by σ. It is clear that Ω_σ is a shift-invariant subset of $\mathcal{A}^{\mathbb{Z}}$ and the reader can check that it is also closed in the metric topology defined in Sect. 2.2.1.

Example 2 (A Constant-Length Substitution) Let $\mathcal{A} = \{a, b\}$ and let $\sigma(a) = abb$ and $\sigma(b) = aaa$. The first few level-n blocks of type a are:

$$a \to abb \to abb\,aaa\,aaa \to abb\,aaa\,aaa\;abb\,abb\,abb\;abb\,abb\,abb \to \cdots ,$$

where the spaces are there to help the reader see the level-n subblocks within. In this example each letter is substituted by a block of the same length (in this case 3), which is why the substitution is known as having constant length.

Clearly one could construct a substitution of constant length on any size of alphabet and any given integer length. The family in the next example contains the most famous and well-studied example of a substitution of *non-constant length*.

Example 3 ('Noble Means' Substitutions and the Fibonacci Substitution) Let $\mathcal{A} = \{a, b\}$ and choose a positive integer k. Define $\sigma(a) = a^k b$ and $\sigma(b) = a$, where by 'a^k' we mean the word composed of k consecutive 'a's. For example, let $k = 1$. In this case the first several level-n blocks of type a are:

$$a \to ab \to ab\,a \to ab\,a\;ab \to ab\,a\;ab\;ab\,a \to ab\,a\,ab\;ab\,a\;ab\,a\;ab \to \cdots ,$$

where again we've included spaces to help the reader distinguish the level-n subblocks. Note that the superword lengths are Fibonacci numbers, and in fact all of the superwords for $k = 1$ share this property. That is why this case is called the *Fibonacci substitution*.

Of course Fibonacci numbers are closely related to the golden mean, and in fact it is the larger eigenvalue of a matrix associated with the substitution (see Sect. 2.5.1). When k is changed we obtain other 'noble means' (silver if $k = 2$) from this matrix. All noble means substitutions have dynamical, spectral, and geometric properties in common with one another and therefore with the Fibonacci tiling, which is well-studied (see [11, 42, 89] and references therein).

The next family also contains the Fibonacci substitution, but in this class Fibonacci is the outlier, having few properties in common with the other elements.

Example 4 Let $\mathcal{A} = \{a, b\}$, choose a positive integer k, and let $\sigma(a) = ab^k$ and $\sigma(b) = a$. The first few supertiles in the case where $k = 3$ are

$$a \to abbb \to abbb\,a\,a\,a \to abbb\,a\,a\,a\;abbb\,abbb\,abbb \to \cdots .$$

A corresponding tiling of the line that yields well to spectral and dynamical analysis is discussed next.

2.4.2 Generalization: One-dimensional Self-Similar Tilings

We know from Sect. 2.2.3 that we can take any sequence in $\mathcal{A}^{\mathbb{Z}}$ and convert it into a tiling of $\mathbb{R}$ by choosing interval lengths for each element of $\mathcal{A}$. We certainly can do this for substitution sequences, and if we do it artfully we get tilings with some geometry to exploit. The process of doing it artfully leads naturally to the idea of inflation rules and self-similar tilings. It is worth doing in the context of an example first.

Example 5 Consider tiles t_a and t_b that are intervals of length $|t_a|$ and $|t_b|$ labelled by a and b. For a positive integer k, we can use the symbolic substitution $\sigma(a) = ab^k$ and $\sigma(b) = a$ to define a tile substitution $\mathcal{S}$ so that the patch $\mathcal{S}(t_a)$ is the tile t_a followed by k t_b's and the patch $\mathcal{S}(t_b)$ is just t_a. In that case the lengths of the supertiles are $|\mathcal{S}(t_a)| = |t_a| + k|t_b|$ and $|\mathcal{S}(t_b)| = |t_a|$.

The ideal situation, geometrically, would be if there was an "inflation factor" $\lambda > 1$ such that $|\mathcal{S}(t_a)| = \lambda|t_a|$ and $|\mathcal{S}(t_b)| = \lambda|t_b|$. A quick calculation yields that this λ would have to satisfy the equation $k = \lambda^2 - \lambda$. As expected, in the Fibonacci case when $k = 1$, we obtain that λ is the golden mean. For larger values of k we find that λ is either 'strongly non-Pisot' or, occasionally, an integer. Later we will discuss how the algebraic properties of λ affect the dynamics of the system. The case where $k = 3$ yields $\lambda = \frac{1+\sqrt{13}}{2}$ and the rule $\mathcal{S}$ is depicted in Fig. 2.7.

The $k = 3$ case is fully analyzed from a diffraction standpoint in [9] and as the basis for a two-dimensional tiling with infinite local complexity in [48]. The diffraction spectrum for all values of k is given a preliminary analysis in [13] and thorough treatment in [14].

Suppose now that σ is a symbolic substitution on a general finite alphabet $\mathcal{A}$. As before it is possible to find the expansion factor and natural tile lengths (more on that later). Suppose t_e is the tile corresponding to the symbol $e \in \mathcal{A}$. Then we define $\mathcal{S}(t_e)$ to be the patch of tiles corresponding to $\sigma(e)$, supported on the interval $\lambda \operatorname{supp}(t_e)$. Often, $\mathcal{S}$ is referred to as an 'inflation rule' or an 'inflate-and-subdivide rule'.

We can extend $\mathcal{S}$ to be a map on the space of all tilings $\Omega_\mathcal{P}$ as follows. Let $\mathcal{T} \in \Omega_\mathcal{P}$ be a tiling and let $t \in \mathcal{T}$ be any tile. We define $\mathcal{S}(t)$ to be the patch given by the substitution of the prototile of t, translated so that it occupies the set $\lambda \operatorname{supp}(t)$. Then $\mathcal{S}(\mathcal{T})$ is the tiling obtained by applying $\mathcal{S}$ to all tiles in $\mathcal{T}$ simultaneously. For most $\mathcal{T} \in \Omega_\mathcal{P}$, $\mathcal{S}(\mathcal{T})$ is not equal to $\mathcal{T}$. However there will be fixed or periodic points for $\mathcal{S}$. A fixed point for $\mathcal{S}$ is known as a *self-similar tiling*.

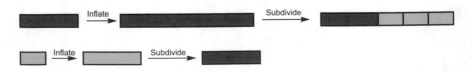

Fig. 2.7 Inflation and subdivision for the case $k = 3$

2.4.3 More Tricky Generalization: Multidimensional Constant-Length Substitutions in $\mathbb{Z}^d$

Symbolic substitutions of constant length generalize directly to substitutions of constant size in $\mathbb{Z}^d$. We choose a 'rectangular' shape in d dimensions, and every letter of the alphabet is substituted with a block of letters in that shape. There is no problem with iteration of the substitution because all of the blocks fit together along every dimension so concatenation of the blocks happens naturally. To generalize symbolic substitutions of non-constant length to $\mathbb{Z}^d$ we will use the fusion paradigm.

Fix *lengths* $l_1, l_2, \ldots l_d$, positive integers with each $l_i > 1$, and define the *location set* $\mathcal{I}^d$ to be the 'rectangle' given by

$$\mathcal{I}^d = \{J = (j_1, \ldots j_d) \text{ such that } j_i \in 0, 1, \ldots, l_i - 1 \text{ for all } i = 1, ..d\}. \quad (2.1)$$

A *block substitution* S is defined to be a map from $\mathcal{A} \times \mathcal{I}^d$ into $\mathcal{A}$. Then for any $e \in \mathcal{A}$ we denote by $S(e)$ a block of letters; we call it a *1-superblock* or *1-supertile*.

In any particular example it is not hard to build an n-superblock through concatenation, but notation describing it precisely obscures this fact. Since the notation is not needed elsewhere in these notes we omit it. Instead we define n-superblocks inductively, using the relative positions of the letters in $S(e)$ to determine the relative positions of their S^{n-1} blocks in $S^n(e)$. Because all of the substituted blocks have the same dimensions, if two letters were adjacent it is clear that their substitutions will fit next to one another properly.

Any position $\mathbf{k} \in \mathcal{I}^d$ represents a location in a 1-superblock and we can think of S restricted to $\mathbf{k}$ as a map from $\mathcal{A}$ to itself. Indeed it can be useful to think of S as a block of maps $(p_{\mathbf{k}})_{\mathbf{k} \in \mathcal{I}^d}$. The nature of these maps is key to the dynamics of the system and they are used to compute the cocycle for the skew product representation of the system [45, 89]. An important subclass is defined as follows.

Definition 2.4.1 Let the substitution S as defined in this section be written as $S = (p_{\mathbf{k}})_{\mathbf{k} \in \mathcal{I}^d}$. We say S is *bijective* if and only if each $p_{\mathbf{k}}$ is a bijection from $\mathcal{A}$ to itself.

Example 6 A two-dimensional version of the Thue-Morse substitution uses $\mathcal{A} = \{0, 1\}$ with $d = 2$ and $l_1 = l_2 = 2$.

$$S(0) = \begin{smallmatrix} 1 & 0 \\ 0 & 1 \end{smallmatrix}, \qquad S(1) = \begin{smallmatrix} 0 & 1 \\ 1 & 0 \end{smallmatrix}, \quad (2.2)$$

where both blocks are located in $\mathbb{Z}^2$ with their lower left corners at the origin. If instead we wish to see S as a matrix $(p_{\mathbf{k}})_{\mathbf{k} \in \mathcal{I}^2}$ of maps on $\mathcal{A}$, denote by g_0 the identity map and g_1 the map switching 0 and 1, we obtain:

$$S(*, \mathcal{I}^2) = (p_{\mathbf{k}})_{\mathbf{k} \in \mathcal{I}^2} = \begin{smallmatrix} g_1 & g_0 \\ g_0 & g_1 \end{smallmatrix}. \quad (2.3)$$

Fig. 2.8 The first three
superblocks of type 0. The
lines emphasize
$(n-1)$-superblocks inside
the n-superblocks

$$
0 \to \begin{array}{cc} 1 & 0 \\ 0 & 1 \end{array} \to \begin{array}{c|c} 0\ 1 & 1\ 0 \\ \hline 1\ 0 & 0\ 1 \\ 1\ 0 & 0\ 1 \\ \hline 0\ 1 & 1\ 0 \end{array} \to \begin{array}{c|c} 1\ 0\ 0\ 1 & 0\ 1\ 1\ 0 \\ 0\ 1\ 1\ 0 & 1\ 0\ 0\ 1 \\ 0\ 1\ 1\ 0 & 1\ 0\ 0\ 1 \\ \hline 1\ 0\ 0\ 1 & 0\ 1\ 1\ 0 \\ 0\ 1\ 1\ 0 & 1\ 0\ 0\ 1 \\ \hline 1\ 0\ 0\ 1 & 0\ 1\ 1\ 0 \\ 1\ 0\ 0\ 1 & 0\ 1\ 1\ 0 \\ 0\ 1\ 1\ 0 & 1\ 0\ 0\ 1 \end{array}
$$

For example we see that $p_{(0,0)} = g_0$ and $p_{(0,1)} = g_1$. Also we note that this example is bijective. The first few superblocks of type 0 are shown in Fig. 2.8.

2.4.4 Geometric Generalization: Self-Affine, Self-Similar, and Pseudo-Self-Similar Tilings

The geometric structure evident in the tilings in this section is governed by expanding linear maps. This makes it particularly amenable to study from a number of viewpoints, and therefore these are the most widely studied of the tilings considered in these notes. There are many examples in Chapter 6 of [11] and we try not to repeat too many of those here. Sometimes tilings created using other supertile methods can be transformed into self-similar tilings and the results that exist for them can be used. Sometimes they can't.

The earliest definition of self-similar tilings that seems to appear in print is in [102], which is a set of AMS colloquium lecture notes by William Thurston. However, the author tells us the ideas in the lectures are not all his own and refers in an informal way to a number of places where the subject was beginning to be studied.

2.4.4.1 Self-Affine and Self-Similar Tilings

We first follow the definitions laid out in [100], and use terminology from there, [46], and [11]. We also give a simpler but more restrictive version of the definition that the reader will find in [11] and lots of other places.

Definition 2.4.2 Let $\phi : \mathbb{R}^d \to \mathbb{R}^d$ be a linear transformation all of whose eigenvalues are greater than one in modulus. A tiling $\mathcal{T}$ is called *self-affine with expansion map ϕ* if

1. for each tile $t \in \mathcal{T}$, $\phi(\mathrm{supp}(t))$ is the support of a union of $\mathcal{T}$-tiles, and
2. t and t' are equivalent up to translation if and only if $\phi(\mathrm{supp}(t))$ and $\phi(\mathrm{supp}(t'))$ support equivalent patches of tiles in $\mathcal{T}$.

If ϕ is a similarity, the tiling is called *self-similar*. For self-similar tilings of $\mathbb{R}$ or $\mathbb{R}^2 \cong \mathbb{C}$ there is an *inflation constant* λ for which $\phi(z) = \lambda z$.[5]

There are a few differences between our definition and the one in [100] upon which it is based. One is that ϕ is not required to be diagonalizable, and the other is that a self-affine tiling is not required to be repetitive. Proofs on the algebraic nature of the expansion constant originally required diagonalizability [66] but the condition was recently removed in [68]. A nonrepetitive tiling satisfying our definition of self-affine would be called "ϕ-subdividing" in [100]. Although we have taken finite local complexity to be a blanket assumption throughout this paper, note that our definition can be used in the infinite local complexity case as well.

What is inconvenient about this definition is the fact that one must begin by already having the self-affine tiling at hand. To be more consistent with the way we think about symbolic substitutions we can define an inflation rule on prototiles first.

Definition 2.4.3 Let $\mathcal{P}$ be a finite prototile set in $\mathbb{R}^d$ and let $\phi : \mathbb{R}^d \rightarrow \mathbb{R}^d$ be a diagonalizable linear transformation all of whose eigenvalues are greater than one in modulus. A function $S : \mathcal{P} \rightarrow \mathcal{P}^*$ is called a *tiling inflation rule*[6] *with inflation map* ϕ if for every $p \in \mathcal{P}$,

$$\phi(\mathrm{supp}(p)) = \mathrm{supp}(S(p)).$$

The linear map ϕ makes it easy to extend S to tiles, patches, and tilings. The substitution of a tile $t = p + x$, for $p \in \mathcal{P}$ and $x \in \mathbb{R}^d$, is the patch $S(t) := S(p) + \phi(x)$. The substitution of a patch or tiling is the substitution applied to each of its tiles. This means that we can consider S as a self-map on the full tiling space $\Omega_{\mathcal{P}}$. If a tiling $\mathcal{T}$ is invariant under S we call it a *self-affine tiling*. We use the term *n-supertile* to mean a patch of the form $S^n(t)$.

We can use either of the methods in Sect. 2.3.1.2 to construct a tiling space for S. If there is a self-similar tiling $\mathcal{T}$, we can make its hull by taking the orbit closure under translation. Or, we could consider the set $\mathcal{R}$ of all n-supertiles, for all n and all prototile types, and use that as our set of admissible patches. Often the resulting spaces are identical, though not in the following example.

Example 7 (Danzer's T2000 Tiling) Figure 2.9 gives an example from the tilings encyclopedia [1] attributed to Ludwig Danzer. It uses a total of 24 tiles, two sizes of triangles in twelve orientations. We show the inflation rule on the two sizes only; the inflation rule of the rotations are the corresponding rotations of these. The expansion map is $\phi(x, y) = (\sqrt{3}x, \sqrt{3}y)$.

---

[5]In the literature (notably [100, 102]) it is taken as given that ϕ is orientation preserving, which can be assumed by squaring any substitution that is not.

[6]These are also known as inflate-and-subdivide rules and tiling substitution rules.

Fig. 2.9 The T2000 inflate-and-subdivide rule

Fig. 2.10 2- and 3-supertiles

Fig. 2.11 A patch from a T2000 tiling

Figure 2.10 shows the second and third iteration of the larger triangle, and Fig. 2.11 shows a large patch of an infinite tiling. The sharp-eyed viewer will notice that the tiling appears to use only 6 rotations each of the small and large triangles, not 12. This is because the substitution is not 'primitive' in the sense of Sect. 2.5.2: there is no number N such that each N-supertile contains all 24 tile types. They will always have exactly 12 when N is sufficiently large. A side effect is that the two methods for producing tiling spaces are different in that the hull of any self-similar tiling is a connected space, while the space of admissible tilings has two connected components, one a rotation of the other by 60°. In this example we could "fix" this problem by restricting our attention to an appropriate 12-prototile set and using two iterations of the substitution.[7]

For the same reason there will be no self-similar tilings for the substitution as it is shown in Fig. 2.9, but there are period-two tilings. By replacing the substitution as shown with its square, we obtain self-similar tilings with expansion factor 3 instead of $\sqrt{3}$.

2.4.4.2 Pseudo-Self-Similar Tilings

In this situation we still have an expanding linear map ϕ acting on our tilings, but we no longer have that $\phi(\text{supp}(p))$ is exactly the support of a patch of tiles. Instead it may only approximate the shape of $\text{supp}(p)$. Well-known examples of such substitutions are the Penrose tilings using rhombuses and/or 'kites and darts' [54, 83] (or also [46, 58, 90]) and the "binary" tilings [55] (or see [46, p. 307] or [11, p. 217]). In these examples there is a substitution rule that still 'fits' to ultimately form a tiling, but not exactly on top of the expanded tiles. Examples of this nature appear in abundance in the Tilings Encyclopedia [1] as they occur in projection tilings constructed in a similar way to Penrose tilings.

For the definition we must make precise what we mean by expanding a tiling $\mathcal{T}$ to obtain the tiling $\psi(\mathcal{T})$. For every tile t in $\mathcal{T}$, $\phi(t)$ is defined to be a tile supported on $\phi(\text{supp}(t))$ that carries the label of t. We define $\phi(\mathcal{T}) := \bigcup_{t \in \mathcal{T}} \phi(t)$.

Definition 2.4.4 (See [88, 99]) Let $\mathcal{P}$ be a finite prototile set in $\mathbb{R}^d$ and let $\phi : \mathbb{R}^d \to \mathbb{R}^d$ be a diagonalizable linear transformation all of whose eigenvalues are greater than one in modulus. We say a tiling $\mathcal{T} \in \Omega_{\mathcal{P}}$ is *pseudo-self-similar with expansion ϕ* if $\mathcal{T}$ is locally derivable from $\phi(\mathcal{T})$.

Example 8 (Variation on Thurston's Hexagonal Example) In [102] a fractal "reptile" is constructed that makes a periodic tiling on the hexagonal lattice. The example is based on the observation that a regular hexagon is approximated by a patch of seven hexagons. To make a nonperiodic version we use two colors of hexagons in our inflation rule.

The inflation map ϕ is given by the matrix $\begin{pmatrix} 5/2 & \sqrt{3}/2 \\ -\sqrt{3}/2 & 5/2 \end{pmatrix}$. Pictured in Fig. 2.12 is what happens to the hexagonal prototiles when inflated by this map and 'subdivided' into a patch of tiles at the original scale. The location of the origin is marked with a point in each tile and patch. To see how each supertile is spatially related to the inflated prototile we have shown their overlap in Fig. 2.13. Figure 2.14 shows the 1-, 2-, and 3-supertiles for the blue hexagon.

---

[7]In general, non-primitive substitutions can have more complicated structure.

Fig. 2.12 The
inflate-and-subdivide rule for
a hexagonal
pseudo-self-similar tiling

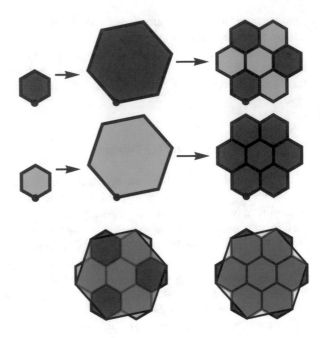

Fig. 2.13 The inflated blue
tile and its patch, left; the
inflated green tile and its
patch, right

First in [88] for $\mathbb{R}^2$ and later [99] for $\mathbb{R}^d$ it is proved that every pseudo-self-similar tiling is mutually locally derivable from a self-similar tiling. In $\mathbb{R}^2$ the argument ends up using ideas from iterated function systems, but in $\mathbb{R}^d$ other methods are required.

2.4.5 Fusion: A General Viewpoint

Symbolic substitutions and tiling inflation rules can be seen as a sort of 'cellular' model: the supertiles grow, level by level, because each symbol or tile within them has expanded to become a word or patch. Fusion takes an 'atomic' model: symbols or tiles are like atoms that come together to form 'molecules' (our 1-supertiles) that then assemble themselves into larger structures (2-supertiles) that continue to merge into larger supertiles level by level.

The sets of n-supertiles obtained by symbolic or tiling substitutions can be seen as fusion rules since it is possible (and natural) to see an n-supertile as being a union of $(n-1)$-supertiles just as easily as seeing it as the union of lots and lots of 1-supertiles. In this viewpoint the $(n-1)$-supertiles in $\mathcal{S}^n(a)$ are concatenated as prescribed by the original substitution rule on a. One could also imagine creating supertiles by applying different substitutions or tile inflations at each stage (if geometry permits); this is a more general situation captured in the symbolic case by S-adic systems and in the tiling case by fusion rules.

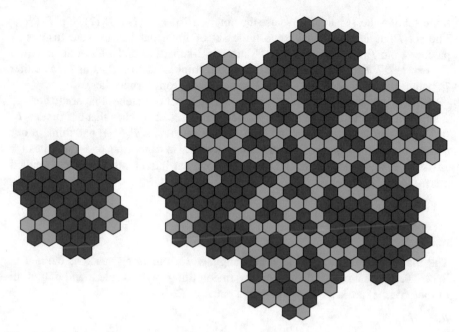

Fig. 2.14 2- and 3-supertiles for the blue prototile

Fusion does not require an underlying linear map ϕ but instead takes a combinatorial approach. There have been other combinatorial approaches to generalizing
symbolic substitutions. An early attempt to understand substitutions on graphs and
in particular the dual graph for the Penrose inflation appears in [86]. A definition of
substitution for the dual graphs of planar tilings is discussed in [46]. Combinatorial
substitutions are defined a little bit differently in [40], and a related notion
called "local rule" substitutions are defined in [39]. A definition of "topological
substitutions" is given in [20]. Separately, an extremely successful program on
"generalized" or "dual" substitutions began with [8]; see [7] for results tying
substitutions to Markov partitions of hyperbolic toral automorphisms, complex
numeration systems and β-expansions. We follow [50] for the fusion definition we
give here and note that although we use the context of tilings in $\mathbb{R}^d$ the definitions
are appropriate for (multidimensional) sequences as well.

Suppose that we have a finite prototile set $\mathcal{P}$ in $\mathbb{R}^d$. Given two $\mathcal{P}$-patches P_1
and P_2 and two translations x_1 and x_2, if the patches $(P_1 - x_1)$ and $(P_2 - x_2)$
intersect only on their boundaries to form a patch with connected support we call
$(P_1 - x_1) \cup (P_2 - x_2)$ a *fusion* of P_1 to P_2. Of course there could be many different
ways two given patches can be fused, and we could make the fusion of any finite
number of patches inductively. Patch fusion is our tiling analogue to concatenation
for symbols.

We form our fusion rule by defining the sets of supertiles as follows. The set $\mathcal{P}_0$
is just the prototile set $\mathcal{P}$. The set $\mathcal{P}_1$ is our set of 1-supertiles and is defined to be a

finite set of finite $\mathcal{P}$-patches. We use the notation $\mathcal{P}_1 = \{P_1(1), P_1(2), \ldots, P_1(j_1)\}$. The set $\mathcal{P}_2$ is defined to be some finite set of finite patches that are fusions of patches from $\mathcal{P}_1$. The elements of $\mathcal{P}_2$ are our 2-supertiles and we use the notation $\mathcal{P}_2 = \{P_2(1), P_2(2), \ldots, P_2(j_2)\}$. One could think of the patches in $\mathcal{P}_2$ as either $\mathcal{P}$-patches, or as $\mathcal{P}_1$-patches (i.e., as patches made from 1-supertiles).

We continue inductively, forming $\mathcal{P}_3$ as a finite set of patches that are fusions of 2-supertiles, and in general letting $\mathcal{P}_n$ be a finite set of patches that are fusions of $(n-1)$-supertiles. We use the notation $\mathcal{P}_n = \{P_n(1), \ldots, P_n(j_n)\}$ and think of our n-supertiles both as patches of ordinary tiles and also as patches of k-supertiles for any $k < n$. We collect all of our supertiles together into an atlas of patches called our *fusion rule* $\mathcal{R}$, that is

$$\mathcal{R} = \bigcup_{n \in \mathbb{N}} \mathcal{P}_n = \{P_n(j) \mid n \in \mathbb{N} \text{ and } 1 \le j \le j_n\}.$$

The fusion rule can be used as a pre-language for our tiling space as defined in Sect. 2.3.1.2. That means that tilings are fusion tilings by this rule if and only if all of their patches are seen somewhere in a patch in $\mathcal{R}$.

Remarks

1. In general we will assume that some sequence of n-supertiles grows to cover $\mathbb{R}^d$ so that there are tilings of $\mathbb{R}^d$ that are allowed by the fusion rule. That is, we take as a standing assumption that our fusion tiling spaces are nonempty.
2. The number j_n of supertiles can vary from level to level.
3. When $d = 1$, if we consider translations by elements of $\mathbb{Z}$ with all tiles having unit length, fusion tilings correspond to Bratteli-Vershik systems (except for edge sequences that have no predecessors or no successors). See [21] for more about the relationship between tilings and Bratteli-Vershik systems.
4. As stated currently, the definition of fusion rule allows for every $\mathcal{P}$-tiling to be seen as a fusion tiling. Construct the fusion rule $\mathcal{R}$ by letting the set of n-supertiles contain every possible patch of $\mathcal{P}$-tiles containing n or fewer tiles. All of $\Omega_{\mathcal{P}}$ is contained in this fusion tiling space.

Example 9 (The Chacon transformation [29]) This example was the first to show that there exist transformations with weakly but not strongly mixing dynamical systems. It was originally constructed using the "cut-and-stack" method[8] and can be seen as a substitution as well as a fusion tiling.

To see the Chacon transformation as a fusion rule for tilings of the line, let l_a and l_b to be two positive numbers, let a denote a prototile with support $[0, l_a]$, and let b denote a prototile with support $[0, l_b]$. We let the symbols a and b also serve as the labels of the tiles if those are needed.

---

[8] Actually, it is possible to see the process of fusion as a cutting and stacking process.

For each n there are two n-supertiles, which we consider being of types a and b. We define $P_1(a) = a \cup (a + l_a) \cup (b + 2l_a) \cup (a + 2l_a + l_b)$ and $P_1(b) = b$. We think of $P_1(a)$ as "$aaba$", and it has length $3l_a + l_b$. The length of $P_1(b)$ is just l_b, and we will let $P_n(b) = b$ for all n.

To make $P_2(a)$ we fuse three copies of $P_1(a)$ and one copy of $P_1(b)$ together in the same order as we did for $P_1(a)$, and we let $P_2(b) = b$. The length of the new a supertile is three times that of the previous a supertile plus the length of b.

In general, we have:

$$\mathcal{P}_{n+1} = \{P_{n+1}(a), P_{n+1}(b)\}$$
$$= \{P_n(a)P_n(a)P_n(b)P_n(a), P_n(b)\}$$
$$= \{P_n(a)P_n(a)\, b\, P_n(a), b\}.$$

This is an example where not all supertiles expand. In the original formulation b is seen as a 'spacer', and the offsets between a's it provides are the cause of the weak but not strong mixing.

Example 10 (A Direct Product Substitution) Let $\mathcal{A} = \{a, b\}$ and define $\sigma(a) = abb, \sigma(b) = aa$. We take the direct product of this substitution with itself, with alphabet $(a, a), (a, b), (b, a), (b, b)$. We use the convention that the substitution on the first letter runs horizontally and the substitution on the second letter goes upwards. With that we obtain

$$S((a,a)) = \begin{array}{l} (a,b)\ (b,b)\ (b,b) \\ (a,b)\ (b,b)\ (b,b), \\ (a,a)\ (b,a)\ (b,a) \end{array} \qquad S((a,b)) = \begin{array}{l} (a,a)\ (b,a)\ (b,a) \\ (a,a)\ (b,a)\ (b,a) \end{array},$$

$$S((b,a)) = \begin{array}{l} (a,b)\ (a,b) \\ (a,b)\ (a,b), \\ (a,a)\ (a,a) \end{array} \qquad S((b,b)) = \begin{array}{l} (a,a)\ (a,a) \\ (a,a)\ (a,a) \end{array}.$$

It is better to visualize the substitution as a tiling, so we show the prototiles and 1-supertiles in Fig. 2.15. The first row is $\mathcal{P}_0 = \{(a,a), (a,b), (b,a), (b,b)\}$ and the second is

$$\mathcal{P}_1 = \{P_1((a,a)), P_1((a,b)), P_1((b,a)), P_1((b,b))\}.$$

It is possible to iterate this as a substitution, concatenating in two dimensions in much the same way as we would do in one dimension. We choose instead to think of it as a fusion, where the $n + 1$-supertiles are constructed using the same concatenation of n-supertiles at every level. This concatenation is diagrammed in Fig. 2.16 and the 2-supertiles are shown in Fig. 2.17.

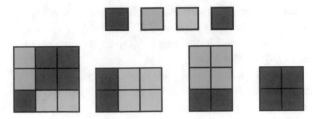

Fig. 2.15 The prototiles and 1-supertiles for the direct product substitution

$P_n(b)$	$P_n(d)$	$P_n(d)$
$P_n(b)$	$P_n(d)$	$P_n(d)$
$P_n(a)$	$P_n(c)$	$P_n(c)$

$P_n(a)$	$P_n(c)$	$P_n(c)$
$P_n(a)$	$P_n(c)$	$P_n(c)$

$P_n(b)$	$P_n(b)$
$P_n(b)$	$P_n(b)$
$P_n(a)$	$P_n(a)$

$P_n(a)$	$P_n(a)$
$P_n(a)$	$P_n(a)$

Fig. 2.16 Direct product fusion rule making $(n + 1)$-supertiles from n-supertiles

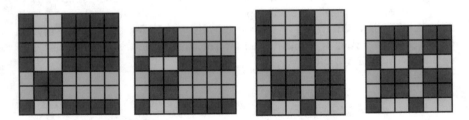

Fig. 2.17 The 2-supertiles for the direct product fusion

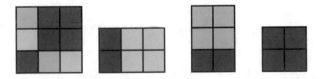

Fig. 2.18 The 1-supertiles for the DPV

The $\mathbb{Z}^2$ dynamical system associated with the direct product acts the same as the direct product of the one-dimensional systems. To get something new we rearrange the substitution on some of the letters to break the direct product structure, obtaining "Direct Product Variation" (DPV) tilings. Varying the structure is easily done but care must be taken so that the DPV substitution can be iterated to form legitimate patches and tilings.

Example 11 (A Variation on the Direct Product) In this example we have chosen to rearrange only the tiles in the first supertile (compare Figs. 2.15 and 2.18). The requisite care was taken to ensure that the 1-supertiles fit together to form 2-supertiles supported on topological disks, and that this nice situation will continue in perpetuity.

Unlike the DP case, if we try to iterate it as a substitution it becomes problematic: it is not clear how to concatenate the substituted tiles. Each supertile may be in a different relationship to its neighbors than the original tile was. In some examples it is possible to determine how to fit the supertiles together by looking at bounded patches around the original tiles. Those are the kinds of examples that have prompted definitions 'combinatorial' or 'local rules' substitutions [39, 40, 43, 86]. For this example, however, concatenation of individual 1-supertiles inside large patches cannot be determined by local information and so the fusion paradigm is necessary. (See [46] for a discussion of the origin of these nonlocal problems and their consequences.)

Figure 2.19 gives the general template for concatenating the n-supertiles to make the $(n+1)$-supertiles, and Fig. 2.20 shows us the set of 2-supertiles. Already we can see that the direct product structure has been disrupted.

For your entertainment we include a comparison of the 3-supertiles of type (a, a) for the DP and DPV substitution in Fig. 2.21. Direct product tilings have a distinct appearance with horizontal and vertical bands clearly visible. The DPV can be compared to the introductory Fig. 2.6, which is a version of the DPV with 'natural' tile sizes.

The topology, and in particular the cohomology, of a DPV based on a strongly non-Pisot substitution in product with the substitution $1 \rightarrow 11$ (which gives a

$P_n(b)$	$P_n(d)$	$P_n(d)$
$P_n(c)$	$P_n(a)$	$P_n(d)$
$P_n(d)$	$P_n(b)$	$P_n(c)$

$P_n(a)$	$P_n(c)$	$P_n(c)$
$P_n(a)$	$P_n(c)$	$P_n(c)$

$P_n(b)$	$P_n(b)$
$P_n(b)$	$P_n(b)$
$P_n(a)$	$P_n(a)$

$P_n(a)$	$P_n(a)$
$P_n(a)$	$P_n(a)$

Fig. 2.19 DPV fusion rule making $(n + 1)$-supertiles from n-supertiles

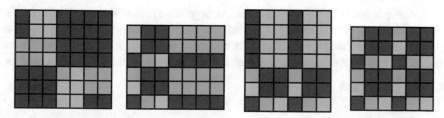

Fig. 2.20 The 2-supertiles for the DPV fusion

Fig. 2.21 The 3-supertiles of type (a, a) for the DP (left) and DPV (right)

"solenoid" system) is analyzed in [49]. In that example the DPV uses natural tile sizes and has infinite local complexity.

2.4.5.1 Special Classes of Fusion Tilings

A fusion rule $\mathcal{R}$ is called *prototile-regular* if each $\mathcal{P}_n$ has the same number of elements. A prototile-regular fusion rule is called *transition-regular* if the number of each tile type in each supertile type doesn't change from level to level. The DP and DPV examples shown were both prototile- and transition-regular, and they are also 'algorithmic' in the sense of the next example.

Example 12 (A "Uniform Shape Substitution") This is an example of the type of substitution found in [39]. We call it *algorithmic* because a simple computer algorithm can be written to describe the formation of the n-supertiles. The algorithm is iterative, accepting n-supertiles and fundamental level-n translations and returning $(n + 1)$ versions of these. Interesting for these examples is that there may be more than one possible input (prototile set and fundamental translations) that leads to a tiling of $\mathbb{R}^2$.

Because it isn't obvious how to make a simplified figure like we did for DPVs that describe the combinatorics of how to put the n-supertiles together to make the $(n+1)$-supertiles, we give the algorithm instead. Let A_n and B_n denote n-supertiles and let $\mathbf{k}_n$ and $\mathbf{l}_n$ represent fundamental translation vectors at the nth level, and let

$L = \begin{pmatrix} 2 & 1 \\ -1 & 1 \end{pmatrix}$. Then

$$A_{n+1} = A_n \cup (B_n + \mathbf{k}_n) \cup (B_n + \mathbf{l}_n) \quad \text{and} \quad B_{n+1} = B_n \cup (A_n + \mathbf{k}_n) \cup (A_n + \mathbf{l}_n).$$

The level-$(n + 1)$ translations are $\mathbf{k}_{n+1} = L\mathbf{k}_n$ and $\mathbf{l}_{n+1} = L\mathbf{l}_n$.

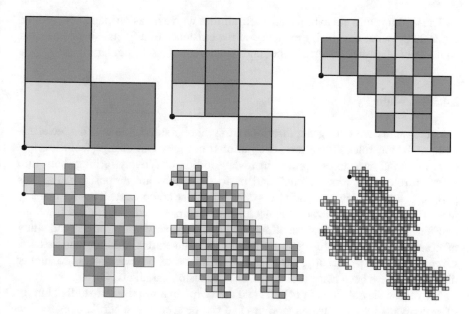

Fig. 2.22 Level-n tiles of type A for $n = 1, \ldots 6$, beginning with square tiles. Each square tile is the same size despite the image rescaling, which is there solely for display purposes

To run the algorithm, put in a prototile set and some initial vectors and see what happens when the algorithm is iterated. There are at least three distinctly-shaped prototile sets and corresponding initial vectors that lead to tilings of $\mathbb{R}^2$; we include only one here.

One possible prototile set is colored unit squares with their lower left endpoints at the origin. With this input set it is necessary to set $\mathbf{k}_0 = (1, 0)$ and $\mathbf{l}_0 = (0, 1)$. Figure 2.22 shows the first six supertiles of the blue (A) prototile. It is essential to note that each successive supertile is shown at a scale smaller than it actually is: all the tiles should be the same size as the first one. The rescaling is just there to display the supertiles together. Also note the dot in each image: it represents the location of the origin and allows us to see the rotational aspect of this fusion rule.

It is the 'shape' of the substitution that matters: it is 'uniform' in the sense that the shapes of the n-supertiles of either type are the same and it is only the coloring that differs. In our example we have chosen a bijective coloring, using the word 'bijective' in the same way as we used it for substitutions in $\mathbb{Z}^d$.

Another prototile set that works is a pair of colored hexagons with vertex set

$$\{(1, 0), (0, 1), (-1, 1), (-1, 0), (0, -1), (1, -1)\},$$

in which case it is necessary to set $\mathbf{k}_0 = (2, -1)$ and $\mathbf{l}_0 = (1, 1)$.

These tilings turn out to be pseudo-self-similar with expansion map L. Moreover, Fig. 2.22 provides convincing evidence of the existence of a fractal-shaped tile that could be used as the 'uniform shape' and which makes a self-similar tiling.

2.4.5.2   S-adic Systems

S-adic systems are both generalizations of symbolic substitutions and specializations of fusion rules. There are substitutions, but they can change from level to level and thus have to be applied in reverse order, effectively turning them into a fusion rule. The term "S-adic" and basic definitions are proposed in [38], as part of a larger study of symbolic systems of low complexity. There are many reasons why this generalization is useful, as it intersects with continued fractions and interval exchange transformations, and has been very interesting in the study of combinatorics on words. The topic and its connections to numerous areas is surveyed in [103]. Recently, a few generalizations of S-adic systems to higher dimensions have been made that do not use the fusion paradigm [51, 53].

We follow the notation of [24] first and then explain how this well-studied family of systems fits into the fusion paradigm and can be seen as a supertile construction method. Let $\mathcal{A}_0, \mathcal{A}_1, \mathcal{A}_2, \ldots$ be a family of finite alphabets, and, for each n, let $\sigma_n : \mathcal{A}_{n+1} \to \mathcal{A}_n^*$ be a map taking an element from $\mathcal{A}_{n+1}$ to a nonempty word in the alphabet $\mathcal{A}_n$.[9] Let $\{a_n\}_{n=0}^{\infty}$ represent a sequence for which $a_n \in \mathcal{A}_n$ for all $n \in \mathbb{N}$. An infinite word $\mathbf{x} \in \mathcal{A}_0^{\mathbb{N}}$ admits the S-adic expansion $\{(\sigma_n, \mathcal{A}_n)\}_{n=0}^{\infty}$ if

$$\mathbf{x} = \lim_{n \to \infty} \sigma_0 \sigma_1 \cdots \sigma_{n-1}(a_n).$$

We have the usual two options to make a sequence space, either as the hull of $\mathbf{x}$ or as the set of all sequences admitted by the set of n-superwords.

(The theoretical computer science community has established terminology for the set of all allowed words: a 'language'. In this terminology the directive sequence $\{\sigma_n\}$ has a language associated with it given by

$$L = \bigcap_{n \in \mathbb{N}} \overline{\sigma_0 \sigma_1 \cdots \sigma_{n-1}(\mathcal{A}_n^*)},$$

where the notation $\overline{M}$ denotes the smallest language containing the set M. The S-adic system given by $\{\sigma_n\}$ can then be studied through the shift space admitted by this language.)

---

[9]There is a separate lexicon in which what are known to some as "substitutions" are known to others as "non-erasing morphisms" and the set $\mathcal{A}^*$ is called the "free monoid" instead of the set of all finite words from $\mathcal{A}$ [6, 76].

Let us see how this fits into the fusion paradigm. The prototile set is $\mathcal{A}_0$, which could be seen as labelled unit intervals if we prefer a tiling to a sequence. The 1-supertiles are constructed using the map $\sigma_0 : \mathcal{A}_1 \to \mathcal{A}_0$. For each $a \in \mathcal{A}_1$, $\sigma_0(a)$ is a word in $\mathcal{A}_0$ which by abuse of notation we might think of as a patch instead. In either case we call it a 1 supertile. The 1-supertiles are given by

$$\mathcal{P}_1 = \{\sigma_0(a) \text{ such that } a \in \mathcal{A}_1\}$$

The 2-supertiles are given by $\sigma_0(\sigma_1(a))$, where a is now an element of $\mathcal{A}_2$, and we need to see why those are fusions of 1-supertiles. Notice that $\sigma_1(a)$ is a word in $\mathcal{A}_1^*$, and so we can apply σ_0 to each of its letters. Thus one can see $\sigma_0(\sigma_1(a))$ as the fusion of blocks of the form $\sigma_0(a')$ in the order prescribed by $\sigma_1(a)$. Thus the set of 2-supertiles is

$$\mathcal{P}_2 = \{\sigma_0\sigma_1(a) \text{ such that } a \in \mathcal{A}_2\}$$

Now the 3-supertiles are given by $\sigma_0\sigma_1\sigma_2(a)$, where now $a \in \mathcal{A}_3$. To see these are fusions of 2-supertiles, suppose $\sigma_2(a) = b_1b_2\cdots b_k$, which is in $\mathcal{A}_2^*$. Then $\sigma_0\sigma_1(\sigma_2(a)) = \sigma_0\sigma_1(b_1)\sigma_0\sigma_1(b_2)\cdots\sigma_0\sigma_1(b_k)$, which is a fusion of 2-supertiles. Clearly, then, the n-supertiles take the form

$$\mathcal{P}_n = \{\sigma_0\sigma_1\cdots\sigma_{n-1}(a) \text{ such that } a \in \mathcal{A}_n\},$$

and can be seen as fusions of $(n-1)$-supertiles as desired. So in this work we consider S-adic constructions to be supertile constructions as well.

2.4.6 Tiling Spaces from Supertile Methods

As we discussed in Sect. 2.3.1.2, tiling spaces are often given as either the hull of a specific tiling or as the tilings admitted by a specific language. It is reasonable to try both approaches for supertile methods.

In the case of general symbolic substitutions in $\mathbb{Z}$, constant-length substitutions in $\mathbb{Z}^d$, and self-affine or -similar tilings of $\mathbb{R}^d$, we have noted that we can consider the substitution as an action on the sequence or tiling space. Thus if $\mathcal{T}$ is a fixed point of the substitution we can study the supertile rule by studying the hull of $\mathcal{T}$. This doesn't work for arbitrary fusion rules, since they don't define actions on the full tiling space.

For fusion rules (and therefore all supertile methods) one can consider the set of all n-supertiles to be the set $\mathcal{R}$. Thinking of $\mathcal{R}$ as a pre-language (which may be an abuse of terminology), we see that if there are infinite tilings in $\Omega_\mathcal{P}$ that are admitted by $\mathcal{R}$, then the space $\Omega_\mathcal{R}$ is nontrivial. We do not attempt to determine precise conditions that enforce nontriviality, but that is easy to check in examples. In [50] a blanket assumption is that the boundaries of the n-supertiles become arbitrarily small compared to their interiors as $n \to \infty$. (Such a sequence of sets is called a "van Hove sequence", and we call fusions of this sort van Hove.) We clearly need some sort of growth condition for supertiles in order for the tiling space to be nontrivial.

Often the Ωs you get by either method are identical, but there are exceptions. One notable exception is if the invariant $\mathcal{T}$ contains a 'defective' patch that is not allowed by the substitution but is stable under the substitution rule. The Danzer T2000 example was another exception arising from not having primitivity.

2.4.7   Recognizability or the Unique Composition Property

Suppose you are given an element of a tiling (or sequence) space given by some supertiling method. All you see are the tiles in the tiling. Can you determine uniquely how the tiles group into supertiles? If so, the tiling (or sequence) is recognizable.

To make this definition it is convenient to introduce the notion of "supertiling spaces". Let Ω be a (nonempty) tiling space defined for some supertile method. Fix an n and choose some $\mathcal{T} \in \Omega$. For any $x \in \mathbb{R}^d$ the patch $\mathcal{T} \cap \{x\}$ must be contained in some n-supertile, either from the generating tiling or from $\mathcal{R}$. The n-supertile might not be unique, but there are only finitely many possibilities. A diagonalization argument can be made to extrapolate that all tiles in $\mathcal{T}$ itself can be composed into n-supertiles that overlap only on their boundary. A tiling $\mathcal{T}_n$ obtained by this composition, i.e. where the prototile set is considered at $\mathcal{P}_n$ rather than $\mathcal{P}$, is called an n-supertiling of $\mathcal{T}$. The space of all n-supertilings of Ω is denoted Ω_n and is a translation-invariant subspace of the tiling space $\Omega_{\mathcal{P}_n}$.

Since each n-supertile is constructed from $(n-1)$-supertiles by definition, there is a unique *decomposition map* f_n taking Ω_n to Ω_{n-1}. It is possible that the tiling $\mathcal{T}$ can be composed in more than one way into a tiling in Ω_n. In this case the supertile rule is "not recognizable" or does not have the "unique composition property".

Definition 2.4.5 (See [50]) A supertile rule is said to be *recognizable* if the decomposition map from Ω_n to Ω_{n-1} is invertible for all n.

This definition looks at recognizability as a sort of global property determined by the connection between supertiling spaces. It can, however, be convenient to think of it locally as converting patches of n-supertiles into $(n+1)$-supertiles. If a supertile rule is recognizable then every tiling in Ω can be unambiguously expressed as a tiling with n-supertiles for every n. It is not difficult to show that the decomposition maps are uniformly continuous, and if they are invertible the inverse is also uniformly continuous. Thus there exists a family of *recognizability radii* r_n ($n = 1, 2, \ldots$), such that, whenever two tilings in Ω have the same patch of radius r_n around a point $\mathbf{v} \in \mathbb{R}^d$, then the n-supertiles intersecting $\mathbf{v}$ in those two tilings are identical.

The terminology original to the symbolic substitutions case is 'recognizability', and it is shown in [81] that recognizability and nonperiodicity are equivalent in that case. Solomyak [100] gave the concept the name 'unique composition property' for self-affine tilings. He proved in [98] that unique composition and nonperiodicity are equivalent. In [50] the notion is defined for fusions, where it is shown by example that there are no general results connecting nonperiodicity to recognizability.

Recognizability turns out to be essential for many arguments and is almost always assumed. It is central in the following construction.

Example 13 (How to Make a Nonlocal Homeomorphism) We use two tiling spaces associated with the Fibonacci substitution of Example 3. Let Ω be the space with labelled unit-length tiles and Ω' to be the space with natural tile lengths, normalized so that the n-supertiles in both spaces asymptotically converge in length. These spaces are easily shown to be recognizable.

We can make an invertible local map Q by requiring that $\mathcal{T}$ and $Q(\mathcal{T})$ have the same underlying sequence of a's and b's and then determining the precise location of 0 in $Q(\mathcal{T})$. This location is determined from the location of 0 in $\mathcal{T}$ using the supertile structure: one considers the sequence of n-supertiles in $\mathcal{T}$ containing 0, and translates $Q(\mathcal{T})$ a little bit for each n so that, say, the left endpoints of the n-supertiles line up. Since the lengths converge asymptotically the adjustments at each stage will go to 0 and the precise location of $Q(\mathcal{T})$ can be determined.

2.5 Ergodic-Theoretic and Dynamical Analysis of Supertile Methods

2.5.1 Transition (a.k.a. Incidence, Substitution, Abelianization, or Subdivision) Matrices

We can obtain basic geometric and statistical information by associating a matrix or matrices to a supertile rule. Transition matrices keep track of how many n-supertiles of each type there are in each $(n + 1)$-supertile. They go by many names in the literature but we will use 'transition' as our terminology.

These matrices are fundamental to their supertile rules, among other things helping to compute frequencies and ergodic measures. Since we can count how many times p_i appears in $\mathcal{S}^n(p_j)$, and we can count the total number of tiles in $\mathcal{S}^n(p_j)$, we can estimate the relative frequency of p_i. The Perron-Frobenius theory of matrices allows us to draw conclusions about the frequency statistics in our hulls accordingly. In particular we will see how to use transition matrices to construct ergodic measures.

The matrices for self-similar tilings of $\mathbb{R}^d$, substitutions in $\mathbb{Z}$, constant-length substitutions of $\mathbb{Z}^d$, and stationary (or even transition-regular) fusion rules in $\mathbb{R}^d$ or $\mathbb{Z}^d$ are all obtained the same way, so we use tiling terminology to refer to all cases. We assume the prototile set has been given some arbitrary order $\mathcal{P} = \{p_1, \ldots p_{|\mathcal{P}|}\}$, which we shall keep fixed. Then the transition matrix for S is the $|\mathcal{P}| \times |\mathcal{P}|$ matrix M whose (i, j) entry M_{ij} is the number of tiles of type p_i in $S(p_j)$. It is not hard to check that the tile population information for S^n is given by M^n.

Fusion rules have a somewhat more complicated transition matrix situation. This is due to two facts: first, that the fusion rules used to construct the n-supertiles may be completely unrelated to the rules used to construct the m-supertiles when $m \neq n$, and second, that the number of supertiles can vary from level to level. Thus we need an infinite family of (possibly non-square) transition matrices in order to give us the population information we seek.

Recall that $\mathcal{P}_n$ is the set of n-supertiles. We define the *transition matrix $M_{n-1,n}$* to be the $|\mathcal{P}_{n-1}| \times |\mathcal{P}_n|$ matrix whose (k, l) entry is the number of supertiles equivalent to $P_{n-1}(k)$ (i.e. the number of $(n-1)$ supertiles of type k) in the supertile $P_n(l)$. Notice that the matrix product $M_{n,N} = M_{n,n+1}M_{n+1,n+2} \cdots M_{N-1,N}$ is well-defined when $N > n$. The entries of $M_{n,N}$ reveal the number of n-supertiles of every type of N-supertile. If there is more than one fusion of $\mathcal{P}_{n-1}$-supertiles that can make $P_n(l)$, we fix a preferred one to be used in all computations.

2.5.2 Primitivity

Generally speaking, a supertile method is primitive if one finds every type of n-supertile in every type of N-supertile, provided N is sufficiently large. This assumption is useful in obtaining minimality, repetitivity, and unique ergodicity results and ensures that our hulls have a certain level of homogeneity.

For this definition, suppose M or $M_{n,n+N}$ are the transition matrices for the supertile rules.

Definition 2.5.1 A symbolic or tiling substitution rule is defined to be *primitive* if and only if there is an N such that all of the entries of M^N are strictly positive. A fusion rule is defined to be *primitive* if and only if for every $n \in \mathbb{N}$ there exists an N such that the entries of $M_{n,n+N}$ are strictly positive.

In this latter situation it is possible that N varies depending on n.

2.5.2.1 General Result: Primitivity Implies Minimality

Recall that a topological dynamical system (Ω, G) is said to be *minimal* if and only if Ω is the orbit closure of any of its elements.

Proposition 2.5.1 [10] *Let Ω be the space of tilings allowed by a supertile construction and let G be its group of translations. If the supertile construction is primitive, then (Ω, G) is minimal.*

Note that this proposition is not an if and only if: the Chacon substitution is an example of a supertile construction that is minimal even though it is not primitive. Although the most study has been done on primitive systems, progress has been made in the non-primitive case as well [25, 26, 78]. We should note, however, that primitivity and minimality are 'morally' the same in the sense that minimal nonprimitive systems in one dimension can be transformed into primitive ones using a return word procedure [78]. A way to do it for a Chacon DPV in two dimensions is shown in [46].

2.5.2.2   Result for Substitution Systems: Primitivity Implies Unique Ergodicity

Because it is easy to write down the transition matrix for a substitution rule, it can often easily be determined that a substitution dynamical system is uniquely ergodic. The general theorem that makes this possible in all cases is the Perron-Frobenius theorem. The situation for fusions is more subtle and will be discussed later.

Part of the Perron-Frobenius theorem requires matrices that are *irreducible* in the sense that for each index (i, j) there is an $n \in \mathbb{N}$ such that $M_{ij}^n > 0$. This condition is weaker than primitivity because the entries are not required to be simultaneously positive for any n. Clearly, a primitive matrix is irreducible. We cite the portions of the Perron-Frobenius theorem that are relevant to our study as a combination of statements from [74, p. 109], [67, p. 16], and [100, p. 704].

Theorem 2.5.2 (Perron-Frobenius Theorem) *Let M be an irreducible matrix. Then M has positive left and right eigenvectors $\mathbf{l}$ and $\mathbf{r}$ with corresponding eigenvalue $\theta > 0$ that is both geometrically and algebraically simple. If θ' is another eigenvalue for M then $|\theta'| \leq \theta$. Any positive left or right eigenvector for M is a multiple of $\mathbf{l}$ or $\mathbf{r}$.*

Moreover, if M is primitive and $\mathbf{l}$ and $\mathbf{r}$ are normalized so that $\mathbf{l} \cdot \mathbf{r} = 1$, it is true that

$$\lim_{n \to \infty} \frac{M^n}{\theta^n} = \mathbf{rl}.$$

The eigenvector $\mathbf{l}$ and $\mathbf{r}$ are ordinarily called the Perron eigenvectors and θ is always called the *Perron eigenvalue* for M.

---

[10]See [64, 89] for symbolic substitutions, [87] for self-affine tilings, and [50] for fusions.

In order to find the result that primitivity implies unique ergodicity for symbolic substitutions, a good reference is [89, Ch. V.4]. There we find this stronger result:

Theorem 2.5.3 ([89], Theorem V.13) *If a symbolic substitution system is minimal, then it is uniquely ergodic.*

It is stronger because primitivity is not the only way for a symbolic substitution system to be minimal. For example, the Chacon system is minimal and therefore uniquely ergodic despite it not being primitive.

The situation for self-affine tilings is as follows.

Corollary 2.5.4 (Folklore; Corollary 2.4 of [100]) *Suppose $\mathcal{T}$ is a self-affine tiling with expansion map ϕ for which the transition matrix M is primitive. Then the Perron eigenvalue of M is $|det\phi|$. Writing the prototile set as $\mathcal{P} = \{p_1, p_2, \ldots, p_m\}$, the left eigenvector can be obtained by $\mathbf{l} = (Vol(p_j))_{j=1}^m$. Moreover,*

$$\lim_{n \to \infty} |det\phi|^{-n} M_{ij}^n = r_i Vol(p_j).$$

The last equation proves particularly useful in computing frequencies and ergodic measures as we will show in the case study, next. Moreover, one can show that the n-supertile frequencies are given by $\frac{1}{|det(\phi)|^n}\mathbf{r}$ and we can get the frequencies of everything else from that information.

The adaptation of the previous result to multidimensional constant-length symbolic substitutions is carried out in [45].

2.5.2.3 Case Study: Ergodic Measures for Constant-Length $\mathbb{Z}^d$ Substitutions

By results in [45], we can use Corollary 2.5.4 to make an instructive example. We consider a tiling model of sequences in Ω where each tile is a unit 'cube' in $\mathbb{R}^d$ labeled by the appropriate element of $\mathcal{A}$.

If $\mathcal{S}$ is a primitive, nonperiodic substitution with size $l_1 \cdot l_2 \cdots l_d = K$ and ϕ is its natural expanding map, Corollary 2.5.4 implies that the largest eigenvalue of M must be equal to $|det \phi| = K$. The left Perron eigenvector $\mathbf{l}$ must the tile volumes, which are all 1. This implies that $\sum_{i=1}^{|\mathcal{A}|} r_i = 1$ and that

$$\lim_{n \to \infty} K^{-n} M^n = \begin{pmatrix} r_1 & r_1 & \cdots & r_1 \\ r_2 & r_2 & \cdots & r_2 \\ \vdots & \vdots & & \vdots \\ r_{|\mathcal{A}|} & r_{|\mathcal{A}|} & \cdots & r_{|\mathcal{A}|} \end{pmatrix}.$$

We know that the dynamical system $(\Omega, \mathbb{Z}^d)$ is uniquely ergodic, and by our discussion of the connection between ergodic measures and frequencies in

Sect. 2.3.4 we know that frequencies exist and must be equal for almost every element of Ω. One can show that it suffices to compute frequencies in larger and larger supertiles, rather than in arbitrary balls of expanding radius. In fact, by primitivity it doesn't matter which type of supertiles we look at, so we will just look at $\mathcal{S}^n(a_1)$ as $n \to \infty$. We will use the notation $N_{a_i}(B)$ to denote the number of occurrences of the letter a_i in a block B. Then

$$freq(a_i) = \lim_{n \to \infty} \frac{N_{a_i}(\mathcal{S}^n(a_1))}{K^n}, \tag{2.4}$$

since K^n is the volume of the substituted block $\mathcal{S}^n(a_1)$. The numerator is easily computed since it is simply M_{i1}^n. Thus we have that $freq(a_i) = \lim_{n \to \infty} K^{-n} M_{i1}^n = r_i$, and so computation of $freq(a_i)$ reduces to computation of the right eigenvector for M.

Sometimes the computation comes out particularly nice. For instance, we have the following proposition:

Proposition 2.5.5 *Let $\mathcal{S}$ be a primitive and nonperiodic substitution of constant length $l_1 \cdot l_2 \cdots l_d = K$ in $\mathbb{Z}^d$. Then M has the property that $\sum_{j=1}^{|\mathcal{A}|} M_{ij} = K$ for all $i \in 1, 2, \ldots |\mathcal{A}|$ if and only if the frequency of any letter $a_i \in \mathcal{A}$ is $1/|\mathcal{A}|$.*

Proof If $\sum_{j=1}^{|\mathcal{A}|} M_{ij} = K$, then a right eigenvector for M is given by $\mathbf{r} = (1/|\mathcal{A}|, \ldots, 1/|\mathcal{A}|)$. Since $\mathbf{l} \cdot \mathbf{r} = 1$ it must be the (unique) right Perron eigenvector for M. Since $freq(a_i) = r_i$ the result follows.

Conversely, the vector $\mathbf{r}$ defined as above again is a right eigenvector and we have that

$$(M\mathbf{r})_i = \sum_{j=1}^{|\mathcal{A}|} M_{ij}/|\mathcal{A}| = (K\mathbf{r})_i = K/|\mathcal{A}|,$$

and the result follows.

Bijective substitutions (defined in Sect. 2.4.3) automatically satisfy the former condition, and so do the Rudin-Shapiro-like substitutions seen in [44].

Corollary 2.5.6 *If $\mathcal{S}$ is a primitive, nonperiodic, bijective substitution of constant length in $\mathbb{Z}^d$, then the frequency of any letter $a_i \in \mathcal{A}$ is $1/|\mathcal{A}|$.*

Proof *(Sketch)* The row sum for row i is the number of times we see a_i in the substitution of any tile. Because the substitution is bijective, for any given location in the substitution we know that a_i appears exactly once. That means that the number of times a_i appears is the number of spots in the substitution, which is K.

2.5.3   Ergodic Measures for Fusions

Let us suppose that $\mathcal{R}$ is a primitive, recognizable, van Hove fusion rule in $\mathbb{R}^d$ that admits a nontrivial tiling space Ω. What are the possibilities for translation-invariant measures? We suppose for convenience[11] that the group action for dynamics is $G = \mathbb{R}^d$ and follow [50], section 3.4.

We cannot use the Perron-Frobenius theorem in this situation because our transition matrices change from level to level, so we adapt it to work here. Because of the relationship between S-adic systems, Bratteli diagrams, and fusion tilings, our analysis is closely related to that in [27, 41] and others. (Our work takes the analysis into the continuous dynamics situation.)

By recognizability we know that every tiling $\mathcal{T} \in \Omega$ has a unique n-supertiling $\mathcal{T}_n \in \Omega_n$. Consider a particular n-supertile $P_n(j) \in \mathcal{P}_n$. We denote its frequency in $\mathcal{T}$ as an n-supertile as $\bar{f}_{P_n(j)}$, if it exists. We know that $P_n(j)$ as a patch of ordinary tiles may have a larger frequency $\bar{f}_{P_n(j)}$ in $\mathcal{T}$. In fact, recognizability gives us a finite list of patches $S_1, S_2, \ldots S_q$ that appear if and only if $P_n(j)$ appears as a supertile. That means that we can compute $\bar{f}_{P_n(j)} = \sum_{i=1}^q \bar{f}_{S_i}$ if the latter frequencies exist in $\mathcal{T}$.

In the symbolic or tiling substitution case we found a right eigenvector $\mathbf{r}$ that represented the prototile frequencies and satisfied $\mathbf{r} \cdot \mathbf{l} = 1$, where $\mathbf{l}$ is the vector of tile volumes. We might say that $\mathbf{r}$ is 'volume-normalized', and the useful thing about that is that it makes the ergodic measure a probability measure. We need to extend this concept to fusion tilings.

In the substitution case, the n-supertile frequencies are all given by the vector $\frac{1}{|det\phi|^n}\mathbf{r}$. For fusion rules the supertile frequencies are not as simple. We let a nonnegative vector $\rho_n = (\rho_n(1), \ldots, \rho_n(j_n)) \in \mathbb{R}^{j_n}$ represent the n-supertile frequencies.

Definition 2.5.2 Let ρ be a sequence of vectors $\{\rho_n\}$ described above. We say that ρ is *volume-normalized* if for all n we have $\sum_{i=1}^{j_n} \rho_n(i)Vol(P_n(i)) = 1$. We say that it has *transition consistency* if $\rho_n = M_{n,N}\rho_N$ whenever $n < N$. A transition-consistent sequence ρ that is normalized by volume is called a sequence of *well-defined supertile frequencies*.

As before, volume-normalization is there to ensure that the measure is a probability measure. The transition-consistency requirement ensures that the measure is additive: it is necessary that the frequency of N-supertiles be related to the frequencies of the n-supertiles they are composed of. This property was automatically satisfied before because $\mathbf{r}$ was an eigenvector. For fusion rules, it turns out that the invariant measures are completely determined by sequences of well-defined supertile frequencies:

---

[11] How to adapt the analysis appears in [50, section 3.7].

Theorem 2.5.7 ([50]) *Let $\mathcal{R}$ be a recognizable, primitive, van Hove fusion rule with tiling dynamical system $(\Omega, \mathbb{R}^d)$. There is a one-to-one correspondence between the set of all invariant Borel probability measures on $(\Omega, \mathbb{R}^d)$ and the set of all sequences of well-defined supertile frequencies with the correspondence that, for all patches P,*

$$freq_\mu(P) = \lim_{n \to \infty} \sum_{i=1}^{j_n} \#(P \text{ in } P_n(i))\, \rho_n(i). \tag{2.5}$$

Thus one could, given a sequence of well-defined supertile frequencies, construct a translation-invariant ergodic measure μ as follows. Given any patch P, get the frequency $freq_\mu(P)$ as in Eq. (2.5). Then, the measure of a cylinder set $\Omega_{P,U}$ will be $freq_\mu(P)Vol(U)$, provided U is not too large. Since the cylinder sets form a basis for the topology, we can now measure any Borel measurable set.

2.5.4   General Result: Substitution Systems are Not Strongly Mixing

Recall that a measure-preserving system is *strongly mixing* if for all measurable sets A, B and for any sequence of vectors $\mathbf{v}_n$ whose lengths increase without bound it is true that $\lim_{n \to \infty} \mu(A \cap (B - \mathbf{v}_n)) = \mu(A)\mu(B)$. There is a standard argument proving that a substitution system isn't strongly mixing, appearing for substitution sequences in [34] and for self-similar tilings in [100]. Here is a general result.

Theorem 2.5.8 ([50]) *The dynamical system of a strongly primitive van Hove fusion rule with a constant number of supertiles at each level and bounded transition matrices, and with group $G = \mathbb{Z}^d$ or $\mathbb{R}^d$, cannot be strongly mixing.*

For tiling or symbolic substitutions neither the transition matrix nor the number of supertiles changes from level to level, so these are never strongly mixing. Indeed, we see that constructing a strongly mixing supertile method requires a certain degree of unboundedness, either in the number of supertiles at each level or in the entries of the transition matrices between consecutive levels.

2.5.5   Fusion Rules with Various Properties

The fusion paradigm can be used to construct interesting examples where the standard results from substitution systems need not apply. Here are some that appear in [50].

1. Example 3.7 provides us with an example of a one-dimensional, prototile- but not transition-regular fusion rule that has a minimal but not uniquely ergodic dynamical system.

2. Again prototile-regular, example 3.8 shows how a measure arising from a sequence of supertiles can fail to be ergodic.

3. Example 4.4, the "scrambled Fibonacci", is based on the Fibonacci substitution/fusion, but is systematically altered at occasional levels. The alterations are enough to eliminate topological point spectrum, but the system still has measurable eigenvalues. Thus we have a system that is topologically weakly mixing but measure-theoretically pure point. It appears in [65], along with a related example where the measurable and topological eigenvalues differ.

4. Example 4.8 gives us an example of a uniquely ergodic symbolic fusion system that has 'coincidence with finite waiting' but is not pure point spectrum. This contrasts with Dekking's classical result on coincidence for constant-length symbolic substitutions [33], where coincidence implies pure pointedness.

5. Example 4.9 provides a one-dimensional symbolic fusion that is not prototile-regular, where not only is the system not uniquely ergodic, but also the ergodic measures can have different spectral types.

6. Example 4.11 is an example of a symbolic fusion rule that is strictly ergodic and yet has positive entropy.

2.6   Spectral Analysis of Supertile Methods: Dynamical Spectrum

For the remainder of these lectures we discuss the two main spectral methods for analyzing tiling spaces: dynamical and diffraction. Spectral theory of dynamical systems is widely used and provides a measure-theoretic tool that standardizes spaces acted on by a group $G \subseteq \mathbb{R}^d$ by comparing them to Lebesgue measure on the dual group of G. This is achieved using what is now called the Koopman representation, which represents G as unitary operators on $L^2(\Omega, \mu)$, and applying the spectral theorem for unitary operators. A nice development of the subject for $G = \mathbb{Z}$ or $\mathbb{Z}^+$ can be found in [84, 104]; also for $G = \mathbb{Z}$ [56] provides a well-contextualized historic overview and survey of results up to 1999. The book [89] is entirely devoted to analyzing symbolic substitutions in one dimension using spectral theory, as is Chapter 7 of [42]. The paper [19] extends many of these results to the multidimensional constant-length $\mathbb{Z}^d$ substitution case. The dynamical spectrum framework for self-similar tilings of $\mathbb{R}^d$ is laid out in [100]. One can see [11, Appendix B] for a brief discussion of how the theory for $\mathbb{Z}$ actions can be extended to the higher-dimensional and/or continuous case, but the author is not aware of any survey of the spectral theory of tiling dynamical systems.

Strong motivation for studying the diffraction spectrum of tilings comes from the quasicrystal model. Indeed, Shechtman's Nobel prize-winning discovery of quasicrystals arose from a diffraction experiment [97]. A lovely and somewhat underappreciated early book on crystals, quasicrystals, tilings, diffraction, and the history of the subject is [96], which bridges the gap between physics and mathematics for those looking for perspective. Fourier analysis is used to define virtual diffraction experiments on tilings, and [11] is an excellent source to learn about how it works. The method was originally proposed in [60], and there is a fundamental argument in [37] that allows us to see that the diffraction spectrum is related to the dynamical spectrum. There has been special emphasis on the discrete (a.k.a. point or atomic) part of the diffraction spectrum of tilings because it represents the bright spots appearing on a diffraction image known as "Bragg peaks". In particular, extensive work has been done to determine conditions under which there exists such point spectrum, and when the spectrum is composed solely of it. We will discuss a selection of results in this direction as well as what is known about the connection between the two types of spectral analysis.

2.6.1   The Koopman Representation

Let (Ω, G) be a dynamical system with G representing a group of translations. Suppose μ is a translation-invariant ergodic Borel probability measure for the system. The function space $L^2(\Omega, \mu)$ is a Hilbert space and one often looks at it when trying to analyze a system. From a physics perspective one could consider a function as taking measurements or running experiments on the tilings in the tiling space.

For each $J \in G$ there is a unitary operator $U^J : L^2(\Omega, \mu) \rightarrow L^2(\Omega, \mu)$ defined by

$$U^J(f)(\mathcal{T}) = f(\mathcal{T} - J).$$

This family of operators is sometimes called the Koopman operator and it is a representation of G. Since $L^2(\Omega, \mu)$ is a separable Hilbert space the tools of operator theory are available. The spectrum of the Koopman operator is called the *dynamical spectrum* of Ω (or of the supertile method that generated it).

Every $f \in L^2(\Omega, \mu)$ has a spectral measure associated with it. No matter which construction method was used, all of the cases look like this: for $J \in G$ we define $\hat{f}(J) = \int_\Omega f(\mathcal{T} - J)\overline{f(\mathcal{T})}d\mu(\mathcal{T})$. One can think of comparing the values of f at two spots in $\mathcal{T}$, separated by J, and averaging the result over all of Ω. In each of our situations this satisfies the appropriate notion of positive definiteness so that the appropriate version of Bochner's theorem[12] guarantees us the existence of a positive

---

[12]The general results on spectral theory of dynamical systems in this section can be found in many places, for instance [56, 71]; specialization to the tiling case first appears in [100].

real-valued measure σ_f on $\mathbb{T}^d$ with these same Fourier coefficients. That is,

$$\hat{f}(\jmath) = \int_\Omega f(\mathcal{T} - \jmath)\overline{f(\mathcal{T})}d\mu(\mathcal{T}) = \int_{\hat{G}} z^{\jmath} d\sigma_f(z), \qquad (2.6)$$

where $z^{\jmath} := z_1^{j_1} \cdots z_d^{j_d}$ and the dual group to G is denoted $\hat{G}$.

The *spectral type* of a function f, then, is determined by how σ_f decomposes with respect to Lebesgue measure. Is it discrete, singular continuous, or absolutely continuous? Or perhaps a mix? The supertile rule that determines Ω ultimately determines what is possible for these spectral types and is our primary interest in this topic.

Each $f \in L^2(\Omega, \mu)$ generates a cyclic subspace of L^2 given by its closed linear span:

$$Z(f) = \overline{\text{span}\{U^{\jmath} f \text{ such that } \jmath \in G\}}.$$

Part of the spectral analysis of operators involves finding generating functions $f_i, i = 1, 2, \ldots$ for which

$$L^2(\Omega, \mu) = \bigoplus Z(f_i).$$

This decomposition is not exactly unique but the number and spectral types of the functions are. It is possible to find functions $f_1, f_2, f_3, \ldots$ such that $L^2(\Omega, \mu) = \bigoplus Z(f_i)$ and for which $\sigma_{f_1} \gg \sigma_{f_2} \gg \sigma_{f_3} \gg \ldots$. Again the functions are not unique, but their spectral types are and the spectral type of f_1 is known as the *maximal spectral type* of the system. This decomposition determines the Koopman operator up to unitary equivalence.

2.6.2 Eigenfunctions

To begin thinking more about the spectrum of $U^{\jmath}$ we can investigate its eigenvalues and eigenvectors. The easiest case is when $G = \mathbb{Z}$. In that case we are looking at any functions $f \in L^2(\Omega, \mu)$ for which there is some λ such that $Uf = \lambda f$. In general f is known as a *measurable eigenfunction*, and if it happens to be continuous then it is called a *topological eigenfunction*. Notice that since U is unitary it must be that $|\lambda| = 1$ and we write $\lambda = e^{2\pi i \alpha}$ for some $\alpha \in \mathbb{R}$. Obviously once we have an eigenvalue/eigenfunction pair then for any $n \in \mathbb{Z}$ we have that $U^n f = \lambda^n f = e^{2\pi i \alpha n} f$.

Now let's generalize to a continuous one-dimensional action, i.e. when $G = \mathbb{R}$. An *eigenfunction* is a function for which there exists an $\alpha \in \mathbb{R}$ such that for all $x \in \mathbb{R}$ it is true that $U^x f = e^{2\pi i \alpha x} f$. That is, for all $\mathcal{T} \in \Omega$ it is true that $f(\mathcal{T} - x) = e^{2\pi i \alpha x} f(\mathcal{T})$.

When G is $\mathbb{Z}^d$ or $\mathbb{R}^d$ with $d > 1$, the eigenvalues themselves live in higher dimensions and the inner product becomes necessary. In this situation we say f is an eigenfunction if there exists an $\boldsymbol{\alpha} \in \mathbb{R}^d$ for which $U^{\boldsymbol{J}} f = \exp(2\pi i \boldsymbol{\alpha} \cdot \boldsymbol{J}) f$ for all $\boldsymbol{J} \in G$. That is, for all $\mathcal{T} \in \Omega$ and all $\boldsymbol{J} \in G$ we find $f(\mathcal{T} - \boldsymbol{J}) = \exp(2\pi i \boldsymbol{\alpha} \cdot \boldsymbol{J}) f(\mathcal{T})$.

An important technical point, which seems to be a point of contention, is whether it is λ or α that is considered the eigenvalue for f. Both perspectives have merit. The argument for using λ is that it is standard usage in functional analysis for the spectrum of the unitary operator. The argument for α is that it resides in the dual group of G and therefore is more directly relevant to diffraction analysis and abstract harmonic analysis. We will allow either to count as the eigenvalue, being more precise when necessary.

There is a difference between the situation where $G = \mathbb{Z}^d$ and $G = \mathbb{R}^d$ that also manifests itself when we allow $G = \mathbb{R}^d$ but the tiling is a suspension of a $\mathbb{Z}^d$ action. When $G = \mathbb{Z}^d$ or the tiling space is a suspension of a $\mathbb{Z}^d$ action, then every $\boldsymbol{\alpha} \in \mathbb{Z}^d$ is an eigenvalue of the system; that is, in these cases there is always some discrete spectrum. In fact when $G = \mathbb{Z}^d$, the dual group is $\mathbb{T}^d$ and we have that if $\boldsymbol{\alpha}$ is any eigenvalue, then $\boldsymbol{\alpha} + \boldsymbol{J}$ is also an eigenvalue for the same eigenfunction, so the spectrum is only considered on $\mathbb{T}^d$.

When $G = \mathbb{R}^d$ but the system is a suspension of a $\mathbb{Z}^d$-action, all elements of $\mathbb{Z}^d$ continue to be eigenvalues but we get that $\boldsymbol{\alpha}$ and $\boldsymbol{\alpha} + \boldsymbol{J}$ have different eigenfunctions for $\boldsymbol{J} \in \mathbb{Z}^d$. However, the eigenfunctions are closely related: if we let f denote the eigenfunction for $\boldsymbol{\alpha}$ and $f_{\boldsymbol{J}}$ that of $\boldsymbol{\alpha} + \boldsymbol{J}$, then $f_{\boldsymbol{J}}(\mathcal{T}) = \exp(2\pi i \boldsymbol{J} \cdot \mathbf{x}(\mathcal{T})) f(\mathcal{T})$, where $\mathbf{x}(\mathcal{T})$ is the location in $\mathbb{R}^d$ of any vertex[13] of $\mathcal{T}$. Put another way, the eigenfunction for $\boldsymbol{\alpha} + \boldsymbol{J}$ is the product of the eigenfunction for $\boldsymbol{\alpha}$ with a function that keeps track of where the tiling is relative to the suspension.

When $G = \mathbb{R}^d$ and the space cannot be seen as the suspension over a lattice we must consider all possible values of the dual group of $\mathbb{R}^d$, which is still $\mathbb{R}^d$. Thus spectral images in this case are not restricted to a torus.

Example 14 Let's compute the spectral measure of an eigenfunction f of (Ω, μ) with eigenvalue α. For $\boldsymbol{J} \in G$ we have

$$\hat{f}(\boldsymbol{J}) = \int_{\Omega} \exp(2\pi i \boldsymbol{\alpha} \cdot \boldsymbol{J}) \overline{f(\mathcal{T})} f(\mathcal{T}) d\mu(\mathcal{T}) = \exp(2\pi i \boldsymbol{\alpha} \cdot \boldsymbol{J}),$$

since eigenfunctions are of almost everywhere constant modulus that can be taken to be 1. Thus the spectral measure of f is a measure σ_f on $\mathbb{T}^d$ with these Fourier coefficients. One can check that the measure on $\mathbb{T}^d$ with these coefficients is the atomic measure supported on α, and so $\sigma_f = \delta_\alpha$.

---

[13] One can show that this term is independent of the choice of vertex.

2.6.2.1 Conditions for Presence of Discrete Spectrum

It is not surprising that spectral properties were investigated soon after the dynamical systems approach to substitution sequences was introduced. Coven and Keane [32] investigated a class of examples and their approach was generalized by Martin [80]. Dekking [33] generalized these results using different methods, completely determining the point part of the dynamical spectrum of constant-length substitutions in one dimension.

Theorem 2.6.1 ([33], Quoted from Section 7.3 of [42]) *Let σ be a non-periodic (symbolic) substitution of constant length n. Let u be a periodic point for σ. We call the* height *of the substitution the greatest integer m which is coprime with n and divides all the strictly positive ranks of occurrence of the letter u_0 in u. The height is less than the cardinality of the alphabet. The maximal equicontinuous factor[14] of the substitutive dynamical system associated with σ is the addition of $(1, 1)$ on the abelian group $\mathbb{Z}_n \times \mathbb{Z}/m\mathbb{Z}$.*

One would expect the discrete spectrum of a symbolic substitution of non-constant length to have a connection to the expansion factor of the system, and Host's result [62] gives a criterion that we quote here, following [42, Chapter 7]. We leave undefined the term "coboundary", which we will not be using again.

Theorem 2.6.2 ([62], Quoted from Section 7.3 of [42]) *Let σ be a not shift-periodic and primitive substitution over the alphabet $\mathcal{A}$. A complex number λ of modulus one is an eigenvalue of $(\Omega, \mathbb{Z})$ if and only if there exists $p > 0$ such that for every $a \in \mathcal{A}$, the limit $h(a) = \lim_{n \to \infty} \lambda^{|\sigma^{p^n}(a)|}$ is well defined, and h is a coboundary of σ.*

The constant function 1 is always a coboundary, making it simpler to check:

Theorem 2.6.3 (Corollary 7.3.17 of [42]) *Let σ be a not shift-periodic and primitive substitution over the alphabet $\mathcal{A}$. If there exists p such that $\lambda \in \mathbb{C}$ satisfies $\lim_{n \to \infty} \lambda^{|\sigma^{p^n}(a)|} = 1$ for every $a \in \mathcal{A}$, then λ is an eigenvalue of the substitutive dynamical system associated with σ.*

One of the main results in [100] is the characterization of eigenvalues and eigenfunctions for self-similar tiling systems. The method of proof is constructive in that given the eigenvalue condition, an eigenfunction is constructed whose value for a tiling $\mathcal{T}$ depends on special points derived from the supertile structure of $\mathcal{T}$.

In [100], general results on the presence or absence of eigenfunctions and therefore on weak mixing are determined. There are results for self-affine tilings of $\mathbb{R}^d$ that are made stronger in the $d = 1$ and 2 cases. The statement presented here is essentially quoted from that paper, except that the result of [98] is taken into account. Note that the set of translations between tiles in $\mathcal{T}$ is given by $\Xi(\mathcal{T})$, where

---

[14]Recall that this is the largest topological factor of the dynamical system that is a rotation of a compact group.

$x \in \Xi(\mathcal{T})$ if and only if x is a translation taking a tile $t \in \mathcal{T}$ to an equivalent tile in $\mathcal{T}$, i.e. $t - x \in \mathcal{T}$.

Theorem 2.6.4 (Theorem 5.1 of [100])

(i) *Let $\mathcal{T}$ be a nonperiodic self-affine tiling of $\mathbb{R}^d$ with expansion map ϕ. Then $\alpha \in \mathbb{R}^d$ is an eigenvalue of the measure-preserving system (Ω, μ) if and only if*

$$\lim_{n \to \infty} e^{2\pi i (\phi^n(x) \cdot \alpha)} = 1 \quad for\ all \quad x \in \Xi(\mathcal{T}). \tag{2.7}$$

Moreover, if Eq. (2.7) holds, the eigenfunction can be chosen continuous.[15]

(ii) *Let $\mathcal{T}$ be a self-similar tiling of $\mathbb{R}$ with expansion constant λ. The tiling dynamical system is not weakly mixing if and only if $|\lambda|$ is a real Pisot number. If λ is real Pisot and $\mathcal{T}$ is nonperiodic, there exists nonzero $a \in \mathbb{R}$ such that the set of eigenvalues contains $a\mathbb{Z}[\lambda^{-1}]$.*

(iii) *Let $\mathcal{T}$ be a nonperiodic self-similar tiling of $\mathbb{R}^2 \equiv \mathbb{C}$ with expansion constant $\lambda \in \mathbb{C}$. Then the tiling dynamical system is not weakly mixing if and only if λ is a complex Pisot number. Moreover, if λ is a non-real Pisot number there exists nonzero $a \in \mathbb{C}$ such that the set of eigenvalues contains $\{(\alpha_1, \alpha_2) : \alpha_1 + i\alpha_2 \in a\mathbb{Z}[\lambda^{-1}]\}$.*

2.6.3 Pure Discrete Dynamical Spectrum

Dekking's work in [33] was the first to define a notion now known as the "coincidence condition" for a symbolic substitution σ. It is that there are numbers k and l such that the image of any letter of the alphabet under σ^k has the same lth letter. This combinatorial condition is easy to check in any given example and eliminates any spectrum that is not discrete. The idea has been generalized to non-constant length symbolic substitutions and to self-similar tiling systems with the goal of characterizing purely discrete spectrum in those cases. Algorithms have been developed such as the "balanced pair algorithm" for substitution sequences (see [79] and references therein or the original source [75], stated for adic transformations on Markov compacta). For the multidimensional case there is a series of papers [3–5] on the "overlap algorithm" making checkable conditions; the original version of this for tilings appears in [100]. In terms of spectral analysis of supertile systems, spaces with a purely discrete dynamical spectrum are the most well understood.

---

[15]This is also true for substitution sequences [62], but not for general fusions.

2.6.3.1 Symbolic Substitutions and the Pisot Substitution Conjecture

Let us begin with Dekking's original result.

Theorem 2.6.5 ([33], Quoted from Section 7.3 of [42]) *Let σ be a substitution of constant length and of height 1. The substitutive dynamical system associated with σ has a purely discrete spectrum if and only if the substitution σ satisfies the condition of coincidence.*

The situation is not settled in the non-constant length substitution case. A substitution σ satisfies the *strong coincidence condition* if there are integers k and l such that for every $a, b \in \mathcal{A}$, the substitutions $\sigma^k(a)$ and $\sigma^k(b)$ not only have the same lth letter, but the prefixes of length $l - 1$ in each have the same number of letters of each type. The latter part of this condition ensures that $\sigma^{(k+j)}(e)$ will have coinciding j-supertiles at the corresponding location for all $j \geq 1$.

It is thought, but not known, that algebraic properties of the transition matrix can determine coincidences. In particular, a substitution is said to be of *Pisot type* if all of the eigenvalues of its transition matrix except for the Perron-Frobenius eigenvalue have modulus strictly between 0 and 1. It is said to be *irreducible* if the characteristic polynomial is irreducible.

There are a family of conjectures collectively known as "Pisot substitution conjectures", that all more or less say, "the substitution dynamical system has pure discrete spectrum if it is of irreducible Pisot type". These conjectures are in place not only for one-dimensional symbolic substitutions, but also for one-dimensional tiling substitutions and a few other sorts of substitutions; the situation is nicely summarized in [2]. Immediately relevant to our work following two conjectures cited there that are equivalent by Clark and Sadun [31].

Conjecture 2.6.1 (Pisot Substitution Conjecture: Symbolic Substitutive Case) If σ is an irreducible Pisot substitution then the substitutive system $(\Omega, \mathbb{Z})$ has pure discrete spectrum.

Conjecture 2.6.2 (Pisot Substitution Conjecture: One-dimensional Tiling Case) If S is an irreducible Pisot substitution for one-dimensional tilings, then its tiling dynamical system $(\Omega, \mathbb{R})$ has pure discrete spectrum.

Progress has recently been made to settle the second conjecture: in [17] the conjecture is confirmed for substitutions that are injective on initial letters and constant on final letters. Closely related is the question of whether the substitution satisfies the strong coincidence condition. Again from [2]:

Conjecture 2.6.3 (Strong Coincidence Conjecture) Every irreducible Pisot substitution satisfies the strong coincidence condition.

The following two theorems together settle the Pisot substitution conjecture for two-letter substitutions:

Theorem 2.6.6 ([61]) *Let σ be a substitution of Pisot type over a two-letter alphabet which satisfies the coincidence condition. Then the substitution dynamical system associated with σ has a purely discrete spectrum.*

Theorem 2.6.7 ([18]) *Any substitution of Pisot type over a two-letter alphabet satisfies the coincidence condition.*

The question of whether a substitution sequence has purely discrete dynamical system if and only if its expansion factor is Pisot remains open for alphabets of size 3 or higher and has proved remarkably difficult to resolve. The methods used in the two-letter case don't apply in these cases. A survey of the state of results as of 2015 (not including [17]) appears in [2].

2.6.3.2 Purely Discrete Spectrum for Supertile Methods in $\mathbb{R}^d$

There is a sufficient condition given in [100] that works in all dimensions, and then a specialized version for $\mathbb{R}^2$ we will talk about.

Theorem 2.6.8 ([100, Theorem 6.1]) *Let $\mathcal{T}$ be a self-affine tiling of $\mathbb{R}^d$ with expansion map ϕ. If there exists a basis B for $\mathbb{R}^d$ such that for all $x \in B$,*

$$\sum (1 - dens(D_{\phi^n(x)})) < \infty, \qquad (2.8)$$

then the tiling dynamical system $(\Omega_{\mathcal{T}}, \mathbb{R}^d, \mu)$ has pure discrete spectrum.

The strongest result is obtained for $\mathbb{R}$ and $\mathbb{R}^2$.

Theorem 2.6.9 ([100, Theorem 6.2]) *Let $\mathcal{T}$ be a self-similar tiling of $\mathbb{R}^d$, $d \leq 2$, with expansion constant λ. The tiling dynamical system $(\Omega_{\mathcal{T}}, \mathbb{R}^d, \mu)$ has pure discrete spectrum if and only if λ is Pisot (real or non-real) and*

$$\lim_{n \to \infty} dens(D_{\lambda^n x}) = 1, \qquad x \in \Xi[\mathcal{T}]. \qquad (2.9)$$

An *overlap algorithm* is defined in [100, p. 721-724] that determines whether a self-similar tiling has a pure discrete spectrum. We give only the general idea of the algorithm here. One makes a graph $G_O(\mathcal{T}, x)$ out of all overlaps one can see by comparing the tile(s) in $\mathcal{T}$ at some $z \in \mathbb{R}^2$, with the tile(s) in $\mathcal{T} - x$ at z. Most of those overlaps will be two tiles of different types, but sometimes they are of the same type, and on occasion x may be a return vector for the tiles at this particular z. When that happens we are said to have a *coincidence*, properly defined as $\mathcal{T} \cap \{z\} = (\mathcal{T} - x) \cap \{z\}$. An overlap is a vertex in the overlap graph, and edges connect overlaps that are related via substitution.

Theorem 2.6.10 ([100, Proposition 6.7]) *Suppose that $\mathcal{T}$ is a self-similar tiling of the plane with expansion constant λ a non-real Pisot number, and $x \in \Xi[\mathcal{T}]$. The following are equivalent:*

(i) *the tiling dynamical system $(\Omega_{\mathcal{T}}, \mathbb{R}^2, \mu)$ has pure discrete spectrum,*

(ii) *from any vertex of $G_0(\mathcal{T}, x)$ there is a path leading to a coincidence, and*

(iii) *$dens(D_{\lambda^n x}) \to 1$ as $n \to \infty$.*

2.6.4 The Continuous Part of the Spectrum

The results given above mean that we have a pretty good understanding of the discrete part of the spectrum for supertile systems. In particular we understand when there is point spectrum, and where the point masses are. We also understand when there must be a continuous component to the spectrum. What remains is to understand the nature of the continuous part. For instance, is it singular or absolutely continuous with respect to Lebesgue measure?

There are two classic examples of one-dimensional constant-length substitution sequences[16] with mixed spectrum: the Thue-Morse sequence and the Rudin-Shapiro sequence. Thue-Morse is given by the bijective substitution $0 \to 01, 1 \to 10$ which, lacking coincidences, was long known to have some continuous spectrum. A variety of results over the years proved that the continuous part is singular with respect to Lebesgue measure. It has been thought that the continuous part of the spectrum of a bijective substitution is always singular, but that question remains open. However, if the alphabet only has two symbols then singularity has been proved in [12], in one and several dimensions.

The original Rudin-Shapiro sequence is not bijective but it lacks coincidences and therefore has a continuous spectral component that has been known to be absolutely continuous for some time (see [42, 89]). Generalizations to higher dimensions were developed in [44], where the continuous part of the dynamical spectrum was shown to be absolutely so. Recent work [30] gives a different generalization and shows the continuous part of the diffraction spectrum to be absolutely continuous. The "twisted silver mean" substitution, introduced in [10], uses a 'bar swap' method seen in some of the Rudin-Shapiro constructions, but on a non-constant length substitution. The result is a mixed spectrum, which is analysed fully in [10] and found to have a mixed but singular spectrum.

For substitutions of length q, a necessary condition for the presence of absolutely continuous spectrum had been conjectured: The transition matrix should have an eigenvalue of modulus $\sqrt{q}$. This conjecture was verified in [23] with the result that if the transition matrix has no eigenvalue of modulus $\sqrt{q}$ then the dynamical spectrum is singular.

---

[16]These sequences were actually defined using number-theoretic constructions, but have simple substitution rules also.

The results and techniques of [89] are given an important generalization to multidimensional constant-length substitutions in [19]. Without requirements on primitivity or height, Bartlett is able to show that "abelian" bijective substitutions have only singular continuous spectrum, settling the longstanding conjecture in these cases. A general algorithm for computing the spectrum for constant-length multidimensional substitutions in $\mathbb{Z}^d$ is also given.

Example 15 (Explicit Computations for the Thue-Morse Substitution) Let Ω_{TM} be the subshift for the Thue-Morse symbolic substitution $0 \to 01$, $1 \to 10$. We will show how to make all of the eigenfunctions for the dynamical system, and also how to make a function whose measure is (singular) continuous with respect to Lebesgue measure. Together these functions generate all of L^2.

To make an example of a nontrivial eigenfunction we consider the 2-supertiles in $\mathcal{T}$. Since 2-supertiles have length 4 in this example, there are four locations the origin can occupy in either type of 2-supertile. Let us call them 0, 1, 2, 3 as we go from left to right. Define $\mathcal{O}(\mathcal{T}) = i$ if the origin occupies the ith location of its 2-supertile. Now define $f(\mathcal{T}) = \exp(2\pi i \mathcal{O}(\mathcal{T})/4)$.

Notice that if $\mathcal{O}(\mathcal{T}) = i$ and $i < 3$ then $\mathcal{O}(\mathcal{T} - 1) = i + 1$. If $\mathcal{O}(\mathcal{T}) = 3$ then $\mathcal{O}(\mathcal{T} - 1) = 0$.[17] Thus if $\mathcal{O}(\mathcal{T}) < 3$ we get that $U(f)(\mathcal{T}) = f(\mathcal{T} - 1) = \exp(2\pi i (\mathcal{O}(\mathcal{T}) + 1)/4)$ and if $\mathcal{O}(\mathcal{T}) = 3$ then $U(f)(\mathcal{T}) = 1$. This means that $U(f) = \exp(2\pi i/4)f$, and so f is an eigenfunction with eigenvalue $\alpha = 1/4$.

We compute the Fourier coefficient $\hat{f}(n)$, as defined in Eq. (2.6). We know

$$U^n(f)(\mathcal{T}) = \exp(2\pi i (\mathcal{O}(\mathcal{T} - n))/4)$$

$$= \exp(2\pi i (\mathcal{O}(\mathcal{T}) + n)/4) = \exp(2\pi i n/4) f(\mathcal{T}),$$

and so

$$\hat{f}(n) = (U^n(f), f) = \int_{\Omega_{TM}} \exp(2\pi i n/4) f(\mathcal{T}) \overline{f(\mathcal{T})} d\mu(\mathcal{T})$$

$$= \int_{\Omega_{TM}} \exp(2\pi i n/4) d\mu(\mathcal{T}) = \exp(2\pi i n/4)$$

for all n. These are the Fourier coefficients of a Dirac δ-function with its peak at $\exp(2\pi i n/4)$. Since σ_f is unique this makes it equal to this Dirac delta function.

Variations of the function f are easy to construct by looking at different sizes of supertiles. If f is based on the location of the origin in an n-supertile, it will have an eigenvalue with a 2^n in the denominator. Since this substitution has height 1, this implies that the eigenvalues of the Koopman operator for the Thue-Morse substitution are $\mathbb{Z}[1/2]$.

---

[17]This is indicative of the 'odometer'-like structure of constant-length substitutions: the shift map augments until a fixed number and then resets to 0, augmenting elsewhere. In general the supertile structure of any constant-length substitution looks like an odometer, see for example [45].

None of the functions we just constructed depend on the actual letter at the origin. We can supplement these with a function that only knows what letter is at the origin in $\mathcal{T}$: Define $g(\mathcal{T}) = \begin{cases} 1 & \text{if } \mathcal{T}(0) = 1 \\ -1 & \text{if } \mathcal{T}(0) = 0 \end{cases}$. It is shown in [45] that together this and the eigenfunctions span all of $L^2(\Omega_{TM}, \mu)$.

Notice that g is orthogonal to the eigenfunction f since

$$< f, g > = \int_{\Omega_{TM}} f(\mathcal{T})\overline{g(\mathcal{T})}d\mu(\mathcal{T}) = \int_{\Omega_{TM}^+} f(\mathcal{T})d\mu(\mathcal{T}) - \int_{\Omega_{TM}^-} f(\mathcal{T})d\mu(\mathcal{T}),$$

where Ω_{TM}^+ (resp. Ω_{TM}^-) are the set of all tilings with a 1 (resp. 0) at the origin. Each of the two integrals on the right are equal because they depend only on the supertile structure of $\mathcal{T}$ and not the letter at the origin. Thus the inner product of g and f is 0. We have obtained:

$$L^2(\Omega_{TM}, \mu) = H_0 \bigoplus Z(g), \tag{2.10}$$

where H_0 denotes the span of the eigenfunctions.

The function space L^2 of general bijective substitutions in $\mathbb{Z}^d$ breaks into a direct sum of pieces that are discrete or continuous analogously to this example. If the bijections comprising the substitution commute with translation, it is possible to explicitly define the generators of the continuous spectral pieces [45, Theorem 4.2]. The nature of the continuous part of the spectrum continues to be investigated.

2.7   Spectral Analysis of Supertile Methods: Diffraction Spectrum

The diffraction spectrum of tilings is motivated by physics. In this viewpoint we consider the tiling as representing the atomic structure of a solid and we wish to mathematically simulate what happens in a diffraction experiment on the solid. That is, one passes x-rays or electrons through the solid, where they will bounce off atoms and interfere constructively and destructively, ultimately creating an image that represents something about the structure they passed through. Fourier analysis turns out to be the right mathematical analogue for this. We describe the situation for symbolic dynamics first, then generalize to $\mathbb{R}^d$.

2.7.1 Autocorrelation for Symbolic Sequences

Consider a sequence $\mathcal{T}_0 \in \Omega \subset \mathcal{A}^{\mathbb{Z}}$, where for convenience we assume that $\mathcal{A}$ is a finite subset of the complex numbers. We know that constructive and destructive interference depends on the repetition at various distances in $\mathcal{T}_0$. For instance, if $\mathcal{T}_0$ was periodic then there would be total agreement at distances that are multiples of the period, leading to strong constructive interference at those distances.

A reasonable way to measure the extent to which $\mathcal{T}_0$ agrees with itself at a distance $k \in \mathbb{Z}$ is to consider the global average of $\mathcal{T}_0(n-k)\overline{\mathcal{T}_0(n)}$ over all n. Thus we define a *correlation function* $C : \mathbb{Z} \to \mathbb{C}$ to be a cluster point of the sequences

$$\left\{ \frac{1}{N} \sum_{n=0}^{N-1} \mathcal{T}_0(n-k)\overline{\mathcal{T}_0(n)} \right\}.$$

A diagonalization argument shows that for a given C there will be some sequence $\{N_j\}$ such that

$$C(k) = \lim_{j\to\infty} \frac{1}{N_j} \sum_{n=0}^{N_j-1} \mathcal{T}_0(n-k)\overline{\mathcal{T}_0(n)}.$$

For most examples of interest from primitive supertile methods, notably the uniquely ergodic ones, it turns out that $C(k)$ is unique, and it is useful to assume that. (The general situation is described in detail starting on page 74 of [89]).

Suppose μ is an ergodic measure on Ω and consider the continuous function $\mathcal{O} : \Omega \to \mathbb{C}$ given by $\mathcal{O}(\mathcal{T}) = \mathcal{T}(0)$. Then for μ-almost every $\mathcal{T} \in \Omega$ we have that

$$C(k) = \lim_{j\to\infty} \frac{1}{N_j} \sum_{n=0}^{N_j-1} \mathcal{T}(n-k)\overline{\mathcal{T}(n)} = \int_{\Omega} \mathcal{O}(\mathcal{T}-k)\overline{\mathcal{O}(\mathcal{T})}d\mu(\mathcal{T}).$$

That means $C(k) = \hat{\mathcal{O}}(k)$ from a dynamical spectrum perspective.

On the other hand, $C(k)$ can be shown to be positive definite and so there is a positive measure on the torus that has $C(k)$ as its Fourier coefficients. This measure, which we denote $\hat{\gamma}$, is known in [89] as the correlation measure of the sequence $\mathcal{T}$; it is the analogue of the diffraction measure (In general the diffraction measure is the Fourier transform of the autocorrelation, which is defined similar to $C(k)$). One can see from this analysis that the diffraction spectrum should be subordinate to the dynamical spectrum.

2.7.2 Diffraction in $\mathbb{R}^d$

2.7.2.1 Overview

A tiling $\mathcal{T}$ is a model for the atomic structure of matter, where the atoms or molecules occupy locations given by the tiles. In our simulation of a diffraction experiment, we imagine that waves of some appropriate wavelength are sent through the tiling, where they interfere constructively and destructively as determined by relative distances between the tiles. The diffracted waves form an image where we see bright spots of intense constructive interference (our "Bragg peaks") and a greyscale spectrum were the interference ranges from constructive to destructive.

The mathematics of diffraction has a long development that is based on Fourier analysis. Because our tilings are infinite there are technicalities that have to be handled using tempered distributions and translation-bounded measures. It was Hof in [60] who first advocated using this overall method to approach the diffraction of aperiodic structures and Dworkin [37] who noticed the connection between diffraction and dynamical spectrum; [15] provides a recent and quite accessible survey. An early computation for self-similar tilings is [52]. A serious treatment of the details as well as the history behind mathematical diffraction appears in Chapters 8 and 9 of [11], along with numerous references. A more condensed and self-contained description of the diffraction spectrum appears in [69], and we loosely follow that development here.

Before we begin, consider this intuitive description of the mathematics of diffraction that appears in [72, Section 5]. It clearly shows why the Fourier transform is central to the theory.

> When modeling diffraction, the two basic principles are the following: Firstly, each point x in the solid gives rise to a wave $\xi \mapsto \exp(-ix\xi)$. The overall wave w is the sum of the single waves. Secondly, the quantity measured in an experiment is the intensity given as the square of the modulus of the wave function.
>
> We start by implementing this for a finite set $F \subset \mathbb{R}^d$. Each $x \in F$ gives rise to a wave $\xi \mapsto \exp(-ix\xi)$ and the overall wavefunction w_F induced by F is accordingly
>
> $$w_F(\xi) = \sum_{x \in F} \exp(-ix\xi).$$
>
> Thus, the intensity I_F is
>
> $$I_F(\xi) = \sum_{x,y \in F} \exp(-i(x-y)\xi) = \widehat{\sum_{x,y \in F} \delta_{x-y}}.\text{''}$$

2.7.2.2 Diffraction via Delone Sets

It is natural to consider diffraction theory on discrete sets in $\mathbb{R}^d$ called Delone sets, so we need to convert our tiling $\mathcal{T}$ into a point set Λ that represents the locations and types of atoms in the solid $\mathcal{T}$ represents.[18] We recall that a subset $\Lambda \subset \mathbb{R}^d$ is called a Delone set if there exist $0 < r \leq R$ such that every ball of radius r contains at most one point of Λ and every ball of radius R contains at least one point of Λ.

A *Delone multiset* is a set $\boldsymbol{\Lambda} = \Lambda_1 \times \Lambda_2 \times \cdots \times \Lambda_m$, where each Λ_i is a Delone set in $\mathbb{R}^d$ and the set $\bigcup_{i \leq m} \Lambda_i$, which by abuse of notation we also denote by $\boldsymbol{\Lambda}$, is Delone. An obvious way to turn $\mathcal{T}$ into a Delone multiset is to mark a special point in the prototile of type i for each $i = 1, .., m$, and let Λ_i be the Delone set of all copies of that point in $\mathcal{T}$.

So $\boldsymbol{\Lambda}$ represents our set of scatterers from $\mathcal{T}$ and we have kept track of the type of each scatterer. To account for different scattering strengths choose $a_i \in \mathbb{C}$ for $i \leq m$. Using the notation δ_x to represent the Dirac delta function at x thought of as a probability measure with support concentrated at x, we have the *weighted Dirac comb*

$$\omega = \sum_{i \leq m} a_i \delta_{\Lambda_i} = \sum_{i \leq m} a_i \sum_{x \in \Lambda_i} \delta_x.$$

This is a point measure on $\mathbb{R}^d$ that is not bounded, but is *translation bounded* in the sense that $\sup_{x \in \mathbb{R}^d} |\omega|(x + K) < \infty$ for all compact K.

The autocorrelation is defined to be the convolution of ω with the weighted Dirac comb $\tilde{\omega} = \sum_{i \leq m} \overline{a_i} \delta_{-\Lambda_i}$. Because convolutions are necessarily defined on measures with bounded support we end up with a limit that yields the *autocorrelation measure*[19]

$$\gamma_\omega = \lim_{R \to \infty} \frac{1}{Vol(B_R(0))} \left(\omega|_{B_R(0)} * \tilde{\omega}|_{B_R(0)} \right) = \sum_{i,j \leq m} a_i \overline{a_j} \sum_{z \in \Lambda_i - \Lambda_j} freq(z) \delta_z,$$

where the frequency is computed as the limit, if it exists, as the average number of times z is a return vector per unit area:

$$freq(z) = \lim_{R \to \infty} \frac{1}{Vol(B_R(0))} \#\{x \in \Lambda_i \cap B_R(0) \text{ and } x - z \in \Lambda_j\}.$$

Since $\mathcal{T}$ has finite local complexity, $\Lambda_i - \Lambda_j$ is a discrete set and that makes γ_ω a point measure also.

Definition 2.7.1 If the autocorrelation measure γ_ω exists, the *diffraction measure* of $\mathcal{T}$ is the Fourier transform $\widehat{\gamma_\omega}$.

---

[18]This volume contains an overview of the history and development of tilings and Delone sets in [101].

[19]This is also known as the "natural" autocorrelation measure because the averaging sets used are balls centered at the origin as opposed to an arbitrary van Hove sequence.

From a physical perspective where we are running a diffraction experiment on a solid modeled by $\mathcal{T}$, the measure $\widehat{\gamma_\omega}$ tells us how much intensity is scattered into a given volume. We decompose $\widehat{\gamma_\omega}$ into its pure point, singular continuous, and absolutely continuous parts with respect to Lebesgue measure on $\mathbb{R}^d$:

$$\widehat{\gamma_\omega} = (\widehat{\gamma_\omega})_{pp} + (\widehat{\gamma_\omega})_{sc} + (\widehat{\gamma_\omega})_{ac}.$$

The pure point part tells us the location of the Bragg peaks that are so characteristic of the diffraction images of crystals and quasicrystals. The degree of disorder in the solid is quantified by the continuous parts. The singular continuous part is rarely observed in physical experiments [11, Remark 9.3].

Example 16 The left column of Fig. 2.23 is a series of increasingly complex examples of self-similar tilings with two colors of square tiles. These tilings are examples of the sort analyzed in [44, 45]. Simulations of the corresponding diffraction images are also shown. Each tiling is a substitution of constant length 2×2 or 4×4. The tiling in the last row is a two-letter factor of a substitution on 8 letters; the other three are simple two-letter substitutions. In all cases there are point measures concentrated on $\mathbb{Z}[1/2] \times \mathbb{Z}[1/2]$.

The top tiling has a purely discrete spectrum because its substitution satisfies the strong coincidence condition. The tiling in the middle of the figure is made from a bijective substitution and thus has a continuous component to its spectral measure. Because it is a constant-length symbolic substitution on two letters the continuous portion of the measure is singularly continuous with respect to Lebesgue measure [19].

The tiling on the right is a generalized Rudin-Shapiro tiling [44]. The original substitution is on eight letters and although it has no coincidence, it is not bijective. The tiling shown in the figure is a two-tile factor that is locally derived from the 8-letter substitution (and in fact the local derivability is mutual, so the factor makes no difference dynamically). The continuous portion of the spectral measure for this tiling is absolutely continuous.

It is interesting to simulate the diffraction images of these tilings in light of these theoretical results. Anyone who produces sample images of any sort probably knows that there are usually parameters that can be altered to enhance the images. In our case such parameters include the weights on the dirac comb, the maximum intensity, and scaling functions. Tinkering with the parameters on a local scale does not change the overall qualitative appearance too much, and the apparent difference between the absolutely continuous diffraction (on the right) and the other two is persistent. The diffraction images for tilings with pure discrete spectrum and those with a singular component consistently appear similar throughout a wide range of parameters, with areas of extreme brightness and darkness. The absolutely continuous spectral images are notable for their lack of these extremes.

Fig. 2.23 The top tiling is substitutive with coincidence and it has a purely discrete spectrum. The substitution for the middle tiling is bijective and the spectrum is mixed, with a singular continuous part. The bottom tiling is not bijective but has a mixed spectrum with absolutely continuous part

2.7.3 Intensities

How bright are the Bragg peaks? A preliminary formula was asserted in [28] and became known after a while as the "Bombieri/Taylor conjecture" (see also [60]). The formula, given below, is in terms of a limit. The convergence of this limit has been studied in many different situations, surveyed in [72]. In that paper Lenz shows that the formula is correct for a wide swath of aperiodic structures, including tilings generated through substitution and through projection, as well as those that are linearly repetitive. The setting in [72] is as follows.

Given a Delone set Λ, an element $\xi \in \mathbb{R}^n$, and a subset $B \subset \mathbb{R}^n$

$$c_B^\xi(\Lambda) = \frac{1}{Vol(B)} \sum_{x \in \Lambda \cap B} \exp(-2\pi i \xi \cdot x).$$

For the cases under consideration the intensity at $\xi \in \mathbb{R}^d$ is shown to exist and is given by

$$\widehat{\gamma(\xi)} = \lim_{n \to \infty} |c_{C_n}^\xi(\Lambda)|^2,$$

where C_n is the cube of side length $2n$ centered at the origin and $\xi \in \mathbb{R}^d$. In many of the situations discussed in [72] it is also proved that the eigenfunctions for the Koopman operator are continuous.

2.8 Connection Between Diffraction and Dynamical Spectrum

A recent survey of this topic is [15], which unifies the various notions of diffraction and dynamical spectrum, explains what was known up until 2016, and provides numerous references. Done in the context of Delone sets with finite local complexity, it applies to tilings of $\mathbb{R}^d$ and their dynamical systems. In particular it explains the notions of diffraction for individual sets Λ as well as their hulls, and explicitly shows how to map from the Schwarz space of test functions under the diffraction measure to the Koopman representation of the dynamical system. Through this mapping they note that "the diffraction measure completely controls a subrepresentation of T", thus making explicit the connection between dynamical and diffraction spectrum.

The original paper connecting diffraction to dynamical spectrum is [37]. In it, Dworkin makes an argument showing how to deduce pure point diffraction spectrum if pure point dynamical spectrum has been established.

For a general system it may be that the diffraction spectrum does not contain as much information as the dynamical spectrum, but in the case of pure point spectrum it is known that the two classes are identical as long as there is unique ergodicity. The result, proved in [69], is in the context of Delone multisets.

Theorem 2.8.1 ([69]) *Suppose that a Delone multiset Λ has finite local complexity and uniform cluster frequencies. Then the following are equivalent:*

(i) *Λ has pure point dynamical spectrum;*
(ii) *The measure $\nu = \sum_{i \leq m} a_i \delta_{\Lambda_i}$ has pure point diffraction spectrum, for any choice of complex numbers $(a_i)_{i \leq m}$;*
(iii) *The measures δ_{Λ_i} have pure point spectrum, for $i \leq m$.*

The condition of Λ having "uniform cluster frequencies" is equivalent to the fact that its hull is uniquely ergodic, which we know is the case for many tilings constructed using supertile methods. It would be remiss not to mention [70], the companion work to [69]. It includes the result that for lattice substitution multiset systems,[20] being a regular model set is equivalent to having pure point spectrum.

2.8.1 When the Diffraction is Not Pure Point

Recent work in [16] attempts to understand the dynamical spectrum when it is larger than the diffraction spectrum. An idea has been around for a while that factors of a system can give nuance to the diffraction spectrum. That is, "the missing parts of the dynamical spectrum could be reconstructed from the diffraction measures of suitable factors of the original system". In the uniquely ergodic case, the authors of [16] are able to show (see Corollary 9 for technical details) that (1) the diffraction measure of a factor is a spectral measure for the Koopman operator, and (2) the set of diffraction measures of factors of a system are dense in the set of all spectral measures for the system.

In [59] it is shown that there exist substitutions which require infinitely many factors to reconstruct the pure point dynamical spectrum from the respective diffraction. There it is noted that it is not true that the maximal spectral measure of a subshift can be realized as the fundamental diffraction of a subshift factor.

As is true for the dynamical spectrum, one of the major areas of study is to determine the nature of the continuous part of the diffraction spectrum. In [9] it is shown that the continuous part of the spectrum of $a \to abbb, b \to a$ is singularly continuous with respect to Lebesgue measure. The general case $a \to ab^k, b \to a$ is considered in [14]. The analysis is based on a 'renormalization' process wherein the substitution structure of the self-similar tiling is used to find recursion relations for the autocorrelation measure. This method was also applied to the twisted silver mean in [10].

---

[20]Not a particularly restrictive subclass according to Section 5.1 of [70].

2.9 For Further Reading

A good primary source for fundamental results on tiling dynamical systems is B. Solomyak's "The dynamics of self-similar tilings" [100]. This paper lays out the basic definitions and takes an ergodic theoretic approach to the systems. A fundamental resource in elementary tiling theory is B. Grunbaum and G. C. Shephard's *Tilings and Patterns* [58], which catalogs nearly everything that is known about periodic tilings and more. It contains an enormous number of examples, and does include a few nonperiodic tilings such as the Penrose, Robinson, and Ammann tilings. Good general ergodic theory references for $\mathbb{Z}$-actions are K. Petersen's *Ergodic Theory* and P. Walters' *An Introduction to Ergodic Theory* [84, 104]. Fundamental symbolic dynamics references are D. Lind and B. Marcus' *An Introduction to Symbolic Dynamics and Coding* and Bruce Kitchens' *Symbolic Dynamcs* [67, 74]. Symbolic substitutions are surveyed up to 2002 in the collectively written *Substitutions in Dynamics, Arithmetics, and Combinatorics* [42]. A recent survey of S-adic expansions appears in V. Berthé and V. Delecroix's "Beyond substitutive dynamical systems: S-adic expansions" [24]. The definitive volume for the study of aperiodic order is M. Baake and U. Grimm's *Aperiodic Order* [11]. It takes a physical perspective and is full of examples of every sort, many analyzed fully.

There are a few other expositions of tilings and tiling spaces that are worth mentioning here. For a rigorous dynamical introduction to the theory, with multidimensional actions surveyed up to 2004 see E. A. Robinson, Jr.'s "Symbolic dynamics and tilings of $\mathbb{R}^d$" [93]. Radin's AMS Student Mathematical Library notes *Miles of Tiles* [90] introduces readers to the dynamics and ergodic theory with a strong physical motivation. At a university student level, it carries the additional interest of treating tilings with infinitely many tile rotations such as the pinwheel tiling. Substitutions on the graphs of tilings are considered in the author's "A primer on substitution tilings of Euclidean space" [46], which includes several examples of such combinatorial substitutions and their associated self-similar tilings. The topology of tiling spaces is the subject of L. Sadun's *Topology of Tiling Spaces* [95], which takes the reader through self-similar tiling constructions with and without rotations, shows tiling spaces are inverse limits, and does cohomology in the tiling context. There are many more topics we have not even mentioned, so the reader is encouraged to find a compelling topic to pursue.

Acknowledgments The author would like to thank Michael Baake, Franz Gähler, E. A. Robinson, Jr., Dan Rust, Lorenzo Sadun, and Boris Solomyak for their comments on drafts of this work.

References

1. Tilings encyclopedia (2018). http://tilings.math.uni-bielefeld.de/. Accessed Dec 2018
2. S. Akiyama, M. Barge, V. Berthé, J.-Y. Lee, A. Siegel, *On the Pisot Substitution Conjecture*. Mathematics of Aperiodic Order, Progress in Mathematics, vol. 309 (Birkhäuser/Springer, Basel, 2015), pp. 33–72, MR 3381478

3. S. Akiyama, Strong coincidence and overlap coincidence. Discrete Contin. Dyn. Syst. **36**(10), 5223–5230 (2016). MR 3543543
4. S. Akiyama, J.-Y. Lee, Algorithm for determining pure pointedness of self-affine tilings. Adv. Math. **226**(4), 2855–2883 (2011). MR 2764877
5. S. Akiyama, J.-Y. Lee, Overlap coincidence to strong coincidence in substitution tiling dynamics. Eur. J. Combin. **39**, 233–243 (2014). MR 3168528
6. J.-P. Allouche, J. Shallit, *Automatic Sequences*. Theory, Applications, Generalizations (Cambridge University Press, Cambridge, 2003), MR 1997038
7. P. Arnoux, M. Furukado, E. Harriss, S. Ito, Algebraic numbers, free group automorphisms and substitutions on the plane. Trans. Amer. Math. Soc. **363**(9), 4651–4699 (2011). MR 2806687
8. P. Arnoux, S. Ito, Pisot substitutions and Rauzy fractals. Bull. Belg. Math. Soc. Simon Stevin **8**(2), 181–207 (2001). Journées Montoises d'Informatique Théorique (Marne-la-Vallée, 2000). MR 1838930
9. M. Baake, N.P. Frank, U. Grimm, E.A. Robinson Jr., Geometric properties of a binary non-pisot inflation and absence of absolutely continuous diffraction. Studia Math. **247**, 109–154 (2019)
10. M. Baake, F. Gähler, Pair correlations of aperiodic inflation rules via renormalisation: some interesting examples. Topology Appl. **205**, 4–27 (2016). MR 3493304
11. M. Baake, U. Grimm, *Aperiodic Order. Vol. 1*. Encyclopedia of Mathematics and its Applications, vol. 149 (Cambridge University Press, Cambridge, 2013), A Mathematical Invitation, With a foreword by Roger Penrose. MR 3136260
12. M. Baake, U. Grimm, Squirals and beyond: substitution tilings with singular continuous spectrum. Ergodic Theory Dynam. Syst. **34**(4), 1077–1102 (2014). MR 3227148
13. M. Baake, U. Grimm, Diffraction of a binary non-pisot inflation tiling. J. Phys. Conf. Ser. **809**(1), 012026 (2017)
14. M. Baake, U. Grimm, N. Mañibo, Spectral analysis of a family of binary inflation rules. Lett. Math. Phys. **108**(8), 1783–1805 (2018). MR 3814725
15. M. Baake, D. Lenz, Spectral notions of aperiodic order. Discrete Contin. Dyn. Syst. Ser. S **10**(2), 161–190 (2017). MR 3600642
16. M. Baake, D. Lenz, A. van Enter, Dynamical versus diffraction spectrum for structures with finite local complexity. Ergodic Theory Dyn. Syst. **35**(7), 2017–2043 (2015). MR 3394105
17. M. Barge, Pure discrete spectrum for a class of one-dimensional substitution tiling systems. Discrete Contin. Dyn. Syst. **36**(3), 1159–1173 (2016). MR 3431249
18. M. Barge, B. Diamond, Coincidence for substitutions of Pisot type. Bull. Soc. Math. France **130**(4), 619–626 (2002). MR 1947456
19. A. Bartlett, Spectral theory of $\mathbb{Z}^d$ substitutions. Ergodic Theory Dyn. Syst. **38**(4), 1289–1341 (2018). MR 3789166
20. N. Bédaride, A. Hilion, Geometric realizations of two-dimensional substitutive tilings. Q. J. Math. **64**(4), 955–979 (2013). MR 3151599
21. J. Bellissard, A. Julien, J. Savinien, Tiling groupoids and Bratteli diagrams. Ann. Henri Poincaré **11**(1–2), 69–99 (2010). MR 2658985
22. R. Berger, The undecidability of the domino problem. Mem. Amer. Math. Soc. No. **66**, 72 (1966). MR 0216954
23. A. Berlinkov, B. Solomyak, Singular substitutions of constant length (2017). ArXiv:1705.00899v2
24. V. Berthé, V. Delecroix, *Beyond Substitutive Dynamical Systems: S-adic Expansions (Numeration and Substitution 2012)*. RIMS Kôkyûroku Bessatsu, vol. B46, (Research Institute for Mathematical Sciences (RIMS), Kyoto, 2014), pp. 81–123. MR 3330561
25. S. Bezuglyi, J. Kwiatkowski, K. Medynets, Aperiodic substitution systems and their Bratteli diagrams. Ergod. Theory Dyn. Syst. **29**(1), 37–72 (2009). MR 2470626
26. S. Bezuglyi, J. Kwiatkowski, K. Medynets, B. Solomyak, Invariant measures on stationary Bratteli diagrams. Ergod. Theory Dyn. Syst. **30**(4), 973–1007 (2010). MR 2669408

27. S. Bezuglyi, J. Kwiatkowski, K. Medynets, B. Solomyak, Finite rank Bratteli diagrams: structure of invariant measures. Trans. Amer. Math. Soc. **365**(5), 2637–2679 (2013). MR 3020111
28. E. Bombieri, J.E. Taylor, Which distributions of matter diffract? An initial investigation. J. Phys. **47**(7) , Suppl. Colloq. C3, C3–19–C3–28 (1986). International workshop on aperiodic crystals (Les Houches, 1986). MR 866320
29. R.V. Chacon, A geometric construction of measure preserving transformations, in *Proceedings of the Fifth Berkeley Symposium Mathematical. Statistics and Probability (Berkeley, California, 1965/66), Vol. II: Contributions to Probability Theory, Part 2* (University of California Press, Berkeley, 1967), pp. 335–360. MR 0212158
30. L. Chan, U. Grimm, Spectrum of a Rudin-Shapiro-like sequence. Adv. Appl. Math. **87**, 16–23 (2017). MR 3629260
31. A. Clark, L. Sadun, When size matters: subshifts and their related tiling spaces. Ergod. Theory Dyn. Syst. **23**(4), 1043–1057 (2003). MR 1997967
32. E.M. Coven, M.S. Keane, The structure of substitution minimal sets. Trans. Amer. Math. Soc. **162**, 89–102 (1971). MR 0284995
33. F.M. Dekking, The spectrum of dynamical systems arising from substitutions of constant length. Z. Wahrscheinlichkeitstheorie und Verw. Gebiete **41**(3), 221–239 (1977/1978). MR 0461470
34. F.M. Dekking, M. Keane, Mixing properties of substitutions. Z. Wahrscheinlichkeitstheorie und Verw. Gebiete **42**(1), 23–33 (1978). MR 0466485
35. F. Durand, Linearly recurrent subshifts have a finite number of non-periodic subshift factors. Ergod. Theory Dyn. Syst. **20**(4), 1061–1078 (2000). MR 1779393
36. F. Durand, Corrigendum and addendum to: "Linearly recurrent subshifts have a finite number of non-periodic subshift factors" [Ergodic Theory Dynam. Systems **20**(4), 1061–1078 (2000); MR1779393 (2001m:37022)]. Ergod. Theory Dyn. Syst. **23**(2), 663–669 (2003). MR 1972245
37. S. Dworkin, Spectral theory and x-ray diffraction. J. Math. Phys. **34**(7), 2965–2967 (1993). MR 1224190
38. S. Ferenczi, Rank and symbolic complexity. Ergod. Theory Dyn. Syst. **16**(4), 663–682 (1996). MR 1406427
39. T. Fernique, Local rule substitutions and stepped surfaces. Theoret. Comput. Sci. **380**(3), 317–329 (2007). MR 2331001
40. T. Fernique, N. Ollinger, Combinatorial substitutions and sofic tilings. Journées Automates Cellulaires (2010) (Turku). arXiv:1009.5167
41. A.M. Fisher, Nonstationary mixing and the unique ergodicity of adic transformations. Stoch. Dyn. **9**(3), 335–391 (2009). MR 2566907
42. N.P. Fogg, *Substitutions in Dynamics, Arithmetics and Combinatorics*, ed. by V. Berthé, S. Ferenczi, C. Mauduit, A. Siegel. Lecture Notes in Mathematics, vol. 1794 (Springer, Berlin, 2002). MR 1970385
43. N.P. Frank, Detecting combinatorial hierarchy in tilings using derived Voronoï tesselations. Discrete Comput. Geom. **29**(3), 459–476 (2003). MR 1961011
44. N.P. Frank, Substitution sequences in $\mathbb{Z}^d$ with a non-simple Lebesgue component in the spectrum. Ergod. Theory Dyn. Syst. **23**(2), 519–532 (2003). MR 1972236
45. N.P. Frank, Multidimensional constant-length substitution sequences. Topol. Appl. **152**(1–2), 44–69 (2005). MR 2160805
46. N.P. Frank, A primer of substitution tilings of the Euclidean plane. Expo. Math. **26**(4), 295–326 (2008). MR 2462439
47. N.P. Frank, *Tilings with Infinite Local Complexity*. Mathematics of Aperiodic Order, Progress in Mathematics, vol. 309 (Birkhäuser/Springer, Basel, 2015), pp. 223–257. MR 3381483
48. N.P. Frank, E.A. Robinson, Jr., Generalized β-expansions, substitution tilings, and local finiteness. Trans. Amer. Math. Soc. **360**(3), 1163–1177 (2008). MR 2357692
49. N.P. Frank, L. Sadun, Topology of some tiling spaces without finite local complexity. Discrete Contin. Dyn. Syst. **23**(3), 847–865 (2009). MR 2461829

50. N.P. Frank, L. Sadun, Fusion: a general framework for hierarchical tilings of $\mathbb{R}^d$. Geom. Dedicata **171**, 149–186 (2014). MR 3226791

51. D. Frettlöh, More Inflation Tilings, in *Aperiodic Order, Vol. 2*. Encyclopedia of Mathematics and its Applications, vol. 166 (Cambridge University Press, Cambridge, 2017), pp. 1–37. MR 3791847

52. F. Gähler, R. Klitzing, The diffraction pattern of self-similar tilings, in *The Mathematics of Long-Range Aperiodic Order (Waterloo, ON, 1995)*. NATO Advanced Science Institutes Series C: Mathematical and Physical Sciences, vol. 489 (Kluwer Academic Publishers, Dordrecht, 1997), pp. 141–174. MR 1460023

53. F. Gähler, E.E. Kwan, G.R. Maloney, A computer search for planar substitution tilings with n-fold rotational symmetry. Discrete Comput. Geom. **53**(2), 445–465 (2015). MR 3316232

54. M. Gardner, Mathematical games. Sci. Amer. **236**(1), 110–121 (1977)

55. C. Godrèche, F. Lançon, A simple example of a non-Pisot tiling with five-fold symmetry. J. Physique I **2**(2), 207–220 (1992). MR 1185612

56. G.R. Goodson, A survey of recent results in the spectral theory of ergodic dynamical systems. J. Dyn. Control Syst. **5**(2), 173–226 (1999). MR 1693318

57. W.H. Gottschalk, Substitution minimal sets. Trans. Amer. Math. Soc. **109**, 467–491 (1963). MR 0190915

58. B. Grünbaum, G.C. Shephard, *Tilings and Patterns* (W. H. Freeman and Company, New York, 1987). MR 857454

59. J.L. Herning, Spectrum and factors of substitution dynamical systems, ProQuest LLC, Ann Arbor, MI. Thesis (Ph.D.)–The George Washington University, 2013. MR 3167382

60. A. Hof, On diffraction by aperiodic structures. Comm. Math. Phys. **169**(1), 25–43 (1995). MR 1328260

61. M. Hollander, B. Solomyak, Two-symbol Pisot substitutions have pure discrete spectrum. Ergod. Theory Dyn. Syst. **23**(2), 533–540 (2003). MR 1972237

62. B. Host, Valeurs propres des systèmes dynamiques définis par des substitutions de longueur variable. Ergod. Theory Dyn. Syst. **6**(4), 529–540 (1986). MR 873430

63. E. Jeandel, P. Vanier, *The Undecidability of the Domino Problem*. This volume, Lecture Notes in Mathematics (Springer, Berlin, 2019)

64. T. Kamae, Spectrum of a substitution minimal set. J. Math. Soc. Japan **22**, 567–578 (1970). MR 0286092

65. J. Kellendonk, L. Sadun, Meyer sets, topological eigenvalues, and Cantor fiber bundles. J. Lond. Math. Soc. (2) **89**(1), 114–130 (2014). MR 3174736

66. R. Kenyon, B. Solomyak, On the characterization of expansion maps for self-affine tilings. Discrete Comput. Geom. **43**(3), 577–593 (2010). MR 2587839

67. B.P. Kitchens, *Symbolic Dynamics: One-sided, Two-sided and Countable State Markov Shifts*. Universitext (Springer, Berlin, 1998). MR 1484730

68. J. Kwapisz, Inflations of self-affine tilings are integral algebraic Perron. Invent. Math. **205**(1), 173–220 (2016). MR 3514961

69. J.-Y. Lee, R.V. Moody, B. Solomyak, Pure point dynamical and diffraction spectra. Ann. Henri Poincaré **3**(5), 1003–1018 (2002). MR 1937612

70. J.-Y. Lee, R.V. Moody, B. Solomyak, Consequences of pure point diffraction spectra for multiset substitution systems. Discrete Comput. Geom. **29**(4), 525–560 (2003). MR 1976605

71. M. Lemańczyk, *Spectral Theory of Dynamical Systems*. Mathematics of Complexity and Dynamical Systems, Vol. 1–3 (Springer, New York, 2012), pp. 1618–1638. MR 3220776

72. D. Lenz, Continuity of eigenfunctions of uniquely ergodic dynamical systems and intensity of Bragg peaks. Comm. Math. Phys. **287**(1), 225–258 (2009). MR 2480747

73. S. Lidin, *The Discovery of Quasicrystals: Scientific Background on the Nobel Prize in Chemistry 2011* (The Royal Swedish Academy of Sciences, Stockholm, 2011)

74. D. Lind, B. Marcus, *An Introduction to Symbolic Dynamics and Coding* (Cambridge University Press, Cambridge, 1995). MR 1369092

75. A.N. Livshits, On the spectra of adic transformations of Markov compact sets. Uspekhi Mat. Nauk **42**(3(255)), 189–190 (1987). MR 896889

76. M. Lothaire, *Algebraic Combinatorics on Words*. Encyclopedia of Mathematics and its Applications, vol. 90 (Cambridge University Press, Cambridge, 2002), A collective work by J. Berstel, D. Perrin, P. Seebold, J. Cassaigne, A. De Luca, S. Varricchio, A. Lascoux, B. Leclerc, J.-Y. Thibon, V. Bruyere, C. Frougny, F. Mignosi, A. Restivo, C. Reutenauer, D. Foata, G.-N. Han, J. Desarmenien, V. Diekert, T. Harju, J. Karhumaki, W. Plandowski, With a preface by Berstel and Perrin. MR 1905123

77. A.L. Mackay, Crystallography and the Penrose pattern. Phys. A **114**(1–3), 609–613 (1982). MR 678468

78. G.R. Maloney, D. Rust, Beyond primitivity for one-dimensional substitution subshifts and tiling spaces. Ergod. Theory Dyn. Syst. **38**(3), 1086–1117 (2018). MR 3784255

79. B.F. Martensen, Generalized balanced pair algorithm. Topology Proc. **28**(1), 163–178 (2004). Spring Topology and Dynamical Systems Conference. MR 2105455

80. J.C. Martin, Substitution minimal flows. Amer. J. Math. **93**, 503–526 (1971). MR 0300261

81. B. Mossé, Puissances de mots et reconnaissabilité des points fixes d'une substitution. Theoret. Comput. Sci. **99**(2), 327–334 (1992). MR 1168468

82. R. Penrose, The rôle of aesthetics in pure and applied mathematical research. Bull. Inst. Math. Appl. **10**, 266–271 (1974)

83. R. Penrose, Pentaplexity: a class of nonperiodic tilings of the plane. Math. Intell. **2**(1), 32–37 (1979/1980). MR 558670

84. K. Petersen, *Ergodic Theory*. Cambridge Studies in Advanced Mathematics, vol. 2 (Cambridge University Press, Cambridge, 1989, Corrected Reprint of the 1983 Original). MR 1073173

85. K. Petersen, Factor maps between tiling dynamical systems. Forum Math. **11**(4), 503–512 (1999). MR 1699171

86. J. Peyrière, Frequency of patterns in certain graphs and in Penrose tilings. J. Physique **47**(7), Suppl. Colloq. C3, C3–41–C3–62 (1986). International workshop on aperiodic crystals (Les Houches, 1986). MR 866322

87. B. Praggastis, Numeration systems and Markov partitions from self-similar tilings. Trans. Amer. Math. Soc. **351**(8), 3315–3349 (1999). MR 1615950

88. N. Priebe, B. Solomyak, Characterization of planar pseudo-self-similar tilings. Discrete Comput. Geom. **26**(3), 289–306 (2001). MR 1854103

89. M. Queffélec, *Substitution Dynamical Systems—Spectral Analysis*. Lecture Notes in Mathematics, vol. 1294 (Springer, Berlin, 1987). MR 924156

90. C. Radin, *Miles of Tiles*. Student Mathematical Library, vol. 1 (American Mathematical Society, Providence, 1999). MR 1707270

91. C. Radin, L. Sadun, Isomorphism of hierarchical structures. Ergod. Theory Dyn. Syst. **21**(4), 1239–1248 (2001). MR 1849608

92. C. Radin, M. Wolff, Space tilings and local isomorphism. Geom. Dedicata **42**(3), 355–360 (1992). MR 1164542

93. E.A. Robinson, Jr., *Symbolic Dynamics and Tilings of $\mathbb{R}^d$s*. Symbolic Dynamics and Its Applications, Proceedings of Symposia in Applied Mathematics, vol. 60 (American Mathematical Society, Providence, 2004), pp. 81–119. MR 2078847

94. R.M. Robinson, Undecidability and nonperiodicity for tilings of the plane. Invent. Math. **12**, 177–209 (1971). MR 0297572

95. L. Sadun, *Topology of Tiling Spaces*. University Lecture Series, vol. 46 (American Mathematical Society, Providence, 2008). MR 2446623

96. M. Senechal, *Quasicrystals and Geometry* (Cambridge University Press, Cambridge, 1995). MR 1340198

97. D. Shechtman, I. Blech, D. Gratias, J.W. Cahn, Metallic phase with long-range orientational order and no translational symmetry. Phys. Rev. Lett. **53**, 1951–1953 (1984)

98. B. Solomyak, Nonperiodicity implies unique composition for self-similar translationally finite tilings. Discrete Comput. Geom. **20**(2), 265–279 (1998). MR 1637896

99. B. Solomyak, Pseudo-self-affine tilings in $\mathbb{R}^d$. Zap. Nauchn. Sem. S.-Peterburg. Otdel. Mat. Inst. Steklov. **326** (2005), Teor. Predst. Din. Sist. Komb. i Algoritm. Metody. **13**, 198–213, 282–283 (2005); Translation in J. Math. Sci. **140**(3), 452–460 (2007). MR 2183221
100. B. Solomyak, Dynamics of self-similar tilings. Ergodic Theory Dynam. Syst. **17**(3), 695–738 (1997). MR 1452190
101. B. Solomyak, *Lecture on Delone Sets and Tilings*. This Volume, Lecture Notes in Mathematics (Springer, Berlin, 2019)
102. W. Thurston, *Groups, Tilings, and Finite State Automata*. AMS Colloquium Lecture Notes (American Mathematical Society, Providence, 1989)
103. J. Thuswaldner, *S-adic Sequences. A Bridge Between Dynamics, Arithmetic, and Geometry*. This Volume, Lecture Notes in Mathematics (Springer, Berlin, 2019)
104. P. Walters, *An Introduction to Ergodic Theory*. Graduate Texts in Mathematics, vol. 79 (Springer, New York, 1982). MR 648108
105. H. Wang, Proving theorems by pattern recognition, II. Bell Syst. Tech. J. **40**, 1–41 (1961).

Chapter 3
S-adic Sequences: A Bridge Between Dynamics, Arithmetic, and Geometry

Jörg M. Thuswaldner

Abstract A Sturmian sequence is an infinite nonperiodic string over two letters with minimal subword complexity. In two papers, the first written by Morse and Hedlund in 1940 and the second by Coven and Hedlund in 1973, a surprising correspondence was established between Sturmian sequences on one side and rotations by an irrational number on the unit circle on the other. In 1991 Arnoux and Rauzy observed that an induction process (invented by Rauzy in the late 1970s), related with the classical continued fraction algorithm, can be used to give a very elegant proof of this correspondence. This process, known as the Rauzy induction, extends naturally to interval exchange transformations (this is the setting in which it was first formalized). It has been conjectured since the early 1990s that these correspondences carry over to rotations on higher dimensional tori, generalized continued fraction algorithms, and so-called *S*-adic sequences generated by substitutions. The idea of working towards such a generalization is known as Rauzy's program. Recently Berthé, Steiner, and Thuswaldner made some progress on Rauzy's program and were indeed able to set up the conjectured generalization of the above correspondences. Using a generalization of Rauzy's induction process in which generalized continued fraction algorithms show up, they proved that under certain natural conditions an *S*-adic sequence gives rise to a dynamical system which is measurably conjugate to a rotation on a higher dimensional torus. Moreover, they established a metric theory which shows that counterexamples like the one constructed in 2000 by Cassaigne, Ferenczi, and Zamboni are rare. It is the aim of the present chapter to survey all these ideas and results.

J. M. Thuswaldner (✉)
Department Mathematics and Information Technology, University of Leoben, Leoben, Austria
e-mail: joerg.thuswaldner@unileoben.ac.at

© The Editor(s) (if applicable) and The Author(s), under exclusive license
to Springer Nature Switzerland AG 2020
S. Akiyama, P. Arnoux (eds.), *Substitution and Tiling Dynamics: Introduction
to Self-inducing Structures*, Lecture Notes in Mathematics 2273,
https://doi.org/10.1007/978-3-030-57666-0_3

3.1 Introduction

A *Sturmian sequence* is an infinite string over two letters with low subword complexity. In particular, it has exactly $n + 1$ different subwords of a given length $n \in \mathbb{N}$. Sturmian sequences have been studied extensively in the literature from various points of view and we refer to Lothaire [101, Chapter 2] or Pytheas Fogg [82, Chapter 6] for detailed accounts. The history of the research surveyed in the present chapter starts with two papers written by Morse and Hedlund [107] as well as Coven and Hedlund [67] in 1940 and 1973, respectively. In these papers the authors established a surprising correspondence between Sturmian sequences and rotations by an irrational number α on the torus $\mathbb{T} = \mathbb{R}/\mathbb{Z}$. In their proof "balance properties" of Sturmian sequences play a prominent role. Several decades later, Arnoux and Rauzy [22] observed that an induction process in which the classical continued fraction algorithm appears can be used to give another very elegant proof of this correspondence (see also Rauzy's earlier papers [112, 113] on this induction process). Their proof also shows how arithmetic and Diophantine properties of an irrational number α are encoded in the corresponding Sturmian sequence.

It has been conjectured since the early 1990s that these correspondences between rotations on $\mathbb{T}$, continued fractions, and Sturmian sequences carry over to rotations on higher dimensional tori, generalized continued fraction algorithms, and so-called S-adic sequences generated by substitutions. The idea of working towards such a generalization is known as *Rauzy's program* and starting with Rauzy [114] a number of examples which hint at such a generalization was devised. A natural class of S-adic sequences to study in this context are so-called *Arnoux-Rauzy sequences* which go back to Arnoux and Rauzy [22]. These are sequences over three letters that behave analogously to Sturmian sequences in many regards. However, in 2000 Cassaigne et al. [63] could construct Arnoux-Rauzy sequences with strong "imbalance", a property which cannot occur for a Sturmian sequence. Cassaigne et al. [62] even constructed Arnoux-Rauzy sequences that give rise to weakly-mixing dynamical systems which are far from rotations in their dynamical behavior. All this shows the limitations of Rauzy's program and indicates that the situation in the general setting is more complicated than it is in the classical case.

Nevertheless, recently Berthé et al. [52] made some progress on Rauzy's program and were indeed able to set up the conjectured generalization of the above correspondences. Using a generalization of Rauzy's induction process in which generalized continued fraction algorithms show up, they proved that under certain natural conditions an S-adic sequence gives rise to a dynamical system which is measurably conjugate to a rotation on a higher dimensional torus. Moreover, they established a metric theory which shows that exceptional cases like the ones constructed in [62] and [63] are rare. A prominent role in this generalization is played by tilings induced by generalizations of the classical *Rauzy fractal* introduced by Rauzy [114].

Another idea which can be linked to the above results goes back to Artin [26], who observed that the classical continued fraction algorithm and its natural extension can be viewed as a Poincaré section of the geodesic flow on the space of two-dimensional lattices $SL_2(\mathbb{Z}) \setminus SL_2(\mathbb{R})$. Arnoux and Fisher [14] revisited Artin's idea and showed that the correspondence between continued fractions, rotations, and Sturmian sequences can be interpreted in a very nice way in terms of an extension of this geodesic flow to pointed lattices which is called the *scenery flow*. Currently, Arnoux et al. [12] are setting up a generalization of this connection between continued fraction algorithms and geodesic flows. In particular, they code the *Weyl Chamber Flow*, a diagonal $\mathbb{R}^{d-1}$-action on the space of d-dimensional lattices $SL_d(\mathbb{Z}) \setminus SL_d(\mathbb{R})$, arithmetically and geometrically by generalized continued fraction algorithms. In this coding, which provides a new view of the relation between S-adic sequences and rotations on higher dimensional tori, non-stationary Markov partitions defined in terms of generalized Rauzy fractals are of great importance.

It is the aim of the present chapter to survey all these ideas and results. In Sect. 3.2 we deal with the case of Sturmian sequences and Sect. 3.3 discusses the problems with the extension of the theory to the more general situation. From Sect. 3.4 onwards we set up the general theory of S-adic sequences and their relation to generalized continued fraction algorithms and rotations on higher dimensional tori.

3.2 The Classical Case

We start our journey by giving some elements of the interaction between Sturmian sequences, the classical continued fraction algorithm, and irrational rotations on the circle. After that we discuss natural extensions of continued fractions and show how all these objects turn up in the study of the geodesic flow acting on the space $SL_2(\mathbb{Z}) \setminus SL_2(\mathbb{R})$ of lattices and its extension to pointed lattices. We will prove most of the results that we state and although our exposition is self-contained we recommend the reader to have a look at the survey [82, Chapter 6] in order to find more background information on the subject of this section.

3.2.1 Sturmian Sequences and Their Basic Properties

For a finite set $\{1, 2, \ldots, d\}$ denote by $\{1, 2, \ldots, d\}^*$ the set of all finite *words* $v_0 \ldots v_{n-1}$ whose *letters* v_i, $0 \leq i < n$, are contained in $\{1, 2, \ldots, d\}$. Moreover, let $\{1, 2, \ldots, d\}^{\mathbb{N}}$ be the space of *(right-infinite) sequences* $w = w_0 w_1 \ldots$ whose letters w_i, $i \in \mathbb{N}$, are elements of $\{1, 2, \ldots, d\}$. The *shift* $\Sigma : \{1, 2, \ldots, d\}^{\mathbb{N}} \to \{1, 2, \ldots, d\}^{\mathbb{N}}$ on this space of sequences is defined by $\Sigma(w_0 w_1 \ldots) = w_1 w_2 \ldots$ Let $w = w_0 w_1 \ldots \in \{1, 2, \ldots, d\}^{\mathbb{N}}$ be a sequence. A *factor* (or *subword*) of w is a word $v_0 \ldots v_{n-1} \in \{1, 2, \ldots, d\}^*$ for which there is $k \geq 0$ such that $w_k \ldots w_{k+n-1} = v_0 \ldots v_{n-1}$. In this case we say that v occurs in w at position k.

The *complexity function* $p_w : \mathbb{N} \to \mathbb{N}$ of w assigns to each integer n the number of words $v_0 \ldots v_{n-1} \in \{1, 2, \ldots, d\}^*$ that are factors of w. If w is *ultimately periodic* in the sense that there exist $k > 0$ and $N \geq 0$ with $w_n = w_{n+k}$ for each $n \geq N$ then p_w is a bounded function. On the other hand, a result by Coven and Hedlund [67] which is not hard to prove states that a sequence $w \in \{1, 2, \ldots, d\}^{\mathbb{N}}$ that admits the inequality $p_w(n) \leq n$ for a single choice of n is ultimately periodic (see also [82, Proposition 1.1.1]). It is the class of not ultimately periodic sequences with smallest complexity function that we are interested in.

Definition 3.2.1 (Sturmian Sequence) A sequence $w \in \{1, 2\}^{\mathbb{N}}$ is called a *Sturmian sequence* if its complexity function satisfies $p_w(n) = n + 1$ for all $n \in \mathbb{N}$.

It is a priori not clear that Sturmian sequences exist at all. However, we will see in Theorem 3.2.11 below that they can be characterized as so-called *natural codings* of irrational rotations which are easy to construct (and will be defined in Sect. 3.2.4).

A detailed account on the early history of Sturmian sequences, which goes back to Bernoulli [39], is given in [101, Notes to Chapter 2]. The name "Sturmian sequence" was coined in 1940 by Morse and Hedlund [107]. Sturmian sequences have been studied extensively. For an overview on fundamental properties of Sturmian sequences we refer in particular to Lothaire [101, Chapter 2], Pytheas Fogg [82, Chapter 6], or Allouche and Shallit [6, Chapters 9 and 10]. Belov et al. [38] discuss some aspects of Sturmian sequences which are related to the present survey.

We start with the discussion of basic properties of Sturmian sequences. The fact that $p_w(n) = n + 1$ holds for a Sturmian sequence entails that for each n there is only one factor $v_0 \ldots v_{n-1}$ of w with the property that both words $v_0 \ldots v_{n-1}1$ and $v_0 \ldots v_{n-1}2$ are factors of w. Such a word $v_0 \ldots v_{n-1}$ is called *right special factor* of w. Left special factors are defined analogously.

Our first lemma deals with *recurrence* of Sturmian sequences. Recall that a sequence $w \in \{1, 2\}^{\mathbb{N}}$ is called *recurrent* if each factor of w occurs infinitely often, i.e., at infinitely many positions, in w.

Lemma 3.2.2 (Cf. e.g. [82, Proposition 6.1.2]) *A Sturmian sequence is recurrent.*

Proof Suppose that this is wrong and let w be a nonrecurrent Sturmian sequence. Then there exists a factor v of length n, say, that occurs only finitely many times in w. Then there exists $k \in \mathbb{N}$ such that $w' = \Sigma^k w$ does not contain v as a factor. However, as $p_w(n) = n + 1$ this implies that $p_{w'}(n) \leq n$ and, hence, w' is ultimately periodic. However, then also w is ultimately periodic, a contradiction.   □

Next we discuss *balance*. To give a formal definition we introduce some notation. For a word $v \in \{1, 2\}^*$ we denote by $|v|$ its *length*, i.e., the number of letters of v. Moreover, for $i \in \{1, 2\}$, we write $|v|_i$ for the number of occurrences of the letter i in v.

Definition 3.2.3 (Balanced Sequence) A sequence $w \in \{1, 2\}^{\mathbb{N}}$ is called *balanced* if each pair of factors (v, v') of w with $|v| = |v'|$ satisfies $\big||v|_1 - |v'|_1\big| \leq 1$.

As was observed already in [107], there is a tight relation between Sturmian sequences and balance.

Proposition 3.2.4 *Let $w \in \{1, 2\}^{\mathbb{N}}$ be given. Then w is a Sturmian sequence if and only if w is not ultimately periodic and balanced.*

The proof of this result is combinatorial. It is based on the observation that for a sequence w which is not balanced there is a word $v \in \{1, 2\}^*$ such that $1v1$ and $2v2$ are factors of w. Since the details are a bit tricky we do not give them here and refer the reader to [107] or [82, Chapter 6, p. 147*ff*].

The fact that Sturmian sequences are balanced will now be exploited in order to prove that they can be *coded* using the *Sturmian substitutions*

$$\sigma_1 : \begin{cases} 1 \mapsto 1, \\ 2 \mapsto 21, \end{cases} \qquad \sigma_2 : \begin{cases} 1 \mapsto 12, \\ 2 \mapsto 2. \end{cases} \tag{3.1}$$

The domain of these substitutions can naturally be extended from $\{1, 2\}$ to $\{1, 2\}^*$ and $\{1, 2\}^{\mathbb{N}}$ by concatenation. The next statement essentially says that balance is maintained by "desubstitution".

Lemma 3.2.5 (See e.g. [14, Lemma 4.2]) *If a sequence $w \in \{1, 2\}^{\mathbb{N}}$ is not balanced, then for each $a \in \{1, 2\}$ the sequence $\sigma_1(aw)$ is not balanced.*

Proof If w is not balanced it is easy to see that there are words u and v with $|u| = |v|$ and $|u|_1 = |v|_1$ such that $1u1$ and $2v2$ are factors of w. Since $1u1$ occurs in w there is $b \in \{1, 2\}$ such that $b1u1$ occurs in aw (we need a in case $1u1$ is the initial word of w). As $\sigma_1(b)$ always ends with 1 and $\sigma_1(2)$ begins with 2, the words $11\sigma_1(u)1$ and $21\sigma_1(v)2$ have the same length and both occur in $\sigma_1(aw)$. As the number of 1s in these two words clearly differs by 2 the lemma follows. $\square$

Let $w = w_0 w_1 \ldots \in \{1, 2\}^{\mathbb{N}}$ be given. If w is a Sturmian sequence, it contains exactly three of the four factors 11, 12, 21, 22. Since it clearly contains 12 and 21 as factors, it either doesn't contain 22, in which case we say that w is of *type 1*, or it doesn't contain 11, in which case we say it is of *type 2*. Using recurrence one can easily see that for each Sturmian sequence $w \in \{1, 2\}^{\mathbb{N}}$ at least one of the sequences $1w$ and $2w$ is Sturmian as well. A Sturmian sequence $w \in \{1, 2\}^{\mathbb{N}}$ is called *special* if $1w$ as well as $2w$ are both Sturmian sequences. With these notions we get the following "desubstitution" of Sturmian sequences (see also [22, Section 1] where an analog of this was proved along somewhat different lines).

Lemma 3.2.6 (See e.g. [14, Proposition 4.3]) *Let u be a Sturmian sequence of type 1.*

(i) *If u is not special then either $u = \sigma_1(v)$ with v Sturmian, or $u = \Sigma \sigma_1(v)$ with v Sturmian starting with 2 (but not both).*

(ii) *If u is special then $u = \sigma_1(v_1) = \Sigma \sigma_1(v_2)$ where $\Sigma v_1 = \Sigma v_2$ is a special Sturmian sequence.*

If u is of type 2 the same statement with the symbols 1 and 2 interchanged holds.

Proof Since u is of type 1 it is immediate that it can be written as $u = \sigma_1(v)$ for some $v \in \{1, 2\}^{\mathbb{N}}$.

To prove (i) suppose that u is not special. Then either $1u$ or $2u$ is Sturmian, but not both.

If $1u$ is Sturmian, $1u = \sigma_1(v')$ with v' starting with 1 and, hence, by Lemma 3.2.5 and Proposition 3.2.4, $v = \Sigma v'$ is Sturmian, and $u = \sigma_1(v)$. If u starts with 2 then $u \neq \Sigma \sigma_1(v')$ for v' starting with 2. If u starts with 1 then also v starts with 1. If we replace the first letter of v by 2 this yields a sequence w satisfying $u = \Sigma \sigma_1(w)$. However, if w is also Sturmian $\Sigma v = \Sigma w$ is special and, hence, one easily checks that u is special, a contradiction and we are done.

If $2u$ is Sturmian then $12u$ has to be Sturmian (since 22 is forbidden) and thus $12u = \sigma_1(1v)$ with v Sturmian and beginning with 2. Thus $u = \Sigma \sigma_1(v)$. As before, we can write $u = \sigma_1(w)$ where w is the word obtained from v by replacing the first letter by 1. This leads again to the contradiction of u being special.

To show (ii) assume u is special. Then, as u has to start with 1 the sequences $12u = \sigma_1(12v)$ and $21u = \sigma_1(21v)$ are Sturmian ($11u$ cannot be Sturmian for imbalance reasons, see [82, Proposition 6.1.23]). By Lemma 3.2.5 and Proposition 3.2.4 the sequences $1v$ and $2v$ are Sturmian, so v is special and $u = \sigma_1(1v) = \Sigma \sigma_1(2v)$.

The proof of the type 2 case is analogous.   □

From the proof of Lemma 3.2.6 we see that for a special sequence u of type 1 there exists a special sequence v such that $21u = \sigma_1(21v)$ and $12u = \sigma_1(12v)$ are Sturmian sequences. If u is special of type 2 we get the existence of a special sequence v with $21u = \sigma_2(21v)$ and $12u = \sigma_2(12v)$ Sturmian by analogous reasoning. If u is a special Sturmian sequence then the two Sturmian sequences $12u$ and $21u$ are called *limit sequences* or *fixed sequences*. By the above arguments they can be desubstituted to sequences that are limit sequences as well. This process can be iterated: let w be a limit sequence. Then there is a sequence $(w^{(n)})_{n \geq 0}$ of limit sequences with

$$w = w^{(0)} \quad \text{and} \quad w^{(n)} = \sigma_{i_n}(w^{(n+1)}) \text{ for } n \geq 0.$$

This can be rewritten as

$$w = \sigma_{i_0} \circ \cdots \circ \sigma_{i_n}(w^{(n+1)}). \tag{3.2}$$

As w is Sturmian, the sequence $(i_n) \in \{1, 2\}^{\mathbb{N}}$ has to change its value infinitely often because otherwise w would be ultimately constant. Now observe that a sequence $w^{(n)}$ starting with a letter a results in a sequence $w^{(0)}$ also starting with a. Moreover, since the sequence (i_n) changes its value infinitely often we see that the first letter of $w^{(n)}$ determines a prefix of w whose length tends to infinity with n. Thus, equipping $\{1, 2\}^{\mathbb{N}}$ with the product topology of the discrete topology yields

$$w = \lim_{n \to \infty} \sigma_{i_0} \circ \cdots \circ \sigma_{i_n}(a), \tag{3.3}$$

where a is the first letter of w (note that we slightly abuse notation here: to be exact the argument of σ_{i_n} should be $aa\ldots \in \mathcal{A}^{\mathbb{N}}$ since the limit is not defined for finite words). We could also group the blocks of the sequence (i_n). So if it starts with a block of a_0 times the symbol 1 followed by a block of a_1 times the symbol 2 and so on we can rewrite (3.3) as

$$w = \lim_{k\to\infty} \sigma_1^{a_0} \circ \sigma_2^{a_1} \circ \sigma_1^{a_2} \circ \cdots \circ \sigma_1^{a_{2k}}(a). \tag{3.4}$$

A sequence w that can be represented by iteratively composing substitutions as in (3.2) is called an *S-adic sequence*.

Note that for arbitrary Sturmian sequences a similar coding as in (3.2) is possible, however, in the general case shifts have to be inserted between the composed substitutions on the appropriate places according to Lemma 3.2.6. Inserting these shifts does not change the collection of factors (called *language*) of the sequence. Thus each Sturmian sequence w is associated with a sequence (σ_{i_m}) which determines its language. We call this sequence the *coding sequence* of w. Summing up we proved the following proposition.

Proposition 3.2.7 (See [22, Section 1]) *Let σ_1, σ_2 be the Sturmian substitutions. Then for each Sturmian sequence w there exists a coding sequence $\boldsymbol{\sigma} = (\sigma_{i_n})$, where (i_n) takes each symbol in $\{1, 2\}$ an infinite number of times, such that w has the same language as*

$$u = \lim_{n\to\infty} \sigma_{i_0} \circ \sigma_{i_1} \circ \cdots \circ \sigma_{i_n}(a).$$

Here $a \in \{1, 2\}$ can be chosen arbitrarily.

Since it will turn out that (3.3) and (3.4) are nonabelian versions of the classical continued fraction algorithm we will now review the basics of this well-known concept.

3.2.2 The Classical Continued Fraction Algorithm

The "*S*-adic" representations of a Sturmian sequence given in (3.3) and (3.4) are related to continued fraction expansions of irrational numbers. For this reason we provide a brief discussion of the classical continued fraction algorithm (see e.g. [76, Chapter 3] for an introduction to continued fractions of a dynamical flavor or [41] for a discussion of continued fractions in a context related to the present paper).

We start with the well-known additive Euclidean algorithm. Given a pair of two nonnegative real numbers $(a, b) \neq (0, 0)$ we define the mapping $F : \mathbb{R}^2_{\geq 0} \setminus \{0\} \to \mathbb{R}^2_{\geq 0} \setminus \{0\}$ by

$$F(a, b) = \begin{cases} (a - b, b), & \text{if } a > b, \\ (a, b - a), & \text{if } a \leq b. \end{cases}$$

If we iterate this mapping starting with $(a, b) \in \mathbb{R}^2_{\geq 0}$ we see that we reach a pair of the form $(0, c)$ or $(c, 0)$ with $c > 0$ if and only if the ratio a/b is rational. If $a/b \notin \mathbb{Q}$ the iterations of F on (a, b) produce an infinite sequence of pairs of strictly positive numbers. Setting

$$M_1 = \begin{pmatrix} 1 & 1 \\ 0 & 1 \end{pmatrix} \quad \text{and} \quad M_2 = \begin{pmatrix} 1 & 0 \\ 1 & 1 \end{pmatrix} \tag{3.5}$$

we see that $F(a, b)^t = M_1^{-1}(a, b)^t$ if $a > b$ and $F(a, b)^t = M_2^{-1}(a, b)^t$ if $a \leq b$. Thus iterating F on a pair (a, b) with $a/b \notin \mathbb{Q}$ produces an infinite sequence $(M_{i_n})_{n \in \mathbb{N}} \in \{M_1, M_2\}^{\mathbb{N}}$ defined by

$$(a, b)^t = M_{i_0} F(a, b)^t = M_{i_0} M_{i_1} F^2(a, b)^t = M_{i_0} M_{i_1} M_{i_2} F^3(a, b)^t = \cdots. \tag{3.6}$$

This sequence (M_{i_n}) is called the *additive continued fraction expansion* of (a, b). In (3.14) we will see that, up to a scalar factor, (a, b) is determined by the sequence (i_n).

Since the sequence (M_{i_n}) is invariant under the multiplication of (a, b) by a scalar, we may use projective coordinates. This motivates the following definition. Let $\mathbb{P}$ be the projective line and $X = \{[a : b] \in \mathbb{P} : a \geq 0, b \geq 0\}$. Define $M : X \to \{M_1, M_2\}$ by $M([a : b]) = M_1$ if $a > b$ and $M([a : b]) = M_2$ if $a \leq b$. Then the mapping

$$F : X \to X; \quad \mathbf{x} \mapsto M(\mathbf{x})^{-1}\mathbf{x} \tag{3.7}$$

is called the *linear additive continued fraction mapping*.

Since $(a, b) \neq (0, 0)$ we can define a *projective* version of (3.7). Indeed, we can write $[a : b] = [1, b/a]$ if $a > b$ and $[a : b] = [a/b, 1]$ if $a \leq b$ and the mapping F can be written as $(c \in [0, 1])$

$$F[1 : c] = \begin{cases} [1 - c : c] = [\frac{1-c}{c} : 1], & \text{if } c > \frac{1}{2}, \\ [1 - c : c] = [1 : \frac{c}{1-c}], & \text{if } c \leq \frac{1}{2}, \end{cases}$$

$$F[c : 1] = \begin{cases} [1 : \frac{1-c}{c}], & \text{if } c > \frac{1}{2}, \\ [\frac{c}{1-c} : 1], & \text{if } c \leq \frac{1}{2}. \end{cases} \tag{3.8}$$

Fig. 3.1 The Farey map

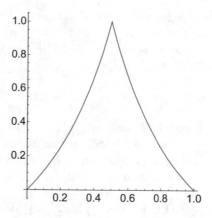

Since the coordinate 1 contains no information in (3.8) and $c \in [0, 1]$, this defines a mapping $f : [0, 1] \to [0, 1]$ by

$$f(x) = \begin{cases} \frac{1-x}{x}, & \text{if } x > \frac{1}{2}, \\ \frac{x}{1-x}, & \text{if } x \leq \frac{1}{2}. \end{cases}$$

The mapping f is called *projective additive continued fraction mapping* or *Farey map*. It is visualized in Fig. 3.1.

The additive continued fraction algorithm can be "accelerated" in the following way. Assume that $a, b > 0$ are given. If $a > b$ we do not just subtract b from a. We subtract it m times where m is chosen in a way that $0 \leq a - mb < b$. If $a \leq b$ we proceed analogously. This results in the *multiplicative Euclidean algorithm* $G :$ $\mathbb{R}^2_{>0} \to \mathbb{R}^2_{\geq 0} \setminus \{\mathbf{0}\}$ with

$$G(a, b) = \begin{cases} (a - \lfloor \frac{a}{b} \rfloor b, b), & \text{if } a > b, \\ (a, b - \lfloor \frac{b}{a} \rfloor a), & \text{if } a \leq b. \end{cases}$$

As in (3.6), iterating G on a pair $(a, b) \in \mathbb{R}^2_{>0}$ yields a sequence of matrices $M_1^{a_0}, M_2^{a_1}, M_1^{a_2}, \ldots$ with positive integers $a_0, a_1, \ldots$ satisfying (we assume $a > b$ here; otherwise the sequence would start with a power of M_2)

$$(a, b)^t = M_1^{a_0} G(a, b)^t = M_1^{a_0} M_2^{a_1} G^2(a, b)^t = M_1^{a_0} M_2^{a_1} M_1^{a_2} G^3(a, b)^t = \cdots .$$
$$(3.9)$$

However, contrary to (3.6) this sequence stops if the iteration runs into a vector one of whose coordinates is 0 because G is not defined for such vectors. Indeed, as can easily be verified, we run into such a vector if and only if $a/b \in \mathbb{Q}$.

Again we move to the projective line and set $X = \{[a : b] \in \mathbb{P} : a > 0, b > 0\}$. Define $M : X \to \{M_1^m, M_2^m : m \geq 1\}$ by $M([a : b]) = M_1^m$ if $a > b$ and $0 \leq a - mb < b$ and $M([a : b]) = M_2^m$ if $a \leq b$ and $0 \leq b - ma < b$. Then the mapping

$$G : X \to X; \quad \mathbf{x} \mapsto M(\mathbf{x})^{-1}\mathbf{x} \tag{3.10}$$

is called the *linear multiplicative continued fraction mapping*.

Similar to the additive case assume that $a, b > 0$ and choose the representatives $[a : b] = [1, b/a]$ if $a > b$ and $[a : b] = [a/b, 1]$ if $a \leq b$. The mapping G can then be written as ($c \in (0, 1]$)

$$G[1 : c] = [1 - \lfloor \tfrac{1}{c} \rfloor c : c] = [\{\tfrac{1}{c}\}c : c] = [\{\tfrac{1}{c}\} : 1], \qquad G[c : 1] = [1 : \{\tfrac{1}{c}\}]. \tag{3.11}$$

As the coordinate 1 contains no information in (3.11) this defines a mapping $g : (0, 1] \to [0, 1)$ by

$$g(x) = \left\{\frac{1}{x}\right\}. \tag{3.12}$$

The mapping g is called *projective multiplicative continued fraction mapping* or *Gauss map*. It is visualized in Fig. 3.2.

By direct calculation (see e.g. [76, Chapter 3]) it follows from the definition that for each irrational $x \in (0, 1)$ the Gauss map g can be iterated infinitely often. This iteration process determines a sequence (a_n) of positive integers defined by

Fig. 3.2 The Gauss map $x \mapsto \{\tfrac{1}{x}\}$

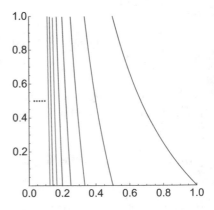

$a_n = \lfloor \frac{1}{g^n(x)} \rfloor$ which admits to develop x in its *(multiplicative) continued fraction expansion*

$$x = \cfrac{1}{a_0 + \cfrac{1}{a_1 + \cfrac{1}{a_2 + \cfrac{1}{a_3 + \ddots}}}}$$

(which will be denoted by $x = [a_0, a_1, \ldots]$). By definition this is the same sequence (a_n) as the one we obtain in the exponents of the matrices in (3.9) when setting $(a, b) = (1, x)$. One can show that this sequence is ultimately periodic if and only if x is a quadratic irrational. If x is rational one can associate a finite sequence with x in this way.

Continued fractions play an eminent role in Diophantine approximation. It is therefore of special interest that they will appear in our theory of Sturmian sequences naturally without being presupposed.

3.2.3 Dynamical Properties of Sturmian Sequences

We want to have a look at the "abelianized" version of (3.3) and (3.4) in order to get a link between Sturmian sequences and the classical continued fraction algorithm. For a word $v \in \{1, 2\}^*$ define the *abelianization* $\mathbf{l}(v) = (|v|_1, |v|_2)^t$, and for $i \in \{1, 2\}$ associate to the Sturmian substitution σ_i from (3.1) the *incidence matrix* $M_i = (|\sigma_i(k)|_j)_{1 \le j, k \le 2}$. Then M_1 and M_2 are the matrices defined in (3.5) which were used to define the linear version of the classical additive continued fraction algorithm in (3.7). Indeed, since $\mathbf{l}\sigma_i(v) = M_i \mathbf{l}(v)$ we see that the vectors (here $\mathbf{e}_1, \mathbf{e}_2$ are the standard basis vectors)

$$M_{i_0} \cdots M_{i_n} \mathbf{e}_a \tag{3.13}$$

form an abelianized version of the expression in the limit of (3.3). Since (i_n) changes its value infinitely often, $M_{i_n} M_{i_{n+1}}$ is a positive matrix for infinitely many n (in particular, $M_{i_n} M_{i_{n+1}} = M_1 M_2$ for infinitely many n; we therefore call the whole sequence (M_{i_n}) a *primitive* sequence of matrices). This property entails that the positive cone $\mathbb{R}^2_{\ge 0}$ is shrunk to a line by these matrices, more precisely, there exists a vector $\mathbf{u} \in \mathbb{R}^2_{>0}$ such that

$$\bigcap_{n \ge 0} M_{i_0} \cdots M_{i_n} \mathbb{R}^2_{\ge 0} = \mathbb{R}_+ \mathbf{u} \tag{3.14}$$

(see [84, pp. 91–95], [125, Chapter 26], or Proposition 3.5.5 below). This says that the additive continued fraction algorithm defined by (3.6) is *weakly convergent* (as is well known, this algorithm is even strongly convergent which is related to the balance property of Sturmian sequences). We call **u**, which is uniquely defined up to scalar factors by the sequence (M_{i_n}), a *generalized right eigenvector* of (M_{i_n}). We also see from (3.14) that the vector $(a, b)^t$ in (3.6) is defined by the sequence (M_{i_n}) up to a scalar factor.

We go back to the (nonabelian) S-adic setting. Assume that a Sturmian sequence $w = w_0 w_1 \ldots$ has a coding sequence (σ_{i_n}) whose associated sequence of incidence matrices (M_{i_n}) satisfies (3.14). We will now prove that in this case w has *uniform letter frequencies*, i.e., the limit

$$f_i(w) = \lim_{\ell \to \infty} \frac{|w_k \ldots w_{k+\ell-1}|_i}{\ell}$$

exists uniformly in k for each $i \in \{1, 2\}$. We get even more, namely, the following lemma holds. In its proof and in all the remaining part of this section we use the abbreviations

$$\sigma_{i_{[m,n)}} = \sigma_{i_m} \circ \cdots \circ \sigma_{i_{n-1}} \quad \text{and} \quad M_{i_{[m,n)}} = M_{i_m} \cdots M_{i_{n-1}}.$$

Lemma 3.2.8 *Let* $w = w_0 w_1 \ldots$ *be a Sturmian sequence with coding sequence* (σ_{i_n}) *whose associated sequence of incidence matrices* (M_{i_n}) *has a generalized right eigenvector* **u**. *Then* w *has uniform letter frequencies and* $(f_1(w), f_2(w))^t = \frac{\mathbf{u}}{\|\mathbf{u}\|_1}$.

Proof Let $\mathbf{u}/\|\mathbf{u}\|_1 = (u_1, u_2)^t$. By Proposition 3.2.7 for all $k, \ell, n \in \mathbb{N}$ we can write

$$w_k \ldots w_{k+\ell-1} = p \sigma_{i_{[0,n)}}(v) s$$

for some $p, v, s \in \{1, 2\}^*$, where the lengths of p, s are bounded by the number $\max\{|\sigma_{i_{[0,n)}}(1)|, |\sigma_{i_{[0,n)}}(2)|\}$.

Now, for each $a \in \{1, 2\}$ we have the inequality

$$\left| \frac{|w_k \ldots w_{k+\ell-1}|_a}{\ell} - u_a \right| \leq \frac{||p|_a - |p|u_a|}{\ell} + $$
$$\frac{||\sigma_{i_{[0,n)}}(v)|_a - |\sigma_{i_{[0,n)}}(v)|u_a|}{\ell} + \frac{||s|_a - |s|u_a|}{\ell}. \tag{3.15}$$

By the convergence of the positive cone to **u** in (3.14) we know that $|\sigma_{i_{[0,n)}}(b)|_a / |\sigma_{i_{[0,n)}}(b)|$ is close to u_a for all $a, b \in \{1, 2\}$ if n is large. Thus for each $\varepsilon > 0$ there is $N \in \mathbb{N}$ such that whenever $\ell \geq N$ we can choose n in a way that $|p|, |s| \leq \varepsilon \ell$ and $||\sigma_{i_{[0,n)}}(b)|_a - |\sigma_{i_{[0,n)}}(b)|u_a| < \varepsilon|\sigma_{i_{[0,n)}}(b)|$ for all letters a and b. This proves that the right hand side of (3.15) is bounded by 3ε and thus $\lim_{\ell \to \infty} |w_k \ldots w_{k+\ell-1}|_a / \ell = u_a$ uniformly in k. $\square$

For a proof of Lemma 3.2.8 along similar lines in a more general setting we refer to Lemma 3.5.10 (see also Berthé and Delecroix [44, Theorem 5.7]; a proof using balance, which also gives irrationality of the frequencies, is contained in [82, Proposition 6.1.10]).

In the same way as for letters, we can define uniform frequencies for factors of an infinite sequence $w \in \{1, 2\}^{\mathbb{N}}$. Let w be a Sturmian sequence with coding sequence (σ_{i_n}). The sequence is the shifted image of another Sturmian sequence under an arbitrary large block $\sigma_{i_{[0,n)}}$ of substitutions. This enables one to show that for the words $\sigma_{i_{[0,n)}}(a)$ there exist uniform frequencies in w. Since (i_n) changes its value infinitely often, the length of the words $\sigma_{i_{[0,n)}}(a)$ tends to infinity for each letter a if $n \to \infty$. Using this fact one can prove the following result along similar lines as Lemma 3.2.8 (for details we refer to the proof of Lemma 3.5.10 below; see also [44, Theorem 5.7]).

Lemma 3.2.9 *Let* $w = w_0 w_1 \ldots \in \{1, 2\}^{\mathbb{N}}$ *be a Sturmian sequence with coding sequence* (σ_{i_n}) *whose associated sequence of incidence matrices* (M_{i_n}) *has a generalized right eigenvector* **u**. *Let* $v \in \{1, 2\}^*$ *be given, and let* $|w_k w_{k+1} \ldots w_{k+\ell-1}|_v$ *be the number of occurrences of the factor* v *in the factor* $w_k w_{k+1} \ldots w_{k+\ell-1}$ *of* w. *Then* $|w_k w_{k+1} \ldots w_{k+\ell-1}|_v / \ell$ *tends to a limit* $f_v(w)$ *for* $\ell \to \infty$ *uniformly in* k.

We can associate a dynamical system with a Sturmian sequence w in a very natural way. Let $X_w = \overline{\{\Sigma^k w : k \in \mathbb{N}\}}$ be the closure of the shift orbit of w. Alternatively, X_w can be viewed as the set of all sequences u whose *language* $L(u)$ (i.e., its set of factors) satisfies $L(u) \subseteq L(w)$. Thus if $\sigma = (\sigma_{i_n})$ is the coding sequence of w, Proposition 3.2.7 implies that X_w contains all Sturmian sequences with coding sequence σ. Since X_w is shift invariant the shift Σ acts on X_w and the dynamical system (X_w, Σ) is well defined. We call (X_w, Σ) a *Sturmian system*. From what we know about Sturmian sequences we can derive a number of properties for these dynamical systems. The notions of *minimality* and *unique ergodicity* of a dynamical system used in the following lemma are defined precisely in Definitions 3.5.2 and 3.5.7, respectively.

Proposition 3.2.10 *A Sturmian system* (X_w, Σ) *has the following properties.*

  (i) *The system* (X_w, Σ) *is minimal.*
 (ii) *The set* X_w *is the set of all Sturmian sequences having the same language.*
(iii) *The set* X_w *is the set of all Sturmian sequences having the same coding sequence* σ.
 (iv) *The system* (X_w, Σ) *is uniquely ergodic.*
  (v) *We have* $X_w = X_{w'}$ *for any* $w' \in X_w$.

Proof Let (σ_{i_n}) be the coding sequence of w with (M_{i_n}) being the associated sequence of matrices.

We start with (i). By Proposition 3.2.7 we may assume w.l.o.g. that $w = \lim_{n \to \infty} \sigma_{i_{[0,n)}}(1)$. Let $v \in X_w$ be given. To prove minimality it suffices to show that $L(v) = L(w)$. Since $L(v) \subseteq L(w)$ is true by definition we need to prove the reverse inclusion. Let $u \in L(w)$. By the definition of w and the primitivity of the sequence (M_{i_n}) there is $m \in \mathbb{N}$ such that u occurs in $\sigma_{i_{[0,m)}}(1)$. However, there is a Sturmian word $w^{(m)}$ satisfying $w = \sigma_{i_{[0,m)}}(w^{(m)})$. Since $w^{(m)}$ is balanced by Proposition 3.2.4, the letter 1 occurs in $w^{(m)}$ with bounded gaps. This implies that $\sigma_{i_{[0,m)}}(1)$ and, hence, u occurs in w with bounded gaps. Thus u occurs in each element of the orbit closure X_w of w, hence, also in v. Thus $L(v) = L(w)$ is established.

Since $L(v) = L(w)$ holds for each $v \in X_w$ according to the previous paragraph we have $p_v(n) = p_w(n) = n + 1$ for all $n \in \mathbb{N}$, hence, v is Sturmian with the same language as w. This proves (ii).

To prove (iii) we follow the proof of [82, Lemma 6.3.12]. Assume w.l.o.g. that the elements of X_w are of type 1 and let $u, u' \in X_w$. Then according to Lemma 3.2.6 there exist Sturmian words v, v' such that $u = \sigma_1(v)$ or $u = \Sigma\sigma_1(v)$ as well as $u' = \sigma_1(v')$ or $u' = \Sigma\sigma_1(v')$. We first prove that v, v' belong to the same Sturmian system. By (ii) we have to show that $L(v) = L(v')$. Suppose that $x \in L(v)$. Since x occurs infinitely often in v by recurrence, there is $y \in L(v)$ starting with the letter 2 such that x is a subword of y. The word $\sigma_1(y)$ occurs in u and by (ii) it occurs also in u' and because $\sigma_1(y)$ begins with 2 and ends with 1 it can be desubstituted in only one way by σ_1, namely to y. This proves that y and, hence, also x occurs in v'. Thus $L(v) \subseteq L(v')$. The other inclusion follows by interchanging the roles of v and v'. Iterating this argument yields that u and u' have the same coding sequence. Thus all elements of X_w have the same coding sequence. As Sturmian sequences with the same coding sequence have the same language by Proposition 3.2.7, X_w contains all Sturmian sequences having the same coding sequence as w.

Item (iv) follows immediately by combining Lemma 3.2.9 with [82, Proposition 5.1.21] (see also Proposition 3.5.9 below) which states that the existence of uniform word frequencies implies unique ergodicity. Alternatively, one can use Boshernitzan [56].

Finally, (v) follows from (ii).   □

We emphasize on the fact that for minimality and unique ergodicity of (X_w, Σ) the recurrence of w as well as the primitivity of the sequence (M_{i_n}) is of importance. This will be the same in the general case (see Sect. 3.5 below). In view of assertion (iii) of the previous lemma we will write X_σ instead of X_w, where σ is the coding sequence of w.

3.2.4   Sturmian Sequences Code Rotations

It was observed already by Morse and Hedlund [107] and Coven and Hedlund [67] that each Sturmian sequence is a *natural coding* of a rotation by some irrational number α. We now sketch a proof of this fact which goes back to Rauzy and in which the multiplicative continued fraction expansion of α pops up when we represent such a coding in an S-adic fashion. For proofs of this kind we refer to [13, 14, 46, 47]; a

Fig. 3.3 Two iterations of
the irrational rotation R_α on
$\mathbb{T}$ which is subdivided into
the two intervals I_1 and I_2

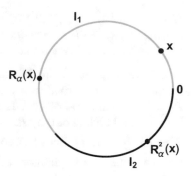

different, combinatorial proof along the lines of the original proof by Morse, Coven, and Hedlund is presented in [101, Theorem 2.1.13] and [6, Section 10.5].

Before we give the main result of this section we provide some definitions. Let $\mathbb{T}$ be the 1-torus, i.e., the unit interval $[0, 1]$ with its end points glued together. A *rotation* or *translation* on $\mathbb{T}$ by a real number α is a mapping $R_\alpha : \mathbb{T} \to \mathbb{T}$ with $x \mapsto x + \alpha$ (mod 1). If $\alpha \notin \mathbb{Q}$ this gives a minimal dynamical system. Moreover, observe that R_α can be regarded as a *two interval exchange* of the intervals $I_1 = [0, 1 - \alpha)$ and $I_2 = [1 - \alpha, 1)$ or of the intervals $I_1' = (0, 1 - \alpha]$ and $I_2' = (1 - \alpha, 1]$, see Fig. 3.3. We say that a sequence $w = w_0 w_1 \ldots \in \{1, 2\}^{\mathbb{N}}$ is a *natural coding* of R_α if there is $x \in \mathbb{T}$ such that $R_\alpha^k(x) \in I_{w_k}$ for each $k \in \mathbb{N}$ or $R_\alpha^k(x) \in I'_{w_k}$ for each $k \in \mathbb{N}$.

Theorem 3.2.11 *A sequence $w \in \{1, 2\}^{\mathbb{N}}$ is Sturmian if and only if there exists $\alpha \in \mathbb{R} \setminus \mathbb{Q}$ such that w is a natural coding of the rotation R_α.*

The sufficiency part of the theorem is easy. Indeed, it just follows from the observation that

$$v_0 \ldots v_{n-1} \text{ is a factor of a natural coding of } R_\alpha \quad \Longleftrightarrow \quad \bigcap_{k=0}^{n-1} R_\alpha^{-k} I_{v_k} \neq \emptyset,$$

$$(3.16)$$

whose proof is an easy exercise (see [46, Lemma 2.7]).

The proof of the necessity part of Theorem 3.2.11 needs more work and we will see that the classical continued fraction algorithm pops up along the way without being presupposed. We need the following key lemma.

Lemma 3.2.12 *For $\alpha \in (0, 1)$ irrational let u be the coding of the point $1 - \alpha/(\alpha + 1)$ under the irrational rotation $R_{\alpha/(\alpha+1)}$. Then there is a sequence (σ_{i_n}) of substitutions such that*

$$u = \lim_{n \to \infty} \sigma_{i_0} \circ \cdots \circ \sigma_{i_n}(2).$$

The sequence $(i_n) \in \{1, 2\}^{\mathbb{N}}$ is of the form $1^{a_0} 2^{a_1} 1^{a_2} 2^{a_3} \ldots$ where the sequence $[a_0, a_1, a_2, a_3, \ldots]$ is the continued fraction expansion of α. For $\alpha > 1$ a similar result with switched symbols holds.

Proof We assume $\alpha < 1$ ($\alpha > 1$ can be treated in a similar way). For computational reasons consider the rotation R by α on the interval $J = [-1, \alpha)$ with the partition $P_1 = [-1, 0)$ and $P_2 = [0, \alpha)$. The natural coding u of $1 - \alpha/(\alpha + 1)$ by $R_{\alpha/(\alpha+1)}$ is the natural coding of 0 by R. Let R' be the first return map of R to the interval $J' = \left[\alpha \lfloor \frac{1}{\alpha} \rfloor - 1, \alpha \right)$. Let v be a coding of the orbit of 0 for R'. As can be seen from Fig. 3.4, after each occurrence of 2 in u we leave the interval J' and there follows a block of 1s of length $\lfloor \frac{1}{\alpha} \rfloor$ before we enter the interval J' again. Thus v emerges from u by removing such a block of 1s after each letter 2 occurring in u. By the definition of σ_1 this just means that $u = \sigma_1^{\lfloor 1/\alpha \rfloor}(v)$. We can now renormalize the interval J' by dividing it by $-\alpha$ and, as illustrated in Fig. 3.4, then R' is conjugate to a rotation (called R' again) by $\left\{ \frac{1}{\alpha} \right\}$ on the interval $\left(-1, \left\{ \frac{1}{\alpha} \right\} \right]$, where v is the natural coding of the partition $P_2' = (-1, 0]$ and $P_1' = \left(0, \left\{ \frac{1}{\alpha} \right\} \right]$. Note that the *Gauss map* $\alpha \mapsto \left\{ \frac{1}{\alpha} \right\}$ from (3.12) comes up here without being presupposed. Since we are in the same setting as before (just with the letters 1 and 2 interchanged), we

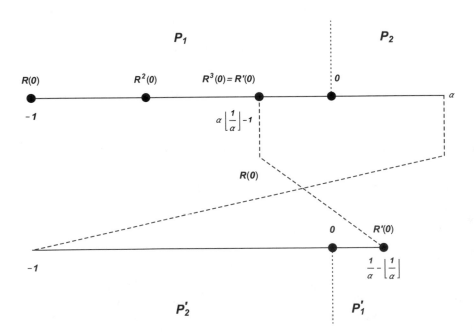

Fig. 3.4 The rotation R' induced by R

can iterate this process and thereby obtain a sequence $(u^{(n)})_{n\geq 0}$ of natural codings such that

$$u = u^{(0)} \quad \text{and} \quad u^{(n)} = \sigma_{i_n}(u^{(n+1)}) \text{ for } n \geq 0$$

for some sequence (σ_{i_n}) with $(i_n) \in \{1, 2\}^{\mathbb{N}}$ having infinitely many changes between the letters 1 and 2. Arguing in the same way as in Sect. 3.2.1 we gain that

$$u = \lim_{n \to \infty} \sigma_{i_0} \circ \cdots \circ \sigma_{i_n}(a)$$

where $a = 2$ is the first letter of u. The assertion on the continued fraction expansion follows from the above proof as well. Just note that the interval we use has length $\alpha + 1$ so that the rotation by α on this interval is conjugate to $R_{\alpha/(\alpha+1)}$. $\qquad\square$

Proof (Conclusion of the Proof of Theorem 3.2.11) The sufficiency assertion has been treated in (3.16). The necessity part of the theorem can now be obtained as follows. Let w be a Sturmian sequence. Consider its coding sequence (σ_{i_n}) and write $(i_n) \in \{1, 2\}^{\mathbb{N}}$ as $1^{a_0}2^{a_1}1^{a_2}2^{a_3} \ldots$ Then $u = \lim_{n\to\infty} \sigma_{i_0} \circ \cdots \circ \sigma_{i_n}(2)$ is a natural coding of $R_{\alpha/(1+\alpha)}$ where $\alpha = [a_0, a_1, a_2, \ldots]$. By Proposition 3.2.7 the sequence w has the same language as u and (3.16) together with an approximation argument implies that w is a natural coding of $R_{\alpha/(1+\alpha)}$ (it is easy to verify that there are limit cases where we really need the intervals I'_1, I'_2 to define the natural coding for w). $\qquad\square$

The fact that Sturmian sequences have irrational uniform letter frequencies is an immediate consequence of Theorem 3.2.11. Moreover, we have the following corollary of Theorem 3.2.11 for Sturmian systems.

Corollary 3.2.13 *A Sturmian system* (X_σ, Σ, μ) *is measurably conjugate to an irrational rotation* $(\mathbb{T}, R_\alpha, \lambda)$. *Here* μ *is the unique* Σ-*invariant measure on* X_σ *and* λ *is the Haar measure on* $\mathbb{T}$.

Proof Let $\varphi : X_\sigma \to \mathbb{T}$ be defined by $\varphi(w_0 w_1 \ldots) = x$ if $R^k_\alpha(x) \in I_{w_k}$ for each $k \in \mathbb{N}$ or $R^k_\alpha(x) \in I'_{w_k}$ for each $k \in \mathbb{N}$. Using Theorem 3.2.11 and the minimality of R_α it is easy to check that this is well defined. Surjectivity of φ follows immediately from Theorem 3.2.11. To investigate injectivity let $u = u_0 u_1 \ldots$ and $v = v_0 v_1 \ldots$ be distinct elements of X_σ with $\varphi(u) = \varphi(v)$. By the minimality of R_α this is only possible if the orbit of $\varphi(u)$ passes through 0 and u is naturally coded by I_1, I_2 while v is naturally coded by I'_1, I'_2 (or vice versa).[1] Since the set of such elements u and v is countable, φ is bijective everywhere save for a countable set. Moreover, φ is easily seen to be continuous and $\varphi \circ \Sigma = R_\alpha \circ \varphi$ holds by the definition of φ. This implies the result. $\qquad\square$

---

[1] This implies that u and v have Σx and Σy in their orbit where x and y are the two limit sequences of σ. This interesting fact, which is not needed in this proof, should be proved by the reader.

We illustrate the concepts of this section by a classical example.

Example 3.2.14 (A Variant of the Fibonacci Sequence) Let σ be given by

$$\sigma = \sigma_1 \circ \sigma_2 : \begin{cases} 1 \mapsto 121, \\ 2 \mapsto 21. \end{cases}$$

This is a reordering of the square of the well-known *Fibonacci substitution* (which is defined by $1 \mapsto 12, 2 \mapsto 1$; see for instance in [82, Section 1.2.1]). Consider the coding sequence $\sigma = (\sigma)$. In this case the associated limit sequences are "purely substitutive". One of the two limit sequences is

$$w = \lim_{n \to \infty} \sigma^n(2) = 211211212112112121121121121 \ldots$$

Since only one substitution plays a role here, the associated "S-adic" system (X_σ, Σ) is called a *substitutive system*. Let $\varphi = \frac{1+\sqrt{5}}{2}$. By the Perron-Frobenius theorem the generalized right eigenvector $\mathbf{u}$ of the sequence of incidence matrices $\mathbf{M}$ of σ is the eigenvector $(\varphi, 1)^t$ corresponding to the dominant eigenvalue φ^2 of the incidence matrix of σ. Let L be the eigenline defined by this eigenvector. Being a Sturmian sequence, w is balanced by Proposition 3.2.4 and has uniform letter frequencies $(f_1(w), f_2(w))^t = \frac{1}{1+\varphi}(\varphi, 1)^t$ by Lemma 3.2.8. This is reflected by the fact that the "broken line"

$$B = \{\mathbf{l}(p) \; : \; p \text{ is a prefix of } w\} \tag{3.17}$$

associated with the sequence w stays at bounded distance from the eigenline L (see Fig. 3.5).

Because $w = \lim_{n \to \infty} \sigma^n(2) = \lim_{n \to \infty} (\sigma_1 \circ \sigma_2)^n(2)$, it has coding sequence $\sigma_1, \sigma_2, \sigma_1, \sigma_2, \ldots$ Since the "run lengths" of σ_i in this sequence are always equal to 1 we set $\alpha = [1, 1, 1, \ldots] = \varphi^{-1}$ and, hence, $\alpha/(\alpha + 1) = \varphi^{-2}$. Thus from Theorem 3.2.11 and its proof we see that w is a natural coding of the rotation by φ^{-2} of the point $1 - \alpha/(\alpha+1) = \varphi^{-1} \in [0, 1)$ with respect to the partition $I_1 = [0, \varphi^{-1})$, $I_2 = [\varphi^{-1}, 1)$ (or the according partition I_1', I_2') of $[0, 1)$. This gives us an easy way to construct w (and the broken line B). Indeed, start at the origin, write out 2 and go up to the lattice point $(0, 1)^t$. After that, inductively proceed as follows: whenever the current lattice point is above L, write out 1 and go right to the next lattice point by adding the vector $(1, 0)^t$ and whenever the current lattice point is below L, write out 2 and go up to the next lattice point by adding the vector $(0, 1)^t$.[2]

Let π be the projection along L to the line $L^\perp$ orthogonal to L. If we project all points on the broken line and take the closure of the image, due to the irrationality

---

[2]We could also have started with writing out 1 and going to the right from the origin. This would have produced the second limit sequence of (σ) which coincides with w save for the first two letters.

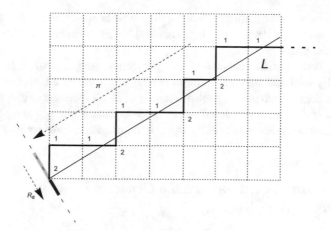

Fig. 3.5 The broken line and its projection to the Rauzy fractal

of **u** we obtain the interval

$$\mathcal{R}_{\mathbf{u}} = \overline{\{\pi \mathbf{l}(p) \ : \ p \text{ is a prefix of } w\}}$$

on $L^\perp$ (the subscript **u** indicates that $\mathcal{R}_{\mathbf{u}}$ lives in the space $L^\perp = \mathbf{u}^\perp$ orthogonal to **u** which is an arbitrary choice; other choices will play a roll in subsequent sections). We color the part of the interval for which we write out 1 at the associated lattice point light grey, the other part dark grey. This subdivides the interval $\mathcal{R}_{\mathbf{u}}$ into two subintervals $\mathcal{R}_{\mathbf{u}}(1)$ and $\mathcal{R}_{\mathbf{u}}(2)$, where

$$\mathcal{R}_{\mathbf{u}}(i) = \overline{\{\pi \mathbf{l}(p) \ : \ pi \text{ is a prefix of } w\}} \qquad (i = 1, 2).$$

Moreover, we see that moving a step along the broken line amounts to exchanging these two intervals in the projection: points in $\mathcal{R}_{\mathbf{u}}(1)$ are moved downwards by a fixed vector, while points in $\mathcal{R}_{\mathbf{u}}(2)$ are moved upwards by a fixed vector.

Thus passing along the broken line each step amounts to exchanging the intervals $\mathcal{R}_{\mathbf{u}}(1)$ and $\mathcal{R}_{\mathbf{u}}(2)$ in the projection. If we identify the end points of $\mathcal{R}_{\mathbf{u}}$ this interval exchange becomes a rotation. This is the rotation which is coded by the Sturmian sequence w. The union $\mathcal{R}_{\mathbf{u}} = \mathcal{R}_{\mathbf{u}}(1) \cup \mathcal{R}_{\mathbf{u}}(2)$ is called the *Rauzy fractal* associated with the substitution σ (or with the sequence $\sigma = (\sigma)$). The reason why we speak about *fractals* here will become apparent in Sect. 3.6.1 when we define the analogs of $\mathcal{R}_{\mathbf{u}}$ in a more general setting.

Suppose we would be given an arbitrary sequence $w \in \{1, 2\}^{\mathbb{N}}$ with letter frequency vector **u** whose broken line stays within bounded distance of the line $L = \mathbb{R}_+\mathbf{u}$. Then we could draw a similar picture as in Fig. 3.5. However, although the projection π would project the vertices of the associated broken line to a bounded set, there is no reason for its closure $\mathcal{R}_{\mathbf{u}}$ to be an interval. Also, if we use two colors as in the example above, it may well happen that the two sets $\mathcal{R}_{\mathbf{u}}(1)$ and $\mathcal{R}_{\mathbf{u}}(2)$ have considerable overlap. This bad behavior prevents us from seeing a rotation in the projections.

Making sure that the closure of the projection of the broken line behaves topologically well and allows a partition whose atoms are essentially different will be our main concern when we establish a theory of S-adic sequences that are codings of rotations on higher dimensional tori in the subsequent sections and, hence, give rise to dynamical systems that are measurably conjugate to torus rotations.

3.2.5 Natural Extensions and the Geodesic Flow on $\mathrm{SL}_2(\mathbb{Z}) \setminus \mathrm{SL}_2(\mathbb{R})$

In this section we talk about natural extensions of the Gauss map and of the coding map of Sturmian sequences by substitutions. Moreover, we show how to relate these natural extensions to the geodesic flow on the space $\mathrm{SL}_2(\mathbb{Z}) \setminus \mathrm{SL}_2(\mathbb{R})$ of unimodular two-dimensional lattices.

So far we could relate Sturmian sequences to rotations on the circle by using the classical continued fraction algorithm. In our discussion we coded a Sturmian sequence w by a sequence of substitutions (σ_{i_n}) as

$$w = \lim_{k \to \infty} \sigma_1^{a_0} \circ \sigma_2^{a_1} \circ \sigma_1^{a_2} \circ \cdots \circ \sigma_1^{a_{2k}}(a)$$

(see (3.4)). In the induction process used in the proof of Theorem 3.2.11 we recoded w by a "desubstitution" process. If we look at the first step of this process we produce the sequence

$$u = \lim_{k \to \infty} \sigma_2^{a_1} \circ \sigma_1^{a_2} \circ \cdots \circ \sigma_1^{a_{2k}}(a).$$

However, the mapping $w \mapsto u$ cannot be inverted since it is not possible to reconstruct a_0 from u. Similarly, the Gauss map g cannot be inverted since $g([a_0, a_1, \ldots]) = [a_1, a_2, \ldots]$, and a_0 cannot be reconstructed from the image $[a_1, a_2, \ldots]$.

In this section we want to make both of these mappings bijective by constructing a geometric model for their *natural extensions* (in the sense of Rohlin [116]). To this matter we look again at the induction used in Lemma 3.2.12 which is visualized once more in Fig. 3.6a. In this figure we see why this induction process cannot be reversed: the intervals $[R(0), R^2(0))$ and $[R^2(0), R^3(0))$ get lost during the induction process and cannot be reconstructed.

A first idea on how to mend this is indicated in Fig. 3.6b: one could "stack" the lost intervals on the larger interval of the induced rotation. This would keep the information of the last induction step. However, acting in this way we can go back at most to the setting from which we started but not farther to the "past".

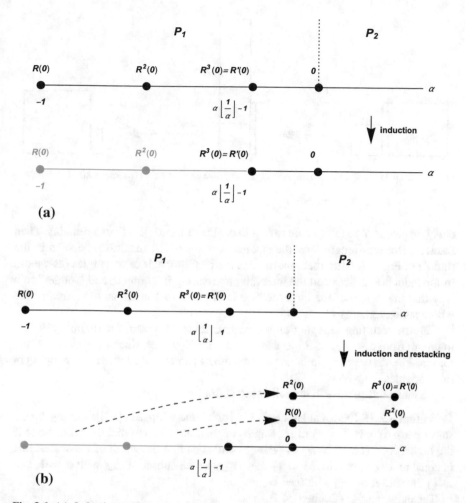

Fig. 3.6 (**a**) Induction without restacking loses some part of the information. The intervals $[R(0), R^2(0))$ and $[R^2(0), R^3(0))$ depicted in light gray are no longer present in the induced rotation. (**b**) Induction together with restacking the intervals keeps all the information. The light gray intervals $[R(0), R^2(0))$ and $[R^2(0), R^3(0))$ are stacked on the longer interval of the induced rotation

To make the induction process bijective, it is more convenient to build rectangular boxes above the intervals as indicated in Fig. 3.7 (this approach is extensively exploited in Arnoux and Fisher [14]; we follow here [82, Section 6.6]). The lengths of the boxes are given by the intervals on which the induction process starts: one box is of length 1, the other one has length α for some $\alpha \in (0, 1) \setminus \mathbb{Q}$. The heights are chosen in a way that the longer rectangle is also the higher one and that the total area of the two rectangles is equal to one. The induction process can now be performed on the rectangles as indicated in Fig. 3.7: let $a \times d$ be the size of the left rectangle and $b \times c$ the size of the right one. Slice the larger rectangle by vertical

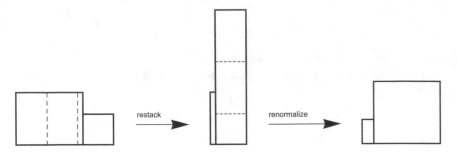

Fig. 3.7 Step 1: restack the boxes. Step 2: renormalize in a way that the larger box has length 1 again

cuts into pieces of lengths equal to b until a slice of length less than b remains. Then stack all slices of length b on the smaller rectangle. The result can be seen in the middle of Fig. 3.7. After that renormalize the resulting pair of rectangles (as we did in the induction process on the intervals) by making it "thinner" and "longer" in a way that the length of the larger rectangle is equal to 1 again and the area of the whole region remains 1.

Call the resulting mapping on the rectangles Ψ. A priori, the mapping Ψ is a mapping from a subset of $\mathbb{R}^4$ to a subset of $\mathbb{R}^4$. However, since $ad + bc = 1$ and $\max\{a, b\} = 1$, we can eliminate two coordinates and we are left with a mapping in two variables.

We make this precise in the following definition.

Definition 3.2.15 (Natural Extension of the Gauss Map, see [14]) Let Δ_m be the set of pairs $(a \times d, b \times c)$ of rectangles of total area 1 such that the widest one is the highest one (i.e., $a > b \Leftrightarrow d > c$) and such that the width of the widest one is equal to 1 (i.e., $\max\{a, b\} = 1$). Let $\Delta_{m,0}$ be the subset of Δ_m with $a = 1$, and $\Delta_{m,1}$ the subset of Δ_m with $b = 1$.

The mapping Ψ is defined on $\Delta_{m,1}$ as

$$(a, d) \mapsto \left(\left\{ \frac{1}{a} \right\}, a - da^2 \right),$$

and similarly on $\Delta_{m,0}$. It is called the *natural extension* of the Gauss map (which is seen in the first coordinate).

Remark 3.2.16 The subscript "m" stands for *multiplicative* since we work here with the multiplicative version of the classical continued fraction algorithm defined by the Gauss map. An analogous theory exists for the additive algorithm as well, see [14].

The mapping Ψ is bijective as becomes clear from its geometric interpretation. Moreover, it is easy to show that Ψ preserves the Lebesgue measure. By integrating away the second coordinate one can show that the invariant measure of the Gauss map is $\frac{dx}{\ln 2(1+x)}$ (see e.g. [76, Chapter 3]). We mention that another natural extension of the Gauss map defined on the unit square is provided in [108].

We can also see Sturmian sequences in the rectangular boxes. To this end note first that a pair of boxes $a \times d$ and $b \times c$ is a fundamental domain of the lattice spanned by the vectors $(a, c)^t$ and $(-b, d)^t$. This is illustrated in Fig. 3.8 and has the consequence that the "L-shaped" region formed by this pair of boxes can be used to tile the plane with respect to this lattice as indicated in Fig. 3.9.

Let us mark a point in this tiling. If we start from this point and move upwards and write out 1 whenever we pass through a large rectangle, and 2, whenever we pass through a small one, we get the coding u of a rotation by α on the interval $(-1, \alpha)$ which, by Theorem 3.2.11, is a Sturmian sequence. This is indicated in Fig. 3.9. In the same way we can produce a Sturmian sequence v by moving horizontally.

If we restack each of the fundamental domains, according to the procedure described above, we get a new fundamental domain (indicated by the shaded region in Fig. 3.9). We now code the same vertical line using this restacked region. Doing this we obtain another Sturmian sequence $u^{(1)}$ which, by the definition of the restacking process, satisfies $u = \sigma(u^{(1)})$, where σ is the substitution defining the induction process as in the proof of Lemma 3.2.12. On the other hand, looking at the

Fig. 3.8 A pair of boxes is a fundamental domain of a lattice

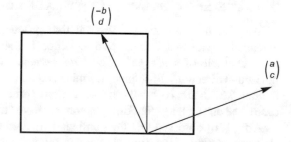

Fig. 3.9 The vertical line is coded by a Sturmian sequence u, the horizontal line by a Sturmian sequence v. The restacking procedure desubstitutes u and substitutes v. The shaded region is a restacked fundamental domain

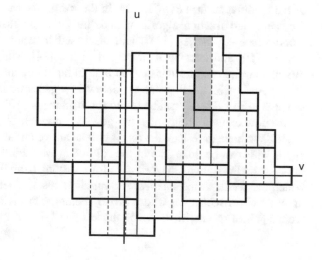

horizontal line we get $v^{(-1)} = \sigma(v)$ as the new coding. Thus the restacking process corresponds to the mapping

$$(u, v) \mapsto (u^{(1)}, v^{(-1)}).$$

As mentioned at the beginning of this section, we cannot reconstruct u from $u^{(1)}$, however we *can* reconstruct (u, v) from $(u^{(1)}, v^{(-1)})$ since the type of the Sturmian sequence $v^{(-1)}$ tells us (which power of) which of the two substitutions σ_1, σ_2 from (3.1) we have to use to get back. This makes the coding process bijective as well. We could mark the pair of rectangles discussed above by a point (x, y) and look at the itinerary of this point under the restacking process. This would give an extension $\tilde{\Psi}$ of the mapping Ψ that is defined on the $\mathbb{T}^2$-fibers over Δ_m (see [14]).

The following remark is of particular importance.

Remark 3.2.17 Regardless of the point in the "L-shaped" region in which we start, the "vertical" Sturmian sequence will always be contained in the same Sturmian system. Thus we can say that the "L-shaped" pairs of rectangles parametrize the Sturmian systems (which are characterized by their coding sequence according to Proposition 3.2.10(iii)), while the (x-coordinates of the) points in a given region parametrize the sequences contained in this system. The same is true for the "vertical" Sturmian sequence w.r.t. the y-coordinates.

We also mention that the vertical line producing the coding u can also be extended downwards. This yields a sequence $\tilde{u} \in \mathcal{A}^{\mathbb{Z}}$ as a coding. Such a sequence is an example of a *bi-infinite Sturmian sequence* (the same can be done in the horizontal direction). Bi-infinite Sturmian sequences are studied for instance in [82, Section 6.2]. It turns out that some of their properties are nicer than in our one-sided case since one no longer has troubles coming from "the beginning" of the sequences.

Artin [26] observed that the continued fraction algorithm can be viewed as a Poincaré section of the geodesic flow on the unit tangent bundle $\mathrm{SL}_2(\mathbb{Z}) \setminus \mathrm{SL}_2(\mathbb{R})$ of the modular surface $\mathrm{SL}_2(\mathbb{Z}) \setminus \mathbb{H}$. In the meantime this correspondence between the continued fraction algorithm and the geodesic flow was studied by many authors (see e.g. Series [121]) and discussed in connection with our setting by Arnoux [10] and later by Arnoux and Fisher [14]. The necessary details on the modular surface and its unit tangent bundle including an explanation why the flow $\mathrm{diag}(e^t, e^{-t})$ which will come up below is a *geodesic flow* on the homogeneous space $\mathrm{SL}_2(\mathbb{Z}) \setminus \mathrm{SL}_2(\mathbb{R})$ can be found for instance in [10] or [76, Chapter 9].

We now explain briefly how the geodesic flow on $\mathrm{SL}_2(\mathbb{Z}) \setminus \mathrm{SL}_2(\mathbb{R})$ enters our model. We have to restack the rectangles as above and then renormalize the lattice again. This can be done also in the following way. First multiply the basis of the lattice from the right by $\mathrm{diag}(e^t, e^{-t})$ for t varying from 0 to the threshold value for which the width of the *smallest* rectangle equals 1. Then restack as above to end up at a pair of rectangles whose *larger* rectangle has width 1. Altogether, starting from a pair of rectangles drawn on the left hand side of Fig. 3.7 we ended up with a pair drawn on its right side. We just did the renormalization smoothly and we did it before the restacking instead of after it.

What we do can be explained more precisely as follows:

- Define the set

$$\Omega_{\mathrm{m}} = \Omega_{\mathrm{m},0} \cup \Omega_{\mathrm{m},1}$$

$$= \left\{ M = \begin{pmatrix} a & c \\ -b & d \end{pmatrix} : 0 < a < 1 \le c, \ 0 < d < b, \ ad + bc = 1 \right\} \cup$$

$$\left\{ M = \begin{pmatrix} a & c \\ -b & d \end{pmatrix} : 0 < c < 1 \le a, \ 0 < b < d, \ ad + bc = 1 \right\}.$$

One can show that a.e. lattice has exactly one basis made of row vectors of a matrix in Ω_{m} (see [10]). Thus Ω_{m} is a (measure theoretic) *fundamental domain* for the action of $SL_2(\mathbb{Z})$ on $SL_2(\mathbb{R})$.

- Start with a lattice, associate with it a basis taken from Ω_{m}.
- Hit this lattice (together with the chosen basis) with the geodesic flow $\mathrm{diag}(e^t, e^{-t})$, $t \ge 0$.
- For increasing t this will eventually deform the basis in a way that the width of the *smaller* rectangle gets equal to 1 (and we would leave Ω_{m} when deforming this basis further). If we restack at this point we end up with a pair of rectangles contained in the *Poincaré section* Δ_{m}: indeed, after restacking the *larger* rectangle will have width 1.
- Change the basis of the lattice to the basis corresponding to the new pair of rectangles according to Fig. 3.8. Note that restacking *does not change the lattice*, so the geodesic flow, which acts on $SL_2(\mathbb{Z}) \setminus SL_2(\mathbb{R})$, is not affected by this base change. However, this restacking has the effect that it creates a new basis of the lattice that remains inside Ω_{m} when it gets further deformed by the action of the flow. Thus we can repeat the procedure.
- Repeating this procedure, the geodesic flow yields a sequence of restackings: any time the width of the smaller rectangle gets equal to 1 by restacking, the according basis gets inside the *Poincaré section* Δ_{m}. This restacking performs one step of the natural extension of the Gauss map.
- Thus the geodesic flow on $SL_2(\mathbb{Z}) \setminus SL_2(\mathbb{R})$ can be regarded as a so-called *suspension flow* of the natural extension of the Gauss map.

This viewpoint has many advantages and one can prove results on continued fractions using the well-developed theory of the geodesic flow on $SL_2(\mathbb{Z}) \setminus SL_2(\mathbb{R})$.

The same procedure can also be performed for *pointed* pairs of rectangles (which we needed to study Sturmian sequences, see Fig. 3.9). This has the effect that the geodesic flow on $SL_2(\mathbb{Z}) \setminus SL_2(\mathbb{R})$ has to be replaced by the so-called *scenery flow* which also takes care of the distinguished point in the "L-shaped" region. All this is described in detail in [14].

We mention that similar results have been obtained for variants of the classical continued fraction algorithm. For instance, Arnoux and Schmidt [23, 24] proved that the α-continued fraction algorithm, Rosen's continued fraction algorithm as well as Veech's continued fraction algorithm can be viewed as Poincaré sections of a geodesic flow. The material presented in this section also forms an easy case of the wide and appealing field of interval exchange transformations and their dynamics (see e.g. Viana [125] for a survey).

3.3 Problems with the Generalization to Higher Dimensions

According to Cassaigne et al. [63] it was conjectured since the beginning of the 1990s that the beautiful correspondence between Sturmian sequences, continued fractions, and irrational rotations on the circle described in Sect. 3.2 can be extended to higher dimensions. The same paper gives strong indications towards the wrongness of this conjecture. Indeed, in [63] Arnoux-Rauzy sequences over a three letter alphabet that are not balanced and that cannot be viewed as natural codings of rotations on the two dimensional torus with finite fundamental domain are constructed. It is the objective of the present section to explain their work and to give an account on further results by Cassaigne et al. [62] concerning weakly mixing Arnoux-Rauzy systems as well as Arnoux-Rauzy systems with nontrivial eigenvalues.

3.3.1 Arnoux-Rauzy Sequences

In an attempt to pave the way for a generalization to higher dimensions of the correspondence between combinatorics, arithmetics, and dynamical systems outlined in Sect. 3.2, Arnoux and Rauzy [22] defined sequences over the alphabet $\{1, 2, 3\}$ whose properties are inspired by Sturmian sequences.

In the following definition a *right special factor* of a sequence $w \in \{1, 2, 3\}^{\mathbb{N}}$ is a factor v of w for which there are distinct letters $a, b \in \{1, 2, 3\}$ such that va and vb both occur in w. A *left special factor* is defined analogously. The definition of several other objects and notations from Sect. 3.2 carry over from two to three letter alphabets without any change and we will use them without defining them again (we will give exact definitions for the general setting from Sect. 3.4 onwards).

Definition 3.3.1 (Arnoux-Rauzy Sequence, see [22]) A sequence $w \in \{1, 2, 3\}^{\mathbb{N}}$ is called *Arnoux-Rauzy sequence* if $p_w(n) = 2n + 1$ and if w has only one right special factor and only one left special factor for each given length n.

Let w be an Arnoux-Rauzy sequence. Let (Γ_n) be a sequence of directed graphs defined in the following way. For each $n \in \mathbb{N}$ the vertices of Γ_n are the factors of length n of w. There is a directed edge from u to v if and only if there are letters

$a, b \in \{1, 2, 3\}$ and a word $x \in \{1, 2, 3\}^*$ such that $u = ax$ and $v = xb$. Inspecting these graphs we see that two cases can occur. If the left special factor v of length n is also the right special factor then Γ_n is a bouquet of three circles whose common vertex is v, otherwise it is a union of three circles that share the line between the vertices corresponding to the right and left special factor. An investigation of these graphs (as done in [22, Section 2]) shows that Arnoux-Rauzy sequences are "S-adic" and we get the following analog of Proposition 3.2.7.

Proposition 3.3.2 (See [22, Section 2]) *Let the* Arnoux-Rauzy *substitutions* $\sigma_1, \sigma_2, \sigma_3$ *be defined by*

$$\sigma_1 : \begin{cases} 1 \mapsto 1, \\ 2 \mapsto 12, \\ 3 \mapsto 13, \end{cases} \qquad \sigma_2 : \begin{cases} 1 \mapsto 21, \\ 2 \mapsto 2, \\ 3 \mapsto 23, \end{cases} \qquad \sigma_3 : \begin{cases} 1 \mapsto 31, \\ 2 \mapsto 32, \\ 3 \mapsto 3. \end{cases} \qquad (3.18)$$

Then for each Arnoux-Rauzy sequence w there exists a sequence $\sigma = (\sigma_{i_n})$, where (i_n) takes each symbol in $\{1, 2, 3\}$ an infinite number of times, such that w has the same language as

$$u - \lim_{n \to \infty} \sigma_{i_0} \circ \sigma_{i_1} \circ \cdots \circ \sigma_{i_n}(1). \qquad (3.19)$$

By this proposition each Arnoux-Rauzy sequence w has a *coding sequence σ* of Arnoux-Rauzy substitutions and we may define the dynamical system $(X_w, \Sigma) = (X_\sigma, \Sigma)$ as the dynamical system associated with w, where $X_w = X_\sigma$ is the set of sequences whose language equals the language of w and which just depends on σ. These dynamical systems are called *Arnoux-Rauzy systems*.

Let w be an Arnoux-Rauzy sequence with coding sequence $\sigma = (\sigma_{i_n})$ and let (M_{i_n}) be the associated sequence of incidence matrices. Since each symbol in $\{1, 2, 3\}$ occurs infinitely often in (i_n) the associated sequence of incidence matrices (M_{i_n}) is easily seen to be primitive in the sense that for each $m \in \mathbb{N}$ there is $n > m$ such that $M_{i_{[m,n)}}$ is a positive matrix. Indeed, a block $M_{i_{[m,n)}}$ is primitive if and only if it contains each of the three matrices M_1, M_2, M_3 at least once.

Lemma 3.3.3 *Let w be an Arnoux-Rauzy sequence with coding sequence σ. Then the dynamical system (X_σ, Σ) is minimal and uniquely ergodic.*

Proof Minimality follows if we can show that $L(v) = L(w)$ for each $v \in X_\sigma$. This in turn holds if each factor of w occurs infinitely often in w with bounded gaps, which we will now prove. Let x be a factor of w. As w has the same language as the sequence u in (3.19), by primitivity of (M_{i_n}) there is $m \in \mathbb{N}$ such that x occurs in $\sigma_{i_{[0,m)}}(1)$. Using primitivity again we see that there exists $n > m$ such that $M_{i_{[m,n)}}$ is a positive matrix. This entails that the word $\sigma_{i_{[m,n)}}(b)$ contains 1 for any $b \in \{1, 2, 3\}$ and, hence, $\sigma_{i_{[0,n)}}(b)$ contains $\sigma_{i_{[0,m)}}(1)$ and, a fortiori, also x for each $b \in \{1, 2, 3\}$. Thus x occurs in w infinitely often with gaps bounded by $2 \max\{|\sigma_{i_{[0,n)}}(b)| : b \in \{1, 2, 3\}\}$.

Unique ergodicity of (X_σ, Σ) can be derived from a general result of Boshernitzan [56] due to the fact that (X_σ, Σ) is minimal and its elements have linear complexity with slope less than 3.   □

This proof implies that each Arnoux-Rauzy sequence is *uniformly recurrent*.

Generalizing an idea of Arnoux [9], in [22] it was shown that each Arnoux-Rauzy sequence w can be viewed as a coding of a 6-interval exchange transformation (by using sequences over an alphabet with only three letters!) and that each Arnoux-Rauzy system can be represented by such a 6-interval exchange. In view of a result by Katok [94] this implies that Arnoux-Rauzy systems cannot be mixing. The incidence matrices of Arnoux-Rauzy substitutions can be used to define a generalized continued fraction algorithm in the sense of Sect. 3.4.2 below. However, this algorithm only works for vectors taken from a set of measure zero, the so-called *Rauzy gasket*. For more on this interesting set we refer to [25, 29, 30, 69, 99].

Another interesting class of sequences of complexity $2n + 1$ over the alphabet $\{1, 2, 3\}$ has been defined recently in [64] and is currently subject to intensive investigation. Compared to Arnoux-Rauzy sequences it has the advantage that it is defined in terms of only two substitutions and gives rise to a continued fraction algorithm that works on a set of full measure.

3.3.2   Imbalanced Arnoux-Rauzy Sequences

To get the perfect analogy with the Sturmian case it would be desirable to represent a given Arnoux-Rauzy sequence w as a natural coding of a rotation on the two-dimensional torus $\mathbb{T}^2$. In the seminal paper of Rauzy [114], this was achieved for the sequence $w = \lim \sigma^n(1)$, where σ is the famous *Tribonacci substitution* defined by

$$\sigma : \begin{cases} 1 \mapsto 12, \\ 2 \mapsto 13, \\ 3 \mapsto 1. \end{cases} \tag{3.20}$$

Since $\sigma^3 = \sigma_1 \circ \sigma_2 \circ \sigma_3$ the sequence w is an example of an Arnoux-Rauzy sequence (with periodic coding sequence). Several years ago Barge, Štimac, and Williams [37] as well as Berthé et al. [48] could generalize this result and proved that each Arnoux-Rauzy sequence w with periodic coding sequence is a natural coding of a rotation on $\mathbb{T}^2$ (a weaker result in this direction is already contained in [18]). A general theory for nonperiodic sequences was established only recently, see Berthé et al. [52], and we will come back to this in later sections.

We recall that a sequence $w = w_0 w_1 \ldots \in \{1, 2, 3\}^{\mathbb{N}}$ is a *natural coding* of a rotation R on $\mathbb{T}^2$ if there exists a fundamental domain Ω of $\mathbb{T}^2$ in $\mathbb{R}^2$ together with a partition $\Omega = \Omega_1 \cup \Omega_2 \cup \Omega_3$ such that on each Ω_i the map R' induced on Ω by the rotation R acts as a translation by a vector $\mathbf{a}_i \in \mathbb{R}^2$ and for some point $x \in \Omega$ we have $R'^k(x) \in \Omega_{w_k}$ for each $k \in \mathbb{N}$ (see also Definition 3.9.2).

An Arnoux-Rauzy sequence is not always a coding of a rotation on $\mathbb{T}^2$ with *bounded* fundamental domain. The reason for this is the lack of balance for some particular instances of such sequences. Following Cassaigne et al. [63] we now sketch the construction of an Arnoux-Rauzy sequence that is not balanced.

Let $C \geq 1$ be an integer. Generalizing the notion of balance from Sect. 3.2.1 we say that a sequence $w \in \{1, 2, 3\}^{\mathbb{N}}$ is *C-balanced* if each pair of factors (u, v) of w having the same length satisfies $\left| |u|_a - |v|_a \right| \leq C$ for each $a \in \{1, 2, 3\}$. The following result implies that there is no uniform C that gives C-balance for each Arnoux-Rauzy sequence. Here a substitution is called primitive if its incidence matrix is primitive.

Lemma 3.3.4 (See [63, Proposition 2.2]) *For each integer $C \geq 1$ there is a finite sequence of Arnoux-Rauzy substitutions $\sigma_{i_1}, \ldots, \sigma_{i_k}$ such that $\sigma = \sigma_{i_1} \circ \cdots \circ \sigma_{i_k}$ is primitive and for each Arnoux-Rauzy sequence w the Arnoux-Rauzy sequence $\sigma(w)$ is not C-balanced.*

Proof We prove by induction that for each $n \geq 2$ there exist $a_n, b_n, c_n \in \mathbb{N}$ and a primitive composition of Arnoux-Rauzy matrices $\sigma^{(n)}$ such that for each Arnoux-Rauzy sequence w the sequence $\sigma^{(n)}(w)$ contains two factors $u^{(n)}$ and $v^{(n)}$ of equal length with

$$\begin{pmatrix} |u^{(n)}|_i \\ |u^{(n)}|_j \\ |u^{(n)}|_k \end{pmatrix} = \begin{pmatrix} a_n \\ b_n + n \\ c_n \end{pmatrix} \quad \text{and} \quad \begin{pmatrix} |v^{(n)}|_i \\ |v^{(n)}|_j \\ |v^{(n)}|_k \end{pmatrix} = \begin{pmatrix} a_n + 1 \\ b_n \\ c_n + n - 1 \end{pmatrix}$$

for some choice i, j, k with $\{i, j, k\} = \{1, 2, 3\}$. This will prove the result because $\left| |u^{(n)}|_j - |v^{(n)}|_j \right| = n$ shows that $\sigma^{(n)}(w)$ is not $(n - 1)$-balanced.

For the induction start take $n = 2$ and $\sigma^{(2)} = \sigma_1 \sigma_2$ with $u^{(2)} = 212$ and $v^{(2)} = 131$.

To perform the induction step assume that the result is true for some n and let $u^{(n)}, v^{(n)}, a_n, b_n, c_n, i, j, k$, and $\sigma^{(n)}$ be as above. Set $\sigma^{(n+1)} = \sigma_k^n \circ \sigma_i^n \circ \sigma^{(n)}$. We now construct $u^{(n+1)}$ and $v^{(n+1)}$. Let u be a nonempty factor of some Arnoux-Rauzy sequence w. Then for each $a \in \{1, 2, 3\}$ the word $\sigma_a(u)a$ is a factor of $\sigma_a(w)$ which begins with a. If we define $\sigma_{(a,+)}(u) = \sigma_a(u)a$ and $\sigma_{(a,-)}(u)$ as the suffix of $\sigma_a(u)$ of length $|\sigma_a(u)| - 1$ (i.e., the first letter of $\sigma_a(u)$ is canceled) we see that $u^{(n+1)} = \sigma_{(k,-)}^n \sigma_{(i,+)}^n(v_n)$ and $v^{(n+1)} = \sigma_{(k,+)}^n \sigma_{(i,-)}^n(v_n)$ are factors of $\sigma^{(n+1)}(w)$. Using the definition of $\sigma_{(a,+)}$ and $\sigma_{(a,-)}$ one can now check directly that

$$\begin{pmatrix} |u^{(n+1)}|_k \\ |u^{(n+1)}|_i \\ |u^{(n+1)}|_j \end{pmatrix} = \begin{pmatrix} a_{n+1} \\ b_{n+1} + n + 1 \\ c_{n+1} \end{pmatrix} \quad \text{and} \quad \begin{pmatrix} |v^{(n+1)}|_k \\ |v^{(n+1)}|_i \\ |v^{(n+1)}|_j \end{pmatrix} = \begin{pmatrix} a_{n+1} + 1 \\ b_{n+1} \\ c_{n+1} + n \end{pmatrix},$$

where

$$\begin{pmatrix} a_{n+1} \\ b_{n+1} \\ c_{n+1} \end{pmatrix} = \begin{pmatrix} c_n + n - 1 + n(a_n + n(b_n + c_n + n) + b_n) \\ a_n + n(b_n + c_n + n - 1) \\ b_n \end{pmatrix}. \qquad \square$$

This lemma can even be sharpened in the following way.

Lemma 3.3.5 (See [63, Proposition 2.3]) *For each integer $C \geq 1$ and each composition of Arnoux-Rauzy substitutions σ there exists a primitive composition of Arnoux-Rauzy sequences σ' such that for each Arnoux-Rauzy sequence w the Arnoux-Rauzy sequence $\sigma \circ \sigma'(w)$ is not C-balanced.*

The proof is technical and we do not provide it here. The idea is to use Lemma 3.3.4 in order to choose σ' in a way that $\sigma'(w)$ is not K-balanced for each Arnoux-Rauzy sequence w, where K, which depends on the incidence matrix of σ, is so large that even after the application of σ we cannot reach C-balance.

We are now able to establish the following result.

Theorem 3.3.6 (See [63, Theorem 2.4]) *There exists an Arnoux-Rauzy sequence which is not C-balanced for any $C \geq 1$.*

Proof By Lemma 3.3.5 one can construct primitive compositions of Arnoux-Rauzy substitutions $\sigma^{(1)}, \ldots, \sigma^{(C)}$ such that $\sigma^{(1)} \circ \cdots \circ \sigma^{(C)}(w)$ is not C-balanced for any Arnoux-Rauzy sequence w. Thus $u = \lim_{C \to \infty} \sigma^{(1)} \circ \cdots \circ \sigma^{(C)}(w)$ is the desired sequence. $\qquad \square$

Using this proposition we are able to establish the following result of [63] which strongly indicates that an unconditional generalization of the theory presented in Sect. 3.2 is not possible.

Corollary 3.3.7 (Cf. [63, Corollary 2.6]) *There exists an Arnoux-Rauzy sequence which is not a natural coding of a minimal rotation on the 2-torus with bounded fundamental domain.*

Proof By Theorem 3.3.6 there is an Arnoux-Rauzy sequence w which is not C-balanced for any $C > 0$. Assume that w is a natural coding of a minimal rotation on $\mathbb{T}^2$ with bounded fundamental domain Ω. Each letter $j \in \{1, 2, 3\}$ corresponds to a translation $\mathbf{a}_j$ on Ω and, hence, to each word $u = u_0 \ldots u_{n-1} \in \{1, 2, 3\}^*$ there corresponds the translation $\mathbf{a}_u = \sum_{k=0}^{n-1} \mathbf{a}_{u_k}$ on Ω. Since Ω is bounded and the rotation is minimal one easily checks that the vectors $\mathbf{a}_1, \mathbf{a}_2, \mathbf{a}_3$ satisfy $\mathbb{R}_+ \mathbf{a}_1 + \mathbb{R}_+ \mathbf{a}_2 + \mathbb{R}_+ \mathbf{a}_3 = \mathbb{R}^2$. This implies that there exists a constant $\gamma > 0$ such that two words $u, v \in \{1, 2, 3\}^*$ with $\big||u|_i - |v|_i\big| \geq C$ for some $i \in \{1, 2, 3\}$ satisfy $\|\mathbf{a}_u - \mathbf{a}_v\|_1 > \gamma C$.

Since w is not balanced there is a letter $i \in \{1, 2, 3\}$ such that for each $C > 0$ there exist two factors $u, v \in \{1, 2, 3\}^*$ of w with $\big||u|_i - |v|_i\big| \geq C$. Thus $\|\mathbf{a}_u - \mathbf{a}_v\|_1 > \gamma C$. Since C can be arbitrarily large, this difference can be made arbitrarily large. Thus one of the two vectors $\mathbf{a}_u, \mathbf{a}_v$ can be made arbitrarily large. Assume w.l.o.g. that this is $\mathbf{a}_u$. Since there is an element $\mathbf{x} \in \Omega$ with $\mathbf{x} + \mathbf{a}_u \in \Omega$, the diameter of Ω is bounded from below by the length of $\mathbf{a}_u$. This contradicts the boundedness of the fundamental domain Ω.   □

Remark 3.3.8 We mention that in [63, Corollary 2.6] it is claimed that Corollary 3.3.7 is true without assuming that the fundamental domain is bounded. However, we were not able to verify this proof.

3.3.3 Weak Mixing and the Existence of Eigenvalues

In Cassaigne et al. [62] the authors give a criterion for *weak mixing* for some class of Arnoux-Rauzy systems. On the other hand they provide a class of Arnoux-Rauzy systems that admit nontrivial *eigenvalues*. Before we give the details, we recall the required terminology from ergodic theory (good references here are for instance Einsiedler and Ward [76] or Walters [126]; we also mention Halmos [85] where some concepts are illustrated in an intuitive way).

Let (X, T, μ) be a dynamical system with invariant measure μ. We say that a complex number λ is a *measurable eigenvalue* of T if there exists $f \in L^1(\mu)$, $f \neq 0$, such that $f(Tx) = \lambda f(x)$ for μ-almost every x. Such an f is called an *eigenfunction* for λ. For topological dynamical systems the notion of *topological eigenvalue* is defined analogously by using continuous eigenfunctions instead of functions from $L^1(\mu)$.

The transformation T is called *weakly mixing* if for each $A, B \subset X$ of positive measure we have

$$\lim_{n \to \infty} \frac{1}{n} \sum_{0 \le k < n} |\mu(T^{-k}(A) \cap B) - \mu(A)\mu(B)| = 0.$$

Weak mixing is equivalent to the fact that 1 is the only measurable eigenvalue of T and the only eigenfunctions are constants (in this case the dynamical system is said to have *continuous spectrum*). We note that rotations are never weakly mixing. They have *pure discrete spectrum* (with will be defined in Definition 3.9.1), meaning that they have "a lot of eigenfunctions" and therefore they have a completely different dynamical behavior. Indeed, from the definition of weak mixing we see that iterated preimages of each set tend to "smear" (or *mix*) over the whole space, this is of course not the case for the iterated preimages of a rotation.

We now come back to the aim of this section and discuss mixing properties of Arnoux-Rauzy systems. Let

$$u = \lim_{n \to \infty} \sigma_{i_1}^{k_1} \circ \sigma_{i_2}^{k_2} \circ \cdots \circ \sigma_{i_n}^{k_n}(1)$$

with $i_n \neq i_{n+1}$ be an Arnoux-Rauzy sequence. We define (n_ℓ) to be the sequence of indices n for which $i_n \neq i_{n+2}$. The sequence u is uniquely defined by the sequences (k_n) and (n_ℓ) (up to permutation of letters). The following result shows a result on weak mixing Arnoux-Rauzy systems for large partial quotients (k_n).

Theorem 3.3.9 (See [62, Theorem 2]) *For an Arnoux-Rauzy sequence w with coding sequence σ and associated sequences (k_n) and (n_ℓ) the system (X_σ, Σ, μ) (with μ being the unique invariant measure) is weakly mixing if the sequence $(k_{n_\ell+2})_{\ell \in \mathbb{N}}$ is unbounded and the sums $\sum_{\ell \geq 1} \frac{1}{k_{n_\ell+1}}$ and $\sum_{\ell \geq 1} \frac{1}{k_{n_\ell}}$ converge.*

This implies that (X_σ, Σ, μ) is not measurably conjugate to a rotation on $\mathbb{T}^2$.

The proof of this result is quite involved. In fact, to get weak mixing, by definition one has to show that there exists no measurable eigenvalue apart from 1 for the system (X_σ, Σ, μ). This is achieved by verifying the following criterion (see [62, Proposition 10]): if ϑ is a measurable eigenvalue of (X_σ, Σ, μ), then $k_{n+1}\{h_n\vartheta\} \to 0$ for $n \to \infty$. Here h_n is the length of $\sigma_{i_1}^{k_1} \circ \cdots \circ \sigma_{i_n}^{k_n}(1)$. This criterion is proved using a sequence of nested Rohlin towers which are naturally built using the coding sequence σ. As mentioned above, because an Arnoux-Rauzy system can be represented by a 6-interval exchange, it cannot be mixing in view of Katok [94].

To give this section a good end we mention that [62] also contains results that support the hope that at least something along the lines of Sect. 3.2 can be done in higher dimensions. Indeed, the authors are able to exhibit criteria for the existence of nontrivial continuous eigenvalues (not equal to 1) for Arnoux-Rauzy systems which implies that these systems have a rotation as a continuous factor. The novelty here is the fact that these systems still have unbounded partial quotients (k_n). For bounded partial quotients criteria for the existence of continuous and measurable eigenvalues are provided in the more general setting of linear recurrent minimal Cantor systems in Cortez et al. [66].

It will be our concern in the subsequent sections to exhibit S-adic sequences that are even measurable conjugates of rotations on tori of dimension greater than or equal to two.

3.4   The General Setting

So far we have seen some elements of the correspondence between Sturmian sequences, the classical continued fraction algorithm, and rotations on the circle. We have also reviewed some results that highlight the problems and limitations of a generalization of this nice interplay between several branches of mathematics to higher dimensions. Nevertheless, we are able to set up a quite general extension of the results contained in Sect. 3.2. Indeed, in the subsequent sections of this chapter we will relate sequences generated by substitutions on alphabets over d letters to generalized continued fraction algorithms and to rotations on the $(d - 1)$-dimensional torus. From this point on we will give exact definitions of all objects we use. This may seem redundant as some objects have already been introduced before but as the subject is quite difficult and a variety of concepts and notations is needed along the way we found it better for the reader to do it that way.

3.4.1 S-adic Sequences

We now define so-called *S-adic sequences* which form analogs of sequences of the form (3.2) and (3.19) for arbitrary "coding sequences" of substitutions over a fixed finite alphabet. To this end we need some notation.

Let $\mathcal{A} = \{1, 2, \ldots, d\}$ be a finite *alphabet* whose elements will be called *letters* or *symbols*. Define $\mathcal{A}^*$ to be the free monoid generated by $\mathcal{A}$ equipped with the operation of concatenation. The elements of $\mathcal{A}^*$, which are of the form $v = v_0 v_1 \ldots v_{n-1}$ with $n \in \mathbb{N}$ and $v_i \in \mathcal{A}$ for $i \in \{0, 1, \ldots, n-1\}$, will be referred to as *words*. The integer n, which is equal to the number of letters in the word v, is called the *length* of v and will be denoted by $|v|$. The unique word of length 0 is called the *empty word*. Let $\mathcal{A}^{\mathbb{N}}$ be the space of *right infinite sequences* $w = w_0 w_1 \ldots$ with $w_i \in \mathcal{A}$ for each $i \in \mathbb{N}$. We equip $\mathcal{A}^{\mathbb{N}}$ with the product topology of the discrete topology on $\mathcal{A}$. To a sequence $w = w_0 w_1 \ldots \in \mathcal{A}^{\mathbb{N}}$ we associate a function $p_w : \mathbb{N} \to \mathbb{N}$ which is defined by

$$n \mapsto |\{v \in \mathcal{A}^* : v = w_k w_{k+1} \ldots w_{k+n-1} \text{ for some } k \in \mathbb{N}\}|.$$

The function p_w is called the *complexity function* of the sequence w. For more on this function we refer for instance to Cassaigne and Nicolas [65].

A *substitution* σ over the alphabet $\mathcal{A}$ is an endomorphism on $\mathcal{A}^*$ that in our setting will always assumed to be *nonerasing* in the sense that the image of each letter is a nonempty word taken from $\mathcal{A}^*$. Being a morphism, a substitution is completely defined by giving its image for each letter. Thus our previous examples of Sturmian substitutions in (3.1) and of Arnoux-Rauzy substitutions in (3.18) are indeed substitutions. We can extend the domain of a substitution σ to $\mathcal{A}^{\mathbb{N}}$ in a natural way by defining it symbol-wise, i.e., by setting $\sigma(w_0 w_1 \ldots) = \sigma(w_0)\sigma(w_1) \ldots$ The mapping σ defined in this way is continuous on $\mathcal{A}^{\mathbb{N}}$.

With each substitution σ over the alphabet $\mathcal{A}$ we associate the $|\mathcal{A}| \times |\mathcal{A}|$ *incidence matrix* M_σ whose columns are the abelianized images of $\sigma(i)$ for $i \in \mathcal{A}$. More precisely, letting $|v|_i$ be the number of occurrences of a given letter $i \in \mathcal{A}$ in a word $v \in \mathcal{A}^*$ this matrix is given by $M_\sigma = (m_{ij}) = (|\sigma(j)|_i)$. The incidence matrix can be seen as the *abelianized* version of σ. If we define the *abelianization mapping* $\mathbf{l} : \mathcal{A}^* \to \mathbb{N}^d$ by $\mathbf{l}(w) = (|w|_1, \ldots, |w|_d)^t$ (here $\mathbf{x}^t$ is the transpose of a vector $\mathbf{x} \in \mathbb{R}^d$) we have the commutative diagram

$$
\begin{array}{ccc}
\mathcal{A}^* & \xrightarrow{\;\sigma\;} & \mathcal{A}^* \\
\downarrow{\scriptstyle \mathbf{l}} & & \downarrow{\scriptstyle \mathbf{l}} \\
\mathbb{N}^d & \xrightarrow{\;M_\sigma\;} & \mathbb{N}^d
\end{array}
\tag{3.21}
$$

which says that $\mathbf{l}\sigma(w) = M_\sigma \mathbf{l}(w)$ holds for each $w \in \mathcal{A}^*$.

We will be interested in special classes of substitutions. Let σ be a substitution. Then σ is called *unimodular* if $|\det M_\sigma| = 1$, it is called *primitive* if M_σ is a primitive matrix (i.e., M_σ has a power each of whose entries is greater than zero), it is called *irreducible* if M_σ has irreducible characteristic polynomial, and it is called *Pisot* if the characteristic polynomial of M_σ is the minimal polynomial of a *Pisot number*. We recall that a Pisot number is an algebraic integer $\beta > 1$ whose Galois conjugates (apart from β itself) are all smaller than 1 in modulus.

In full generality substitutions are studied for instance in [6, 40, 87] and, in a context related to the present chapter, in [82].

We will now define the analogs of the "coding sequences" used in Sects. 3.2 and 3.3 for a more general setting. We go in the reverse direction: in the mentioned earlier sections the sequence (of letters) was there first and we constructed a sequence of substitutions that generates this sequence. Now we start with a sequence of substitutions in order to define a sequence of letters.

Let $\sigma = (\sigma_n)_{n \in \mathbb{N}}$ be a sequence of substitutions over a given finite alphabet $\mathcal{A}$. For convenience, we will set $M_n = M_{\sigma_n}$ for the incidence matrix of σ_n and write $\mathbf{M} = (M_n)$ for the sequence of these incidence matrices. Moreover, as we will often need blocks of substitutions as well as blocks of matrices we set

$$\sigma_{[m,n)} = \sigma_m \circ \sigma_{m+1} \circ \cdots \circ \sigma_{n-1} \quad \text{and} \quad M_{[m,n)} = M_m M_{m+1} \cdots M_{n-1}$$

for positive integers $m \leq n$ (here we set $\sigma_{[n,n)}(a) = a$ for all $a \in \mathcal{A}$ and define $M_{[n,n)}$ to be the $|\mathcal{A}| \times |\mathcal{A}|$ identity matrix).

We associate with σ a sequence of languages, for all $m \in \mathbb{N}$,

$$\mathcal{L}_\sigma^{(m)} = \{v \in \mathcal{A}^* : v \text{ is a factor of } \sigma_{[m,n)}(a) \text{ for some } a \in \mathcal{A}, \, m \leq n\}$$

and call $\mathcal{L}_\sigma = \mathcal{L}_\sigma^{(0)}$ the *language of* σ. Here $u \in \mathcal{A}^*$ is a *factor* of $v \in \mathcal{A}^*$ if $v \in \mathcal{A}^* u \mathcal{A}^*$, or, more informally, if the word u occurs somewhere as *subword* in the word v. We will use this notation also for (right infinite) sequences later. Then $u \in \mathcal{A}^*$ is a *factor* of $v \in \mathcal{A}^{\mathbb{N}}$ if $v \in \mathcal{A}^* u \mathcal{A}^{\mathbb{N}}$. The set of all factors of a sequence v is called the *language* of v. It is denoted by $L(v)$. We also introduce the notion of prefix and suffix that will be used later. A *prefix* of a word $v \in \mathcal{A}^*$ is a word $u \in \mathcal{A}^*$ with $v \in u\mathcal{A}^*$ and a *suffix* of $v \in \mathcal{A}^*$ is a word $u \in \mathcal{A}^*$ with $v \in \mathcal{A}^* u$. A prefix of a sequence $v \in \mathcal{A}^{\mathbb{N}}$ is a word $u \in \mathcal{A}^*$ with $v \in u \mathcal{A}^{\mathbb{N}}$.

After these preparations we can define *S-adic sequences* for a given sequence of substitutions σ. The terminology "*S-adic*" goes back to Ferenczi [77]. In our definition we follow Arnoux et al. [20] (see also [52, Section 2.2]).

Definition 3.4.1 (*S-adic* Sequence) Let $\mathcal{A}$ be a given finite alphabet, let $\sigma = (\sigma_n)_{n \geq 0}$ be a sequence of substitutions over $\mathcal{A}$, and set $S := \{\sigma_n : n \in \mathbb{N}\}$. We call a sequence $w \in \mathcal{A}^{\mathbb{N}}$ an *S-adic sequence* (or a *limit sequence*) for σ if there exists a sequence $(w^{(n)})_{n \geq 0}$ of sequences $w^{(n)} \in \mathcal{A}^{\mathbb{N}}$ with

$$w^{(0)} = w, \quad w^{(n)} = \sigma_n(w^{(n+1)}) \quad \text{(for all } n \in \mathbb{N}). \tag{3.22}$$

In this case we call σ the *coding sequence* or the *directive sequence* for w. (Note that (3.22) says that w can be "desubstituted" infinitely often).

Let S be a finite set of substitutions over a given alphabet $\mathcal{A}$. For this case S-adic sequences have been thoroughly studied in the literature. With Sturmian sequences and Arnoux-Rauzy sequences we already discussed two prominent classes of S-adic sequences. Durand [71, 72] proved that linearly recurrent[3] sequences are S-adic with finite S. Ferenczi [77] and Leroy [97] showed that a uniformly recurrent[4] sequence w with an at most linear complexity function p_w is S-adic with finite S; see also [98]. The so-called *S-adic conjecture* (see e.g. [82, Section 12.1.2] or [73, 97]) is also formulated for a finite set of substitutions S. It asks to what extent a converse of this assertion can be true, i.e., which criteria are needed for an S-adic sequence w to have linear complexity function p_w. Berthé and Labbé [49] show linearity of the complexity of S-adic sequences associated with the Arnoux-Rauzy-Poincaré multidimensional continued fraction algorithm (their bound $p_w(n) \leq \frac{5}{2}n + 1$ is even strong enough to conclude from Boshernitzan [56] that, like Arnoux-Rauzy sequences, these sequences pertain to uniquely ergodic dynamical systems). Arnoux et al. [20] study S-adic sequences in the same context as we will do it. However, they restrict their attention to sets of substitutions S whose elements have a common incidence matrix. If S is a singleton, an S-adic sequence is called *substitutive*. Substitutive sequences are very well studied (see for instance [82]; moreover in the paragraphs following Definition 3.4.2 we review the literature on substitutive sequences related to our subject). They are strongly related to automatic sequences by *Cobham's Theorem*, see e.g. [6, Theorem 6.3.2].

Generalizing Sturmian systems we introduce dynamical systems for S-adic sequences. To this end, for a finite alphabet $\mathcal{A}$ define the *shift* on $\mathcal{A}^{\mathbb{N}}$ as $\Sigma : \mathcal{A}^{\mathbb{N}} \to \mathcal{A}^{\mathbb{N}}$ by $\Sigma(w_0 w_1 \ldots) = w_1 w_2 \ldots$

Definition 3.4.2 (*S*-adic System)   For an S-adic sequence w over a finite alphabet $\mathcal{A}$ we denote by $X_w = \overline{\{\Sigma^k w : k \in \mathbb{N}\}}$ the orbit closure of w under the action of the shift Σ. If we denote the restriction of Σ to X_w by Σ again we call the pair (X_w, Σ) the *S-adic system* (or *S-adic shift*) generated by w.

Alternatively, the set X_w can be defined using languages by setting $X_w = \{v \in \mathcal{A}^{\mathbb{N}} : L(v) \subseteq L(w)\}$. The proof of the fact that both definitions of X_w agree is an easy exercise. Also the set $X_\sigma = \bigcup X_w$, where the union is extended over all S-adic sequences with directive sequence σ, and the associated dynamical system (X_σ, Σ) are of interest.[5] A recent survey on S-adic systems is provided in [44].

---

[3]A sequence is called *linearly recurrent* if there is a constant K such that each of its factors u occurs infinitely often in the sequence with gaps bounded by $K|u|$.

[4]A sequence is called *uniformly recurrent* if each of its factors occurs infinitely often in the sequence with bounded gaps.

[5]If we impose the additional property of *primitivity* on the coding sequence of a sequence $w \in \mathcal{A}^{\mathbb{N}}$ it turns out that X_w depends only on the directive sequence σ defining the S-adic sequence w and we have $X_\sigma = X_w$. This will be worked out precisely in Sect. 3.5.

In all what follows we will assume that all our substitutions and matrices are unimodular.

The case of $\boldsymbol{\sigma} = (\sigma)$, the constant sequence formed by a given unimodular substitution σ over some alphabet $\mathcal{A}$, has been studied extensively. In this case we call $(X_{(\sigma)}, \Sigma)$ a *substitutive system* (see Queffelec [110] for a profound study of dynamical properties of these systems). The theory of Sect. 3.2 can be generalized quite well to substitutive systems if σ is a unimodular Pisot substitution. The seed for such a generalization was planted by Rauzy [114]. Constructing the prototype of what is now called *Rauzy fractal*, he proved that the dynamical system $(X_{\boldsymbol{\sigma}}, \Sigma)$ is measurably conjugate to a rotation on $\mathbb{T}^2$ if $\boldsymbol{\sigma} = (\sigma)$ with σ being the Tribonacci substitution introduced in (3.20). It was conjectured since then that each unimodular Pisot substitution σ gives rise to a substitutive system $(X_{(\sigma)}, \Sigma)$ which is measurably conjugate to a rotation on the torus. This conjecture is still open and known as *Pisot (substitution) conjecture*.

In the meantime, the Pisot conjecture was studied by many people and interesting partial results have been achieved. We mention Arnoux and Ito [18] as well as Ito and Rao [91] who could prove the Pisot conjecture subject to some combinatorial *coincidence conditions*. Conditions of this type will also play an important role in the general theory we will develop here, see Sect. 3.8.2. Recently, Barge [32, 33] made considerable progress on this subject using refinements of the notion of *proximality* (see [27, 35]). For survey papers on the subject we refer e.g. to [4, 51]. For extensions of this theory to the nonunimodular case see [106, 122].

3.4.2   Generalized Continued Fraction Algorithms

We now generalize the concept of continued fraction algorithm defined in Sect. 3.2.2 and introduce *generalized continued fraction algorithms*. Standard references for these objects are Brentjes [58] and Schweiger [119]. Also Labbé's *Cheat Sheets* [95] for 3-dimensional continued fraction algorithms are highly recommended. For discussions of generalized continued fraction algorithms in a context similar to ours we refer e.g. to [11, 19, 21, 41].

Definition 3.4.3 (Generalized Continued Fraction Algorithm)   For $d \geq 2$ let X be a closed subset of the projective space $\mathbb{P}^{d-1}$ and let $\{X_i\}_{i \in I}$ be a partition of X (up to a set of measure 0) indexed by a countable set I. Let $\mathcal{M} = \{M_i : i \in I\}$ be a set of unimodular $d \times d$ integer matrices (that act on $\mathbb{P}^{d-1}$ by homogeneity) satisfying $M_i^{-1} X_i \subset X$ and let $M : X \to \mathcal{M}$ given by $M(\mathbf{x}) = M_i$ whenever $\mathbf{x} \in X_i$. The *generalized continued fraction algorithm* associated with this data is given by the mapping

$$F : X \to X; \quad \mathbf{x} \mapsto M(\mathbf{x})^{-1}\mathbf{x}.$$

If I is a finite set, the algorithm given by F is called *additive*, otherwise it is called *multiplicative*.

Note that F is defined only almost everywhere since $\{X_i\}_{i \in I}$ in general is only a partition up to measure zero. We confine ourselves to unimodular matrices. Thus the algorithms in Definition 3.4.3 are sometimes called *unimodular algorithms*. Interesting examples of nonunimodular continued fraction algorithms are provided by the *N-continued fraction algorithm* introduced by Burger et al. [60] and by the *Reverse algorithm*, a certain "completion" of the Arnoux-Rauzy algorithm studied in [19, Section 4].

We illustrate the definition of generalized continued fraction algorithms by a classical example: *Brun's continued fraction algorithm*.

Example 3.4.4 (Brun's Algorithm)   The linear version of Brun's algorithm is defined on the subset

$$X = \{[w_1 : w_2 : w_3] \ : \ 0 \le w_1 \le w_2 \le w_3\} \subset \mathbb{P}^2.$$

It maps a vector $[w_1 : w_2 : w_3]$ to sort$[w_1 : w_2 : w_3 - w_2]$, i.e., it subtracts the second largest entry from the largest one and sorts the resulting entries in ascending order. By a straightforward calculation we see that $\mathcal{M} = \{M_1, M_2, M_3\}$ with

$$M_1 = \begin{pmatrix} 0 & 1 & 0 \\ 0 & 0 & 1 \\ 1 & 0 & 1 \end{pmatrix}, \quad M_2 = \begin{pmatrix} 1 & 0 & 0 \\ 0 & 0 & 1 \\ 0 & 1 & 1 \end{pmatrix}, \quad M_3 = \begin{pmatrix} 1 & 0 & 0 \\ 0 & 1 & 0 \\ 0 & 1 & 1 \end{pmatrix}, \tag{3.23}$$

and that the partition $X = X_1 \cup X_2 \cup X_3$ is given by Fig. 3.10.

Fig. 3.10 The partition of X induced by Brun's continued fraction algorithm

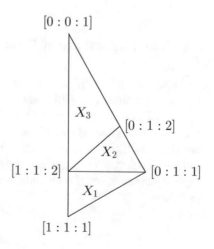

With this data the linear Brun continued fraction mapping can be defined according to Definition 3.4.3 by

$$F_B : X \to X; \quad \mathbf{x} \mapsto M_i^{-1}\mathbf{x} \quad \text{for } \mathbf{x} \in X_i.$$

Since M_1, M_2, and M_3 are unimodular, Brun's algorithm is a unimodular continued fraction algorithm. As we did for the classical continued fraction algorithm in Sect. 3.2.2, we can define a projective version also in the case of Brun's algorithm. This projective version is the original version of this algorithm and goes back to Brun [59]. It is defined on the set

$$\Delta = \{(x_1, x_2) \in \mathbb{R}^2 : 0 \le x_1 \le x_2 \le 1\} \tag{3.24}$$

by

$$f_B : (x_1, x_2) \mapsto \begin{cases} \left(\frac{x_1}{1-x_2}, \frac{x_2}{1-x_2}\right), & \text{for } x_2 \le \frac{1}{2}, \\ \left(\frac{x_1}{x_2}, \frac{1-x_2}{x_2}\right), & \text{for } \frac{1}{2} \le x_2 \le 1 - x_1, \\ \left(\frac{1-x_2}{x_2}, \frac{x_1}{x_2}\right), & \text{for } 1 - x_1 \le x_2. \end{cases} \tag{3.25}$$

To see that f_B is the projective version of F_B we use the same reasoning as in the classical case in Sect. 3.2.2.

We refer to Example 3.5.12 where we provide S-adic sequences associated with Brun's algorithm.

Other well-known generalized continued fraction algorithms include the Jacobi-Perron algorithm [109] and the Selmer algorithm [120].

3.5 The Importance of Primitivity and Recurrence

As indicated in Sect. 3.3 it is not possible to generalize the results of Sect. 3.2 to higher dimensions (or, equivalently, to alphabets of cardinality greater than two) without additional conditions on the sequence of substitutions σ. In this section we will discuss two natural conditions that we will have to impose on our sequences of substitutions. The first one is *primitivity*, the second one is *recurrence*. Both of them will have important consequences for the underlying S-adic system: primitivity will imply minimality, and if we assume recurrence on top of primitivity, the system will be uniquely ergodic.

3.5.1 Primitivity and Minimality

In the following definition a matrix is called *nonnegative* if each of its entries is greater than or equal to zero. In a *positive matrix* each entry is greater than zero.

Definition 3.5.1 (Primitivity) A sequence $\mathbf{M} = (M_n)_{n \geq 0}$ of nonnegative integer matrices is *primitive* if for each $m \in \mathbb{N}$ there is $n > m$ such that $M_{[m,n)}$ is a positive matrix. A sequence σ of substitutions is *primitive* if its associated sequence of incidence matrices is primitive.

Note that primitivity of $(M_n)_{n \geq 0}$ implies primitivity of the "shifted" sequence $(M_{n+k})_{n \geq 0}$ for each $k \in \mathbb{N}$. The same applies for primitive sequences of substitutions.

Our definition of primitivity is taken from [52, Section 2.2]. It coincides with the notion of weak primitivity introduced in [44, Definition 5.1] and with the notion of nonstationary primitivity defined in [81, p. 339]. The more restrictive property of strong primitivity which is also introduced in [44, Definition 5.1] requires that the integer n in Definition 3.5.1 can be chosen in a way that the difference $n - m$ is uniformly bounded in m. In other papers, this stronger property is called primitivity (see e.g. [71–73]).

As we will see in the first result of this section, the assumption of primitivity entails *minimality* of the associated *S*-adic systems. We recall the definition of this basic concept.

Definition 3.5.2 (Minimality) Let (X, T) be a topological dynamical system. (X, T) is called *minimal* if the orbit of each point is dense in X, i.e., if $\overline{\{T^n x : n \in \mathbb{N}\}} = X$ holds for each $x \in X$.

The following lemma summarizes the consequences of primitivity for an *S*-adic system. It is proved for instance in [20, Proposition 2.1 and 2.2]; the minimality assertion can already be found in [71, Lemma 7].

Proposition 3.5.3 *If σ is a primitive sequence of substitutions, the following properties hold.*

  (i) *There exists at least one and at most $|\mathcal{A}|$ limit sequences for σ.*
 (ii) *Let w, w' be two S-adic sequences with directive sequence σ. Then the dynamical systems (X_w, Σ) and $(X_{w'}, \Sigma)$ are equal.*
(iii) *For a limit sequence w of σ the S-adic system (X_w, Σ) is minimal.*

Proof To show (i) let $\sigma = (\sigma_n)$ and for each $n \in \mathbb{N}$ let $\mathcal{A}_n$ be the set of all first letters occurring in the family $\sigma_{[0,n)}(\mathcal{A})$ of words. Then $(\mathcal{A}_n)$ is a decreasing sequence of nonempty subsets of $\mathcal{A}$. Hence, there is a $a \in \bigcap_{n \geq 0} \mathcal{A}_n$. By construction there is a sequence (a_n) with $a_0 = a$ such that a_n is the first letter of $\sigma_n(a_{n+1})$. Moreover, $\sigma_{[0,n)}(a_n)$ is a prefix of $\sigma_{[0,n+1)}(a_{n+1})$. By primitivity, the lengths of these words tend to infinity which implies that $w = \lim_{n \to \infty} \sigma_{[0,n)}(a_n a_n \ldots)$ converges.[6] By the same reasoning (here we use that primitivity also holds for "shifted" sequences), we see that $w^{(m)} = \lim_{n \to \infty} \sigma_{[m,n)}(a_n a_n \ldots)$ converges as well and the sequence $(w^{(m)})$ satisfies the conditions of Definition 3.4.1. Thus $w = w^{(0)}$ is an S-adic sequence with directive sequence σ.

If w is an S-adic sequence with directive sequence σ, by Definition 3.4.1 we can associate a sequence $(w^{(n)})$ with it. For $n \in \mathbb{N}$ let a_n be the first letter of $w^{(n)}$. Primitivity implies that $|\sigma_{[0,n)}(a_n)| \to \infty$ for $n \to \infty$ and, hence, the sequence w is determined by the sequence (a_n). In particular, we can write $w = \lim_{n \to \infty} \sigma_{[0,n)}(a_n a_n \ldots)$. Since a_n uniquely determines a_p for each $p < n$, there are at most $|\mathcal{A}|$ possible different choices for such a sequence.

To prove (ii) let w and w' be two S-adic sequences with directive sequence σ. Associate the sequences (a_n) and (a_n'), respectively, with them as above. If u is a factor of w then u is a factor of $\sigma_{[0,m)}(a_m)$ for some m. By primitivity, there exists $n > m$ such that a_m occurs in $\sigma_{[m,n)}(a_n')$. Thus $\sigma_{[0,m)}(a_m)$ and a fortiori also u is a factor of w' and, hence, $L(w) \subseteq L(w')$. Exchanging the roles of w and w' we can therefore conclude that $L(w) = L(w')$ which implies that $X_w = X_{w'}$.

It remains to prove (iii). This follows if we can show that $L(v) = L(w)$ for each $v \in X_w$. This in turn holds if each factor of w occurs infinitely often in w with bounded gaps, which we will now prove. Let u be a factor of w and let (a_n) be the sequence of letters associated to w as above. Then u is a factor of $\sigma_{[0,m)}(a_m)$ for some m. By primitivity, there exists $n > m$ such that u is a factor of $\sigma_{[0,n)}(a)$ for each $a \in \mathcal{A}$. Since w is an S-adic sequence, $w = \sigma_{[0,n)}(w^{(n)})$ holds for some $w^{(n)} \in \mathcal{A}^{\mathbb{N}}$. Thus u occurs in w infinitely often with gaps bounded by $2 \max\{|\sigma_{[0,n)}(a)| : a \in \mathcal{A}\}$. □

If σ is a primitive sequence of substitutions, assertion (ii) of this proposition implies that $X_\sigma = X_w$ and, hence, $(X_\sigma, \Sigma) = (X_w, \Sigma)$ for w being an arbitrary S-adic sequence with directive sequence σ. Since we will assume primitivity throughout the remaining part of the paper we will always work with X_σ.

3.5.2 Recurrence, Weak Convergence, and Unique Ergodicity

The next concept we introduce is *recurrence*. Let S be a finite set of substitutions. If we take a random sequence of substitutions $\sigma \in S^{\mathbb{N}}$ whose elements are taken from a finite set S we will almost always (w.r.t. any natural measure on the space $S^{\mathbb{N}}$) get a sequence σ each of whose patterns occurs infinitely often. This infinite repetition of patterns is made precise in the following definition.

---

[6]Only the first letter a_n in the argument of $\sigma_{[0,n)}$ is relevant for the limit. However, since we use the topology on $\mathcal{A}^{\mathbb{N}}$ and $\sigma_{[0,n)}(a_n) \notin \mathcal{A}^{\mathbb{N}}$ we have to write $\sigma_{[0,n)}(a_n a_n \ldots)$.

Definition 3.5.4 (Recurrence) A sequence $\mathbf{M} = (M_n)$ of integer matrices is called *recurrent* if for each $m \in \mathbb{N}$ there is $n \geq 1$ such that $(M_0, \ldots, M_{m-1}) = (M_n, \ldots, M_{n+m-1})$. A sequence $\sigma = (\sigma_n)$ of substitutions is called *recurrent* if for each $m \in \mathbb{N}$ there is $n \geq 1$ such that $(\sigma_0, \ldots, \sigma_{m-1}) = (\sigma_n, \ldots, \sigma_{n+m-1})$.

Note that recurrence of a sequence of substitutions σ implies that each block of substitutions that occurs once in σ must occur infinitely often (the same is true for sequences of matrices). Thus recurrence of $(\sigma_n)_{n\in\mathbb{N}}$ implies recurrence of $(\sigma_{m+n})_{n\in\mathbb{N}}$ for each $m \in \mathbb{N}$ and an analogous statement holds for sequences of matrices. We also emphasize that a nonrecurrent sequence of substitutions may well have a recurrent sequence of incidence matrices. This is due to the fact that two different substitutions can have the same incidence matrix.

We now study consequences of primitivity and recurrence. We start with the following result which follows from contraction properties of the *Hilbert metric*, a metric on projective space that goes back to Birkhoff [55] and Furstenberg [84, pp. 91–95] (we mention [81, Appendix A] and [125, Chapter 26] as more recent references). A special case of this result is stated in Sect. 3.2, see (3.14).

Proposition 3.5.5 *Let* $\mathbf{M} = (M_n)$ *be a primitive and recurrent sequence of nonnegative integer matrices. Then there is a vector* $\mathbf{u} \in \mathbb{R}_{>0}^d$ *satisfying*

$$\bigcap_{n\geq 0} M_{[0,n)}\mathbb{R}_{\geq 0}^d = \mathbb{R}_+\mathbf{u}. \tag{3.26}$$

Proof To prove this result we define a metric on the space $\mathcal{W} = \{\mathbb{R}_+\mathbf{w} : \mathbf{w} \in \mathbb{R}_{\geq 0}^d \setminus \{\mathbf{0}\}\}$ of nonnegative rays through the origin by (see [81, Appendix A])

$$d_{\mathcal{W}}(\mathbb{R}_+\mathbf{v}, \mathbb{R}_+\mathbf{w}) = \max_{1\leq i,j\leq d} \log \frac{v_i w_j}{v_j w_i},$$

where $\mathbf{v} = (v_1, \ldots, v_d)$ and $\mathbf{w} = (w_1, \ldots, w_d)$. It can be checked by direct calculation that this is a metric on $\mathcal{W}$ which is the so-called *Hilbert Metric* (cf. e.g. [81, Lemma A.5] or [125, Chapter 26]). Let $\mathrm{diam}_{\mathcal{W}}(A)$ be the diameter of a set $A \subset \mathcal{W}$ w.r.t. this metric. Then $\mathrm{diam}_{\mathcal{W}}(\mathcal{W}) = \infty$ and $\mathrm{diam}_{\mathcal{W}}(M\mathcal{W}) < \infty$ for every positive matrix M. It follows from the definitions that a nonnegative matrix M is nonexpanding in the sense that $d_{\mathcal{W}}(M\mathbb{R}_+\mathbf{v}, M\mathbb{R}_+\mathbf{w}) \leq d_{\mathcal{W}}(\mathbb{R}_+\mathbf{v}, \mathbb{R}_+\mathbf{w})$ for all $\mathbb{R}_+\mathbf{v}, \mathbb{R}_+\mathbf{w} \in \mathcal{W}$. Moreover, one can show that each positive matrix M is a contraction, i.e., there is $\kappa < 1$ (depending on M) such that $d_{\mathcal{W}}(M\mathbb{R}_+\mathbf{v}, M\mathbb{R}_+\mathbf{w}) \leq \kappa\, d_{\mathcal{W}}(\mathbb{R}_+\mathbf{v}, \mathbb{R}_+\mathbf{w})$ for all $\mathbb{R}_+\mathbf{v}, \mathbb{R}_+\mathbf{w} \in \mathcal{W}$ (see for instance [55] or [125, Proposition 26.3] for a proof of this).

We now apply these contraction properties to our setting. Since $\mathbf{M}$ is primitive and recurrent, there exists a positive matrix B and an integer $h > 0$ such that $B = M_{[m_i,m_i+h)}$ for a sequence of positive integers $(m_i)_{i\geq 0}$ satisfying $m_i + h \leq m_{i+1}$. By the preceding paragraph we get that $\mathrm{diam}_{\mathcal{W}}(B\mathcal{W}) = \gamma$ for some $\gamma > 0$ and that B is a contraction with some contraction factor $\kappa < 1$. Thus for each $m \in \{m_i + h, m_{i+1} + h - 1\}$ we have

$$\mathrm{diam}_{\mathcal{W}}\left(\bigcap_{0 \leq n \leq m} M_{[0,n)} \mathbb{R}_{\geq 0}^d \right) \leq \gamma \kappa^i.$$

Since $\kappa < 1$ and $i \to \infty$ for $m \to \infty$ this yields the result. Positivity of the entries of $\mathbf{u}$ follows from the primitivity of $\mathbf{M}$. $\square$

This result motivates the following definition.

Definition 3.5.6 (Weak Convergence and Generalized Right Eigenvector) If a sequence of nonnegative integer matrices satisfies (3.26) for some $\mathbf{u} \in \mathbb{R}_{\geq 0}^d \setminus \{\mathbf{0}\}$ we say that $\mathbf{M}$ is *weakly convergent* to $\mathbf{u}$. In this case we call $\mathbf{u}$ a *generalized right eigenvector* of $\mathbf{M}$. If a sequence σ of substitutions has a sequence of incidence matrices $\mathbf{M}$ which is weakly convergent to $\mathbf{u}$, we say that σ is *weakly convergent* to $\mathbf{u}$ and call $\mathbf{u}$ a *generalized right eigenvector* of σ.

Our next goal is to establish *unique ergodicity* of S-adic systems with primitive and recurrent directive sequences. We start with a fundamental definition (and refer to [126, §6.5] for background material on this).

Definition 3.5.7 (Unique Ergodicity) A topological dynamical system (X, T) on a compact space X is said to be *uniquely ergodic* if there is a unique T-invariant Borel probability measure on X.

By a theorem of Krylov and Bogoliubov (see e.g. [126, Corollary 6.9.1]) there always exists an invariant probability measure on (X, T) if X is compact.

A uniquely ergodic dynamical system is ergodic (thus the name) since otherwise there would be a T-invariant set E with $\mu(E) \in (0, 1)$ which could be used to define a second T-invariant Borel probability measure $\nu(B) = \frac{\mu(B \cap E)}{\mu(E)}$ on X. Unique ergodicity is equivalent to the fact that each point is generic in the sense that Birkhoff's ergodic theorem holds everywhere (cf. [126, Theorem 6.19]). Roughly speaking, this is true since nongeneric points (as for instance periodic points) could be used to construct a second invariant measure.

We note that unique ergodicity is close to minimality in the sense that there are many dynamical systems that either enjoy both or none of the two properties. If (X, T) is uniquely ergodic with T-invariant measure μ having full support then minimality follows. However, there are examples of systems that have only one of these two properties. For a discussion of such examples in a context similar to ours see [78] and the references given there. What happens for these examples is that although we have a primitive sequence of matrices (leading to minimality) this primitivity is so weak that it does not make the positive cone converge to a single line as in (3.26). This entails that no letter frequencies exist which permits to construct many invariant measures (see also [53, 54, 81]).

It has been mentioned already in Sect. 3.2 that the existence of uniform frequencies of letters and words in a shift (X_w, Σ) entail unique ergodicity. We want to give the elegant proof of this result here before we use it in order to establish unique ergodicity of primitive and recurrent S-adic systems. To this matter we need the following definition (see Lemma 3.2.9 for the special case of Sturmian sequences).

Definition 3.5.8 (Uniform Word and Letter Frequencies) Let $w = w_0 w_1 \ldots \in \mathcal{A}^{\mathbb{N}}$ be given and for each $k, \ell \in \mathbb{N}$ and each $v \in \mathcal{A}^*$ let $|w_k \ldots w_{k+\ell-1}|_v$ be the number of occurrences of v in $w_k \ldots w_{k+\ell-1}$. We say that w has *uniform word frequencies* if for each $v \in \mathcal{A}^*$ the ratio $|w_k \ldots w_{k+\ell-1}|_v / \ell$ tends to a limit $f_v(w)$ (which does not depend on k) for $\ell \to \infty$ uniformly in k. It has *uniform letter frequencies* if this is true for each $v \in \mathcal{A}$.

Proposition 3.5.9 (See [82, Proposition 5.1.21]) *Let $w \in \mathcal{A}^{\mathbb{N}}$ be a sequence with uniform word frequencies and let $X_w = \overline{\{\Sigma^k w : k \in \mathbb{N}\}}$ be the shift orbit closure of w. Then (X_w, Σ) is uniquely ergodic.*

Proof For every factor v of $w = w_0 w_1 \ldots$ let $[v]$ be the *cylinder* of all sequences in X_w that have v as a prefix. Define a function μ on these cylinders by $\mu([v]) = \mu(\Sigma^{-n}[v]) = f_v(w)$. Since cylinders generate the topology on X_w this defines a Borel measure μ on X_w. Our goal is to show that every element of X_w is generic in the sense of Birkhoff's ergodic theorem. To this end note first that (here $\mathbb{1}_Y$ denotes the characteristic function of a set $Y \subset X_w$)

$$\frac{1}{N} \sum_{n < N} \mathbb{1}_{[v]}(\Sigma^{n+j} w) \to \mu([v]) = \int \mathbb{1}_{[v]} d\mu$$

holds uniformly in $j \in \mathbb{N}$ for every $v \in \mathcal{A}^*$ by the existence of uniform word frequencies for w. Since continuous functions are monotone limits of simple functions this extends to

$$\frac{1}{N} \sum_{n < N} g(\Sigma^{n+j} w) \to \int g \, d\mu \qquad (3.27)$$

uniformly in $j \in \mathbb{N}$ for each $g \in C(X_w)$. By this uniform convergence, in (3.27) we may choose $j = n_k$ with any sequence (n_k) and (3.27) holds uniformly in k. Since each $u \in X_w$ is the limit of $(\Sigma^{n_k} w)$ for some sequence (n_k) this implies that

$$\frac{1}{N} \sum_{n < N} g(\Sigma^n u) \to \int g \, d\mu$$

holds for each $g \in C(X_w)$ and each $u \in X_w$. Thus each point is generic in the sense of Birkhoff's ergodic theorem which is equivalent to unique ergodicity (by [126, Theorem 6.19] which was already mentioned above). □

We now show that the conditions we introduced so far imply unique ergodicity of S-adic systems. In view of Proposition 3.5.9 we will establish the following lemma (see also [44, Theorem 5.7]).

Lemma 3.5.10 *Let σ be a sequence of substitutions with associated sequence of incidence matrices* **M**. *If* **M** *is primitive and recurrent then each sequence $w \in X_\sigma$ has uniform word frequencies.*

Proof Let $w = w_0 w_1 \ldots \in X_\sigma$ be given. We follow the proof of [44, Theorem 5.7] to establish that w has uniform word frequencies.

Part 1: Uniform Letter Frequencies Since **M** satisfies the conditions of Proposition 3.5.5, it admits a generalized right eigenvector **u**. Let $\mathbf{u}/\|\mathbf{u}\|_1 = (u_1, u_2, \ldots, u_d)^t$. Since $w \in X_\sigma$, for all $k, \ell, n \in \mathbb{N}$ we can write

$$w_k \ldots w_{k+\ell-1} = p\sigma_{[0,n)}(v)s$$

for some $p, v, s \in \mathcal{A}^*$, where the lengths of p, s are bounded by the number $\max\{|\sigma_{[0,n)}(a)| : a \in \mathcal{A}\}$. Now for each $a \in \mathcal{A}$

$$\left| \frac{|w_k \ldots w_{k+\ell-1}|_a}{\ell} - u_a \right| \leq \frac{\left| |p|_a - |p|u_a \right|}{\ell} + \\ \frac{\left| |\sigma_{[0,n)}(v)|_a - |\sigma_{[0,n)}(v)|u_a \right|}{\ell} + \frac{\left| |s|_a - |s|u_a \right|}{\ell}. \tag{3.28}$$

By the convergence of the positive cone to **u** in Proposition 3.5.5 we know that $|\sigma_{[0,n)}(b)|_a/|\sigma_{[0,n)}(b)|$ is close to u_a for all $a, b \in \mathcal{A}$ if n is large. Thus for each $\varepsilon > 0$ there is $N \in \mathbb{N}$ such that whenever $\ell \geq N$ we can choose n in a way that $|p|, |s| \leq \varepsilon\ell$ and $\left| |\sigma_{[0,n)}(b)|_a - |\sigma_{[0,n)}(b)|u_a \right| < \varepsilon|\sigma_{[0,n)}(b)|$ for all letters a and b. This proves that the right hand side of (3.28) is bounded by 3ε and, hence, $\lim_{\ell \to \infty} |w_k \ldots w_{k+\ell-1}|_a/\ell = u_a$ uniformly in k. Thus w has uniform letter frequencies.

Part 2: Uniform Word Frequencies For $m \in \mathbb{N}$ let $\mathbf{u}^{(m)}$ be a right eigenvector of the shifted sequence $\sigma^{(m)} = (\sigma_{m+n})_{n\in\mathbb{N}}$ and set $\mathbf{u}^{(m)}/\|\mathbf{u}^{(m)}\|_1 = (u_1^{(m)}, \ldots, u_d^{(m)})$. Such an eigenvector exists by Proposition 3.5.5 since the shifted sequence $\sigma^{(m)}$ has a primitive and recurrent sequence of incidence matrices as well.

Fix $v \in \mathcal{L}_\sigma$. We claim that for each $m \in \mathbb{N}$ and each $w^{(m)} = w_0^{(m)} w_1^{(m)} \ldots \in X_{\sigma^{(m)}}$ we have

$$\lim_{j \to \infty} \frac{\sum_{i=q}^{q+j-1} |\sigma_{[0,m)}(w_i^{(m)})|_v}{|\sigma_{[0,m)}(w_q^{(m)} \ldots w_{q+j-1}^{(m)})|} = \frac{\sum_{a\in\mathcal{A}} u_a^{(m)} |\sigma_{[0,m)}(a)|_v}{\sum_{a\in\mathcal{A}} u_a^{(m)} |\sigma_{[0,m)}(a)|} =: g(v, m) \tag{3.29}$$

uniformly in $q \in \mathbb{N}$. This claim follows because, since $w^{(m)}$ has uniform letter frequencies $(u_1^{(m)}, \ldots, u_d^{(m)})$ by Part 1, we get that

$$\lim_{j \to \infty} \frac{|\sigma_{[0,m)}(w_q^{(m)} \ldots w_{q+j-1}^{(m)})|}{j} = \sum_{a \in \mathcal{A}} u_a^{(m)} |\sigma_{[0,m)}(a)| \quad \text{and}$$

$$\lim_{j \to \infty} \frac{\sum_{i=q}^{q+j-1} |\sigma_{[0,m)}(w_i^{(m)})|_v}{j} = \sum_{a \in \mathcal{A}} u_a^{(m)} |\sigma_{[0,m)}(a)|_v$$

uniformly in $q \in \mathbb{N}$.

Now we proceed similarly to Part 1. First define

$$m_n^+ = \max\{|\sigma_{[0,n)}(a)| : a \in \mathcal{A}\} \quad \text{and} \quad m_n^- = \min\{|\sigma_{[0,n)}(a)| : a \in \mathcal{A}\},$$

and observe that primitivity of σ implies that both of these quantities tend to ∞ for $n \to \infty$. For each $n \in \mathbb{N}$ choose a fixed $w^{(n)} = w_0^{(n)} w_1^{(n)} \ldots \in X_{\sigma^{(n)}}$. As $w \in X_\sigma$, for all $k, \ell \in \mathbb{N}$ we can write

$$w_k \ldots w_{k+\ell-1} = p\sigma_{[0,n)}(w_q^{(n)} \ldots w_{q+r-1}^{(n)})s$$

for some $q, r \in \mathbb{N}$, where the lengths of $p, s \in \mathcal{A}^*$ are bounded by m_n^+. There are three possibilities for an occurrence of v in $w_k \ldots w_{k+\ell-1}$. Firstly, v can overlap with p or s. This can happen at most $2m_n^+$ times. Secondly, v can have nonempty overlap with the images $\sigma_{[0,n)}(w_i^{(n)})$ and $\sigma_{[0,n)}(w_{i+1}^{(n)})$ of two consecutive letters $w_i^{(n)}$ and $w_{i+1}^{(n)}$ of $w_q^{(n)} \ldots w_{q+r-1}^{(n)}$. This can happen at most $|v|(r-1) \le |v|\frac{\ell}{m_n^-}$ times. Thirdly, v can occur as a factor of $\sigma_{[0,n)}(w_i^{(n)})$ for some $i \in \{q, \ldots, q + r - 1\}$ which happens exactly $\sum_{i=q}^{q+r-1} |\sigma_{[0,n)}(w_i^{(n)})|_v$ times. Each of these three possibilities contributes one of the summands of the right hand side of the estimate

$$\left| \frac{|w_k \ldots w_{k+\ell-1}|_v}{\ell} - g(v, n) \right| \le \frac{2m_n^+}{\ell} + \frac{|v|}{m_n^-} + \left| \frac{\sum_{i=q}^{q+r-1} |\sigma_{[0,n)}(w_i^{(n)})|_v}{\ell} - g(v, n) \right|. \tag{3.30}$$

Letting $\ell \to \infty$ and using (3.29) for the third term on the right this yields that

$$\limsup_{\ell \to \infty} \left| \frac{|w_k \ldots w_{k+\ell-1}|_v}{\ell} - g(v, n) \right| \le \frac{|v|}{m_n^-}. \tag{3.31}$$

Since for $n \to \infty$ the quantity $\frac{|w_k \ldots w_{k+\ell-1}|_v}{\ell}$ does not change while $\frac{|v|}{m_n^-} \to 0$ we conclude from (3.31) that $(g(v, n))_{n \in \mathbb{N}}$ is a Cauchy sequence converging to the frequency $f_v(w)$ of v in w. Since $\frac{|v|}{m_n^-}$ does not depend on k and the convergence in (3.29) is uniform in q, the estimate (3.30) implies that $\frac{|w_k \ldots w_{k+\ell-1}|_v}{\ell} \to f_v(w)$ for $\ell \to \infty$ uniformly in k and the proof is finished.   □

The following main result of this section is an immediate consequence of Propositions 3.5.3, 3.5.9, and Lemma 3.5.10.

Theorem 3.5.11 *Let σ be a sequence of substitutions with associated sequence of incidence matrices* **M**. *If* **M** *is primitive and recurrent then* (X_σ, Σ) *is minimal and uniquely ergodic.*

A proof of a similar result as Theorem 3.5.11 is sketched in Berthé and Delecroix [44]. Moreover, we refer to Fisher [81] and Bezuglyi et al. [53, 54], where theorems of this flavor are proved in the context of Bratteli-Vershik systems.

Example 3.5.12 We associate substitutions with the matrices M_1, M_2, and M_3 that came up in (3.23) during the definition of Brun's continued fraction algorithm. Indeed, the substitutions

$$\sigma_1 : \begin{cases} 1 \mapsto 3, \\ 2 \mapsto 1, \\ 3 \mapsto 23, \end{cases} \qquad \sigma_2 : \begin{cases} 1 \mapsto 1, \\ 2 \mapsto 3, \\ 3 \mapsto 23, \end{cases} \qquad \sigma_3 : \begin{cases} 1 \mapsto 1, \\ 2 \mapsto 23, \\ 3 \mapsto 3. \end{cases} \tag{3.32}$$

are called *Brun substitutions* (see [52, Sections 3.3 and 9.2] where also the relation between these substitutions and a slightly different set of "Brun substitutions" studied in [42] is discussed).

It is immediate that $M_1 M_2 M_1 M_2$ is a strictly positive matrix. Thus we get the following result.

Proposition 3.5.13 *Let $S = \{\sigma_1, \sigma_2, \sigma_3\}$ be the set of Brun substitutions and $\sigma \in S^{\mathbb{N}}$. If σ is recurrent and contains the block $(\sigma_1, \sigma_2, \sigma_1, \sigma_2)$ then the associated S-adic system (X_σ, Σ) is minimal and uniquely ergodic.*

Proof Since σ is recurrent it contains the block $(\sigma_1, \sigma_2, \sigma_1, \sigma_2)$ infinitely often. Thus σ is primitive and the result follows from Theorem 3.5.11.   □

3.6 The Importance of Balance and Algebraic Irreducibility

Let σ be a sequence of unimodular substitutions over an alphabet $\mathcal{A} = \{1, 2, \ldots, d\}$ and let (X_σ, Σ) be the S-adic system defined by it. At the end of Sect. 3.2.4 we gave some rough idea on how we want to prove that (X_σ, Σ) is measurably conjugate to a rotation on $\mathbb{T}^{d-1}$. Indeed, we wish to project the broken line (see (3.17) for an example) associated with a limit sequence $w \in X_\sigma$ to a hyperplane in $\mathbb{R}^d$ not containing the frequency vector **u** of the sequences in X_σ. On a natural subdivision $\mathcal{R}(1), \ldots, \mathcal{R}(d)$ of the closure $\mathcal{R}$ of this projection we want to define a domain exchange and a rotation. This is possible only if the sets $\mathcal{R}(i)$, $i \in \mathcal{A}$, have suitable topological properties and the mentioned subdivision has no essential overlaps.

In the present section we will define these sets $\mathcal{R}$ and $\mathcal{R}(i)$, $i \in \mathcal{A}$, and discuss basic properties of them. Besides primitivity and recurrence, the crucial conditions we will have to impose on σ in order to get suitable properties of the sets $\mathcal{R}$ and $\mathcal{R}(i)$ will be *algebraic irreducibility* of the sequence of incidence matrices of σ and *balance* of the language $\mathcal{L}_\sigma$. Both of these conditions will be defined and first consequences of them will be discussed. This paves the way to obtain deeper topological and measure theoretic properties of $\mathcal{R}$ and $\mathcal{R}(i)$ in Sect. 3.7. The theory we will outline in the present as well as in the forthcoming sections is mainly due to Berthé et al. [52] and we refer to this paper for rigorous proofs of the statements we give.

3.6.1 *S-adic Rauzy Fractals*

Following Berthé et al. [52, Section 2.9] we will now define *S*-adic Rauzy fractals. As mentioned before, on these objects we will be able to "see" the rotations to which we want to (measurably) conjugate our *S*-adic systems. In the definition we will use the following notations. For a vector $\mathbf{w} \in \mathbb{R}^d \setminus \{\mathbf{0}\}$ we write $\mathbf{w}^\perp$ for the hyperplane orthogonal to $\mathbf{w}$, i.e., $\mathbf{w}^\perp = \{\mathbf{x} \in \mathbb{R}^d : \langle \mathbf{x}, \mathbf{w} \rangle = 0\}$ with $\langle \cdot, \cdot \rangle$ being the dot product on $\mathbb{R}^d$, and we equip the space $\mathbf{w}^\perp$ with the $(d-1)$-dimensional Lebesgue measure $\lambda_\mathbf{w}$. Since its orthogonal hyperplane will be of special interest later we introduce the vector $\mathbf{1} = (1, \ldots, 1)^t$.

For vectors $\mathbf{u}, \mathbf{w} \in \mathbb{R}^d \setminus \{\mathbf{0}\}$ satisfying $\mathbf{u} \notin \mathbf{w}^\perp$ we denote the projection along $\mathbf{u}$ to $\mathbf{w}^\perp$ by $\pi_{\mathbf{u},\mathbf{w}}$.

Definition 3.6.1 (*S*-adic Rauzy Fractal and Subtiles) Let σ be a sequence of unimodular substitutions over the alphabet $\mathcal{A}$ and assume that σ is weakly convergent to a generalized right eigenvector $\mathbf{u} \in \mathbb{R}^d_{>0}$. The *S-adic Rauzy fractal* in the representation space $\mathbf{w}^\perp$, $\mathbf{w} \in \mathbb{R}^d_{\geq 0} \setminus \{\mathbf{0}\}$, associated with σ is the set

$$\mathcal{R}_\mathbf{w} := \overline{\{\pi_{\mathbf{u},\mathbf{w}}\mathbf{l}(p) \ : \ p \text{ is a prefix of a limit sequence of } \sigma\}}.$$

The set $\mathcal{R}_\mathbf{w}$ can be covered by the *subtiles*

$$\mathcal{R}_\mathbf{w}(i) := \overline{\{\pi_{\mathbf{u},\mathbf{w}}\mathbf{l}(p) \ : \ pi \text{ is a prefix of a limit sequence of } \sigma\}} \qquad (i \in \mathcal{A}). \tag{3.33}$$

For convenience we will use the notation $\mathcal{R}(i) = \mathcal{R}_\mathbf{1}(i)$ and $\mathcal{R} = \mathcal{R}_\mathbf{1}$.

The prototype of a Rauzy fractal goes back to Rauzy [114] and was used there in order to show that a certain substitutive dynamical system is measurably conjugate to a rotation on the torus, see Example 3.6.2 below. In the meantime there exists a vast literature on Rauzy fractals. For constant sequences $\boldsymbol{\sigma} = (\sigma)$ with σ being a Pisot substitution fundamental properties of Rauzy fractals were studied for instance by Ito and Kimura [89], Holton and Zamboni [86], Arnoux and Ito [18], Canterini and Siegel [61], Sirvent and Wang [124], Hubert and Messaoudi [88], and Ito and Rao [91]. Akiyama [2, 3] and Messaoudi [103, 104] consider versions of Rauzy fractals for β-numeration, in Siegel [122], Minervino and Thuswaldner [106], and Minervino and Steiner [105] Rauzy fractals with p-adic factors are related to nonunimodular substitutions. For versions of Rauzy fractals corresponding to substitutions with reducible incidence matrices (whose most prominent representative is the so-called "Hokkaido Fractal" studied by Akiyama and Sadahiro [5]) we refer to [3, 75, 100]. A case of a non-Pisot substitution is treated in [16]. Surveys containing information on Rauzy fractals are provided in [51, 123] (see also [4] for their relation to the Pisot substitution conjecture). An easily accessible treatment of Rauzy fractals intended for a broad audience is given in [17].

Recently, Boyland and Severa [57] considered a particular family of S-adic sequences associated with the so-called *infimax S-adic family* over three letters. These sequences do not fit into our framework. Indeed, they have two "expanding directions" which entails that the authors have to project on a 1-dimensional subspace of $\mathbb{R}^3$ in order to obtain compact Rauzy fractals. Their Rauzy fractals turn out to be Cantor sets which can be subdivided naturally into three subtiles whose convex hulls are intervals that intersect on their boundary points. This fact is used to show that the infimax S-adic systems can be geometrically represented as 3-interval exchange transformations.

In what follows, instead of "S-adic Rauzy fractal" we will often just say "Rauzy fractal". This will cause no confusion. To give the reader a feeling for a Rauzy fractal and its importance in the remaining part of this chapter we provide an example.

Example 3.6.2 (Tribonacci Substitution) We explain the definition of a Rauzy fractal for the constant sequence $\boldsymbol{\sigma} = (\sigma)_{n \in \mathbb{N}}$ with σ being the Tribonacci substitution introduced in (3.20). The sequence $\boldsymbol{\sigma}$ is easily checked to be primitive and obviously it is recurrent. Thus it admits a generalized right eigenvector $\mathbf{u}$ which is just the Perron-Frobenius eigenvector of M_σ. Since each of the words $\sigma(1), \sigma(2)$, and $\sigma(3)$ begins with 1 the only limit sequence of (σ) is given by

$$w = \lim_{n \to \infty} \sigma^n(1) = 1213121121312121312112131213121\ldots$$

and, hence, $\mathcal{R}_{\mathbf{w}} := \overline{\{\pi_{\mathbf{u},\mathbf{w}} \mathbf{l}(p) \ : \ p \text{ is a prefix of } w\}}$ for $\mathbf{w} \in \mathbb{R}^d_{\geq 0} \setminus \{\mathbf{0}\}$. In Fig. 3.11 we illustrate the definition of $\mathcal{R}_{\mathbf{u}}$ and its subtiles (we choose $\mathbf{w} = \mathbf{u}$ in this case so the occurring projection $\pi_{\mathbf{u},\mathbf{u}}$ is an orthogonal projection). As mentioned before, this famous prototype of a Rauzy fractal first appears in Rauzy [114].

For this example it is known since Rauzy [114] that one can define a rotation on the Rauzy fractal using the broken line. This can be used to prove that the substitutive system $(X_{(\sigma)}, \Sigma)$ is measurably conjugate to a rotation on $\mathbb{T}^2$. We want to give an idea on how this works without going into the details. To this end it is

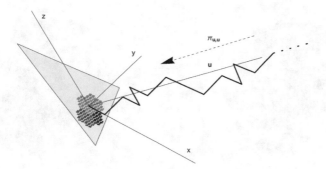

Fig. 3.11 The broken line and its projection to $\mathbf{u}^\perp$ defining the Rauzy fractal $\mathcal{R}_\mathbf{u}$ for the case of the Tribonacci substitution (note that only the vertices of the broken line are projected; not the whole edges). Each of the three subtiles $\mathcal{R}_\mathbf{u}(i)$ is shaded differently. The shaded triangle represents a part of the plane $\mathbf{u}^\perp$ in which $\mathcal{R}_\mathbf{u}$ is situated

convenient to work with $\mathcal{R} = \mathcal{R}_1$ and its subtiles. It was shown in [114] that each of the three subtiles $\mathcal{R}(i)$, $i \in \{1, 2, 3\}$, is a compact subset of the space $\mathbf{1}^\perp$ which is equal to the closure of its interior and has a boundary of λ_1-measure 0. Moreover, it is proved that these subtiles are pairwise disjoint apart from overlaps on their boundaries. Thus we can almost everywhere define a "domain exchange" E in the following way. If we set

$$\tilde{\mathcal{R}}(i) := \overline{\{\pi_{\mathbf{u},1}\mathbf{l}(pi) \; : \; pi \text{ is a prefix of } w\}} \qquad (i \in \{1, 2, 3\})$$

we see from the definition of $\mathcal{R}(i)$ that $\tilde{\mathcal{R}}(i) = \mathcal{R}(i) + \pi_{\mathbf{u},1}\mathbf{l}(i)$ (recall that w is the only limit sequence of σ). As the Lebesgue measure λ_1 doesn't change under translation and we still have that $\mathcal{R} = \tilde{\mathcal{R}}(1) \cup \tilde{\mathcal{R}}(2) \cup \tilde{\mathcal{R}}(3)$ also the translated pieces only overlap on a set of measure 0. The domain exchange

$$E : \mathcal{R} \to \mathcal{R}; \quad \mathbf{x} \mapsto \mathbf{x} + \pi_{\mathbf{u},1}\mathbf{l}(i) \quad \text{for } \mathbf{x} \in \mathcal{R}(i)$$

is thus well defined almost everywhere and it moves $\mathcal{R}(i)$ to $\tilde{\mathcal{R}}(i)$ for each $i \in \{1, 2, 3\}$. By what was said above, E is an almost everywhere bijective symmetry. The effect of E on the points of $\mathcal{R}$ is illustrated in Fig. 3.12. As in the Sturmian case discussed in Sect. 3.2.4, each step on the broken line performs the domain exchange on $\mathcal{R}$.

The problem that remains is the transition from the domain exchange to the rotation. In the Sturmian case this was achieved by identifying the endpoints of an interval. Here things become more complicated as intervals are replaced by fractals and we have to make identifications on $\partial\mathcal{R}$.

To settle this, Rauzy [114] proved that $\mathcal{R}$ forms a fundamental domain of the lattice

$$\Lambda = (\pi_{\mathbf{u},1}\mathbf{l}(1) - \pi_{\mathbf{u},1}\mathbf{l}(2))\mathbb{Z} \oplus (\pi_{\mathbf{u},1}\mathbf{l}(1) - \pi_{\mathbf{u},1}\mathbf{l}(3))\mathbb{Z},$$

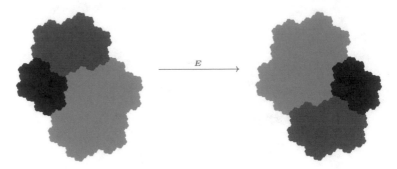

Fig. 3.12 The domain exchange on the classical Rauzy fractal associated with the Tribonacci substitution: the bright domain $\mathcal{R}(1)$ is translated by $\pi_{\mathbf{u},1}\mathbf{l}(1)$, the darker domain $\mathcal{R}(2)$ is translated by $\pi_{\mathbf{u},1}\mathbf{l}(2)$, and finally the darkest domain $\mathcal{R}(3)$ is translated by $\pi_{\mathbf{u},1}\mathbf{l}(3)$. The union of the translated domains gives $\mathcal{R}$ again

i.e., it forms a tiling of $\mathbf{1}^{\perp}$ when translated by elements of Λ. Thus $\mathcal{R}$ can be seen as a subset of the 2-torus $\mathbf{1}^{\perp}/\Lambda$ and since it is a fundamental domain of Λ it covers the torus without overlaps (apart from the boundary). This gives the desired identifications on $\partial\mathcal{R}$. If we look at the domain exchange on this torus we see that $\pi_{\mathbf{u},1}\mathbf{l}(i) \equiv \pi_{\mathbf{u},1}\mathbf{l}(j) \pmod{\Lambda}$ holds for $i, j \in \{1, 2, 3\}$. Thus on this torus all the translations performed by the domain exchange E become the same and, hence, on the torus the mapping E induces a rotation by $\pi_{\mathbf{u},1}\mathbf{l}(1)$. One can show (by defining a suitable "representation map" for the elements of $X_{(\sigma)}$ on the torus $\mathbf{1}^{\perp}/\Lambda$) that $(X_{(\sigma)}, \Sigma)$ is measurably conjugate to $(\mathbf{1}^{\perp}/\Lambda, +\pi_{\mathbf{u},1}\mathbf{l}(1))$, which is a rotation on the 2-torus. We also refer to [52, Section 8] where rigorous arguments are given in a general context (a sketch of these arguments is provided in Sect. 3.9.2 below).

In the preceding example various properties of the Rauzy fractal were needed in order to get the measurable conjugacy between the substitutive system and the rotation. Our aim is to establish these conditions for S-adic Rauzy fractals under a set of natural conditions. Since tiling properties of S-adic Rauzy fractals will play an important role we will now define some collections of Rauzy fractals that will later be shown to provide *tilings* in the following sense.

Definition 3.6.3 (Multiple Tiling and Tiling) A collection $\mathcal{K}$ of subsets of a Euclidean space $\mathcal{E}$ is called a *multiple tiling* of $\mathcal{E}$ if each element of $\mathcal{K}$ is a compact set which is equal to the closure of its interior, and if there is $m \in \mathbb{N}$ such that almost every point (w.r.t. Lebesgue measure) of $\mathcal{E}$ is contained in exactly m elements of $\mathcal{K}$. If $m = 1$ then a multiple tiling is called a *tiling*.

The collections of tiles we need in our setting are defined in terms of so-called *discrete hyperplanes*. These objects were first defined and studied in the context of theoretical computer science (see [115] and later [7, 93]) and have interesting connections to generalized continued fraction algorithms (cf. e.g. [45, 79, 80, 90, 92]). The formal definition reads as follows. Pick $\mathbf{w} \in \mathbb{R}^d_{\geq 0} \setminus \{\mathbf{0}\}$, then (setting $\mathbf{e}_i = \mathbf{l}(i)$ for $i \in \mathcal{A}$)

$$\Gamma(\mathbf{w}) = \{[\mathbf{x}, i] \in \mathbb{Z}^d \times \mathcal{A} \; : \; 0 \leq \langle \mathbf{x}, \mathbf{w} \rangle < \langle \mathbf{e}_i, \mathbf{w} \rangle\}.$$

This has a geometrical meaning: if we interpret the symbol $[\mathbf{x}, i] \in \mathbb{Z}^d \times \mathcal{A}$ as the hypercube or "face"

$$[\mathbf{x}, i] = \left\{ \mathbf{x} + \sum_{j \in \mathcal{A} \setminus \{i\}} \lambda_j \mathbf{e}_j \; : \; \lambda_j \in [0, 1] \right\}, \tag{3.34}$$

the set $\Gamma(\mathbf{w})$ turns into a "stepped hyperplane" that approximates $\mathbf{w}^{\perp}$ by hypercubes. In Fig. 3.13 this is illustrated for two cases: for a rational vector $\mathbf{w}$, which leads to a periodic pattern and for an irrational vector $\mathbf{w}$ which yields an aperiodic one. A finite subset of a discrete hyperplane will often be called a *patch*.

Using the concept of discrete hyperplane we define the following collections of Rauzy fractals. Let σ be a sequence of substitutions with generalized right eigenvector $\mathbf{u} \in \mathbb{R}^d_{>0}$ and choose $\mathbf{w} \in \mathbb{R}^d_{\geq 0} \setminus \{\mathbf{0}\}$. Then, following [52, Section 2.10], we set

$$\mathcal{C}_\mathbf{w} = \{\pi_{\mathbf{u}, \mathbf{w}} \mathbf{x} + \mathcal{R}_\mathbf{w}(i) \; : \; [\mathbf{x}, i] \in \Gamma(\mathbf{w})\}. \tag{3.35}$$

As mentioned above, we will see that each of these collections forms a tiling of the space $\mathbf{w}^{\perp}$ under natural conditions. A special role will be played by the collection $\mathcal{C}_\mathbf{1}$ which will give rise to a periodic tiling of $\mathbf{1}^{\perp}$ by lattice translates of the Rauzy fractal $\mathcal{R}$.

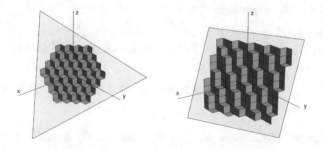

Fig. 3.13 Examples of stepped planes. On the left hand side the stepped plane $\Gamma(\mathbf{1})$, on the right hand side $\Gamma(\mathbf{u})$ with $\mathbf{u}$ as in Example 3.6.2. Since $\mathbf{1}$ is rational the stepped plane $\Gamma(\mathbf{1})$ is periodic, while the irrationality of $\mathbf{u}$ leads to an aperiodic structure in $\Gamma(\mathbf{u})$

3.6.2 Balance, Algebraic Irreducibility, and Strong Convergence

Let (X_σ, Σ) be an S-adic system. As we mentioned already, the associated Rauzy fractals can be used to prove that (X_σ, Σ) is measurably conjugate to a rotation on a torus provided that they have suitable properties. In the present section we will discuss two conditions that have to be imposed on σ in order to guarantee that each of the associated Rauzy fractals $\mathcal{R}_\mathbf{w}$, $\mathbf{w} \in \mathbb{R}_{\geq 0} \setminus \{\mathbf{0}\}$, as well as each of their subtiles $\mathcal{R}_\mathbf{w}(i)$, $i \in \mathcal{A}$, is a compact set that is the closure of its interior and has a boundary of zero measure $\lambda_\mathbf{w}$.

The first property is *balance* and as we will see immediately it entails compactness of $\mathcal{R}_\mathbf{w}$ and its subtiles (see e.g. [1, 44] or [52, Section 2.4] for similar definitions).

Definition 3.6.4 (Balance) Let $\mathcal{A}$ be an alphabet and consider a pair of words $(u, v) \in \mathcal{A}^* \times \mathcal{A}^*$ of the same length. If there is $C > 0$ such that $\big||v|_i - |u|_i\big| \leq C$ holds for each letter $i \in \mathcal{A}$, the pair (u, v) is called *C-balanced*. A language $\mathcal{L} \subset \mathcal{A}^*$ is called *C-balanced* if each pair $(u, v) \in \mathcal{L} \times \mathcal{L}$ with $|u| = |v|$ is C-balanced. It is called *finitely balanced* if it is C-balanced for some $C > 0$.

In Definition 3.2.3 and in Sect. 3.3.2 we defined balance of an infinite sequence and applied this notion to Sturmian sequences as well as to Arnoux-Rauzy sequences. For a general S-adic system (X_σ, Σ) it is more convenient to look at balance of the associated language $\mathcal{L}_\sigma$ since there might be more than one limit sequence associated with the given directive sequence σ. Of course, by Proposition 3.5.3(ii) primitivity of σ implies that each of these limit sequences has the language $\mathcal{L}_\sigma$ of factors.

The following result goes back essentially to [1, Proposition 7] and, in the form we present it here, is contained in [52, Lemma 4.1] (in fact, the conditions that are imposed on σ in that paper are slightly weaker than ours).

Proposition 3.6.5 *Let σ be a primitive and recurrent sequence of unimodular substitutions. Then $\mathcal{R}_\mathbf{w}$ and each of its subtiles is compact for each $\mathbf{w} \in \mathbb{R}_{\geq 0}^d \setminus \{\mathbf{0}\}$ if and only if $\mathcal{L}_\sigma$ is finitely balanced.*

Proof Since $\mathcal{R}_\mathbf{w}$ as well as each of its subtiles is closed by definition it suffices to prove that

$$\mathcal{R}_\mathbf{w} \text{ is bounded for each } \mathbf{w} \in \mathbb{R}_{\geq 0}^d \setminus \{\mathbf{0}\} \quad \Longleftrightarrow \quad \mathcal{L}_\sigma \text{ is finitely balanced.}$$
$$(3.36)$$

We start with proving (3.36) for the case $\mathbf{w} = \mathbf{1}$ and follow [52]. Let $\mathbf{u}$ be a generalized right eigenvector for σ which exists by Proposition 3.5.5.

If $\mathcal{R}$ is bounded then there is $C > 0$ such that $\|\pi_{\mathbf{u},\mathbf{1}}\mathbf{l}(p)\|_\infty \leq C$ for each prefix of a limit sequence of σ. Let $u, v \in \mathcal{L}_\sigma$ be of equal length. Then, by primitivity these words are factors of a limit sequence which entails that

$\|\pi_{\mathbf{u},1}\mathbf{l}(u)\|_\infty, \|\pi_{\mathbf{u},1}\mathbf{l}(v)\|_\infty \le 2C$. As $\mathbf{l}(u) - \mathbf{l}(v) \in \mathbf{1}^\perp$ this yields $\|\mathbf{l}(u) - \mathbf{l}(v)\|_\infty = \|\pi_{\mathbf{u},1}\mathbf{l}(u) - \pi_{\mathbf{u},1}\mathbf{l}(v)\|_\infty \le 4C$ and, hence, $\mathcal{L}_\sigma$ is $4C$-balanced.

Assume now that $\mathcal{L}_\sigma$ is C-balanced and let w be a limit sequence of σ. Let p be a prefix of w and write $w = v_0 v_1 \dots$ where $v_k \in \mathcal{A}^*$ with $|v_k| = |p|$ for each $k \ge 0$. By C-balance, $\|\pi_{\mathbf{u},1}\mathbf{l}(v_k) - \pi_{\mathbf{u},1}\mathbf{l}(p)\|_\infty \le C$ for each $k \in \mathbb{N}$ and, hence, $\left\| \frac{1}{n} \sum_{k=0}^{n-1} \pi_{\mathbf{u},1}\mathbf{l}(v_k) - \pi_{\mathbf{u},1}\mathbf{l}(p) \right\|_\infty \le C$ for each $n \in \mathbb{N}$. By Lemma 3.5.10 (see proof of Part 1), the letter frequencies of w are given by the entries of the vector $\mathbf{u}/\|\mathbf{u}\|_1$ which implies that $\lim_{n\to\infty} \frac{1}{n} \sum_{k=0}^{n-1} \pi_{\mathbf{u},1}\mathbf{l}(v_k) = \mathbf{0}$ and thus

$$\|\pi_{\mathbf{u},1}\mathbf{l}(p)\|_\infty = \left\| \lim_{n\to\infty} \frac{1}{n} \sum_{k=0}^{n-1} \pi_{\mathbf{u},1}\mathbf{l}(v_k) - \pi_{\mathbf{u},1}\mathbf{l}(p) \right\|_\infty \le C.$$

This finishes the proof of (3.36) for the case $\mathbf{w} = \mathbf{1}$. The full statement (3.36) follows from this because $\mathcal{R}_\mathbf{w} = \pi_{\mathbf{u},\mathbf{w}}\mathcal{R}$, which implies that $\mathcal{R}$ is bounded if and only if $\mathcal{R}_\mathbf{w}$ is bounded for each $\mathbf{w} \in \mathbb{R}_{\ge 0}^d \setminus \{\mathbf{0}\}$. □

Our next aim is to make sure that $\mathcal{R}_\mathbf{w}(i)$ has nonempty interior for each $\mathbf{w} \in \mathbb{R}_{\ge 0}^d \setminus \{\mathbf{0}\}$ and each $i \in \mathcal{A}$. This will require much more work. In a first step observe that we have no hope to get nonempty interior if $\mathbf{u}$ has coordinates which are *rationally dependent*, i.e., if there is $\mathbf{x} \in \mathbb{Z}^d$ such that $\langle \mathbf{x}, \mathbf{u} \rangle = 0$. Indeed, in this case the set $\mathcal{R}_\mathbf{w}$ is contained in a finite union of proper affine subspaces of $\mathbf{w}^\perp$. We wish to exclude this case first. This is related to an irreducibility property (going back to [52, Section 2.2]) of the underlying set of incidence matrices which we define now.

Definition 3.6.6 (Algebraic Irreducibility) Let $\mathbf{M} = (M_n)$ be a sequence of nonnegative integer matrices. We say that $\mathbf{M}$ is *algebraically irreducible* if for each $m \in \mathbb{N}$ there is $n > m$ such that the characteristic polynomial of $M_{[m,\ell)}$ is irreducible for each $\ell \ge n$.

A sequence σ of substitutions is called *algebraically irreducible* if it has a sequence of incidence matrices which is algebraically irreducible.

Remark 3.6.7 For our purposes we can replace algebraic irreducibility by the weaker condition that for each $m \in \mathbb{N}$ the matrix M_m is regular and there is $n > m$ such that $M_{[m,\ell)}$ does not have 1 as eigenvalue for each $\ell \ge n$. This condition is easier to check than algebraic irreducibility.

However, since we will always have to assume balance in our setting all but the dominant eigenvalue of large blocks $M_{[m,\ell)}$ should be inside the closed unit disk anyway (cf. also the definition of the Pisot condition in (3.59)). Thus this new condition is not essentially weaker than algebraic irreducibility. For this reason we work with algebraic irreducibility in the sequel.

Together with other properties, algebraic irreducibility of σ implies rational independence of the right eigenvector. We announce this in the following lemma, whose elegant proof is taken from [52, Lemma 4.2].

Lemma 3.6.8 *Let σ be an algebraically irreducible sequence of substitutions with finitely balanced language $\mathcal{L}_\sigma$ that admits a generalized right eigenvector $\mathbf{u} \in \mathbb{R}_{\geq 0}^d \setminus \{0\}$. Then $\mathbf{u}$ has rationally independent coordinates.*

Proof The proof is done by contradiction. Assume that $\mathbf{u}$ has rationally dependent coordinates. Then there is $\mathbf{x} \in \mathbb{Z}^d \setminus \{0\}$ such that $\langle \mathbf{x}, \mathbf{u} \rangle = 0$. This implies that $\langle (M_{[0,n)})^t \mathbf{x}, \mathbf{e}_i \rangle = \langle \mathbf{x}, M_{[0,n)} \mathbf{e}_i \rangle = \langle \mathbf{x}, \mathbf{l}\sigma_{[0,n)}(i) \rangle = \langle \mathbf{x}, \pi_{\mathbf{u},1} \mathbf{l}\sigma_{[0,n)}(i) \rangle$ is uniformly bounded in $i \in \mathcal{A}$ and $n \in \mathbb{N}$ by balance of $\mathcal{L}_\sigma$. Thus $(M_{[0,n)})^t \mathbf{x} \in \mathbb{Z}^d$ is bounded and, hence, there exists an integer k and infinitely many $\ell > k$ with $(M_{[0,k)})^t \mathbf{x} = (M_{[0,\ell)})^t \mathbf{x}$. Multiplying by $((M_{[0,k)})^t)^{-1}$ we see that $\mathbf{x}$ is an eigenvector of $(M_{[k,\ell)})^t$ with eigenvalue 1. Since ℓ can be chosen arbitrarily large this contradicts algebraic irreducibility. $\qquad\square$

In Definition 3.5.6 the concept of weak convergence of a sequence of matrices is introduced. In what follows, we will need a stronger form of convergence, viz. *strong convergence*. If we look back to Lemma 3.2.12 we see that the cascade of inductions we perform on the interval leads to smaller and smaller intervals (that are blown up by renormalization) whose lengths tend to 0. To get an analogous behavior on S-adic Rauzy fractals we need to introduce a certain subdivision on them whose pieces have a diameter that tends to zero. It will turn out that strong convergence is the right condition to guarantee this behavior. We thus recall the definition of strong convergence which is well known in the theory of generalized continued fraction algorithms (see e.g. [119, Definition 19]) and then derive it from the conditions we introduced so far.

Definition 3.6.9 (Strong Convergence) We say that a sequence $\mathbf{M} = (M_n)$ of nonnegative integer matrices is *strongly convergent* to $\mathbf{u} \in \mathbb{R}_{\geq 0}^d \setminus \{0\}$ if

$$\lim_{n \to \infty} \pi_{\mathbf{u},1} M_{[0,n)} \mathbf{e}_i = \mathbf{0} \quad \text{for all } i \in \mathcal{A}.$$

If σ has a strongly convergent sequence of incidence matrices we say that σ is *strongly convergent*.

The difference between weak and strong convergence is explained and illustrated in Fig. 3.14: while weak convergence of vectors can be seen on the unit ball, strong convergence takes place at their end points.

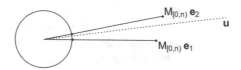

Fig. 3.14 The sequence $\mathbf{M} = (M_n)$ of matrices is weakly convergent, if the intersections of $M_{[0,n)} \mathbf{e}_i$ with the unit ball converge to the intersection of the generalized right eigenvector $\mathbf{u}$ with the unit ball. It is strongly convergent, if the minimal distance of the point $M_{[0,n)} \mathbf{e}_i$ to the ray $\mathbb{R}_+ \mathbf{u}$ converges to zero for each $i \in \mathcal{A}$. Summing up: weak convergence takes place on the unit ball while strong convergence concerns the end points of the vectors

The following result on strong convergence will be needed in the sequel. It is the content of [52, Proposition 4.3].

Proposition 3.6.10 *Let σ be a primitive, algebraically irreducible, and recurrent sequence of substitutions with finitely balanced language $\mathcal{L}_\sigma$. Then*

$$\lim_{n\to\infty} \sup\{\|\pi_{\mathbf{u},1} M_{[0,n)} \mathbf{l}(v)\|_\infty \; : \; v \in \mathcal{L}_\sigma^{(n)}\} = 0.$$

By primitivity this implies that σ is strongly convergent.

The proof of this result is quite tricky. We give a sketch to illustrate the ideas and refer to [52, Proposition 4.3] for details.

Proof (Sketch) Let w be a limit sequence of σ. By primitivity we may apply Proposition 3.5.3(ii) to the shifted sequence $(\sigma_n, \sigma_{n+1}, \sigma_{n+2}, \ldots)$. Thus the language $\mathcal{L}_\sigma^{(n)}$ is equal to the language $L(w^{(n)})$ of factors of the nth "desubstitution" $w^{(n)}$ of w (see (3.22)), i.e., each $v \in \mathcal{L}_\sigma^{(n)}$ satisfies $\mathbf{l}(v) = \mathbf{l}(p) - \mathbf{l}(q)$, where p and q are prefixes of $w^{(n)}$. Thus it suffices to prove

$$\lim_{n\to\infty} \sup\{\|\pi_{\mathbf{u},1} M_{[0,n)} \mathbf{l}(p)\|_\infty \; : \; p \text{ is a prefix of } w^{(n)}\} = 0. \tag{3.37}$$

Let (i_n) be the sequence of first letters of $w^{(n)}$ and choose $\varepsilon > 0$ arbitrary. Define the sets $\mathcal{S}_n = \{\pi_{\mathbf{u},1} \mathbf{l}(p) \; : \; p \text{ is a prefix of } \sigma_{[0,n)}(i_n)\}$ and $\tilde{\mathcal{R}} := \overline{\{\pi_{\mathbf{u},1} \mathbf{l}(p) \; : \; p \text{ is a prefix of } w\}}$.

Then $\mathcal{S}_n \to \tilde{\mathcal{R}}$ for $n \to \infty$ in Hausdorff metric. Since, on the other hand, $\pi_{\mathbf{u},1} M_{[0,n)} \mathbf{l}(p) + \mathcal{S}_n \subset \tilde{\mathcal{R}}$ holds for each $p \in \mathcal{A}^*$ such that $p i_n$ is a prefix $w^{(n)}$ we obtain

$$\|\pi_{\mathbf{u},1} M_{[0,n)} \mathbf{l}(p)\|_\infty < \varepsilon \tag{3.38}$$

for each $p \in \mathcal{A}^*$ such that $p i_n$ is a prefix $w^{(n)}$ for a large enough n. We have to prove (3.38) for arbitrary prefixes p of $w^{(n)}$. If $N(p) = \{n \in \mathbb{N} \; : \; p i_n \text{ is a prefix of } w^{(n)}\}$ is infinite then (3.38) yields

$$\lim_{n \in N(p),\, n\to\infty} \|\pi_{\mathbf{u},1} M_{[0,n)} \mathbf{l}(p)\|_\infty = 0. \tag{3.39}$$

Using algebraic irreducibility and balance by some tricky arguments it is now possible to find a set P of prefixes of w such that the abelianizations $\mathbf{l}(P)$ contain a basis of $\mathbb{R}^d$ and $N(P) = \bigcap_{p \in P} N(p)$ is an infinite set (moreover, the elements of P can by "synchronized" in a certain way by using the recurrence of σ). This implies that (3.39) is true for each $p \in P$ when $N(p)$ is replaced by $N(P)$, i.e.,

$$\lim_{n \in N(P),\, n\to\infty} \|\pi_{\mathbf{u},1} M_{[0,n)} \mathbf{l}(p)\|_\infty = 0 \qquad (p \in P).$$

Since $\mathbf{l}(P)$ contains a basis of $\mathbb{R}^d$ we gain

$$\lim_{n \in N(P),\, n \to \infty} \|\pi_{\mathbf{u},1} M_{[0,n)} \mathbf{x}\|_\infty = 0 \qquad (\mathbf{x} \in \mathbb{R}^d). \tag{3.40}$$

Using primitivity and recurrence again, Eq. (3.37) can be obtained using (3.38) and (3.40). This again requires some work and we omit the details.   □

3.7 Properties of S-adic Rauzy Fractals

Based on the results of the previous section we will now study deeper properties of S-adic Rauzy fractals. In particular, the present section is devoted to the illustration of the proof of the following result from Berthé et al. [52, Theorem 3.1 (ii)].

Theorem 3.7.1 *Let S be a finite set of unimodular substitutions over a finite alphabet $\mathcal{A}$ and let $\sigma = (\sigma_n)$ be a primitive and algebraically irreducible sequence of substitutions taken from the set S. Assume that there is $C > 0$ such that for every $\ell \in \mathbb{N}$ there exists $n \geq 1$ such that $(\sigma_n, \ldots, \sigma_{n+\ell-1}) = (\sigma_0, \ldots, \sigma_{\ell-1})$ and the language $\mathcal{L}_\sigma^{(n+\ell)}$ is C-balanced.*

* *Then each subtile $\mathcal{R}(i)$, $i \in \mathcal{A}$, of the Rauzy fractal $\mathcal{R}$ is a nonempty compact set which is equal to the closure of its interior and has a boundary whose Lebesgue measure λ_1 is zero.*

Remark 3.7.2

(i) We can see that the assumptions of this theorem contain all the properties we discussed in the previous subsections. We could have used the stronger assumption that σ is primitive, recurrent, algebraically irreducible, and has C-balanced language $\mathcal{L}_\sigma^{(n)}$ for each $n \in \mathbb{N}$. However, although this assumption is more handy and holds for many natural examples it would lead to a measure zero subset of the set of "all" sequences σ. The conditions we give in Theorem 3.7.1 will turn out to be "generic" in the sense that they are true for "almost all" sequences σ. All this will be made precise when we develop a metric counterpart of our theory in Sect. 3.9.3.

(ii) Let σ be a substitution on the alphabet $\mathcal{A}$. It is easy to prove that for each $C > 0$ there is $C' > 0$ such that $\sigma(w)$ is C'-balanced for each C-balanced sequence $w \in \mathcal{A}^{\mathbb{N}}$. Applying this to the substitution $\sigma = \sigma_{[0,n+\ell)}$ for some n, ℓ with balanced language $\mathcal{L}_\sigma^{(n+\ell)}$ we see that we can choose the constant C in Theorem 3.7.1 in a way that also $\mathcal{L}_\sigma$ is C-balanced. We will always assume that C is chosen in this way in the sequel.

(iii) We confine ourselves to finite sets S of substitutions to keep things as simple as possible. With a bit more effort it is possible to generalize Theorem 3.7.1 to infinite sets S. This is of interest because infinite sets S correspond to multiplicative continued fraction algorithms like the important Jacobi-Perron algorithm or an acceleration of the Arnoux-Rauzy algorithm proposed recently by Avila et al. [30]. This more general setting is treated in [52].

Theorem 3.7.1 will enable us to study tiling properties of $\mathcal{R}_{\mathbf{w}}$ and its subtiles which will finally lead to the measurable conjugacy of (X_σ, Σ) to a rotation.

The proof of Theorem 3.7.1 is quite long and technical and we refer to [52, Section 6] for details. Our aim here is to illustrate the main ideas in a way that is hopefully more accessible to a broader readership than the original research paper. First we will establish a set equation for the subtiles $\mathcal{R}_{\mathbf{w}}(i)$, $i \in \mathcal{A}$, of $\mathcal{R}_{\mathbf{w}}$ that governs certain subdivisions of $\mathcal{R}_{\mathbf{w}}(i)$. Using this set equation we will be able to establish the properties of S-adic Rauzy fractals stated in Theorem 3.7.1.

Theorem 3.7.1 has a number of predecessors. For instance, Lagarias and Wang [96] proved that each *self-affine tile* $\mathcal{T}$ is the closure of its interior and $\partial\mathcal{T}$ has Lebesgue measure zero. For substitutive Rauzy fractals the according result was proved by Sirvent and Wang [124]. However, in all these cases the sets have strong self-affinity properties which are no longer present in our setting. We therefore need new ideas and more efforts to get the desired results (in particular, the proof of the fact that the boundary of an S-adic Rauzy fractal has measure zero will need quite some work).

3.7.1 Set Equations for S-adic Rauzy Fractals and Dual Substitutions

The first important tool in the proof of Theorem 3.7.1 will be a *set equation* for the subtiles $\mathcal{R}_{\mathbf{w}}(i)$, $\mathbf{w} \in \mathbb{R}_{\geq 0}^d \setminus \{\mathbf{0}\}$ and $i \in \mathcal{A}$, of a sequence σ of unimodular substitutions as well as for related subtiles associated with "shifts" of σ. This set equation equips the sets $\mathcal{R}_{\mathbf{w}}(i)$ with a subdivision structure that is governed by σ. We now give an idea on how this works.

Let $\sigma = (\sigma_n)$ be a primitive and recurrent sequence of unimodular substitutions over the alphabet $\mathcal{A}$ with generalized right eigenvector $\mathbf{u} \in \mathbb{R}_{>0}^d$ and choose $\mathbf{w} \in \mathbb{R}_{\geq 0}^d \setminus \{\mathbf{0}\}$. In all what follows, keep in mind the definition of the subtile $\mathcal{R}_{\mathbf{w}}(i)$ from (3.33). We choose a limit sequence w of σ and associate with it the sequence $(w^{(n)})$ of its "desubstitutions" according to (3.22).

Consider the set $\{\pi_{\mathbf{u},\mathbf{w}}\mathbf{l}(p) \ : \ pi \text{ is a prefix of } w\}$ and observe that by the definition of a limit sequence each $p \in \mathcal{A}^*$ for which pi is a prefix of w can be written as $p = \sigma_0(p')p_0$ with $p_0 i$ a prefix of $\sigma_0(j)$ for some $j \in \mathcal{A}$ and $p'j$ some prefix of $w^{(1)}$. Using this decomposition of p we obtain the decomposition

$$\{\pi_{\mathbf{u},\mathbf{w}}\mathbf{l}(p) \ : \ pi \text{ is a prefix of } w\} =$$

$$\bigcup_{\substack{j\in\mathcal{A},\, p_0\in\mathcal{A}^* \\ \sigma_0(j)=p_0 i\,\mathcal{A}^*}} \{\pi_{\mathbf{u},\mathbf{w}}\mathbf{l}(p_0) + \pi_{\mathbf{u},\mathbf{w}}(\mathbf{l}\sigma_0(p')) \ : \ p'j \text{ is a prefix of } w^{(1)}\}.$$

$$(3.41)$$

From (3.21) we see that $l\sigma_0(p') = M_0 l(p')$. Moreover, direct calculation (see [52, Lemma 5.2]) yields that $\pi_{\mathbf{u},\mathbf{w}} M_0 = M_0 \pi_{M_0^{-1}\mathbf{u}, M_0^t \mathbf{w}}$. Inserting this in (3.41) we gain

$$\{\pi_{\mathbf{u},\mathbf{w}} l(p) \; : \; pi \text{ is a prefix of } w\} =$$

$$\bigcup_{\substack{j \in \mathcal{A},\, p_0 \in \mathcal{A}^* \\ \sigma_0(j) = p_0 i \mathcal{A}^*}} \pi_{\mathbf{u},\mathbf{w}} l(p_0) + M_0 \{\pi_{M_0^{-1}\mathbf{u}, M_0^t \mathbf{w}} l(p') \; : \; p'j \text{ is a prefix of } w^{(1)}\}.$$

Taking the union over all (finitely many, by Proposition 3.5.3) limit sequences of σ and taking the closure we obtain by (3.33) that

$$\mathcal{R}_{\mathbf{w}}(i) = \bigcup_{\substack{j \in \mathcal{A},\, p_0 \in \mathcal{A}^* \\ \sigma_0(j) = p_0 i \mathcal{A}^*}} \pi_{\mathbf{u},\mathbf{w}} l(p_0) +$$

$$M_0 \overline{\{\pi_{M_0^{-1}\mathbf{u}, M_0^t \mathbf{w}} l(p') \; : \; p'j \text{ is a prefix of some limit sequence of } (\sigma_{n+1})\}}.$$

(3.42)

We now inspect the closures in the union in (3.42). Looking at the definition of subtiles in (3.33) we see that these are subtiles of the Rauzy fractal corresponding to the shifted sequence $(\sigma_{n+1})_{n\geq 0}$ of σ. Indeed, it follows from Proposition 3.5.5 that $M_0^{-1}\mathbf{u}$ is the right eigenvector of this shifted sequence. This motivates the following definitions.

For $k \in \mathbb{N}$ let

$$\pi_{\mathbf{u},\mathbf{w}}^{(k)} = \pi_{M_{[0,k)}^{-1}\mathbf{u}, M_{[0,k)}^t \mathbf{w}}, \tag{3.43}$$

denote the subtiles of the shifted sequence of substitutions $(\sigma_{n+k})_{n\in\mathbb{N}}$ which live in the hyperplane $(M_{[0,k)}^t \mathbf{w})^\perp$ by

$$\mathcal{R}_{\mathbf{w}}^{(k)}(i) := \overline{\{\pi_{\mathbf{u},\mathbf{w}}^{(k)} l(p') \; : \; p'j \text{ is a prefix of some limit sequence of } (\sigma_{n+k})_{n\in\mathbb{N}}\}}, \tag{3.44}$$

and set $\mathcal{R}_{\mathbf{w}}^{(k)} = \bigcup_{i\in\mathcal{A}} \mathcal{R}_{\mathbf{w}}^{(k)}(i)$. Together with these notations (3.42) can be generalized by using similar arguments as we used in its proof. The generalized form of (3.42) reads as follows (for a detailed proof see [52, Proposition 5.6]).

Proposition 3.7.3 (The Set Equation) *Let σ be a primitive and recurrent sequence of unimodular substitutions with generalized right eigenvector $\mathbf{u}$. Then for each $[\mathbf{x}, i] \in \mathbb{Z}^d \times \mathcal{A}$ and every $k, \ell \in \mathbb{N}$ with $k < \ell$ we have*

$$\pi_{\mathbf{u},\mathbf{w}}^{(k)} \mathbf{x} + \mathcal{R}_{\mathbf{w}}^{(k)}(i) = \bigcup_{[\mathbf{y},j]\in E_1^*(\sigma_{[k,\ell)})[\mathbf{x},i]} M_{[k,\ell)}(\pi_{\mathbf{u},\mathbf{w}}^{(\ell)} \mathbf{y} + \mathcal{R}_{\mathbf{w}}^{(\ell)}(j)), \tag{3.45}$$

where

$$E_1^*(\sigma)[\mathbf{x}, i] = \{[M_\sigma^{-1}(\mathbf{x} + \mathbf{l}(p)), j] \; : \; j \in \mathcal{A}, \; p \in \mathcal{A}^*, \; pi \; prefix \; of \; \sigma(j)\}. \tag{3.46}$$

The elements in the union on the right hand side of (3.45) are called the *level* $(\ell - k)$ *subtiles* of $\pi_{\mathbf{u},\mathbf{w}}^{(k)}\mathbf{x} + \mathcal{R}_{\mathbf{w}}^{(k)}(i)$. The collection of all the elements in the union is called the $(\ell - k)$*-th subdivision* of $\pi_{\mathbf{u},\mathbf{w}}^{(k)}\mathbf{x} + \mathcal{R}_{\mathbf{w}}^{(k)}(i)$. This will often be applied for the case $k = 0$. In Fig. 3.16 the set equation is illustrated for the situation discussed in Example 3.7.8.

The *dual geometric realization* $E_1^*(\sigma)$ of a substitution σ defined in (3.46) will turn out to be useful when we define so-called *coincidence conditions* in Sect. 3.8.2. If we regard the pairs $[\mathbf{x}, i]$ as hypercubes as we did in (3.34) this dual also has a geometric meaning. We explain this in the following example.

Example 3.7.4 Let σ be the Tribonacci substitution defined in (3.20). Then by direct computation we see that $E_1^*(\sigma)$ is given by

$$E_1^*(\sigma)[\mathbf{0}, 1] = \{[\mathbf{0}, 1], [\mathbf{0}, 2], [\mathbf{0}, 3]\},$$

$$E_1^*(\sigma)[\mathbf{0}, 2] = \{[(0, 0, 1)^t, 1]\},$$

$$E_1^*(\sigma)[\mathbf{0}, 3] = \{[(0, 0, 1)^t, 2]\}$$

together with the obvious fact that $E_1^*(\sigma)[\mathbf{x}, i] = M_\sigma^{-1}\mathbf{x} + E_1^*(\sigma)[\mathbf{0}, i]$. One can extend the definition of $E_1^*(\sigma)$ to subsets of $Y \subset \mathbb{Z}^d \times \mathcal{A}$ in a natural way by setting

$$E_1^*(\sigma)Y = \bigcup_{[\mathbf{x}, i] \in Y} E_1^*(\sigma)[\mathbf{x}, i].$$

Using this extension we can then iterate $E_1^*(\sigma)$. The geometric interpretation of $E_1^*(\sigma)^{12}[\mathbf{0}, 1]$ is depicted in Fig. 3.15. It is not by accident that this image is a good approximation of (an affine image of) the classical Rauzy fractal corresponding to σ depicted in Fig. 3.12. In fact, $E_1^*(\sigma)$ can even be used to give an alternative definition of $\mathcal{R}$, see for example [18, 51].

The dual $E_1^*(\sigma)$ and its higher dimensional generalizations have been investigated thoroughly in connection with the study of substitutive dynamical systems and their Rauzy fractals (see [18, 51, 74, 91, 100, 117]). We need a result of Fernique [79] that shows how $E_1^*(\sigma)$ behaves with respect to discrete hyperplanes. Before we state it we introduce some notation. Let σ be a sequence of substitutions with generalized right eigenvector $\mathbf{u} \in \mathbb{R}_{>0}^d$ and let a fixed vector $\mathbf{w} \in \mathbb{R}_{\geq 0}^d \setminus \{\mathbf{0}\}$ be given (such that the Rauzy fractal $\mathcal{R}_{\mathbf{w}}$ can be defined). Then, motivated by the projections (3.43) we needed in the formulation of the set equation we set

$$\mathbf{u}^{(k)} = (M_{[0,k)})^{-1}\mathbf{u}, \qquad \mathbf{w}^{(k)} = (M_{[0,k)})^t\mathbf{w} \qquad (k \in \mathbb{N}). \tag{3.47}$$

Fig. 3.15 An approximation
of $\mathcal{R}$ using $E_1^*(\sigma)$

Lemma 3.7.5 *Let $\sigma = (\sigma_n)$ be a sequence of unimodular substitutions. Then for all $k < \ell$ the following assertions hold.*

(i) $M_{[k,\ell)}(\mathbf{w}^{(\ell)})^{\perp} = (\mathbf{w}^{(k)})^{\perp}$,

(ii) $E_1^*(\sigma_{[k,\ell)})\Gamma(\mathbf{w}^{(k)}) = \Gamma(\mathbf{w}^{(\ell)})$,

(iii) *for distinct pairs $[\mathbf{x}, i], [\mathbf{x}', i'] \in \Gamma(\mathbf{w}^{(k)})$ the images $E_1^*(\sigma_{[k,\ell)})[\mathbf{x}, i]$ and $E_1^*(\sigma_{[k,\ell)})[\mathbf{x}', i']$ are disjoint patches of $\Gamma(\mathbf{w}^{(\ell)})$.*

Proof Assertion (i) is an immediate consequence of the definition of $\mathbf{w}^{(k)}$, assertions (ii) and (iii) are the content of [79, Theorem 1]. Their proof is a bit tedious, however, it just uses the definition of discrete hyperplane and checks the required conditions (assertion (iii) is essentially already contained in [18, Lemma 3]). □

Combining Proposition 3.7.3 and Lemma 3.7.5 we get the following result in which we use the notation

$$\mathcal{C}_{\mathbf{w}}^{(k)} = \{\pi_{\mathbf{u},\mathbf{w}}\mathbf{x} + \mathcal{R}_{\mathbf{w}}^{(k)}(i) \ : \ [\mathbf{x}, i] \in \Gamma(\mathbf{w}^{(k)})\} \qquad (k \in \mathbb{N})$$

for the collection of subtiles associated with the shifted sequence $(\sigma_{n+k})_{n \in \mathbb{N}}$ of σ.

Proposition 3.7.6 *Let σ be a primitive and recurrent sequence of unimodular substitutions with generalized right eigenvector $\mathbf{u}$. Then for each $[\mathbf{x}, i] \in \mathbb{Z}^d \times \mathcal{A}$ and every $k, \ell \in \mathbb{N}$ with $k < \ell$ we have*

$$\bigcup_{[\mathbf{x},i]\in\Gamma(\mathbf{w}^{(k)})} \pi_{\mathbf{u},\mathbf{w}}\mathbf{x} + \mathcal{R}_{\mathbf{w}}^{(k)}(i) = \bigcup_{[\mathbf{y},j]\in\Gamma(\mathbf{w}^{(\ell)})} M_{[k,\ell)}(\pi_{\mathbf{u},\mathbf{w}}^{(\ell)}\mathbf{y} + \mathcal{R}_{\mathbf{w}}^{(\ell)}(j)).$$

The collection $M_{[k,\ell)}C_{\mathbf{w}}^{(\ell)}$ is a refinement of $C_{\mathbf{w}}^{(k)}$ in the sense that each element of the latter is a finite union of elements of the former.

The following lemma shows that the set equation subdivides Rauzy fractals into sets whose diameter eventually tends to zero (see [52, Lemma 5.5]).

Lemma 3.7.7 *Let $\boldsymbol{\sigma} = (\sigma_n) \in S^{\mathbb{N}}$ be a primitive, algebraically irreducible, and recurrent sequence of unimodular substitutions with balanced language $\mathcal{L}_{\boldsymbol{\sigma}}$, and let $\mathbf{w} \in \mathbb{R}_{\geq 0}^d \setminus \{\mathbf{0}\}$. Then*

$$\lim_{n \to \infty} M_{[0,n)} \mathcal{R}_{\mathbf{w}}^{(n)} = \{\mathbf{0}\}.$$

Proof As $M_{[0,n)} \pi_{\mathbf{u},\mathbf{w}}^{(n)} = \pi_{\mathbf{u},\mathbf{w}} M_{[0,n)}$ and $\pi_{\mathbf{u},\mathbf{w}} = \pi_{\mathbf{u},\mathbf{w}} \pi_{\mathbf{u},\mathbf{1}}$, we conclude that $M_{[0,n)} \pi_{\mathbf{u},\mathbf{w}}^{(n)} \mathbf{l}(v) = \pi_{\mathbf{u},\mathbf{w}} \pi_{\mathbf{u},\mathbf{1}} M_{[0,n)} \mathbf{l}(v)$ for all $v \in \mathcal{L}_{\boldsymbol{\sigma}}^{(n)}$. Now, the result follows from Proposition 3.6.10 and the definition of $\mathcal{R}_{\mathbf{w}}^{(n)}$ in (3.44). $\qquad\square$

We explain the concepts and results of this section in the next example.

Example 3.7.8 Recall the definition of the Arnoux-Rauzy substitutions $\sigma_1, \sigma_2, \sigma_3$ from (3.18) and consider a sequence

$$\boldsymbol{\sigma} = (\sigma_1, \sigma_2, \sigma_3, \sigma_1, \sigma_2, \sigma_3, \sigma_1, \sigma_2, \sigma_3, \dots),$$

where the dots "..." mean that the sequence is continued in a way that $\boldsymbol{\sigma}$ is primitive and recurrent.

If we start with three blocks of the form $\sigma_1, \sigma_2, \sigma_3$ it turns out that $\mathcal{R}_{\mathbf{w}}$ is close to the classical Rauzy fractal studied in Example 3.6.2 in Hausdorff metric, which, of course, doesn't say anything about its topological properties or tiling properties; we just did it this way to get nice pictures in Fig. 3.16.

In Fig. 3.16a we show a patch P_0 of the collection $C_{\mathbf{w}}$ (for some convenient vector $\mathbf{w} \in \mathbb{R}_{\geq 0}^3 \setminus \{\mathbf{0}\}$) of subtiles associated with $\boldsymbol{\sigma}$, while Fig. 3.16b shows a patch P_1 of the collection $C_{\mathbf{w}}^{(1)}$ associated with the shifted sequence

$$\boldsymbol{\sigma}^{(1)} = (\sigma_2, \sigma_3, \sigma_1, \sigma_2, \sigma_3, \sigma_1, \sigma_2, \sigma_3, \dots).$$

Note that, since $\mathbf{w}$ and $\mathbf{w}^{(1)}$ are not collinear, these patches live in two different planes which is illustrated in Fig. 3.16c.

In this setting, the set equation in Proposition 3.7.3 says that each element of the collection $C_{\mathbf{w}}$ can be viewed as the union of elements from $M_0 C_{\mathbf{w}}^{(1)}$. In other words, if we take the patch P_1 depicted in Fig. 3.16b and apply the linear mapping M_0 to it, the resulting patch $M_0 P_1$ lies in the same plane $\mathbf{w}^{\perp}$ as the collection $C_{\mathbf{w}}$ and some elements of P_0 are unions of elements from $M_0 P_1$. In Fig. 3.16d this is illustrated: the image of the patch P_1 from Fig. 3.16b under the mapping M_0 is subdividing some parts of P_0. Figure 3.16e illustrates that, according to Proposition 3.7.6, each element of $C_{\mathbf{w}}$ is a union of elements from $M_0 C_{\mathbf{w}}^{(1)}$.

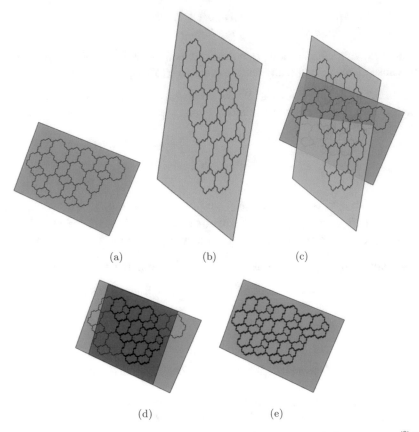

(a)   (b)   (c)

(d)   (e)

Fig. 3.16 Illustration of the set equation. (**a**) shows a patch P_0 of the collection $\mathcal{C}_\mathbf{w} = \mathcal{C}_\mathbf{w}^{(0)}$, (**b**) shows a patch P_1 of $\mathcal{C}_\mathbf{w}^{(1)}$. In (**c**) P_0 and P_1 are drawn together to illustrate that they lie in different planes. In (**d**) the matrix M_0 is applied to P_1: the image $M_0 P_1$ is located in the same plane as P_0 and forms a subdivision of tiles of P_0. The subdivision of P_0 in patches of $M_0 \mathcal{C}_\mathbf{w}^{(1)}$ is shown in (**e**) for the whole patch P_0

Note that in Fig. 3.16 the collections $\mathcal{C}_\mathbf{w}$ and $\mathcal{C}_\mathbf{w}^{(1)}$ are depicted as tilings and the patches of $M_0 \mathcal{C}_\mathbf{w}^{(1)}$ subdivide the elements of $\mathcal{C}_\mathbf{w}$ without overlap. This is the situation we "dream" of. So far, we only know that elements of $\mathcal{C}_\mathbf{w}$ are unions of elements of $M_0 \mathcal{C}_\mathbf{w}^{(1)}$. To realize this ideal situation we need to work more.

3.7.2 An S-adic Rauzy Fractal Is the Closure of Its Interior

The present section is devoted to the interior of the subtiles. We start with a covering result taken from [52, Proposition 6.2]. In its statement we use the following terminology. Let $\mathcal{K}$ be a collection of subsets of a set D. The *covering degree* of $\mathcal{K}$ (in D) is the largest number m having the property that each $x \in D$ is contained in at least m elements of $\mathcal{K}$.

Lemma 3.7.9 *Let* σ *be a sequence of unimodular substitutions and* $\mathbf{w} \in \mathbb{R}_{\geq 0} \setminus \{\mathbf{0}\}$. *If* σ *is primitive, recurrent, algebraically irreducible, and has finitely balanced language* $\mathcal{L}_\sigma$ *then* $C_{\mathbf{w}}^{(n)}$ *covers* $(\mathbf{w}^{(n)})^{\perp}$ *with finite covering degree for each* $n \in \mathbb{N}$. *The covering degree of* $C_{\mathbf{w}}^{(n)}$ *increases monotonically with* n.

Proof We prove the covering property for $C_{\mathbf{w}}$. The covering property for $C_{\mathbf{w}}^{(n)}$ as well as the monotonicity of the covering degree follow from this by the set equation in Proposition 3.7.3.

By Proposition 3.7.6 with the choices $k = 0$ and $\ell = n \geq n_0$ we know that

$$\bigcup_{T \in C_{\mathbf{w}}} T = \bigcup_{n \geq n_0} \bigcup_{T \in C_{\mathbf{w}}^{(n)}} M_{[0,n)} T = \bigcup_{n \geq n_0} \bigcup_{[\mathbf{y},j] \in \Gamma(\mathbf{w}^{(n)})} M_{[0,n)} (\pi_{\mathbf{u},\mathbf{w}}^{(n)} \mathbf{y} + \mathcal{R}_{\mathbf{w}}^{(n)}(j))$$

(3.48)

holds for each $n_0 \in \mathbb{N}$. Because $C_{\mathbf{w}}$ is a locally finite collection of compact sets it suffices to show that $\bigcup_{T \in C_{\mathbf{w}}} T$ is dense in $\mathbf{w}^{\perp}$. To prove this we show that the right hand side of (3.48) is dense in $\mathbf{w}^{\perp}$ for each $n_0 \in \mathbb{N}$. To see this note that by the definition of the discrete hyperplane $\Gamma(\mathbf{w}^{(n)})$ the set of translates in this union satisfies (recall from (3.47) that $\mathbf{w}^{(n)} = (M_{[0,n)})^t \mathbf{w}$)

$$\{M_{[0,n)} \pi_{\mathbf{u},\mathbf{w}}^{(n)} \mathbf{y} \; : \; [\mathbf{y}, j] \in \Gamma(\mathbf{w}^{(n)})\} =$$

$$\{\pi_{\mathbf{u},\mathbf{w}} M_{[0,n)} \mathbf{y} \; : \; \mathbf{y} \in \mathbb{Z}^d, \, 0 \leq \langle \mathbf{y}, (M_{[0,n)})^t \mathbf{w} \rangle \leq \max_{i \in \mathcal{A}} \langle \mathbf{e}_i, (M_{[0,n)})^t \mathbf{w} \rangle\} =$$

$$\{\pi_{\mathbf{u},\mathbf{w}} \mathbf{z} \; : \; \mathbf{z} \in \mathbb{Z}^d, \, 0 \leq \langle \mathbf{z}, \mathbf{w} \rangle \leq \max_{i \in \mathcal{A}} \langle M_{[0,n)} \mathbf{e}_i, \mathbf{w} \rangle\}.$$

As $\mathbf{u}$ has rationally independent coordinates by Lemma 3.6.8, the set

$$\{\pi_{\mathbf{u},\mathbf{w}} \mathbf{z} \; : \; \mathbf{z} \in \mathbb{Z}^d, \, 0 \leq \langle \mathbf{z}, \mathbf{w} \rangle\}$$

is dense in $\mathbf{w}^{\perp}$. Since $\max_{i \in \mathcal{A}} \langle M_{[0,n)} \mathbf{e}_i, \mathbf{w} \rangle \to \infty$ for $n \to \infty$ by primitivity, this yields that

$$\bigcup_{n \geq n_0} \bigcup_{[\mathbf{y},j] \in \Gamma(\mathbf{w}^{(n)})} M_{[0,n)} \pi_{\mathbf{u},\mathbf{w}}^{(n)} \mathbf{y} = \{\pi_{\mathbf{u},\mathbf{w}} \mathbf{z} \; : \; \mathbf{z} \in \mathbb{Z}^d, \, 0 \leq \langle \mathbf{z}, \mathbf{w} \rangle\}. \tag{3.49}$$

is dense in $\mathbf{w}^{\perp}$ for each $n_0 \in \mathbb{N}$. Because n_0 was arbitrary, and the limit $\lim_{n \to \infty} M_{[0,n)} \mathcal{R}_{\mathbf{w}}^{(n)}(i) = \{\mathbf{0}\}$ by Lemma 3.7.7 this implies that the right hand side of (3.48) is dense in $\mathbf{w}^{\perp}$ for each $n_0 \in \mathbb{N}$ and we are done. $\qquad \square$

From this result we get the assertion on the interiors of S-adic Rauzy fractals.

Proposition 3.7.10 *Let σ be a sequence of unimodular substitutions over the alphabet $\mathcal{A}$ and $\mathbf{w} \in \mathbb{R}_{\geq 0} \setminus \{0\}$. If σ is primitive, recurrent, algebraically irreducible, and has finitely balanced language $\mathcal{L}_\sigma$, then $\mathcal{R}(i)$ is the closure of its interior for each $i \in \mathcal{A}$.*

Proof Choose some $\mathbf{w} \in \mathbb{R}_{\geq 0} \setminus \{0\}$. By Lemma 3.7.9 the collection $\mathcal{C}_{\mathbf{w}}^{(n)}$ is a locally finite covering of $(\mathbf{w}^{(n)})^{\perp}$ by compact sets for each $n \in \mathbb{N}$. Thus by Baire's theorem for each $n \in \mathbb{N}$ there is $i_n \in \mathcal{A}$ such that $\text{int}(\mathcal{R}_{\mathbf{w}}^{(n)}(i_n)) \neq \emptyset$. By primitivity of σ the set equation in Proposition 3.7.3 implies that each $\mathcal{R}_{\mathbf{w}}^{(n)}(i)$ contains $\mathcal{R}_{\mathbf{w}}^{(k)}(i_k)$ for some $k > n$. Thus for each $n \in \mathbb{N}$ and each $i \in \mathcal{A}$ we have $\text{int}(\mathcal{R}_{\mathbf{w}}^{(n)}(i)) \neq \emptyset$.

For each $i \in \mathcal{A}$ and each $n \in \mathbb{N}$, Proposition 3.7.3 yields a subdivision of $\mathcal{R}_{\mathbf{w}}(i)$ in translates of sets of the form $M_{[0,n)} \mathcal{R}_{\mathbf{w}}^{(n)}(j)$, $j \in \mathcal{A}$. The diameters of these sets tend to 0 by Lemma 3.7.7. Since they all contain inner points, the set of inner points of $\mathcal{R}_{\mathbf{w}}(i)$ is dense in $\mathcal{R}_{\mathbf{w}}(i)$. In other words, $\mathcal{R}_{\mathbf{w}}(i)$ is the closure of its interior. The result now follows by taking $\mathbf{w} = \mathbf{1}$. $\qquad\square$

3.7.3 The Generalized Left Eigenvector

Let σ be a primitive and recurrent sequence of unimodular substitutions. If we look at the set equation in Proposition 3.7.3 for $k = 0$ and $\ell = n$ we see that it subdivides the sets $\mathcal{R}_{\mathbf{w}}(i)$, $i \in \mathcal{A}$, into translates of sets of the form $\mathcal{R}_{\mathbf{w}}^{(n)}(j)$, $j \in \mathcal{A}$. In the well-studied substitutive case $\mathcal{R}_{\mathbf{w}}^{(n)}(i) = \mathcal{R}_{\mathbf{w}}(i)$ holds for each n, i.e., the sets $\mathcal{R}_{\mathbf{w}}(i)$ are subdivided into small copies of themselves. This fact is crucial in most of the proofs of properties of substitutive Rauzy fractals (see e.g. [123]). In our case, in general the sets $\mathcal{R}_{\mathbf{w}}^{(n)}$ are not only different for each $n \in \mathbb{N}$, but also live in different hyperplanes $(\mathbf{w}^{(n)})^{\perp}$ of $\mathbb{R}^n$.

In what follows we want to deal with this problem by choosing a strictly increasing sequence (n_k) of integers such that $\mathcal{R}_{\mathbf{w}}^{(n_k)}(i)$ is at least getting closer and closer to $\mathcal{R}_{\mathbf{w}}(i)$ in Hausdorff metric when $k \to \infty$.

To this matter let σ be a sequence of substitutions that satisfies the assumptions of Theorem 3.7.1. We now successively choose subsequences of the integers to get the desired properties.

(a) Consider the set equation in Proposition 3.7.3 for the choices $k = 0$, $\ell = m$ and $k = n$, $\ell = n + m$. Look at the subdivision of $\mathcal{R}_{\mathbf{w}}(i)$ and $\mathcal{R}_{\mathbf{w}}^{(n)}(i)$. We can hope to get $\mathcal{R}_{\mathbf{w}}(i)$ and $\mathcal{R}_{\mathbf{w}}^{(n)}(i)$ close to each other in Hausdorff metric if they have the same subdivision structure. From Proposition 3.7.3 we see that these subdivision structures are the same if $(\sigma_0, \ldots, \sigma_{m-1}) = (\sigma_n, \ldots, \sigma_{n+m-1})$. Since σ is recurrent, there exist strictly increasing sequences (n_k) and (ℓ_k) such that

$$(\sigma_0, \ldots, \sigma_{\ell_k - 1}) = (\sigma_{n_k}, \ldots, \sigma_{n_k + \ell_k - 1}). \tag{3.50}$$

By recurrence and primitivity it is possible to choose (n_k) and (ℓ_k) in a way that there is some h such that $M_{[\ell_k - h, \ell_k)}$ is the same primitive matrix for all $k \in \mathbb{N}$.

(b) We know that $M_{[0,\ell_k)}\mathcal{R}_\mathbf{w}^{(\ell_k)}(j)$ tends to $\{\mathbf{0}\}$ in Hausdorff metric for $k \to \infty$ by Lemma 3.7.7 so that the subdivision corresponding to the choice $k = 0$, $\ell = \ell_k$ in the set equation gives a subdivision of $\mathcal{R}_\mathbf{w}(j)$ into sets whose diameter tends to 0 for $k \to \infty$. However, if we consider $\mathcal{R}_\mathbf{w}^{(n_k)}(i)$, there is no reason for $M_{[n_k,n_k+\ell_k)}\mathcal{R}_\mathbf{w}^{(n_k+\ell_k)}(j) = M_{[0,\ell_k)}\mathcal{R}_\mathbf{w}^{(n_k+\ell_k)}(j)$ to tend to $\{\mathbf{0}\}$ unless $\mathcal{R}_\mathbf{w}^{(n_k+\ell_k)}(j)$ is bounded uniformly in k. To this end we need to assume that $\mathcal{L}_\sigma^{(n_k+\ell_k)}$ is C-balanced as this implies that $\mathcal{R}_\mathbf{w}^{(n_k+\ell_k)}(j)$ is indeed bounded by Proposition 3.6.5. In view of the conditions imposed on σ in Theorem 3.7.1 it is, however, possible to change the sequence (n_k) and (ℓ_k) chosen in (a) in a way that also $\mathcal{L}_\sigma^{(n_k+\ell_k)}$ is C-balanced for $C \in \mathbb{N}$ not depending on k.

(c) Still (a) and (b) give us no reason for $\mathcal{R}_\mathbf{w}^{(n_k)}$ living in a hyperplane $\mathbf{w}^{(n_k)}$ close to $\mathbf{w}$ which is needed in order to get $\mathcal{R}_\mathbf{w}^{(n_k)}$ close to $\mathcal{R}_\mathbf{w}$ in Hausdorff metric. By the compactness of the space of directions in $\mathbb{R}^d$, using the Hilbert metric from Proposition 3.5.5 it is possible to exhibit a vector $\mathbf{v} \in \mathbb{R}_{\geq 0} \setminus \{\mathbf{0}\}$ for which there exists subsequences of (n_k) and (ℓ_k) (called (n_k) and (ℓ_k) again) such that $\lim_{k\to\infty} \mathbf{v}^{(n_k)}/\|\mathbf{v}^{(n_k)}\|_1 = \mathbf{v}/\|\mathbf{v}\|_1$. Here we set $\mathbf{v}^{(n)} = (M_{[0,n)})^t\mathbf{v}$.

Summing up, if the conditions of Theorem 3.7.1 are in force we can choose sequences (n_k) and (ℓ_k) satisfying (a), (b), and (c). The vector $\mathbf{v}$ defined in (c) deserves special attention.

Definition 3.7.11 (Generalized Left Eigenvector) A vector $\mathbf{v}$ as in (c) is called a *generalized left eigenvector* of σ.

Sequences (n_k) and (ℓ_k) associated with σ in the above way will just be called *associated sequences* for σ in the sequel (they are related to the property *PRICE* of [52, Definition 5.8]). It turns out that associated sequences are suitable for our purposes. In particular, we get the following result (we refer to [52, Proposition 5.12] for details).

Proposition 3.7.12 *Let σ be a sequence of substitutions that admits associated sequences (n_k) and (ℓ_k) and has a generalized left eigenvector $\mathbf{v}$. Then for each $i \in \mathcal{A}$*

$$\lim_{k\to\infty} \mathcal{R}_\mathbf{v}^{(n_k)}(i) = \mathcal{R}_\mathbf{v}(i)$$

in Hausdorff metric.

Proof *(Sketch)* By (3.50) in (a) the sets $\mathcal{R}_\mathbf{v}(i)$ and $\mathcal{R}_\mathbf{v}^{(n_k)}(i)$ have the same subdivision structure governed by $E_1^*(\sigma_{[0,\ell_k)})$ for $k \in \mathbb{N}$. More precisely,

$$\mathcal{R}_\mathbf{v}(i) = \bigcup_{[\mathbf{y},j]\in E_1^*(\sigma_{[0,\ell_k)})[\mathbf{0},i]} M_{[0,\ell_k)}(\pi_{\mathbf{u},\mathbf{v}}^{(\ell_k)}\mathbf{y} + \mathcal{R}_\mathbf{v}^{(\ell_k)}(j)),$$

$$\mathcal{R}_\mathbf{v}^{(n_k)}(i) = \bigcup_{[\mathbf{y},j]\in E_1^*(\sigma_{[0,\ell_k)})[\mathbf{0},i]} M_{[0,\ell_k)}(\pi_{\mathbf{u},\mathbf{v}}^{(n_k+\ell_k)}\mathbf{y} + \mathcal{R}_\mathbf{v}^{(n_k+\ell_k)}(j)).$$

$$(3.51)$$

By Proposition 3.6.10 the sets $M_{[0,\ell_k)}\mathcal{R}_\mathbf{v}^{(\ell_k)}(j)$ tend to $\{\mathbf{0}\}$ in Hausdorff metric for $k \to \infty$. With more effort, using the balance conditions of (b) and the convergence properties of (c), one can also show that the sets $M_{[0,\ell_k)}\mathcal{R}_\mathbf{v}^{(n_k+\ell_k)}(j)$ tend to $\{\mathbf{0}\}$ in Hausdorff metric for $k \to \infty$. So replacing all these sets by $\{\mathbf{0}\}$ on the right hand side of (3.51) changes the sets on the left hand side of (3.51) only very little in Hausdorff metric for large $k \in \mathbb{N}$. Thus for large $k \in \mathbb{N}$ the Hausdorff distance between $\mathcal{R}_\mathbf{v}^{(n_k)}(i)$ and $\mathcal{R}_\mathbf{v}(i)$ is (up to an error tending to 0 for $k \to \infty$) bounded by

$$\max\left\{\|M_{[0,\ell_k)}(\pi_{\mathbf{u},\mathbf{v}}^{(\ell_k)}\mathbf{y} - \pi_{\mathbf{u},\mathbf{v}}^{(n_k+\ell_k)}\mathbf{y}\|_\infty \ : \ [\mathbf{y}, j] \in E_1^*(\sigma_{[0,\ell_k)})[\mathbf{0}, i]\right\} =$$

$$\max\left\{\|\pi_{\mathbf{u},\mathbf{v}}M_{[0,\ell_k)}\mathbf{y} - \pi_{\mathbf{u},\mathbf{v}}^{(n_k)}M_{[0,\ell_k)}\mathbf{y}\|_\infty \ : \ [\mathbf{y}, j] \in E_1^*(\sigma_{[0,\ell_k)})[\mathbf{0}, i]\right\}.$$

One can now show that the latter maximum tends to 0 for $k \to \infty$. Here one uses that by the definition of the generalized left eigenvector in (c) the hyperplanes $(\mathbf{v}^{(n_k)})^\perp$ converge to $\mathbf{v}^\perp$.   □

3.7.4 An S-adic Rauzy Fractal Has a Boundary of Measure Zero

We now turn to the boundary of an S-adic Rauzy fractal. We start with a result on level ℓ subtiles contained in the interior of a given subtile whose detailed proof is contained in [52, Lemma 6.6].

Lemma 3.7.13 *Let σ be a sequence of unimodular substitutions that satisfies the properties of Theorem 3.7.1 and let associated sequences (n_k), (ℓ_k), and a generalized left eigenvector $\mathbf{v}$ be given.*

Then there is $\ell \in \mathbb{N}$ such that for each $i, j \in \mathcal{A}$ there is $[\mathbf{y}, j] \in E_1^*(\sigma_{[0,\ell)})[\mathbf{0}, i]$ such that

(i) $M_{[0,\ell)}(\pi_{\mathbf{u},\mathbf{v}}^{(\ell)}\mathbf{y} + \mathcal{R}_\mathbf{v}^{(\ell)}(j)) \subset \text{int}(\mathcal{R}_\mathbf{v}(i))$,

(ii) $M_{[0,\ell)}(\pi_{\mathbf{u},\mathbf{v}}^{(n_k+\ell)}\mathbf{y} + \mathcal{R}_\mathbf{v}^{(n_k+\ell)}(j)) \subset \text{int}(\mathcal{R}_\mathbf{v}^{(n_k)}(i))$ for each sufficiently large $k \in \mathbb{N}$.

Moreover, the covering degree of $\mathcal{C}_\mathbf{v}^{(n)}$ does not depend on n.

Proof *(Sketch)* Since the conditions in the lemma imply that $\text{int}(\mathcal{R}_\mathbf{v}(i)) \neq \emptyset$ (see Proposition 3.7.10) and that the diameter of $M_{[0,\ell)}\mathcal{R}_\mathbf{v}^{(\ell)}(j)$ becomes arbitrarily small for large ℓ (see Lemma 3.7.7), assertion (i) follows easily from primitivity.

The fact that ℓ and $\mathbf{y}$ can be chosen in a way that (i) and (ii) hold simultaneously is more difficult to prove. By Proposition 3.7.12 we get that $\mathcal{R}_\mathbf{v}^{(n_k)}(i) \to \mathcal{R}_\mathbf{v}(i)$ in Hausdorff metric. Moreover, $\mathcal{R}_\mathbf{v}(i)$ and $\mathcal{R}_\mathbf{v}^{(n_k)}(i)$ have the same subdivision structure for ℓ_k steps. This implies that the "inner structure" of these tiles is similar for large k. However, as inner points are not respected by the Hausdorff metric, technical difficulties occur and also the "outer structure", i.e., the structure of the collections

$C_\mathbf{v}$ and $C_\mathbf{v}^{(n_k)}$ has to be exploited. One can show that if a patch P occurs in a discrete hyperplane $\Gamma(\mathbf{w})$, then translates of P occur relatively densely in each discrete hyperplane $\Gamma(\tilde{\mathbf{w}})$ provided that $\|\mathbf{w} - \tilde{\mathbf{w}}\|_\infty$ is small enough. In particular, containment of translates of a patch P is an open property of discrete hyperplanes. Thus, if k is large then at level n_k there is a translation $\mathbf{y}_k \in \mathbb{Z}^d$ such that the sets $\Gamma(\mathbf{v})$ and $\Gamma(\mathbf{v}^{(n_k)}) - \mathbf{y}_k$ have a large patch around the origin in common.

Summing up, this means that the collections $C_\mathbf{v}^{(n_k)} - \pi_{\mathbf{u},\mathbf{y}}^{(n_k)} \mathbf{y}_k$ converge[7] to $C_\mathbf{v}$ for $k \to \infty$. This implies that the covering degree of $C_\mathbf{v}^{(n_k)} - \pi_{\mathbf{u},\mathbf{y}}^{(n_k)} \mathbf{y}_k$ is less than or equal to the covering degree of $C_\mathbf{v} = C_\mathbf{v}^{(0)}$ for k large enough. Since the covering degree of $C_\mathbf{v}^{(n)}$ is monotonically increasing in n by Lemma 3.7.9, the last assertion of the lemma follows.

The fact that the collections $C_\mathbf{v}^{(n_k)} - \pi_{\mathbf{u},\mathbf{y}}^{(n_k)} \mathbf{y}_k$ converge to $C_\mathbf{v}$ and have the same covering degree can now be used to show that inner points of elements of $C_\mathbf{v}$ are close to inner points of elements of $C_\mathbf{v}^{(n_k)} - \pi_{\mathbf{u},\mathbf{y}}^{(n_k)} \mathbf{y}_k$ for large k. Using this together with the fact that $\mathcal{R}_\mathbf{v}(i)$ and $\mathcal{R}_\mathbf{v}^{(n_k)}(i)$ have the same subdivision structure for ℓ_k steps, one can show that (i) and (ii) holds simultaneously as claimed. $\square$

After these preparations we can also prove the result on the measure of the boundary of S-adic Rauzy fractals announced in Theorem 3.7.1.

Proposition 3.7.14 *Let σ be a sequence of unimodular substitutions over the alphabet $\mathcal{A}$ that satisfies the assertions of Theorem 3.7.1. Then the Lebesgue measure $\lambda_1(\partial\mathcal{R}(i))$ is zero for each $i \in \mathcal{A}$.*

Proof *(Sketch)* Choose $\ell \in \mathbb{N}$ and the sequences (n_k) and (ℓ_k) as in Lemma 3.7.13 and consider $\mathcal{R}_\mathbf{v}(i)$ for some $i \in \mathcal{A}$ (see Fig. 3.17a), where $\mathbf{v}$ is a generalized left eigenvector of σ. Then subdivide $\mathcal{R}_\mathbf{v}(i)$ into its level ℓ subtiles as shown in Fig. 3.17b. According to Lemma 3.7.13(i) there is at least one level ℓ subtile $M_{[0,\ell)}(\pi_{\mathbf{u},\mathbf{v}}^{(\ell)}\mathbf{y} + \mathcal{R}_\mathbf{v}^{(\ell)}(j))$ which is a subset of $\mathrm{int}(\mathcal{R}_\mathbf{v}(i))$; this is indicated with a black boundary in Fig. 3.17b. Letting $m_{ij} = \lambda_\mathbf{v}(M_{[0,\ell)}\mathcal{R}_\mathbf{v}^{(\ell)}(j))/\lambda_\mathbf{v}(\mathcal{R}_\mathbf{v}(i))$ and $m = \min\{m_{ij} : i, j \in \mathcal{A}\}$ we therefore gain

$$\lambda_\mathbf{v}(\partial\mathcal{R}_\mathbf{v}(i)) = \lambda_\mathbf{v}(\mathcal{R}_\mathbf{v}(i) \setminus \mathrm{int}(\mathcal{R}_\mathbf{v}(i))) \leq (1 - m)\lambda_\mathbf{v}(\mathcal{R}_\mathbf{v}(i)).$$

Now we subdivide all level ℓ subtiles of $\mathcal{R}_\mathbf{v}(i)$ apart from $M_{[0,\ell)}(\pi_{\mathbf{u},\mathbf{v}}^{(\ell)}\mathbf{y} + \mathcal{R}_\mathbf{v}^{(\ell)}(j))$ in level n_k subtiles where k is chosen in a way that $n_k \geq \ell$. This is illustrated in Fig. 3.17c.

---

[7]In [111] a space of tilings is equipped with a topology by saying that two tilings are close to each other if their tiles are close to each other in Hausdorff metric inside a large ball around the origin. Although $C_\mathbf{v}$ and $C_\mathbf{v}^{(n_k)}$ are no tilings, an analogous topology can be used here: $C_\mathbf{v}$ and $C_\mathbf{v}^{(n_k)} - \pi_{\mathbf{u},\mathbf{y}}^{(n_k)} \mathbf{y}_k$ are said to be close to each other if $\Gamma(\mathbf{v})$ and $\Gamma(\mathbf{v}^{(n_k)}) - \mathbf{y}_k$ coincide inside a large ball B around the origin and the tiles associated to an element of $[\mathbf{y}, i] \in \Gamma(\mathbf{v}) \cap B$ in each of these two collections are close to each other in Hausdorff metric.

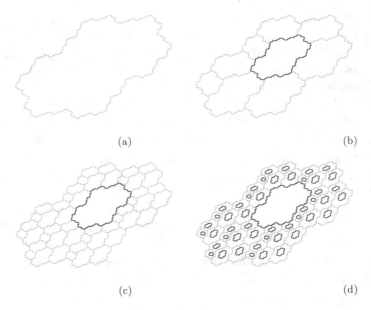

(a)   (b)

(c)   (d)

Fig. 3.17 Illustration of the proof of Proposition 3.7.14. In (**a**) a subtile $\mathcal{R}_{\mathbf{v}}(i)$, $i \in \mathcal{A}$, is shown. In (**b**) we see the ℓth subdivision of $\mathcal{R}_{\mathbf{v}}(i)$. The level ℓ subtile contained in $\mathrm{int}(\mathcal{R}_{\mathbf{v}}(i))$ has black boundary. In (**c**) all other level ℓ subtiles are further subdivided in level n_k subtiles. Each of them contains a level $n_k + \ell$ subtile in its interior. These level $n_k + \ell$ subtiles, which a fortiori are also contained in $\mathrm{int}(\mathcal{R}_{\mathbf{v}}(i))$, are depicted in (**d**) also with black boundary

We iterate this procedure: each level n_k subtile R_{n_k} we got in this way is subdivided in level $n_k + \ell$ subtiles. By Lemma 3.7.13(ii) one of these level $n_k + \ell$ subtiles lies in the interior of R_{n_k} (see Fig. 3.17d for an illustration of this) and, a fortiori, in the interior of $\mathcal{R}_{\mathbf{v}}(i)$. If we set

$$m_{ij}^{(n_k)} = \frac{\lambda_{\mathbf{v}}(M_{[0,n_k+\ell)}\mathcal{R}_{\mathbf{v}}^{(n_k+\ell)}(j))}{\lambda_{\mathbf{v}}(M_{[0,n_k)}\mathcal{R}_{\mathbf{v}}^{(n_k)}(i))} = \frac{\lambda_{\mathbf{v}}(M_{[n_k,n_k+\ell)}\mathcal{R}_{\mathbf{v}}^{(n_k+\ell)}(j))}{\lambda_{\mathbf{v}}(\mathcal{R}_{\mathbf{v}}^{(n_k)}(i))}$$

$$= \frac{\lambda_{\mathbf{v}}(M_{[0,\ell)}\mathcal{R}_{\mathbf{v}}^{(n_k+\ell)}(j))}{\lambda_{\mathbf{v}}(\mathcal{R}_{\mathbf{v}}^{(n_k)}(i))}$$

(note that the last equation follows from recurrence of σ if k is chosen large enough) and $m^{(n_k)} = \min\{m_{ij}^{(n_k)} : i, j \in \mathcal{A}\}$ we obtain

$$\lambda_{\mathbf{v}}(\partial \mathcal{R}_{\mathbf{v}}(i)) \le (1 - m)(1 - m^{(n_k)})\lambda_{\mathbf{v}}(\mathcal{R}_{\mathbf{v}}(i)).$$

Iterating this further we get for some infinite set $K \subset \mathbb{N}$ that

$$\lambda_{\mathbf{v}}(\partial \mathcal{R}_{\mathbf{v}}(i)) \le (1 - m) \prod_{k \in K}(1 - m^{(n_k)})\lambda_{\mathbf{v}}(\mathcal{R}_{\mathbf{v}}(i)). \qquad (3.52)$$

One can show that $m^{(n_k)}$ is uniformly bounded away from 0. To this end one needs Proposition 3.7.12 and the fact that ℓ_k is chosen in a way that there is some h such that $M_{[\ell_k - h, \ell_k)}$ is the same primitive matrix for all $k \in \mathbb{N}$ (see (a) in Sect. 3.7.3). Now (3.52) yields $\lambda_\mathbf{v}(\partial \mathcal{R}_\mathbf{v}(i)) = 0$ and, hence, $\lambda_1(\partial \mathcal{R}(i)) = 0$. $\qquad\square$

Propositions 3.7.10 and 3.7.14 imply Theorem 3.7.1.

3.8 Tilings, Coincidence Conditions, and Combinatorial Issues

We now turn to tiling conditions of Rauzy fractals. Already in the substitutive case combinatorial conditions like the *strong coincidence condition* (see e.g. [18]) or the *super coincidence condition* and its variants (cf. [36, 51, 91]) have to be imposed in order to gain all the tiling results on Rauzy fractals required for our purposes. Here we discuss an S-adic version of these concepts and establish a variety of tiling results. For detailed proofs we refer again to Berthé et al. [52]. As before, our aim is to discuss the main ideas and to make these ideas understandable without going into all the technical details.

3.8.1 Multiple Tiling and Inner Subdivision of the Subtiles

In this section we prove tiling properties of Rauzy fractals that hold without further combinatorial conditions. Our first result contains a multiple tiling property of the collections of Rauzy fractals $\mathcal{C}_\mathbf{v}$ defined in (3.35).

Proposition 3.8.1 *Let $\sigma = (\sigma_n)$ be a primitive and algebraically irreducible sequence of unimodular substitutions. Assume that there is $C > 0$ such that for every $\ell \in \mathbb{N}$ there exists $n \geq 1$ such that $(\sigma_n, \ldots, \sigma_{n+\ell-1}) = (\sigma_0, \ldots, \sigma_{\ell-1})$ and the language $\mathcal{L}_\sigma^{(n+\ell)}$ is C-balanced.*

If $\mathbf{v}$ is a generalized left eigenvector of σ then the collection $\mathcal{C}_\mathbf{v}$ forms a multiple tiling of the hyperplane $\mathbf{v}^\perp$.

Proof *(Sketch)* Let (ℓ_k) and (n_k) be associated sequences for σ. We subdivide the proof in seven observations. In the sequel $B_X(\mathbf{x}, \varepsilon)$ denotes an open ball in a metric space X centered at $\mathbf{x}$ with radius ε.

(i) Let $\mathbf{w} \in \mathbb{R}_{\geq 0}^d \setminus \{\mathbf{0}\}$. As mentioned in the proof of Lemma 3.7.13 one can show that each patch $P \subset \Gamma(\mathbf{w})$ is *repetitive* in the following sense: there exists $\delta_P > 0$ and a radius $r_P > 0$ such that for each $\tilde{\mathbf{w}} \in \mathbb{R}_{\geq 0}^d \setminus \{\mathbf{0}\}$ with $\|\tilde{\mathbf{w}} - \mathbf{w}\|_\infty < \delta_P$ and each $\mathbf{z}$ with $[\mathbf{z}, i] \in \Gamma(\tilde{\mathbf{w}})$ a translate of P occurs in $\Gamma(\tilde{\mathbf{w}}) \cap B_{\mathbb{R}^n}(\mathbf{z}, r_P)$. This means that each patch occurring in a discrete hyperplane $\mathcal{D}$

occurs uniformly repetitively in each hyperplane $\mathcal{D}'$ which is close enough to $\mathcal{D}$. This general property of discrete hyperplanes is proved in [52, Lemma 6.5].

(ii) Let m be the covering degree of $C_{\mathbf{v}}$. Then each point $\mathbf{x} \in \mathbf{v}^\perp$ which is covered exactly m times by elements of $C_{\mathbf{v}}$ is not contained in the boundary of any element of $C_{\mathbf{v}}$. Suppose this was wrong and let $R_1, \ldots, R_m \in C_{\mathbf{v}}$ be the elements containing $\mathbf{x}$. Since $C_{\mathbf{v}}$ is a locally finite union of compact sets there is $\varepsilon > 0$ such that $B_{\mathbf{v}^\perp}(\mathbf{x}, \varepsilon)$ doesn't intersect any $R \in C_{\mathbf{v}} \setminus \{R_1, \ldots, R_m\}$. By assumption $\mathbf{x} \in \partial R_i$ for some $1 \leq i \leq m$. Thus there is $\mathbf{y} \in B_{\mathbf{v}^\perp}(\mathbf{x}, \varepsilon)$ with $\mathbf{y} \notin R_i$ and, hence, $\mathbf{y}$ is covered by at most $m - 1$ elements of $C_{\mathbf{v}}$, a contradiction.

(iii) Choose $\mathbf{x}$ which is covered exactly m times by elements of $C_{\mathbf{v}}$. Since the elements of $C_{\mathbf{v}}$ are uniformly bounded, the set of elements of $C_{\mathbf{v}}$ which contain $\mathbf{x}$ is contained in a set $\{\pi_{\mathbf{u},\mathbf{v}}\mathbf{x} + \mathcal{R}_{\mathbf{v}}(i) : [\mathbf{x}, i] \in P\}$, where P is a patch of $\Gamma(\mathbf{v})$ which is chosen so large that, regardless of how the elements of $\Gamma(\mathbf{v})$ continue outside P, they will not contribute elements of $C_{\mathbf{v}}$ containing $\mathbf{x}$ because they are bounded and located "too far away" from $\mathbf{x}$. Thus, whenever we encounter a translate $P + \mathbf{t}$ of P in $\Gamma(\mathbf{v})$, the point $\mathbf{x} + \pi_{\mathbf{u},\mathbf{v}}\mathbf{t}$ will be covered m times by elements of $C_{\mathbf{v}}$ as well. Thus by (i) and (ii) there exist r_m and r_m' such that in each ball of radius r_m' the hyperplane $\mathbf{v}^\perp$ contains a ball of radius r_m that is covered by exactly m elements of $C_{\mathbf{v}}$.

(iv) $C_{\mathbf{v}}^{(n_k)}$ converges to $C_{\mathbf{v}}$ in a sense described in the proof of Lemma 3.7.13. Thus by (i) the radii r_m and r_m' in (iii) can be chosen in a way that in each ball of radius r_m' the hyperplane $(\mathbf{v}^{(n_k)})^\perp$ contains a ball of radius r_m that is covered by exactly m elements of $C_{\mathbf{v}}^{(n_k)}$ for k large enough.

(v) Suppose that $C_{\mathbf{v}}$ is not a multiple tiling. Then there is a set $X \subset \mathbf{v}^\perp$ with $\lambda_{\mathbf{v}}(X) > 0$ which is covered at least $m + 1$ times. Since the boundaries of the elements of $C_{\mathbf{v}}$ have measure 0 by Proposition 3.7.14, there is $\mathbf{x}$, which is covered a least $m + 1$ times and which is not contained in the boundary of any element of $C_{\mathbf{v}}$. Thus there is $\varepsilon > 0$ such that $B_{\mathbf{v}^\perp}(\mathbf{x}, \varepsilon)$ is covered at least $m + 1$ times.

(vi) Suppose that $C_{\mathbf{v}}$ is not a multiple tiling. By analogous arguments as in (iii), by (v) there exist r_{m+1} and r_{m+1}' such that in each ball of radius r_{m+1}' the hyperplane $\mathbf{v}^\perp$ contains a ball of radius r_{m+1} that is covered by at least $m + 1$ elements of $C_{\mathbf{v}}$.

(vii) By Proposition 3.7.6 each element of $C_{\mathbf{v}}$ can be subdivided into elements of $M_{[0,n_k)} C_{\mathbf{v}}^{(n_k)}$. The diameters of the elements of $M_{[0,n_k)} C_{\mathbf{v}}^{(n_k)}$ tend to 0 for $k \to \infty$ by Lemma 3.7.7 and the balls of radius r_m' occurring in (iv) are shrunk by $M_{[0,n_k)}$ to ellipsoids contained in balls of radius less than r_{m+1}. Thus by (iv) we can chose k so large that in each ball of radius r_{m+1} in $\mathbf{v}^\perp$ there are points which are covered exactly m times by $M_{[0,n_k)} C_{\mathbf{v}}^{(n_k)}$. Thus, by Proposition 3.7.6, in each ball of radius r_{m+1} there are points which are covered at most m times by $C_{\mathbf{v}}$. This contradicts (vi) and the result follows. $\qquad\square$

This result can be generalized to $\mathcal{C}_\mathbf{w}$ for arbitrary $\mathbf{w} \in \mathbb{R}^d_{\geq 0} \setminus \{\mathbf{0}\}$. To establish this generalization one needs to show first that the measures of the subtiles of $\mathcal{R}_\mathbf{v}$ are determined by

$$(\lambda_\mathbf{v}(\mathcal{R}_\mathbf{v}(1)), \ldots, \lambda_\mathbf{v}(\mathcal{R}_\mathbf{v}(d))) = m(\lambda_\mathbf{v}(\pi_{\mathbf{u},\mathbf{v}}[\mathbf{0}, 1]), \ldots, \lambda_\mathbf{v}(\pi_{\mathbf{u},\mathbf{v}}[\mathbf{0}, d])),$$

where m is the covering degree of the multiple tiling $\mathcal{C}_\mathbf{v}$. This can be proved along similar lines as in the substitutive case, see [91, Lemma 2.3]. Using this, measure theoretical considerations lead to the following generalization of Proposition 3.8.1.

Proposition 3.8.2 *Let* $\sigma = (\sigma_n)$ *be a primitive and algebraically irreducible sequence of unimodular substitutions. Assume that there is* $C > 0$ *such that for every* $\ell \in \mathbb{N}$ *there exists* $n \geq 1$ *such that* $(\sigma_n, \ldots, \sigma_{n+\ell-1}) = (\sigma_0, \ldots, \sigma_{\ell-1})$ *and the language* $\mathcal{L}_\sigma^{(n+\ell)}$ *is* C-*balanced.*

Then for each $\mathbf{w} \in \mathbb{R}^d_{\geq 0} \setminus \{\mathbf{0}\}$ *the collection* $\mathcal{C}_\mathbf{w}$ *forms a multiple tiling of the hyperplane* $\mathbf{w}^\perp$.

It remains to show that this multiple tiling is actually a tiling. As we will see later, additional assumptions are needed to prove this. However, there is one tiling result which holds without additional assumptions. This result, which concerns the "inner tiling" of $\mathcal{R}_\mathbf{w}(i)$ by the set equation (3.45) will be proved next.

Proposition 3.8.3 *Let* $\sigma = (\sigma_n)$ *be a primitive and algebraically irreducible sequence of unimodular substitutions. Assume that there is* $C > 0$ *such that for every* $\ell \in \mathbb{N}$ *there exists* $n \geq 1$ *such that* $(\sigma_n, \ldots, \sigma_{n+\ell-1}) = (\sigma_0, \ldots, \sigma_{\ell-1})$ *and the language* $\mathcal{L}_\sigma^{(n+\ell)}$ *is* C-*balanced.*

Then the unions in the set equation (3.45) of Proposition 3.7.3 are disjoint in measure.

Proof From Proposition 3.8.2 we know that $\mathcal{C}_\mathbf{w}$ is a multiple tiling for each $\mathbf{w} \in \mathbb{R}^d_{\geq 0} \setminus \{\mathbf{0}\}$ with multiplicity m not depending on $\mathbf{w}$. Together with Proposition 3.7.6 this implies that the $(\ell - k)$th subdivisions of all the tiles in the multiple tiling $\mathcal{C}_{\mathbf{w}^{(k)}}$ form a multiple tiling $M_{[k,\ell)}\mathcal{C}_{\mathbf{w}^{(\ell)}}$ of the same covering degree (for all $k, \ell \in \mathbb{N}$ with $k < \ell$). This is possible only if each tile of $\mathcal{C}_{\mathbf{w}^{(k)}}$ is tiled without overlaps by elements of $M_{[k,\ell)}\mathcal{C}_{\mathbf{w}^{(\ell)}}$. This proves the result. $\square$

3.8.2   Coincidence Conditions and Tiling Properties

Let σ be a sequence of unimodular substitutions over an alphabet $\mathcal{A}$. In view of Example 3.6.2 in order to prove that (X_σ, Σ) is measurably conjugate to a rotation on a torus we need two properties of the associated Rauzy fractal $\mathcal{R}$. Firstly, the subtiles $\mathcal{R}(i)$, $i \in \mathcal{A}$, need to be disjoint in measure and secondly, the Rauzy fractal itself has to be a fundamental domain of a (well-chosen) torus. The latter property is equivalent to the fact that $\mathcal{R}$ admits a lattice tiling of $\mathbf{v}^\perp$. Setting $\mathbf{w} = \mathbf{1}$ in

Proposition 3.8.2 we obtain that $\mathcal{C}_1$ is a multiple tiling of $\mathbf{1}^\perp$. Since the discrete hyperplane $\Gamma(\mathbf{1})$ can be written as $\Gamma(\mathbf{1}) = \{[\mathbf{x}, i] : \langle \mathbf{x}, \mathbf{1} \rangle = 0, \ i \in \mathcal{A}\}$ we see that

$$\bigcup_{R \in \mathcal{C}_1} R = \bigcup_{[\mathbf{x},i] \in \Gamma(\mathbf{1})} \pi_{\mathbf{u},\mathbf{1}}\mathbf{x} + \mathcal{R}(i) = \bigcup_{\mathbf{x} \in \mathbb{Z}^d \,:\, \langle \mathbf{x}, \mathbf{1} \rangle = 0} \pi_{\mathbf{u},\mathbf{1}}\mathbf{x} + \mathcal{R}.$$

Thus $\mathcal{R}$ is a covering of $\mathbf{1}^\perp$ w.r.t. the lattice $\{\mathbf{x} \in \mathbb{Z}^d \ : \ \langle \mathbf{x}, \mathbf{1} \rangle = 0\}$ and we have to prove that the elements of the union on the right hand side are measure disjoint to get tiling properties of $\mathcal{R}$.

We have therefore three types of unions which we want to be disjoint in measure:

(i) The unions of subtiles on the right hand side of the set equation (3.45).
(ii) The union $\mathcal{R} = \mathcal{R}(1) \cup \cdots \cup \mathcal{R}(d)$.
(iii) The union $\mathbf{1}^\perp = \bigcup_{\mathbf{x} \in \mathbb{Z}^d \,:\, \langle \mathbf{x}, \mathbf{1} \rangle = 0} \pi_{\mathbf{u},\mathbf{1}}\mathbf{x} + \mathcal{R} = \bigcup_{[\mathbf{x},i] \in \Gamma(\mathbf{1})} \pi_{\mathbf{u},\mathbf{1}}\mathbf{x} + \mathcal{R}(i)$.

The elements of the unions in (i) are disjoint in measure by Proposition 3.8.3. One can use this fact in order to prove that the unions in (ii) are disjoint in measure as well. However, to make this proof work we need an additional assumption on σ.

Definition 3.8.4 (Strong Coincidence Condition) A sequence σ of substitutions over an alphabet $\mathcal{A}$ satisfies the *strong coincidence condition* if there is $\ell \in \mathbb{N}$ such that for each pair $(j_1, j_2) \in \mathcal{A}^2$ there are $i \in \mathcal{A}$ and $p_1, p_2 \in \mathcal{A}^*$ with $\mathbf{l}(p_1) = \mathbf{l}(p_2)$ such that $\sigma_{[0,\ell)}(j_1) \in p_1 i \mathcal{A}^*$ and $\sigma_{[0,\ell)}(j_2) \in p_2 i \mathcal{A}^*$.

This definition has an easy geometric meaning: it says that the broken lines associated with $\sigma_{[0,\ell)}(j_1)$ and $\sigma_{[0,\ell)}(j_2)$ have at least one line segment in common for each pair $(j_1, j_2) \in \mathcal{A}^2$.

Example 3.8.5 Figure 3.18 shows that the strong coincidence condition is satisfied for the constant sequence $\sigma = (\sigma)$ with $\sigma(1) = 121, \sigma(2) = 21$. Because we are in a case with a two letter alphabet we only have to deal with the instance $(j_1, j_2) = (1, 2)$.

Using the strong coincidence condition we get the following result.

Proposition 3.8.6 *Let* $\sigma = (\sigma_n)$ *be a primitive and algebraically irreducible sequence of unimodular substitutions. Assume that there is $C > 0$ such that for every $\ell \in \mathbb{N}$ there exists $n \geq 1$ such that $(\sigma_n, \ldots, \sigma_{n+\ell-1}) = (\sigma_0, \ldots, \sigma_{\ell-1})$ and the language $\mathcal{L}_\sigma^{(n+\ell)}$ is C-balanced.*

Fig. 3.18 The broken lines associated with i, $\sigma_{[0,1)}(i)$, and $\sigma_{[0,2)}(i)$ for $i \in \{1, 2\}$. Coincidence is indicated by the bold line

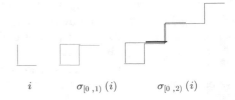

i $\sigma_{[0,1)}(i)$ $\sigma_{[0,2)}(i)$

If the strong coincidence condition holds then the subtiles $\mathcal{R}(i)$, $i \in \mathcal{A}$, are disjoint in measure.

Proof *(Sketch)* Let (n_k) and (ℓ_k) be the associated sequences of σ. Let $\mathcal{R}(j_1)$ and $\mathcal{R}(j_2)$ be two subtiles with $j_1, j_2 \in \mathcal{A}$ distinct and assume that the strong coincidence condition holds with $\ell \in \mathbb{N}$. By the definition of the dual E_1^* in (3.46) this implies that for k satisfying $n_k \geq \ell$ there is $\mathbf{z}_k \in \mathbb{Z}^d$ and a letter $i \in \mathcal{A}$ such that $[\mathbf{z}_k, j_1], [\mathbf{z}_k, j_2] \in E_1^*(\sigma_{[0,n_k)})[\mathbf{0}, i]$. Thus the set equation

$$\mathcal{R}(i) = \bigcup_{[\mathbf{y},j] \in E_1^*(\sigma_{[0,n_k)})[\mathbf{0},i]} M_{[0,n_k)}(\pi_{\mathbf{u},\mathbf{1}}^{(n_k)}\mathbf{y} + \mathcal{R}^{(n_k)}(j)),$$

(see (3.45)) contains $M_{[0,n_k)}(\mathbf{z}_k + \mathcal{R}^{(n_k)}(j_1))$ and $M_{[0,n_k)}(\mathbf{z}_k + \mathcal{R}^{(n_k)}(j_2))$ in the union on the right hand side. Proposition 3.8.3 now implies that $\mathcal{R}^{(n_k)}(j_1)$ and $\mathcal{R}^{(n_k)}(j_2)$ are disjoint in measure. Since this is true for arbitrarily large k, using results along the line of Proposition 3.7.12 (in particular, [52, Lemma 6.8]) this implies that $\mathcal{R}(j_1)$ and $\mathcal{R}(j_2)$ are disjoint in measure as well.   □

What we did in the proof of Proposition 3.8.6 can be explained in a simple way. If the strong coincidence condition holds, each intersection of the subtiles $\mathcal{R}(j_1) \cap \mathcal{R}(j_2)$ can be realized as an intersection of two elements in the union on the right hand side of the set equation (3.45). Since we know that the elements in the union of the set equation are measure disjoint, the same is true for $\mathcal{R}(j_1)$ and $\mathcal{R}(j_2)$. More briefly: in case of strong coincidence the elements in the union in (ii) are special cases of the elements in some union in (i).

The same strategy can be used in order to prove that the unions in (iii) are measure disjoint. To this end we need another type of coincidence condition.

Definition 3.8.7 (Geometric Coincidence Condition)   A sequence σ of unimodular substitutions over an alphabet $\mathcal{A}$ satisfies the *geometric coincidence condition* if the following is true. For each $r > 0$ there is $n_0 \in \mathbb{N}$ such that for each $n \geq n_0$ the set $E_1^*(\sigma_{[0,n)})[\mathbf{0}, i_n]$ contains a ball of radius r of the discrete hyperplane $\Gamma((M_{[0,n)})^t\mathbf{1})$ for some $i_n \in \mathcal{A}$.

Along similar lines as Proposition 3.8.6 one can prove the following tiling criterion for Rauzy fractals (see [52, Proposition 7.9]).

Proposition 3.8.8 *Let $\sigma = (\sigma_n)$ be a primitive and algebraically irreducible sequence of unimodular substitutions. Assume that there is $C > 0$ such that for every $\ell \in \mathbb{N}$ there exists $n \geq 1$ such that $(\sigma_n, \ldots, \sigma_{n+\ell-1}) = (\sigma_0, \ldots, \sigma_{\ell-1})$ and the language $\mathcal{L}_\sigma^{(n+\ell)}$ is C-balanced. Then the following assertions are equivalent.*

   (i)   *The collection $\mathcal{C}_1$ forms a tiling of $\mathbf{1}^\perp$.*
   (ii) *The sequence σ satisfies the geometric coincidence condition.*
   (iii) *The sequence σ satisfies the strong coincidence condition and for each $r > 0$ there exists $n_0 \in \mathbb{N}$ such that $\bigcup_{i \in \mathcal{A}} E_1^*(\sigma_{[0,n)})[\mathbf{0}, i]$ contains a ball of radius r of $\Gamma((M_{[0,n)})^t \mathbf{1})$ for all $n \geq n_0$.*

(iv) The sequence σ satisfies the following effective condition: There are $n \in \mathbb{N}$, $i \in \mathcal{A}$, and $\mathbf{z} \in \mathbb{R}^d$, such that

$$\left\{[\mathbf{y}, j] \in \Gamma((M_{[0,n)})^t \mathbf{1}) : \|\pi_{(M_{[0,n)})^{-1}\mathbf{u},\mathbf{1}}(\mathbf{y} - \mathbf{z})\| \leq C\right\} \subset E_1^*(\sigma_{[0,n)})[\mathbf{0}, i],$$

where $C \in \mathbb{N}$ is chosen in a way that $\mathcal{L}_\sigma^{(n)}$ is C-balanced.

An (essentially) more restrictive condition than the geometric coincidence condition and its variants in Proposition 3.8.8 is the following one.

Definition 3.8.9 (Geometric Finiteness Property) A sequence σ of unimodular substitutions over an alphabet $\mathcal{A}$ satisfies the *geometric finiteness property* if for each $r > 0$ there is $n_0 \in \mathbb{N}$ such that $\bigcup_{i \in \mathcal{A}} E_1^*(\sigma_{[0,n)})[\mathbf{0}, i]$ contains the ball $\{[\mathbf{x}, i] \in \Gamma((M_{[0,n)})^t \mathbf{1}) : \|\mathbf{x}\| \leq r\}$ for all $n \geq n_0$.

The geometric finiteness property implies that $\bigcup_{i \in \mathcal{A}} E_1^*(\sigma_{[0,n)})[\mathbf{0}, i]$ generates a whole discrete plane for $n \to \infty$, and that $\mathbf{0}$ is an inner point of the Rauzy fractal $\mathcal{R}$ (as is proved in [52, Proposition 7.10]). It is immediate that together with the strong coincidence condition the geometric finiteness property is more restrictive than the condition in Proposition 3.8.8(iii). The name *geometric finiteness property* comes from the fact that it is related to certain finiteness properties in number representations w.r.t. positional number systems (see for instance Barat et al. [31] for a survey on these objects). By Proposition 3.8.8(iii) strong coincidence plus geometric finiteness imply that $\mathcal{C}_1$ forms a tiling of $\mathbf{1}^\perp$.

3.8.3   How to Check Geometric Coincidence and Geometric Finiteness?

In most cases it is easy to check strong coincidence of a sequence $\sigma = (\sigma_n)$ of substitutions over an alphabet $\mathcal{A}$. For instance, this property trivially holds if $\sigma_0(i)$ starts with the same letter for each $i \in \mathcal{A}$. However, it is a priori not so clear how to check geometric coincidence or geometric finiteness and although there is an effective criterion for geometric coincidence contained in Proposition 3.8.8(iv) this is only suitable for checking single instances. Geometric coincidence asserts that a large piece of a discrete hyperplane can be generated by the dual substitution $E_1^*(\sigma_{[0,n)})$ acting on $[\mathbf{0}, i_n]$ if n is large. If geometric finiteness holds, even a whole discrete hyperplane can be generated by the patches $E_1^*(\sigma_{[0,n)}) \bigcup_{i \in \mathcal{A}}[\mathbf{0}, i]$ for $n \to \infty$. The idea of generating discrete hyperplanes in this way using sequences of substitutions coming from generalized continued fraction algorithms goes back to Ito and Ohtsuki [90]. More recently, Berthé et al. [42, 48] provide a systematic study on how to check geometric coincidence as well as geometric finiteness. While [48] concentrates on Arnoux-Rauzy substitutions, the more general treatment in [42] uses substitutions related to the Brun as well as the Jacobi-Perron algorithm as guiding examples. In this section we give a brief discussion of their ideas which are centered around an "annulus property" of stepped hyperplanes generated by $E_1^*(\sigma_{[0,n)})$.

Let $\sigma = (\sigma_n)$ be a sequence of unimodular substitutions over an alphabet $\mathcal{A}$ and let $S = \{\sigma_n : n \in \mathbb{N}\}$. The fact that σ satisfies the geometric coincidence condition in Definition 3.8.7 roughly says that the patch $E_1^*(\sigma_{[0,n)})[\mathbf{0}, i_n]$ contains a larger and larger ball when n is growing. In this section, for the sake of simplicity, we will deal with the geometric finiteness property. Indeed, we will assume that this ball is centered at the origin and instead of $[\mathbf{0}, i_n]$ we will use $\mathcal{U} = \bigcup_{i \in \mathcal{A}}[\mathbf{0}, i]$ as our "seed". So we want to show that for each $R > 0$ there is $n_0 \in \mathbb{N}$ such that $E_1^*(\sigma_{[0,n)})\mathcal{U}$ contains the ball $\{[\mathbf{x}, i] \in \Gamma((M_{[0,n)})^t \mathbf{1}) : \|\mathbf{x}\| \leq R\}$ for all $n \geq n_0$.

Following [42] we shall reformulate the geometric finiteness property in a more combinatorial way. Let P be a patch of a discrete hyperplane containing $\mathcal{U}$ and interpret its elements as faces as in (3.34). Then the *minimal combinatorial radius* rad(P) of P is equal to the length ℓ of the shortest sequence of faces $[\mathbf{x}_1, j_1], \ldots, [\mathbf{x}_\ell, j_\ell] \in P$ satisfying $[\mathbf{x}_1, j_1] \in \mathcal{U}$, $[\mathbf{x}_\ell, j_\ell]$ contains a part of the boundary of P (regarded as a topological manifold), and $[\mathbf{x}_k, j_k] \cap [\mathbf{x}_{k+1}, j_{k+1}] \neq \emptyset$ for $1 \leq k \leq \ell - 1$. Intuitively, rad(P) is the minimal distance between $\mathbf{0}$ and the boundary of P. For instance, one easily checks that the minimal combinatorial radius of the patch on the left hand side of Fig. 3.13 is equal to six. Clearly a sequence σ enjoys the geometric finiteness property if and only if rad$\left(E_1^*(\sigma_{[0,n)})\mathcal{U}\right)$ tends to ∞ for $n \to \infty$.

Let $P_{[m,n)} = E_1^*(\sigma_{[m,n)})\mathcal{U}$. We have to show that the minimal combinatorial radii of the patches $P_{[0,n)}$ tend to ∞ for $n \to \infty$. Since the patches $P_{[0,n)}$ can have complicated shapes there is no obvious way to do this. One approach to prove this property goes back to Ito and Ohtsuki [90] and makes use of "annuli". Let $\ell < m < n$ and suppose that $\mathcal{U} \subset E_1^*(\sigma)\mathcal{U}$ holds for each $\sigma \in S$ (this is not a crucial assumption and, if it is not true, can often be gained by blocking the substitutions of the sequence σ). Then $P_{[m,n)} \subset P_{[\ell,n)}$ holds by the definition of $E_1^*(\sigma)$ (note in particular that $E_1^*(\tau)E_1^*(\sigma) = E_1^*(\sigma\tau)$ for $\sigma, \tau \in S$). The idea is to make sure that whenever $(\sigma_\ell, \ldots, \sigma_{m-1})$ is of a certain shape then $P_{[\ell,n)} \setminus P_{[m,n)}$ contains an annulus of positive width. One can then show that if $(\sigma_0, \ldots, \sigma_n)$ contains the block $(\sigma_\ell, \ldots, \sigma_{m-1})$ for k times, the patch $P_{[0,n)}$ contains k "concentric" annuli and has a minimal combinatorial radius greater than or equal to k.

To achieve this we first search for a block $(\sigma_0, \ldots, \sigma_{m-1})$ such that $A = P_{[0,m)} \setminus \mathcal{U}$ contains an annulus of positive width, i.e., $\partial P_{[0,m)} \cap \mathcal{U} = \emptyset$. If σ is recurrent, the block $(\sigma_0, \ldots, \sigma_{m-1})$ occurs infinitely often in σ. Let (n_j) with $n_0 = 0$

and $n_j \geq n_{j-1} + m$ be an increasing sequence such that $(\sigma_{n_j}, \ldots, \sigma_{n_j+m-1}) = (\sigma_0, \ldots, \sigma_{m-1})$. Fix $k \in \mathbb{N}$ and set $A_0 = P_{[0,n_k+m)} \setminus P_{[m,n_k+m)}$ and $A_j := P_{[n_{j-1}+m,n_k+m)} \setminus P_{[n_j+m,n_k+m)}$ for $j \geq 1$. Then

$$
\begin{aligned}
P_{[0,n_k+m)} &= (P_{[0,n_k+m)} \setminus P_{[m,n_k+m)}) \cup P_{[m,n_k+m)} \\
&= A_0 \cup P_{[m,n_k+m)} \\
&= A_0 \cup (P_{[m,n_k+m)} \setminus P_{[n_1+m,n_k+m)}) \cup P_{[n_1+m,n_k+m)} \\
&= A_0 \cup A_1 \cup P_{[n_1+m,n_k+m)} \qquad\qquad (3.53)\\
&= A_0 \cup A_1 \cup (P_{[n_1+m,n_k+m)} \setminus P_{[n_2+m,n_k+m)}) \cup P_{[n_2+m,n_k+m)} \\
&= A_0 \cup A_1 \cup A_2 \cup P_{[n_2+m,n_k+m)} \\
&= \cdots = A_0 \cup \cdots \cup A_k \cup \mathcal{U}.
\end{aligned}
$$

Because

$$
\begin{aligned}
A_j &= P_{[n_{j-1}+m,n_k+m)} \setminus P_{[n_j+m,n_k+m)} \\
&\supset P_{[n_j,n_k+m)} \setminus P_{[n_j+m,n_k+m)} \\
&= E_1^*(\sigma_{[n_j+m,n_k+m)})(P_{[n_j,n_j+m)} \setminus \mathcal{U}) \\
&= E_1^*(\sigma_{[n_j+m,n_k+m)})A
\end{aligned}
$$

for $j \geq 1$ (the last step comes from the recurrence property; the case $j = 0$ follows along similar lines) each A_j contains some image of A under E_1^*. If the annulus A has certain "covering properties" that are described in detail in [42, 48], one can show that images of A under E_1^* are annuli of positive width as well. Thus such an annulus of positive width is contained in each of the pairwise disjoint subsets $A_0, \ldots, A_k$ of $P_{[0,n_k+m)}$ and therefore (3.53) implies that the patch $P_{[0,n_k+m)}$ contains a "concentric" annulus for each of the $k + 1$ (non overlapping) occurrence of the block $(\sigma_0, \ldots, \sigma_{m-1})$ in $(\sigma_0, \ldots, \sigma_{n_k+m-1})$. Since an application of E_1^* maps disjoint annuli to disjoint annuli also $P_{[0,n)} = E_1^*(\sigma_{[n_k+m,n)})P_{[0,n_k+m)}$ with $n_k + m \leq n < n_{k+1} + m$ contains $k + 1$ such "concentric" annuli. Thus if $n \to \infty$, the number of such annuli in $P_{[0,n)}$ tends to ∞. Since the above-mentioned covering properties of A imply that $A_0 \cup \cdots \cup A_k \cup \mathcal{U} = P_{[0,n_k+m)}$ is simply connected for each $k \in \mathbb{N}$ and that the same is true for all the patches $P_{[0,n)}$ (see [48]), we gain that the minimal combinatorial radii of the patches $P_{[0,n)}$ tend to ∞ for $n \to \infty$.

The following example shows that this method can be used in order to prove geometric finiteness for large classes of sequences of substitutions.

Example 3.8.10 We want to illustrate the construction of the annulus A around $\mathcal{U}$ for the case of sequences of Arnoux-Rauzy substitutions $\sigma = (\sigma_n)$ (all details for

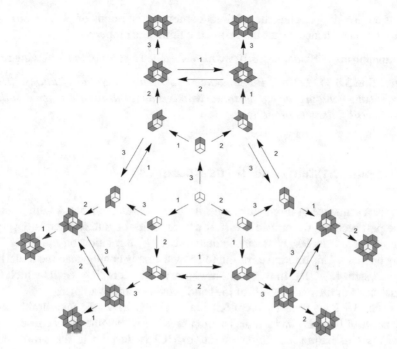

Fig. 3.19 An illustration of the annulus property for sequences of Arnoux-Rauzy substitutions

this case can be found in [48]). Suppose that σ is a recurrent sequence of Arnoux-Rauzy substitutions which contains each of the three Arnoux-Rauzy substitutions (3.18). Then, by recurrence, σ contains a block $(\sigma_0, \ldots, \sigma_{m-1})$ in which each Arnoux Rauzy substitution occurs at least twice. In the graph depicted in Fig. 3.19 the action of E_1^* on $\mathcal{U}$ is illustrated.[8] The vertices of this graph are patches and there is an edge $P_1 \xrightarrow{i} P_2$ if $P_2 \subset E_1^*(\sigma_i)P_1$. Thus each vertex has an outgoing edge for each $i \in \{1, 2, 3\}$ (loops and outgoing edges of patches that contain an annulus of positive width around $\mathcal{U}$ are suppressed). Examining the graph we see that $E_1^*(\sigma_{[k,n)})\mathcal{U}$ contains an annulus around $\mathcal{U}$ of positive width whenever the block $(\sigma_k, \ldots, \sigma_{n-1})$ contains at least two occurrences of each Arnoux-Rauzy substitution. Thus, $P_{[0,m)}$ is a patch which contains $\mathcal{U}$ together with an annulus A of positive width around it.

If one proves that the annulus A has the above-mentioned covering properties (which was done in [48]) one can iterate this procedure as indicated above and prove that $P_{[0,n)}$ is simply connected and contains a growing number of "concentric"

---

[8]We note that in [48] the dual $E_1^*(\sigma)$ is defined using suffixes of the images of σ instead of prefixes. Nevertheless, this difference does not change the behavior of $E_1^*(\sigma)$ significantly and in Fig. 3.19 we get the same image as the authors obtained in [48, Figure 1].

annuli for growing n. Thus the minimal combinatorial radius of $P_{[0,n)}$ tends to ∞ for $n \to \infty$ and, hence, σ has the geometric finiteness property.

Summing up, in Example 3.8.10 we have sketched a proof of the following result.

Proposition 3.8.11 *Let σ be a sequence of Arnoux-Rauzy substitutions. If σ is recurrent and contains each of the three Arnoux-Rauzy substitutions then σ satisfies the geometric finiteness property.*

3.9   *S*-adic Systems and Torus Rotations

Let σ be a sequence of unimodular substitutions over an alphabet $\mathcal{A}$ with d letters. In the past sections we proved a variety of properties of Rauzy fractals. Using all these results makes Rauzy fractals suitable to "see" a rotation on the torus $\mathbb{T}^{d-1}$ acting on them. This rotation turns out to be measurably conjugate to the underlying S-adic system (X_σ, Σ). In this section we prove the according results which form special cases of the main results of [52] and provide some examples.

In Sect. 3.9.1 we state Theorem 3.9.4, a result that gives the measurable conjugacy between (X_σ, Σ) and a torus rotation together with some of its consequences under a set of natural conditions. Section 3.9.2 is devoted to the proof of this result. In Sect. 3.9.3 we formulate a metric version of Theorem 3.9.4. In particular, for a finite set S of substitutions we consider the shift[9] $(S^{\mathbb{N}}, \Sigma, \nu)$ acting on all infinite sequences of substitutions taken from S. The measure ν is chosen in a way that this shift becomes ergodic. We prove that the conditions of Theorem 3.9.4 are "generic" w.r.t. the measure ν if the *Pisot condition* (3.59) on the Lyapunov exponents associated with a linear cocycle of $(S^{\mathbb{N}}, \Sigma, \nu)$ is in force. Thus under this Pisot condition we gain that ν-almost all $\sigma \in S^{\mathbb{N}}$ give rise to an S-adic system (X_σ, Σ) that is measurably conjugate to a torus rotation. This result is the content of Theorem 3.9.5. Section 3.9.4 is devoted to the proof of this result. Finally, Sect. 3.9.5 gives examples for S-adic systems associated with Arnoux-Rauzy and Brun substitutions. This shows that the Pisot condition is satisfied in many natural situations.

---

[9]Note that there are two kinds of shifts: the one just defined acts on the sequence of substitutions $S^{\mathbb{N}}$, the other one (the S-adic shift) acts on the set of sequences X_σ which is defined in terms of a single sequence of substitutions $\sigma \in S^{\mathbb{N}}$. It should cause no confusion that both of these shift mappings are denoted by Σ.

3.9.1   Statement of the Conjugacy Result

Before we state the first main result of this survey we give some terminology.
We start with a spectral property of a measurable dynamical system that is "the
opposite" of continuous spectrum (see Sect. 3.3.3; we refer to this section also for
the definition of an *eigenfunction*).

Definition 3.9.1 (Pure Discrete Spectrum, see [126, Defintion 3.2])   An ergodic
dynamical system (X, T, μ) on a probability space X has *pure discrete spectrum* if
there exists an orthonormal basis of $L^2(\mu)$ which consists of eigenfunctions of T.

It is well known that an ergodic dynamical system on a probability space that has
pure discrete spectrum is measurably conjugate to a rotation on a compact abelian
group. On the other hand, each ergodic rotation on a compact abelian group has
pure discrete spectrum (see for instance [126, Theorems 3.5 and 3.6]; these results
can be proved by using character theory and Pontryagin duality for compact abelian
groups).

The notion of *natural coding* came up already in Sects. 3.2.4 and 3.3.2 in the
framework of Sturmian sequences and Arnoux-Rauzy sequences. Sloppily speaking
a natural coding is a coding of a torus rotation that induces translations on the
atoms of the partition that was used to define the coding. We give a precise general
definition of this concept.

Definition 3.9.2 (Coding and Natural Coding)   Let Λ be a full-rank lattice in $\mathbb{R}^d$
and $T_{\mathbf{t}} : \mathbb{R}^d/\Lambda \to \mathbb{R}^d/\Lambda, \mathbf{x} \mapsto \mathbf{x} + \mathbf{t}$ a rotation on the torus $\mathbb{R}^d/\Lambda$. Let $\Omega \subset \mathbb{R}^d$ be a
fundamental domain for the lattice Λ and $\tilde{T}_{\mathbf{t}} : \Omega \to \Omega$ the mapping induced by $T_{\mathbf{t}}$
on Ω. Assume that $\Omega = \Omega_1 \cup \cdots \cup \Omega_k$ is a (measure theoretic w.r.t. the Lebesgue
measure) partition of Ω.

A sequence $w = w_0 w_1 \ldots \in \{1, \ldots, k\}^{\mathbb{N}}$ is the *coding* of a point $\mathbf{x} \in \Omega$ with
respect to this partition if $\tilde{T}_{\mathbf{t}}^{\,j}(\mathbf{x}) \in \Omega_{w_j}$ holds for each $j \in \mathbb{N}$. If, in addition, for
each $1 \le i \le k$ the restriction $\tilde{T}_{\mathbf{t}}|_{\Omega_i}$ is given by the translation $\mathbf{x} \mapsto \mathbf{x} + \mathbf{t}_i$ for
some $\mathbf{t}_i \in \mathbb{R}^d$ we call w a *natural coding* of $T_{\mathbf{t}}$.

For the sake of completeness we give the definition of *bounded remainder set*.

Definition 3.9.3 (Bounded Remainder Set)   Let Λ be a full-rank lattice in $\mathbb{R}^d$.
A subset A of $\mathbb{R}^d/\Lambda$ is called a *bounded remainder set* for the rotation $T_{\mathbf{t}} : \mathbb{R}^d/\Lambda \to$
$\mathbb{R}^d/\Lambda, \mathbf{x} \mapsto \mathbf{x} + \mathbf{t}$ if there exist $\gamma, C > 0$ such that, for a.e. $\mathbf{x} \in \mathbb{R}^d/\Lambda$,

$$|\#\{n < N : T_{\mathbf{t}}^n(\mathbf{x}) \in A\} - \gamma N| < C$$

holds for all $N \in \mathbb{N}$.

The following result gives sufficient conditions for an S-adic system (X_σ, Σ) to
be measurably conjugate to an irrational rotation on a torus. The subtiles $\mathcal{R}(i)$ of the
Rauzy fractal $\mathcal{R}$ turn out to be bounded remainder sets for this rotation and induce
natural codings of the elements of (X_σ, Σ).

Theorem 3.9.4 (See [52, Theorem 3.1]) *Let S be a finite set of unimodular substitutions over a finite alphabet $\mathcal{A} = \{1, 2, \ldots, d\}$ and let $\sigma = (\sigma_n)$ be a primitive and algebraically irreducible sequence of substitutions taken from the set S. Assume that there is $C > 0$ such that for every $\ell \in \mathbb{N}$ there exists $n \geq 1$ such that $(\sigma_n, \ldots, \sigma_{n+\ell-1}) = (\sigma_0, \ldots, \sigma_{\ell-1})$ and the language $\mathcal{L}_\sigma^{(n+\ell)}$ is C-balanced.*

If the collection $\mathcal{C}_1$ forms a tiling of $\mathbf{1}^\perp$ then the following results hold.

1. *The S-adic shift (X_σ, Σ, μ), with μ being the unique Σ-invariant Borel probability measure on X_σ, is measurably conjugate to a rotation T on the torus $\mathbb{T}^{d-1}$; in particular, its measure-theoretic spectrum is purely discrete.*
2. *Each element of X_σ is a natural coding of the torus rotation T with respect to the partition $\{\mathcal{R}(i) : i \in \mathcal{A}\}$ of the fundamental domain $\mathcal{R}$.*
3. *The subtile $\mathcal{R}(i)$ is a bounded remainder set for the torus rotation T for each $i \in \mathcal{A}$.*

For the special case of two letter alphabets the tiling condition does not have to be assumed. It can be derived from the remaining assumptions of Theorem 3.9.4. The corresponding result is proved in [50] and generalizes an analogous result for substitutive systems from [34].

3.9.2 Proof of the Conjugacy Result

In this section we illustrate the proof of Theorem 3.9.4 given in [52]. We assume throughout this section that the sequence σ satisfies the conditions of Theorem 3.9.4. The main part is the proof of the measurable conjugacy between (X_σ, Σ, μ) and a rotation on the torus $\mathbb{T}^{d-1}$, where d is the cardinality of the underlying alphabet. Here μ is the unique Σ-invariant Borel probability measure on X_σ (see Theorem 3.5.11).

Our first aim is to set up the *representation map* from X_σ to the Rauzy fractal. We define this map using a nested sequence of the subsets

$$\mathcal{R}(u) := \overline{\{\pi_{\mathbf{u}, \mathbf{w}} \mathbf{l}(p) \; : \; pu \text{ is a prefix of a limit sequence of } \sigma\}} \qquad (u \in \mathcal{A}^*)$$

of the Rauzy fractal $\mathcal{R}$. In particular, we set

$$\varphi : X_\sigma \to \mathcal{R}; \quad v_0 v_1 v_2 \ldots \mapsto \bigcap_{n \in \mathbb{N}} \mathcal{R}(v_0 v_1 \ldots v_{n-1}). \qquad (3.54)$$

To show that φ is a well-defined continuous surjection one has to prove that the intersection on the right-hand side of (3.54) is a single point. Using the minimality of (X_σ, Σ) and the strong convergence property from Proposition 3.6.10 this is done in [52, Section 8].

In the next step one proves that (X_σ, Σ, μ) is measurably conjugate to the *domain exchange* $(\mathcal{R}, E, \lambda_1)$, where E is given by

$$E : \mathcal{R} \to \mathcal{R}; \quad \mathbf{x} \mapsto \mathbf{x} + \pi_{\mathbf{u},1}\mathbf{l}(i) \quad \text{for } \mathbf{x} \in \mathcal{R}(i) \setminus \bigcup_{j \neq i} \mathcal{R}(j)$$

which is illustrated in Fig. 3.12. Since $\mathcal{C}_1$ is a tiling, the overlaps of the subtiles $\mathcal{R}(i)$ have measure 0 and, hence, E is well defined a.e. w.r.t. the measure λ_1 on $\mathcal{R}$. To prove the asserted conjugacy, we have to show that φ is bijective μ-a.e. and that the diagram

$$
\begin{array}{ccc}
X_\sigma & \xrightarrow{\ \Sigma\ } & X_\sigma \\
\downarrow{\scriptstyle \varphi} & & \downarrow{\scriptstyle \varphi} \\
\mathcal{R} & \xrightarrow{\ E\ } & \mathcal{R}
\end{array}
\tag{3.55}
$$

commutes. Since

$$E \circ \varphi = \varphi \circ \Sigma \tag{3.56}$$

follows easily by direct calculation it remains to prove the bijectivity assertion. This runs as follows (all statements are true up to measure zero). First observe that, for all $i \in \mathcal{A}$, E satisfies

$$E(\mathcal{R}(i)) = \overline{\{\pi_{\mathbf{u},1}\mathbf{l}(pi) : p \in \mathcal{A}^*, \ pi \text{ is a prefix of a limit word of } \sigma\}}.$$

Therefore, we have $\bigcup_{i \in \mathcal{A}} E(\mathcal{R}(i)) = \mathcal{R}$ and, hence, E is a surjective piecewise isometry. Therefore, E is bijective. Since the subtiles $\mathcal{R}(i)$, $i \in \mathcal{A}$, are disjoint in measure and

$$\mathcal{R}(w_0 w_1 \cdots w_{n-1}) = \bigcap_{\ell=0}^{n-1} E^{-\ell} \mathcal{R}(w_\ell), \tag{3.57}$$

the injectivity of E implies that also the elements of the collection of "length n subtiles"[10] $\mathcal{K}_n = \{\mathcal{R}(u) : u \in \mathcal{L}_\sigma \text{ with } |u| = n\}$ are disjoint in measure. By (3.56) the measure $\lambda_1 \circ \varphi$ is a shift invariant probability measure on X_σ. As by Theorem 3.5.11 there is only one such measure, $\mu = \lambda_1 \circ \varphi$. Now, essential disjointness of the elements of $\mathcal{K}_n$ implies that $\varphi(\mathbf{x}) \neq \varphi(\mathbf{y})$ for all distinct $\mathbf{x}, \mathbf{y}$ satisfying $\varphi(\mathbf{x}), \varphi(\mathbf{y}) \in \mathcal{R} \setminus \bigcup_{n \in \mathbb{N}, K \in \mathcal{K}_n} \partial K$. As, by (3.57) and Theorem 3.7.1, $\lambda_1(\partial K) = \mu(\varphi^{-1}(\partial K)) = 0$ for all $K \in \mathcal{K}_n$, $n \in \mathbb{N}$, the map φ is μ-a.e. injective.

---

[10]Not to be confused with the level n subtiles introduced in Sect. 3.7.1.

Since surjectivity follows from the definition of φ this proves μ-a.e. bijectivity. Finally, using (3.56), the commutativity of the diagram (3.55) follows from the bijectivity of φ.

Since $\mathcal{C}_1$ forms a tiling of $\mathbf{1}^{\perp}$ by assumption, the Rauzy fractal $\mathcal{R}$ is a fundamental domain of the lattice $\Lambda = \mathbf{1}^{\perp} \cap \mathbb{Z}^d$ spanned by $\mathbf{e}_1 - \mathbf{e}_i$, $i \in \mathcal{A} \setminus \{1\}$. But as $\pi_{\mathbf{u},1} \mathbf{e}_i \equiv \pi_{\mathbf{u},1} \mathbf{e}_1 \pmod{\Lambda}$ holds for each $i \in \mathcal{A}$, the canonical projection of E onto the torus $\mathbf{1}^{\perp}/\Lambda \simeq \mathbb{T}^{d-1}$ is equal to the rotation $T : \mathbb{T}^{d-1} \to \mathbb{T}^{d-1}$, $\mathbf{x} \mapsto \mathbf{x} + \pi_{\mathbf{u},1} \mathbf{e}_1$. Thus, if we denote by $\overline{\varphi}$ the canonical projection of φ to the torus $\mathbf{1}^{\perp}/\Lambda$, the diagram

$$
\begin{array}{ccc}
X_\sigma & \xrightarrow{\ \Sigma\ } & X_\sigma \\
\downarrow{\overline{\varphi}} & & \downarrow{\overline{\varphi}} \\
\mathbf{1}^{\perp}/\Lambda & \xrightarrow{\ +\pi_{\mathbf{u},1}\mathbf{e}_1\ } & \mathbf{1}^{\perp}/\Lambda
\end{array}
$$

commutes. Note that $\overline{\varphi}$ is m to 1 onto, where m is the covering degree of $\mathcal{C}_1$, and, hence, a bijection as $\mathcal{C}_1$ forms a tiling. This proves the first assertion of Theorem 3.9.4.

The second assertion of Theorem 3.9.4 follows from the definition of a natural coding because the rotation T was defined in terms of an exchange of domains. Finally, due to [1, Proposition 7], the C-balance of $\mathcal{L}_\sigma$ implies that $\mathcal{R}(i)$ is a bounded remainder set for each $i \in \mathcal{A}$, which also proves the last assertion.

3.9.3   A Metric Result

As mentioned already in Remark 3.7.2(i), the assumptions of Theorems 3.7.1 and 3.9.4 allow for a metric version of these results. To be more precise, let S be a finite set of substitutions and consider the full shift $(S^{\mathbb{N}}, \Sigma, \nu)$, where ν is an ergodic Σ-invariant probability measure satisfying some mild conditions. Our aim is to state a version of Theorems 3.5.11, 3.7.1, and 3.9.4 that is valid for ν-a.e. $\sigma \in S^{\mathbb{N}}$. This second main result of the present survey is also a special case of a result from Berthé et al. [52].

To state our result we need to introduce some new concepts. Let S be a finite set of substitutions over the alphabet $\mathcal{A} = \{1, 2, \ldots, d\}$ and consider the shift $(S^{\mathbb{N}}, \Sigma, \nu)$, where ν is some Σ-invariant probability measure on $S^{\mathbb{N}}$. With each $\sigma = (\sigma_n)_{n \geq 0}$ we associate the *linear cocycle operator* $A(\sigma) = (M_0)^t$ (recall that M_0 is the incidence matrix of σ_0) and define the *Lyapunov exponents* $\vartheta_1, \ldots, \vartheta_d$ of

$(S^{\mathbb{N}}, \Sigma, \nu)$ iteratively by

$$\vartheta_1 + \vartheta_2 + \cdots + \vartheta_k = \lim_{n \to \infty} \frac{1}{n} \int_{S^{\mathbb{N}}} \log \| \wedge^k \left(A(\Sigma^{n-1}(\sigma)) \cdots A(\sigma) \right) \|_{\infty} \, d\nu(\sigma)$$

$$= \lim_{n \to \infty} \frac{1}{n} \int_{S^{\mathbb{N}}} \log \| \wedge^k (M_{[0,n)})^t \|_{\infty} \, d\nu \qquad (3.58)$$

$$= \lim_{n \to \infty} \frac{1}{n} \int_{S^{\mathbb{N}}} \log \| \wedge^k M_{[0,n)} \|_{\infty} \, d\nu$$

for $1 \le k \le d$, where $\wedge^k$ denotes the k-fold wedge product. We say that $(S^{\mathbb{N}}, \Sigma, \nu)$ satisfies the *Pisot condition* if

$$\vartheta_1 > 0 > \vartheta_2 \ge \cdots \ge \vartheta_d \qquad (3.59)$$

(cf. [44, §6.3]). Using these definitions we get the following metric version of Theorems 3.5.11, 3.7.1, and 3.9.4.

Theorem 3.9.5 (See [52, Theorem 3.3]) *Let S be a finite set of unimodular substitutions and assume that the shift $(S^{\mathbb{N}}, \Sigma, \nu)$ is ergodic and satisfies the Pisot condition. Assume further that ν assigns positive measure to every cylinder and that there exists a cylinder corresponding to a substitution with positive incidence matrix. Then, for ν-almost every $\sigma \in S^{\mathbb{N}}$ the following assertions hold.*

1. *(X_σ, Σ) is minimal and uniquely ergodic (denote the unique Σ-invariant measure by μ).*
2. *Each subtile $\mathcal{R}(i)$, $i \in \mathcal{A}$, is equal to the closure of its interior and satisfies $\lambda_1(\partial \mathcal{R}(i)) = 0$.*
3. *If the collection $\mathcal{C}_1$ associated with σ forms a tiling of $\mathbf{1}^{\perp}$ then (X_σ, Σ, μ) is measurably conjugate to a rotation T on $\mathbb{T}^{d-1}$, each element of X_σ is a natural coding of T w.r.t. the partition $\{\mathcal{R}(i) : i \in \mathcal{A}\}$ of $\mathcal{R}$, and each $\mathcal{R}(i)$, $i \in \mathcal{A}$, is a bounded remainder set for T.*

3.9.4 Proof of the Metric Result

In the present section we give a quite complete proof of Theorem 3.9.5. The idea is to show that each of the conditions posed in Theorem 3.9.4 is *generic*. A prominent tool in this proof is the *Multiplicative Ergodic Theorem* (also called *Oseledec Theorem*; see for instance [8, 3.4.1 Theorem]). Also the famous *Poincaré Recurrence Theorem* (cf. e.g. [126, Theorem 1.4]), which states that a.e. orbit in a measurable dynamical system (X, T, μ) starting in a set of positive measure E hits E infinitely often, will be used. In our setting, the Oseledec theorem has the following consequence.

Proposition 3.9.6 *Let S be a finite set of unimodular substitutions over the alphabet $\mathcal{A} = \{1, 2, \ldots, d\}$ and assume that the shift $(S^{\mathbb{N}}, \Sigma, \nu)$ is ergodic with Lyapunov exponents $\vartheta_1, \ldots, \vartheta_d$ satisfying the Pisot condition (3.59). Assume further that ν assigns positive measure to every cylinder and that there exists a cylinder corresponding to a substitution with positive incidence matrix. Then for ν-a.e. $\sigma \in S^{\mathbb{N}}$ the following assertions hold.*

(i) *The sequence σ is primitive and recurrent, thus the letter frequency vector $\mathbf{u} = \mathbf{u}(\sigma)$ exists.*

(ii) *For each $\varepsilon > 0$ there exists $n_0 = n_0(\varepsilon, \sigma)$ such that the sequence of incidence matrices $\mathbf{M} = (M_n) = (M_n(\sigma))$ satisfies*[11]

$$\|(M_{[0,n)})^t|_{\mathbf{u}^\perp}\|_2 < e^{(\vartheta_2+\varepsilon)n}$$

for each $n \geq n_0$.

Proof Since ν puts positive mass on each cylinder, ν-a.e. σ is recurrent by Poincaré recurrence. Together with the fact that there is a cylinder corresponding to a positive incidence matrix Poincaré recurrence also implies primitivity for ν-a.e. σ. Thus ν-a.e. σ has a letter frequency vector $\mathbf{u}$ by Proposition 3.5.5. This proves (i).

In order to apply the Multiplicative Ergodic Theorem [8, 3.4.1 Theorem] we need to assure *log-integrability* of the cocycle which, in our case, means that

$$\max\{0, \log \|M_0(\sigma)\|_2\} \in L^1(S^{\mathbb{N}}, \nu). \tag{3.60}$$

Since S finite, the quantity $\max\{0, \log \|M_0(\sigma)\|_2\}$ is bounded and therefore (3.60) always holds. Thus, because ϑ_1 is a simple Lyapunov exponent, [8, 3.4.1 Theorem] implies that for ν-a.e. σ there is a hyperplane $\mathcal{H} = \mathcal{H}(\sigma) \subset \mathbb{R}^d$ such that $\lim_{n\to\infty} \frac{1}{n} \log \|M_{[0,n)}(\sigma)^t|_{\mathcal{H}}\|_2 \leq \vartheta_2$. This implies that for each $\varepsilon > 0$ there is $n_0 = n_0(\varepsilon, \sigma)$ such that

$$\|M_{[0,n)}(\sigma)^t|_{\mathcal{H}}\|_2 < e^{(\vartheta_2+\varepsilon)n} \tag{3.61}$$

holds for $n \geq n_0$. It remains to show that $\mathcal{H} = \mathbf{u}^\perp$. However, this follows because for $\mathbf{x} \notin \mathbf{u}^\perp$ we have that $\langle M_{[0,n)}(\sigma)^t \mathbf{x}, \mathbf{1}\rangle = \langle \mathbf{x}, M_{[0,n)}(\sigma)\mathbf{1}\rangle$ is unbounded because for large n the vector $M_{[0,n)}(\sigma)\mathbf{1}$ is a large vector close to the line $\mathbb{R}_+\mathbf{u}$. Thus the only hyperplane for which (3.61) can possibly hold is $\mathcal{H} = \mathbf{u}^\perp$ and (ii) follows. $\square$

Proposition 3.9.6 is now used in order to show that balance is generic for elements of a shift $(S^{\mathbb{N}}, \Sigma, \nu)$ satisfying the Pisot condition.

Lemma 3.9.7 *Let S be a finite set of unimodular substitutions over the alphabet $\mathcal{A} = \{1, 2, \ldots, d\}$ and assume that the shift $(S^{\mathbb{N}}, \Sigma, \nu)$ is ergodic and satisfies the Pisot condition (3.59). Assume further that ν assigns positive measure to every*

---

[11] Here $\|\cdot\|_2$ is the operator norm w.r.t. the Euclidean norm on $\mathbb{R}^d$.

cylinder and that there exists a cylinder corresponding to a substitution with positive incidence matrix. Then the sets

$$S(C) = \{\sigma \in S^{\mathbb{N}} : \mathcal{L}_\sigma \text{ is } C\text{-balanced}\} \qquad (C \in \mathbb{N})$$

satisfy

$$\lim_{C \to \infty} \nu(S(C)) = 1,$$

i.e., balance of $\mathcal{L}_\sigma$ is a generic property of $\sigma \in S^{\mathbb{N}}$.

Proof By Proposition 3.9.6 we see that for ν-a.e. $\sigma \in S^{\mathbb{N}}$ the sequence is primitive and recurrent, and for the letter frequency vector $\mathbf{u} = (u_1, \dots, u_d)^t$ (with $\|\mathbf{u}\|_1 = 1$) we have

$$\sum_{n \geq 0} \|(M_{[0,n)})^t|_{\mathbf{u}^\perp}\|_2 < \infty. \tag{3.62}$$

We assume that $\sigma \in S^{\mathbb{N}}$ has all these properties and follow the proof of [44, Theorem 5.8]. Let $w \in X_\sigma$ be arbitrary. Since by the proof of Proposition 3.5.3(iii) each element of X_σ has the same language, each factor v of w is a factor of a limit sequence of σ and, hence, by (3.4.1) can be written as

$$v = p_0 \sigma_0 (p_1 \dots \sigma_{N-2}(p_{N-1} \sigma_{N-1}(x) s_{N-1}) \dots s_1) s_0 \tag{3.63}$$

where p_n and s_n is a prefix and a suffix of $\sigma_n(i)$ for some $i \in \mathcal{A}$, respectively, for each $0 \leq n \leq N - 1$ and x is a factor of $\sigma_N(i)$ for some $i \in \mathcal{A}$. To make the notation easier we set $p_N = x$ and $s_N = \varepsilon$. We mention that (3.63) is the Dumont-Thomas decomposition of v which was first introduced in [70]. Using (3.63) and denoting by $\mathbf{e}_1, \dots, \mathbf{e}_d$ the standard basis vectors of $\mathbb{R}^d$ we have

$$|v|_i - |v| u_i = \sum_{n=0}^{N} (|\sigma_{[0,n)}(p_n)|_i - |\sigma_{[0,n)}(p_n)| u_i + |\sigma_{[0,n)}(s_n)|_i - |\sigma_{[0,n)}(s_n)| u_i)$$

$$= \sum_{n=0}^{N} \langle \mathbf{e}_i - u_i(\mathbf{e}_1 + \dots + \mathbf{e}_d), M_{[0,n)} \mathbf{l}(p_n + s_n) \rangle$$

for each $i \in \mathcal{A}$. Since $u_1 + \cdots + u_d = 1$ we see that $\mathbf{e}_i - u_i(\mathbf{e}_1 + \cdots + \mathbf{e}_d) \in \mathbf{u}^\perp$. This can be used to get

$$
\left| |v|_i - |v| u_i \right| \leq \sum_{n=0}^{N} \left| \langle \mathbf{e}_i - u_i(\mathbf{e}_1 + \cdots + \mathbf{e}_d), M_{[0,n)} \mathbf{l}(p_n + s_n) \rangle \right|
$$

$$
= \sum_{n=0}^{N} \left| \langle (M_{[0,n)})^t (\mathbf{e}_i - u_i(\mathbf{e}_1 + \cdots + \mathbf{e}_d)), \mathbf{l}(p_n + s_n) \rangle \right|
$$

$$
\leq 2\sqrt{d} \sum_{n=0}^{N} \| (M_{[0,n)})^t |_{\mathbf{u}^\perp} \|_2 \| M_n \|_2.
$$

Since S is a finite set, the quantity $\|M_n\|_2$ is uniformly bounded in n. Thus, using (3.62) this implies that w is finitely balanced. Since σ was taken from a set of full measure ν of $S^{\mathbb{N}}$ this finishes the proof.   □

Before we can put everything together we need to deal with the genericness of algebraic irreducibility. This has been done in [52, Lemma 8.7] in the following fashion.

Lemma 3.9.8 *Let S be a finite set of unimodular substitutions over the alphabet $\mathcal{A} = \{1, 2, \ldots, d\}$ and assume that the shift $(S^{\mathbb{N}}, \Sigma, \nu)$ is ergodic and satisfies the Pisot condition (3.59). If ν-a.e. sequence $\boldsymbol{\sigma} \in S^{\mathbb{N}}$ is primitive then ν-a.e. sequence $\boldsymbol{\sigma} \in S^{\mathbb{N}}$ is algebraically irreducible.*

Proof *(Sketch)* Let $\boldsymbol{\sigma}$ be a generic sequence with sequence of incidence matrices $\mathbf{M} = (M_n)$ and fix $k \in \mathbb{N}$. Then for $\ell \to \infty$ the matrix $M_{[k,\ell)}$ maps the unit sphere into an ellipse whose largest semiaxis tends to infinity and all of whose other semiaxes tend to zero by the Pisot condition. We prove that for ℓ large enough there can be only one eigenvalue λ with $|\lambda| \geq 1$.

Indeed, if ℓ is large enough then $M_{[k,\ell)}$ is strictly positive, thus there is a dominant Perron-Frobenius eigenvalue $\lambda_0 > 1$. It corresponds to an eigenvector $\mathbf{w}_0$ with strictly positive entries. Suppose that there is another real eigenvalue λ with $|\lambda| \geq 1$ and corresponding eigenvector $\mathbf{w}$. Since the image of the unit sphere under $M_{[k,\ell)}$ is an ellipse with the above mentioned properties, the corresponding eigenvector has to have a direction close to $\mathbf{w}_0$ for ℓ large, because otherwise its length would be shrunk by the application of $M_{[k,\ell)}$ as can be seen in Fig. 3.20. Thus, if ℓ is large enough then $\mathbf{w}$ must have strictly positive entries. However, such an eigenvector has to belong to the Perron-Frobenius eigenvalue, a contradiction. The case of nonreal eigenvalues can be treated similarly. Thus $M_{[k,\ell)}$ has only one eigenvalue of modulus greater than or equal to 1. Since $M_{[k,\ell)}$ is an unimodular integer matrix, it cannot have 0 as an eigenvalue. This implies that the characteristic polynomial of $M_{[k,\ell)}$ is irreducible and, hence, $\boldsymbol{\sigma}$ is algebraically irreducible. Indeed, we even proved that the characteristic polynomial of $M_{[k,\ell)}$ is the minimal polynomial of the Pisot number λ_0.   □

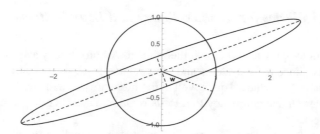

Fig. 3.20 An illustration of the elliptic image of the unit circle under $M_{[k,\ell)}$. The dashed lines are the axes of the ellipse, the largest axis being the direction of the Perron-Frobenius eigenvector $\mathbf{w}_0$. If the indicated vector $\mathbf{w}$ is an eigenvector of $M_{[k,\ell)}$ for another eigenvalue, its direction has to be far from the direction of $\mathbf{w}_0$ (because not all of its entries can be positive). This entails that its length is less than 1 and so it can only correspond to an eigenvalue less than 1 in modulus

We now have all the necessary ingredients to finish the proof of Theorem 3.9.5.

Proof *(Conclusion of the Proof of Theorem 3.9.5)* We show that the conditions of Theorem 3.9.4 are satisfied for ν-a.e. $\sigma \in S^{\mathbb{N}}$. To keep things simple we give the proof only for ν being a Bernoulli measure. Primitivity and algebraic irreducibility hold ν-a.e. by Proposition 3.9.6(i) and Lemma 3.9.8, respectively.

It remains to deal with the condition involving recurrence and balance. We claim that there is $C \in \mathbb{N}$ such that

$$\nu([\sigma_0, \ldots, \sigma_{\ell-1}] \cap \Sigma^{-\ell} S(C)) > 0 \text{ for each } (\sigma_n) \in S^{\mathbb{N}} \text{ and each } \ell \geq 0. \quad (3.64)$$

Indeed, since ν is a Bernoulli measure, $[\sigma_0, \ldots, \sigma_{\ell-1}]$ is independent from $\Sigma^{-\ell} S(C)$. Thus we have

$$\nu([\sigma_0, \ldots, \sigma_{\ell-1}] \cap \Sigma^{-\ell} S(C)) = \nu([\sigma_0, \ldots, \sigma_{\ell-1}])\nu(S(C))$$

and the claim (3.64) follows because $\nu([\sigma_0, \ldots, \sigma_{\ell-1}]) > 0$ by assumption and $\nu(S(C)) > 0$ for C large enough by Lemma 3.9.7. By another application of Poincaré recurrence (3.64) yields that for ν-a.e. $\sigma \in S^{\mathbb{N}}$ and for every $\ell \in \mathbb{N}$ there is $n > 0$ such that $\Sigma^n \sigma \in [\sigma_0, \ldots, \sigma_{\ell-1}]$ and $\Sigma^{n+\ell} \sigma \in S(C)$.

Summing up we see that the assumptions of Theorem 3.9.4 are satisfied for ν-a.e. $\sigma \in S^{\mathbb{N}}$. Thus Theorem 3.9.5 (1) follows from Theorem 3.5.11, Theorem 3.9.5 (2) follows from Theorem 3.7.1, and Theorem 3.9.5 (3) follows from Theorem 3.9.4.

$\square$

Remark 3.9.9 With small amendments in the conclusion of the proof of Theorem 3.9.5 it is possible to prove Theorem 3.9.5 for sofic subshifts (X, Σ, ν) of $(S^{\mathbb{N}}, \Sigma, \nu)$. Even the case of infinitely many substitutions (i.e., $|S| = \infty$) can be treated provided that the log-integrability condition (3.60) is satisfied. In this case one has to deal with the *S-adic graph* introduced in [44]. As mentioned above, the general result is contained in [52].

3.9.5   Corollaries for Arnoux-Rauzy and Brun Systems

We now want to apply the two main theorems to Arnoux-Rauzy as well as Brun S-adic systems. Since these systems and their related generalized continued fraction algorithms have been studied quite well in the literature this will yield unconditional results on measurable conjugacy to a torus rotation, natural codings, and bounded remainder sets.

We start with the case of Arnoux-Rauzy systems. Let $S = \{\sigma_1, \sigma_2, \sigma_3\}$ be the set of Arnoux-Rauzy substitutions defined in (3.18). First we give a version of Theorem 3.9.5 for the S-adic sequences taken from $S^{\mathbb{N}}$.

Corollary 3.9.10 (See [52, Theorem 3.8]) *Let S be the set of Arnoux-Rauzy substitutions defined in (3.18) and consider the full shift $(S^{\mathbb{N}}, \Sigma, \nu)$ equipped with an ergodic invariant measure ν that assigns positive mass to each cylinder. Then ν-a.e. $\sigma \in S^{\mathbb{N}}$ defines an S-adic system (X_σ, Σ) that is measurably conjugate to a rotation T on the 2-torus $\mathbb{T}^2$. Moreover, each element of X_σ forms a natural coding of T w.r.t. the partition $\{\mathcal{R}_i : i \in \mathcal{A}\}$ defined by the subtiles of the Rauzy fractal $\mathcal{R}$. Each of these subtiles is a bounded remainder set of T.*

Proof *(Sketch)* It is easy to see that each cylinder containing each of the three substitutions has positive incidence matrix. Thus the result follows from Theorem 3.9.5 if we can establish that $(S^{\mathbb{N}}, \Sigma, \nu)$ satisfies the Pisot condition and that for ν-a.e. $\sigma \in S^{\mathbb{N}}$ the associated collection $\mathcal{C}_1$ of Rauzy fractals forms a tiling. The fact that the Pisot condition holds was proved by Avila and Delecroix [28]. The tiling property is a consequence of Proposition 3.8.8. Indeed, assertion (iii) of this proposition holds by the following results. Firstly, strong coincidence follows from [37, Proposition 4] (or [52, Section 9] where "negative coincidence" was used). The other assertion from Proposition 3.8.8(iii) is a weaker form of the geometric finiteness property which holds by Proposition 3.8.11 (see also [48, Theorem 4.7]).   □

With help of the balance properties of Arnoux-Rauzy sequences proved in [43] it is possible to use Theorem 3.9.4 in order to show results for concrete Arnoux-Rauzy systems. For instance it is proved in [52, Corollary 3.9] that any linearly recurrent Arnoux-Rauzy sequence with recurrent directive sequence generates an S-adic system (X_σ, Σ) that is measurably conjugate to a rotation on a 2-torus.

For the second class of examples let $S = \{\sigma_1, \sigma_2, \sigma_3\}$ be the set of Brun substitutions defined in (3.32). In this case a version of Theorem 3.9.5 completely analogous to Corollary 3.9.10 holds.

Corollary 3.9.11 (See [52, Theorem 3.10]) *Let S be the set of Brun substitutions defined in (3.32) and consider the full shift $(S^{\mathbb{N}}, \Sigma, \nu)$ equipped with an ergodic invariant measure ν that assigns positive mass to each cylinder. Then ν-a.e. $\sigma \in S^{\mathbb{N}}$ defines an S-adic system (X_σ, Σ) that is measurably conjugate to a rotation T on the 2-torus $\mathbb{T}^2$. Moreover, each element of X_σ forms a natural coding of T w.r.t. the partition $\{\mathcal{R}_i : i \in \mathcal{A}\}$ defined by the subtiles of the Rauzy fractal $\mathcal{R}$. Each of these subtiles is a bounded remainder set of T.*

Proof *(Sketch)* First observe that $\sigma_1\sigma_2\sigma_1\sigma_2$ has positive incidence matrix. One uses again [28] to ensure that the Pisot condition holds (see also [83, 102, 118] for similar results). The tiling property follows from geometric coincidence which is established in [42] for the Brun class.   □

Contrary to the Arnoux-Rauzy continued fraction algorithm, the Brun algorithm can be performed for all elements $(x_1, x_2) \in \Delta$ with Δ as in (3.24). Thus, using Brun systems we get natural codings for a.a. torus rotations $\mathbf{t} \in \mathbb{T}^2$.

Corollary 3.9.12 (See [52, Corollary 3.12]) *Let S be the set of Brun substitutions defined in (3.32). Then for almost every $\mathbf{t} \in \mathbb{T}^2$ (w.r.t. the Haar measure on $\mathbb{T}^2$) there is $\sigma \in S^{\mathbb{N}}$ such that the shift (X_σ, Σ) is measurably conjugate to the rotation $T_{\mathbf{t}}$ by $\mathbf{t}$ on $\mathbb{T}^2$. Moreover, the sequences in X_σ form natural codings of the rotation $T_{\mathbf{t}}$.*

To create concrete examples of Brun S-adic shifts being measurably conjugate to a rotation, one can use Theorem 3.9.4 together with the balance results established in [68].

3.10 Concluding Remarks: Natural Extensions, Flows, and Their Poincaré Sections

It remains to extend the ideas and results presented in Sect. 3.2.5 to generalized continued fraction algorithms and S-adic systems on d letters. This is the subject of the ongoing paper by Arnoux et al. [12].

It is possible to study natural extensions of generalized continued fraction algorithms (see for instance [19, 21]). In the way we do it in [12], the analogs of the L-shaped regions of Sect. 3.2.5 are "Rauzy-Boxes" which are defined as suspensions of S-adic Rauzy fractals. They were introduced in the S-adic setting in [52, Section 2.9] but have been studied earlier in the substitutive case, see for instance Ito and Rao [91]. These Rauzy boxes allow nonstationary Markov partitions for so-called "mapping families" in the sense studied by Arnoux and Fisher [15] that can be visualized by restacking S-adic Rauzy fractals in a suitable way.

Also Artin's idea of viewing continued fraction algorithms as Poincaré sections of the geodesic flow on $SL_2(\mathbb{Z})\backslash SL_2(\mathbb{R})$ can be generalized. In this generalization the role of the geodesic flow is played by the *Weyl Chamber Flow*, a diagonal $\mathbb{R}^{d-1}$-action on the space $SL_d(\mathbb{Z})\backslash SL_d(\mathbb{R})$ of d-dimensional lattices. It turns out that each coordinate direction of this $\mathbb{R}^{d-1}$-action has a Poincaré section which is arithmetically coded by a generalized continued fraction algorithm. Geometrically, this is visualized by deforming a given Rauzy box (one for each coordinate) by the action of the Weyl Chamber Flow and restacking it accordingly as soon as a Poincaré section is reached.

Details of all this will be contained in [12].

Acknowledgments I warmly thank Shigeki Akiyama and Pierre Arnoux for inviting me to contribute to this volume. I very much appreciate their constant encouragement and support, and the many discussions I had with them. Moreover, I am indebted to Valérie Berthé, Sébastien Labbé, and Wolfgang Steiner for their suggestions.

References

1. B. Adamczewski, Balances for fixed points of primitive substitutions. Theor. Comput. Sci. **307**(1), 47–75 (2003). Words
2. S. Akiyama, Self affine tilings and Pisot numeration systems, in *Number Theory and Its Applications*, ed. by K. Győry, S. Kanemitsu (Kluwer, Dordrecht, 1999), pp. 1–17
3. S. Akiyama, On the boundary of self affine tilings generated by Pisot numbers. J. Math. Soc. Japan **54**(2), 283–308 (2002)
4. S. Akiyama, M. Barge, V. Berthé, J.Y. Lee, A. Siegel, On the Pisot substitution conjecture, in *Mathematics of Aperiodic Order*. Progr. Math., vol. 309 (Birkhäuser/Springer, Basel, 2015), pp. 33–72
5. S. Akiyama, T. Sadahiro, A self-similar tiling generated by the minimal Pisot number, in *Proceedings of the 13th Czech and Slovak International Conference on Number Theory (Ostravice, 1997)*, vol. 6 (1998), pp. 9–26
6. J.P. Allouche, J. Shallit, *Automatic Sequences: Theory, Applications, Generalizations*. (Cambridge University Press, Cambridge, 2003)
7. A. Andres, R. Acharya, C. Sibata, Discrete analytical hyperplanes. Graph. Model. Image Process. **59**, 302–309 (1997)
8. L. Arnold, *Random Dynamical Systems*. Springer Monographs in Mathematics (Springer, Berlin, 1998)
9. P. Arnoux, Un exemple de semi-conjugaison entre un échange d'intervalles et une translation sur le tore. Bull. Soc. Math. France **116**(4), 489–500 (1989) (1988)
10. P. Arnoux, Le codage du flot géodésique sur la surface modulaire. Enseign. Math. (2) **40**(1-2), 29–48 (1994)
11. P. Arnoux, Continued fractions: natural extensions and invariant measures, in *Natural Extension of Arithmetic Algorithms and S-adic System*. RIMS Kôkyûroku Bessatsu, vol. B58 (Res. Inst. Math. Sci. (RIMS), Kyoto, 2016), pp. 19–32
12. P. Arnoux, V. Berthé, M. Minervino, W. Steiner, J. Thuswaldner, Nonstationary Markov partitions, flows on homogeneous spaces, and continued fractions (2018, in preparation)
13. P. Arnoux, S. Ferenczi, P. Hubert, Trajectories of rotations. Acta Arith. **87**(3), 209–217 (1999)
14. P. Arnoux, A.M. Fisher, The scenery flow for geometric structures on the torus: the linear setting. Chinese Ann. Math. Ser. B **22**(4), 427–470 (2001)
15. P. Arnoux, A.M. Fisher, Anosov families, renormalization and non-stationary subshifts. Ergodic Theory Dyn. Syst. **25**(3), 661–709 (2005)
16. P. Arnoux, M. Furukado, E. Harriss, S. Ito, Algebraic numbers, free group automorphisms and substitutions on the plane. Trans. Am. Math. Soc. **363**(9), 4651–4699 (2011)
17. P. Arnoux, E. Harriss, What is ... a Rauzy fractal? Not. Am. Math. Soc. **61**(7), 768–770 (2014)
18. P. Arnoux, S. Ito, Pisot substitutions and Rauzy fractals. Bull. Belg. Math. Soc. Simon Stevin **8**(2), 181–207 (2001). Journées Montoises d'Informatique Théorique (Marne-la-Vallée, 2000)
19. P. Arnoux, S. Labbé, On some symmetric multidimensional continued fraction algorithms. Ergodic Theory Dyn. Syst. **38**, 1601–1626 (2018)
20. P. Arnoux, M. Mizutani, T. Sellami, Random product of substitutions with the same incidence matrix. Theor. Comput. Sci. **543**, 68–78 (2014)
21. P. Arnoux, A. Nogueira, Mesures de Gauss pour des algorithmes de fractions continues multidimensionnelles. Ann. Sci. École Norm. Sup. (4) **26**(6), 645–664 (1993)

22. P. Arnoux, G. Rauzy, Représentation géométrique de suites de complexité $2n + 1$. Bull. Soc. Math. France **119**(2), 199–215 (1991)
23. P. Arnoux, T.A. Schmidt, Cross sections for geodesic flows and α-continued fractions. Nonlinearity **26**(3), 711–726 (2013)
24. P. Arnoux, T.A. Schmidt, Commensurable continued fractions. Discrete Contin. Dyn. Syst. **34**(11), 4389–4418 (2014)
25. P. Arnoux, Š. Starosta, The Rauzy gasket, in *Further Developments in Fractals and Related Fields*. Trends Mathematics (Birkhäuser/Springer, New York, 2013), pp. 1–23
26. E. Artin, Ein mechanisches System mit quasiergodischen Bahnen. Abh. Math. Sem. Univ. Hamburg **3**(1), 170–175 (1924)
27. J. Auslander, *Minimal Flows and Their Extensions*. North-Holland Mathematics Studies, vol. 153. (North-Holland Publishing Co., Amsterdam, 1988). Notas de Matemática [Mathematical Notes], 122
28. A. Avila, V. Delecroix, Some monoids of Pisot matrices (2015). Preprint, http://arxiv.org/abs/1506.03692
29. A. Avila, P. Hubert, A. Skripchenko, Diffusion for chaotic plane sections of 3-periodic surfaces. Invent. Math. **206**(1), 109–146 (2016)
30. A. Avila, P. Hubert, A. Skripchenko, On the Hausdorff dimension of the Rauzy gasket. Bull. Soc. Math. France **144**(3), 539–568 (2016)
31. G. Barat, V. Berthé, P. Liardet, J. Thuswaldner, Dynamical directions in numeration. Ann. Inst. Fourier (Grenoble) **56**(7), 1987–2092 (2006). Numération, pavages, substitutions
32. M. Barge, The Pisot conjecture for β-substitutions. Ergodic Theory Dyn. Syst. **38**(2), 444–472 (2016)
33. M. Barge, Pure discrete spectrum for a class of one-dimensional substitution tiling systems. Discrete Contin. Dyn. Syst. **36**(3), 1159–1173 (2016)
34. M. Barge, B. Diamond, Coincidence for substitutions of Pisot type. Bull. Soc. Math. France **130**(4), 619–626 (2002)
35. M. Barge, J. Kellendonk, Proximality and pure point spectrum for tiling dynamical systems. Mich. Math. J. **62**(4), 793–822 (2013)
36. M. Barge, J. Kwapisz, Geometric theory of unimodular Pisot substitutions. Amer. J. Math. **128**(5), 1219–1282 (2006)
37. M. Barge, S. Štimac, R.F. Williams, Pure discrete spectrum in substitution tiling spaces. Discrete Contin. Dyn. Syst. **33**(2), 579–597 (2013)
38. A.Y. Belov, G.V. Kondakov, I.V. Mitrofanov, Inverse problems of symbolic dynamics, in *Algebraic Methods in Dynamical Systems*. Banach Center Publ., vol. 94 (Polish Acad. Sci. Inst. Math., Warsaw, 2011), pp. 43–60
39. J. Bernoulli, Sur une nouvelle espece de calcul. *Recueil pour les Astronomes*, Berlin, vol. 1 (1772), pp. 255–284
40. J. Berstel, Transductions and context-free languages, *Leitfäden der Angewandten Mathematik und Mechanik [Guides to Applied Mathematics and Mechanics]*, vol. 38 (B. G. Teubner, Stuttgart, 1979)
41. V. Berthé, Multidimensional Euclidean algorithms, numeration and substitutions. Integers **11B**, A2 (2011)
42. V. Berthé, J. Bourdon, T. Jolivet, A. Siegel, A combinatorial approach to products of Pisot substitutions. Ergodic Theory Dyn. Syst. **36**(6), 1757–1794 (2016)
43. V. Berthé, J. Cassaigne, W. Steiner, Balance properties of Arnoux-Rauzy words. Int. J. Algebra Comput. **23**(4), 689–703 (2013)
44. V. Berthé, V. Delecroix, Beyond substitutive dynamical systems: S-adic expansions. RIMS Lecture Note 'Kôkyûroku Bessatsu' **B46**, 81–123 (2014)
45. V. Berthé, E. Domenjoud, D. Jamet, X. Provençal, Fully subtractive algorithm, tribonacci numeration and connectedness of discrete planes, in *Numeration and Substitution 2012*. RIMS Kôkyûroku Bessatsu, vol. B46 (Res. Inst. Math. Sci. (RIMS), Kyoto, 2014), pp. 159–174

46. V. Berthé, S. Ferenczi, L.Q. Zamboni, Interactions between dynamics, arithmetics and combinatorics: the good, the bad, and the ugly, in *Algebraic and Topological Dynamics*. Contemp. Math., vol. 385 (American Mathematical Society, Providence, RI, 2005), pp. 333–364

47. V. Berthé, C. Holton, L.Q. Zamboni, Initial powers of Sturmian sequences. Acta Arith. **122**(4), 315–347 (2006)

48. V. Berthé, T. Jolivet, A. Siegel, Substitutive Arnoux-Rauzy sequences have pure discrete spectrum. Unif. Distrib. Theory **7**(1), 173–197 (2012)

49. V. Berthé, S. Labbé, Factor complexity of S-adic words generated by the Arnoux-Rauzy-Poincaré algorithm. Adv. Appl. Math. **63**, 90–130 (2015)

50. V. Berthé, M. Minervino, W. Steiner, J. Thuswaldner, The S-adic Pisot conjecture on two letters. Topology Appl. **205**, 47–57 (2016)

51. V. Berthé, A. Siegel, J.M. Thuswaldner, Substitutions, Rauzy fractals, and tilings, in *Combinatorics, Automata and Number Theory*. Encyclopedia of Mathematics and its Applications, vol. 135 (Cambridge University Press, Cambridge, 2010)

52. V. Berthé, W. Steiner, J.M. Thuswaldner, Geometry, dynamics, and arithmetic of S-adic shifts. Ann. Inst. Fourier (Grenoble) **69**(3), 1347–1409 (2019)

53. S. Bezuglyi, J. Kwiatkowski, K. Medynets, B. Solomyak, Invariant measures on stationary Bratteli diagrams. Ergodic Theory Dyn. Syst. **30**(4), 973–1007 (2010)

54. S. Bezuglyi, J. Kwiatkowski, K. Medynets, B. Solomyak, Finite rank Bratteli diagrams: structure of invariant measures. Trans. Am. Math. Soc. **365**(5), 2637–2679 (2013)

55. G. Birkhoff, Extensions of Jentzsch's theorem. Trans. Am. Math. Soc. **85**, 219–227 (1957)

56. M. Boshernitzan, A unique ergodicity of minimal symbolic flows with linear block growth. J. Anal. Math. **44**, 77–96 (1984/85)

57. P. Boyland, W. Severa, Geometric representation of the infimax S-adic family. Fundam. Math. **240**(1), 15–50 (2018)

58. A.J. Brentjes, Multidimensional continued fraction algorithms, *Mathematical Centre Tracts*, vol. 145 (Mathematisch Centrum, Amsterdam, 1981)

59. V. Brun, Algorithmes euclidiens pour trois et quatre nombres, in *Treizième congrès des mathèmaticiens scandinaves, tenu à Helsinki 18-23 août 1957* (Mercators Tryckeri, Helsinki, 1958), pp. 45–64

60. E.B. Burger, J. Gell-Redman, R. Kravitz, D. Walton, N. Yates, Shrinking the period lengths of continued fractions while still capturing convergents. J. Number Theory **128**(1), 144–153 (2008)

61. V. Canterini, A. Siegel, Geometric representation of substitutions of Pisot type. Trans. Am. Math. Soc. **353**(12), 5121–5144 (2001)

62. J. Cassaigne, S. Ferenczi, A. Messaoudi, Weak mixing and eigenvalues for Arnoux-Rauzy sequences. Ann. Inst. Fourier (Grenoble) **58**(6), 1983–2005 (2008)

63. J. Cassaigne, S. Ferenczi, L.Q. Zamboni, Imbalances in Arnoux-Rauzy sequences. Ann. Inst. Fourier (Grenoble) **50**(4), 1265–1276 (2000)

64. J. Cassaigne, S. Labbé, J. Leroy, A set of sequences of complexity $2n + 1$, in *Combinatorics on Words*. Lecture Notes in Comput. Sci., vol. 10432 (Springer, Cham, 2017), pp. 144–156

65. J. Cassaigne, F. Nicolas, Factor complexity, in: *Combinatorics, Automata and Number Theory*. Encyclopedia Math. Appl., vol. 135 (Cambridge University Press, Cambridge, 2010), pp. 163–247

66. M.I. Cortez, F. Durand, B. Host, A. Maass, Continuous and measurable eigenfunctions of linearly recurrent dynamical Cantor systems. J. Lond. Math. Soc. (2) **67**(3), 790–804 (2003)

67. E.M. Coven, G.A. Hedlund, Sequences with minimal block growth. Math. Systems Theory **7**, 138–153 (1973)

68. V. Delecroix, T. Hejda, W. Steiner, Balancedness of Arnoux-Rauzy and Brun words, in *WORDS*. Lecture Notes in Computer Science, vol. 8079, pp. 119–131 (Springer, Berlin, 2013)

69. R. DeLeo, I.A. Dynnikov, Geometry of plane sections of the infinite regular skew polyhedron {4, 6 | 4}. Geom. Dedicata **138**, 51–67 (2009)

70. J.M. Dumont, A. Thomas, Systemes de numeration et fonctions fractales relatifs aux substitutions. Theor. Comput. Sci. **65**(2), 153–169 (1989)

71. F. Durand, Linearly recurrent subshifts have a finite number of non-periodic subshift factors. Ergodic Theory Dyn. Syst. **20**, 1061–1078 (2000)

72. F. Durand, Corrigendum and addendum to: "Linearly recurrent subshifts have a finite number of non-periodic subshift factors" [Ergodic Theory Dynam. Systems **20** (2000), no. 4, 1061–1078]. Ergodic Theory Dyn. Syst. **23**, 663–669 (2003)

73. F. Durand, J. Leroy, G. Richomme, Do the properties of an S-adic representation determine factor complexity? J. Integer Seq. **16**(2), Article 13.2.6, 30 pp. (2013)

74. H. Ei, Some properties of invertible substitutions of rank d, and higher dimensional substitutions. Osaka J. Math. **40**(2), 543–562 (2003)

75. H. Ei, S. Ito, H. Rao, Atomic surfaces, tilings and coincidences. II. Reducible case. Ann. Inst. Fourier (Grenoble) **56**(7), 2285–2313 (2006). Numération, pavages, substitutions

76. M. Einsiedler, T. Ward, *Ergodic Theory with a View Towards Number Theory*. Graduate Texts in Mathematics, vol. 259 (Springer, London, 2011)

77. S. Ferenczi, Rank and symbolic complexity. Ergodic Theory Dyn. Syst. **16**(4), 663–682 (1996)

78. S. Ferenczi, A.M. Fisher, M. Talet, Minimality and unique ergodicity for adic transformations. J. Anal. Math. **109**, 1–31 (2009)

79. T. Fernique, Multidimensional Sturmian sequences and generalized substitutions. Int. J. Found. Comput. Sci. **17**(3), 575–599 (2006)

80. T. Fernique, Generation and recognition of digital planes using multi-dimensional continued fractions, in *Discrete Geometry for Computer Imagery*. Lecture Notes in Comput. Sci., vol. 4992 (Springer, Berlin, 2008)

81. A.M. Fisher, Nonstationary mixing and the unique ergodicity of adic transformations. Stochastics Dyn. **9**(3), 335–391 (2009)

82. N.P. Fogg, *Substitutions in Dynamics, Arithmetics and Combinatorics*. Lecture Notes in Mathematics, vol. 1794 (Springer, Berlin, 2002)

83. T. Fujita, S. Ito, M. Keane, M. Ohtsuki, On almost everywhere exponential convergence of the modified Jacobi-Perron algorithm: a corrected proof. Ergodic Theory Dyn. Syst. **16**(6), 1345–1352 (1996)

84. H. Furstenberg, *Stationary Processes and Prediction Theory*. Annals of Mathematics Studies, No. 44 (Princeton University Press, Princeton, NJ, 1960)

85. P.R. Halmos, *Lectures on Ergodic Theory* (Chelsea Publishing Co., New York, 1960)

86. C. Holton, L.Q. Zamboni, Geometric realizations of substitutions. Bull. Soc. Math. France **126**(2), 149–179 (1998)

87. J.E. Hopcroft, J.D. Ullman, *Introduction to Automata Theory, Languages, and Computation*. Addison-Wesley Series in Computer Science (Addison-Wesley Publishing Co., Reading, MA, 1979).

88. P. Hubert, A. Messaoudi, Best simultaneous Diophantine approximations of Pisot numbers and Rauzy fractals. Acta Arith. **124**(1), 1–15 (2006)

89. S. Ito, M. Kimura, On Rauzy fractal. Japan J. Ind. Appl. Math. **8**(3), 461–486 (1991)

90. S. Ito, M. Ohtsuki, Parallelogram tilings and Jacobi-Perron algorithm. Tokyo J. Math. **17**(1), 33–58 (1994)

91. S. Ito, H. Rao, Atomic surfaces, tilings and coincidence. I. Irreducible case. Israel J. Math. **153**, 129–155 (2006)

92. D. Jamet, N. Lafrenière, X. Provençal, Generation of digital planes using generalized continued-fractions algorithms, in *Discrete Geometry for Computer Imagery*. Lecture Notes in Comput. Sci., vol. 9647, pp. 45–56 (Springer, Cham, 2016)

93. D. Jamet, J.L. Toutant, On the connectedness of rational arithmetic discrete hyperplanes, in *Discrete Geometry for Computer Imagery*. Lecture Notes in Comput. Sci., vol. 4245, (Springer, Berlin, 2006), pp. 223–234

94. A. Katok, Interval exchange transformations and some special flows are not mixing. Israel J. Math. **35**(4), 301–310 (1980)

95. S. Labbé, 3-dimensional continued fraction algorithms cheat sheets (2015). http://arxiv.org/abs/1511.08399
96. J. Lagarias, Y. Wang, Self-affine tiles in $\mathbb{R}^n$. Adv. Math. **121**, 21–49 (1996)
97. J. Leroy, Some improvements of the S-adic conjecture. Adv. Appl. Math. **48**(1), 79–98 (2012)
98. J. Leroy, An S-adic characterization of minimal subshifts with first difference of complexity $1 \le p(n+1) - p(n) \le 2$. Discrete Math. Theor. Comput. Sci. **16**(1), 233–286 (2014)
99. G. Levitt, La dynamique des pseudogroupes de rotations. Invent. Math. **113**(3), 633–670 (1993)
100. B. Loridant, M. Minervino, Geometrical models for a class of reducible Pisot substitutions. Discrete Comput. Geom. **60**, 981–1028 (2018)
101. M. Lothaire, *Algebraic Combinatorics on Words*. Encyclopedia of Mathematics and its Applications, vol. 90 (Cambridge University Press, Cambridge, 2002). A collective work by J. Berstel, D. Perrin, P. Seebold, J. Cassaigne, A. De Luca, S. Varricchio, A. Lascoux, B. Leclerc, J.-Y. Thibon, V. Bruyere, C. Frougny, F. Mignosi, A. Restivo, C. Reutenauer, D. Foata, G.-N. Han, J. Desarmenien, V. Diekert, T. Harju, J. Karhumaki and W. Plandowski, with a preface by J. Berstel and D. Perrin
102. R. Meester, A simple proof of the exponential convergence of the modified Jacobi-Perron algorithm. Ergodic Theory Dyn. Syst. **19**(4), 1077–1083 (1999)
103. A. Messaoudi, Frontière du fractal de Rauzy et système de numération complexe. Acta Arith. **95**(3), 195–224 (2000)
104. A. Messaoudi, Propriétés arithmétiques et topologiques d'une classe d'ensembles fractales. Acta Arith. **121**(4), 341–366 (2006)
105. M. Minervino, W. Steiner, Tilings for Pisot beta numeration. Indag. Math. (N.S.) **25**(4), 745–773 (2014)
106. M. Minervino, J.M. Thuswaldner, The geometry of non-unit Pisot substitutions. Ann. Inst. Fourier (Grenoble) **64**(4), 1373–1417 (2014)
107. M. Morse, G.A. Hedlund, Symbolic dynamics II. Sturmian trajectories. Am. J. Math. **62**, 1–42 (1940)
108. H. Nakada, S. Ito, S. Tanaka, On the invariant measure for the transformations associated with some real continued-fractions. Keio Eng. Rep. **30**(13), 159–175 (1977)
109. O. Perron, Grundlagen für eine Theorie des Jacobischen Kettenbruchalgorithmus. Math. Ann. **64**(1), 1–76 (1907)
110. M. Queffélec, Substitution Dynamical Systems—Spectral Analysis. Lecture Notes in Mathematics, vol. 1294, 2nd edn. (Springer, Berlin, 2010)
111. C. Radin, M. Wolff, Space tilings and local isomorphism. Geom. Dedicata **42**(3), 355–360 (1992)
112. G. Rauzy, Une généralisation du développement en fraction continue, in *Séminaire Delange-Pisot-Poitou, 18e année: 1976/77, Théorie des nombres*, Fasc. 1, pp. Exp. No. 15, 16 (Secrétariat Math., Paris, 1977)
113. G. Rauzy, échanges d'intervalles et transformations induites. Acta Arith. **34**(4), 315–328 (1979)
114. G. Rauzy, Nombres algébriques et substitutions. Bull. Soc. Math. France **110**(2), 147–178 (1982)
115. J.P. Reveillès, Géométrie discrète, calculs en nombres entiers et algorithmes. Ph.D. thesis, Université Louis Pasteur, Strasbourg (1991)
116. V.A. Rohlin, Exact endomorphism of a Lebesgue space. Magyar Tud. Akad. Mat. Fiz. Oszt. Közl. **14**, 443–474 (1964)
117. Y. Sano, P. Arnoux, S. Ito, Higher dimensional extensions of substitutions and their dual maps. J. Anal. Math. **83**, 183–206 (2001)
118. B.R. Schratzberger, The exponent of convergence for Brun's algorithm in two dimensions. Österreich. Akad. Wiss. Math.-Natur. Kl. Sitzungsber. II **207**, 229–238 (1999) (1998)
119. F. Schweiger, *Multidimensional Continued Fractions* Oxford Science Publications (Oxford University Press, Oxford, 2000)

120. E.S. Selmer, Continued fractions in several dimensions. Nordisk Nat. Tidskr. **9**, 37–43, 95 (1961)
121. C. Series, The modular surface and continued fractions. J. London Math. Soc. (2) **31**(1), 69–80 (1985)
122. A. Siegel, Représentation des systèmes dynamiques substitutifs non unimodulaires. Ergodic Theory Dyn. Syst. **23**(4), 1247–1273 (2003)
123. A. Siegel, J.M. Thuswaldner, Topological properties of Rauzy fractals. Mém. Soc. Math. Fr. (N.S.) **118**, 144pp (2009)
124. V.F. Sirvent, Y. Wang, Self-affine tiling via substitution dynamical systems and Rauzy fractals. Pac. J. Math. **206**(2), 465–485 (2002)
125. M. Viana, Ergodic theory of interval exchange maps. Rev. Mat. Complut. **19**(1), 7–100 (2006)
126. P. Walters, *An Introduction to Ergodic Theory*. Graduate Texts in Mathematics, vol. 79 (Springer, New York, 1982)

Chapter 4
Operators, Algebras and Their Invariants for Aperiodic Tilings

Johannes Kellendonk

Abstract We review the construction of operators and algebras from tilings of Euclidean space. This is mainly motivated by physical questions, in particular after topological properties of materials. We explain how the physical notion of locality of interaction is related to the mathematical notion of pattern equivariance for tilings and how this leads naturally to the definition of tiling algebras. We give a brief introduction to the K-theory of tiling algebras and explain how the algebraic topology of K-theory gives rise to a correspondence between the topological invariants of the bulk and its boundary of a material.

4.1 Tilings and the Topology of Their Hulls

In condensed matter theory tilings are used to describe the spatial arrangement of the constituents which make up a material, for instance a quasicrystal. They describe the spatial structure of the material.

Associated to a tiling are various topological spaces and topological dynamical systems. Their topology is peculiar. It takes into account the topology of the space in which the tiling lies and, at the same time, its pattern structure, that is, the way how finite patterns repeat over the tiling. Continuity in the tiling topology is related to locality in physics.

J. Kellendonk (✉)
Institute Camille Jordan, Université Claude Bernard Lyon1, Villeurbanne, France
e-mail: kellendonk@math.univ-lyon1.fr

© The Editor(s) (if applicable) and The Author(s), under exclusive license
to Springer Nature Switzerland AG 2020
S. Akiyama, P. Arnoux (eds.), *Substitution and Tiling Dynamics: Introduction
to Self-inducing Structures*, Lecture Notes in Mathematics 2273,
https://doi.org/10.1007/978-3-030-57666-0_4

4.1.1 Basic Notions

There exist various introductions to the theory of aperiodic tilings which cover the basic notions we need, for instance [4, 41]. We recall here only the strict necessary.

- A *tiling* is a covering of $\mathbb{R}^d$ by tiles whose interiors do not overlap.
- A *tile* is here a possibly decorated compact convex polyhedron. The polyhedron (up to translational congruence) is its shape. The decoration is by symbols from a symbol set $\mathcal{A}$ and serves to distinguish equal shapes (or enforce matching conditions). We assume that tiles match face to face.
- The tiles of a tiling at different positions are considered different. If the tiling has only finitely many tiles up to translation and only finitely many symbols then it is said to have *finite local complexity* (FLC).

Examples are Voronoi tilings (or their dual) coming from Delone sets.

A tiling $\mathcal{T}$ can be translated by a vector $x \in \mathbb{R}^d$ in the space it lies. By this we mean that the individual tiles are shifted by x. We denote the tiling translated by x as $\mathcal{T} + x$ (the tile which has been on the origin 0 of the space in which $\mathcal{T}$ lies is now on x). A period of a tiling is a vector $x \in \mathbb{R}^d$ such that $\mathcal{T} + x = \mathcal{T}$. Aperiodicity may occur through the shapes of the tiles, like for instance in the Penrose tilings, or through the decorations, or through both. In order not to overburden the notation and complexity we now consider the first two cases separately and only indicate how to do the third.

4.1.2 Undecorated Tilings

An undecorated tiling by compact convex polyhedra without decorations can be equivalently described as the closed subset of $\mathbb{R}^d$ given by the union of the boundaries of its tiles. In the following we attach a language and a hull to any closed subset of $\mathbb{R}^d$.

Let $\mathcal{T}$ be a closed subset of $\mathbb{R}^d$. For $R \geq 0$ and $x \in \mathbb{R}^d$, the *R-patch of $\mathcal{T}$ centered at x* is intersection of the R-ball at x with $\mathcal{T}$, union with the boundary of the R-ball,

$$B_R[\mathcal{T}; x] := (B_R(x) \cap \mathcal{T}) \cup \partial B_R(x).$$

The R-patch class of an R-patch $B_R[\mathcal{T}; x]$ is its translational congruence class. The representative of the R-patch class of $B_R[\mathcal{T}; x]$ at 0 is $B_R[\mathcal{T}; x] - x = B_R[\mathcal{T} - x; 0]$, we simply denote it by $B_R[\mathcal{T} - x]$.

Using R-patches we define a metric-topology on the set of all closed subsets of $\mathbb{R}^d$. Let $\mathcal{T}$, $\mathcal{T}'$ be two closed subsets, their distance is

$$D(\mathcal{T}, \mathcal{T}') = \inf\{\epsilon : d_H(B_{\epsilon^{-1}}[\mathcal{T}], B_{\epsilon^{-1}}[\mathcal{T}']) \leq \epsilon\}$$

where d_H is the Hausdorff distance between the set of all compact subsets of $B_{\epsilon^{-1}}(0)$ which contain $\partial B_{\epsilon^{-1}}(0)$. We call $\mathcal{L}(\mathcal{T}) = \bigcup_{R \geq 0} \mathcal{L}^R(\mathcal{T})$,

$$\mathcal{L}^R(\mathcal{T}) = \overline{\{B_R[\mathcal{T} - x] : x \in \mathbb{R}^d\}}^{d_H}$$

the language of $\mathcal{T}$. We say that a closed subset $\mathcal{T}'$ of $\mathbb{R}^d$ is allowed for the language of $\mathcal{T}$ if each R-patch-class of $\mathcal{T}'$ occurs in $\mathcal{L}(\mathcal{T})$. The set of all closed subsets which are allowed for $\mathcal{T}$ is called the *continuous hull* of $\mathcal{T}$ and denoted by $\Omega(\mathcal{T})$. If $R' \leq R$ we have a continuous map $\mathcal{L}^R(\mathcal{T}) \ni B_R[\mathcal{T}'] \mapsto B_{R'}[\mathcal{T}'] \in \mathcal{L}^{R'}(\mathcal{T})$ associating to the R'-patch of $\mathcal{T}'$ its restriction to the ball of smaller radius R. The hull $\Omega(\mathcal{T})$ can be seen as the inverse limit of these maps. In particular, the metric topology on $\Omega(\mathcal{T})$ is the same as the inverse limit topology, namely it is the smallest topology so that all the maps $\Omega(\mathcal{T}) \ni \mathcal{T}' \mapsto B_R[\mathcal{T}'] \in \mathcal{L}^R(\mathcal{T})$ are continuous.

Examples:

1. If $\mathcal{T}$ is bounded then $\Omega(\mathcal{T})$ is the d-sphere.
2. If Λ is a regular lattice in $\mathbb{R}^d$ and $\mathcal{T}$ has period lattice Λ then $\Omega(\mathcal{T})$ is the torus $\mathbb{R}^d/\Lambda$.
3. If $\mathcal{T}$ is a hyperplane of dimension k then $\Omega(\mathcal{T})$ is the $d - k$-sphere, its south pole corresponds to $\mathcal{T}$ and its north pole to the empty set.

Applied to an undecorated tiling this construction yields its continuous hull. If $\mathcal{T}$ has FLC then $\{B_R[\mathcal{T} - x] : x \in \mathbb{R}^d\}$ is already closed in the Hausdorff topology and so any R-patch of $\mathcal{L}^R(\mathcal{T})$ has a translate occuring in $\mathcal{T}$. Then $\Omega(\mathcal{T})$ is the set of all tilings of $\mathbb{R}^d$ whose R-patches occur somewhere in $\mathcal{T}$.

4.1.3 Wang Tilings

Consider an (undecorated) tiling $\mathcal{T}$ whose vertex set $\mathcal{V}$ happens to lie in a regular lattice $\Lambda \subset \mathbb{R}^d$. Choose a closed fundamental domain I (a parallel epiped) for Λ and superimpose $\mathcal{T}$ with the periodic tiling whose tiles are $I + \lambda$, $\lambda \in \Lambda$. Then the boundary points of the tiles of $\mathcal{T}$ mark on the tiles of the periodic tiling patterns which we can take as symbols, and thus our tiling $\mathcal{T}$ can be alternatively described by a tiling by *decorated* parallel epipeds. The two tilings contain the same information, they can be transformed into each other by inspection of their R-patches. This is an example of mutual local derivability [5] the definition of which we recall further down. The tiling by *decorated* parallel epipeds is a so-called Wang tiling.

A Wang tiling is a tiling in which all tiles are cubes meeting face to face (or a parallel epipeds, but we can always apply a linear transformation to deform it into a cube), but the tiles carry decorations[1] and so Wang tilings may be aperiodic. Once we have fixed which tile contains the origin 0 of our space, the set of tiles of the tiling are in bijective correspondence to $\mathbb{Z}^d$ and their decorations may be described by a map into a space of decorations, or symbols, $\xi : \mathbb{Z}^d \to \mathcal{A}$. This suggests a more symbolic approach to the construction of the hull, and so we start with the description of the *symbolic* hull of a Wang tiling. We suppose that $\mathcal{A}$ is a compact set and if it is finite the Wang tiling is said to have finite local complexity. The (symbolic) N-patch of a Wang tiling at $n \in \mathbb{Z}^d$ is the decoration of the cube of size $2N + 1$ at n, more precisely, this is the subset $\{\xi_{n+m} : \|m\|_{max} \leq N\} \subset \mathcal{A}^{[-N,N]^d}$. The (symbolic) language of ξ is $\ell(\xi) = \bigcup_N \ell^N(\xi)$ where $\ell^N(\xi)$ is the closure of the set of N-patches of ξ (shifted to 0) in $\mathcal{A}^{[-N,N]^d}$. We can make the same inverse limit construction as above to obtain the symbolic hull of the Wang tiling which we denote $\Xi(\xi)$ or simply Ξ if the context is clear.

The additional information which is needed to construct from a map $\xi : \mathbb{Z}^d \to \mathcal{A}$ a tiling of $\mathbb{R}^d$ is the location of the origin, that is, a choice of point x in the cube I. Given such a point the Wang tiling symbolized by ξ is placed in $\mathbb{R}^d$ in such a way that the point x lies on the origin of our space if we identify I with the tile corresponding to $0 \in \mathbb{Z}^d$ (so with symbol ξ_0). The continuous hull Ω of the Wang tiling is then the quotient of $I \times \Xi$ by the following relation: If x belongs to the boundary of I then the origin of our space will be in two or more tiles of the Wang tiling. These possibilities have to be identified. More precisely, if $\alpha : \mathcal{A}^{\mathbb{Z}^d} \to \mathcal{A}^{\mathbb{Z}^d}$ denotes the translation action, $\alpha_m(\xi)(n) = \xi(n - m)$, then

$$\Omega = \mathbb{R}^d \times_{\mathbb{Z}^d} \Xi$$

which is the quotient of the cartesian product w.r.t. the diagonal action, i.e. w.r.t. the relation $(x, \xi) \sim (x + m, \alpha_m(\xi))$. The continuous hull Ω is thus a fibre bundle over the d-torus $\mathbb{R}^d/\mathbb{Z}^d$ with typical fibre Ξ.

4.1.4   Pattern Equivariant Functions and Local Derivability

The notion of pattern equivariant functions has been introduced for tilings of finite local complexity in [27]. We extend here this notion to arbitrary tilings.

Let $\mathcal{T}$ be a closed subset. A map f from $\mathbb{R}^d$ to a topological space Y is called *strongly pattern equivariant* for $\mathcal{T}$ (or *local*) if there exist $R \geq 0$ and a continuous function $b : \mathcal{L}^R(\mathcal{T}) \to Y$ (a sort of continuous sliding block code) such that

$$f(x) = b(B_R[\mathcal{T} - x]).$$

---

[1]Originally the decorations encode matching conditions, but we will not make use of that here.

In particular, if the R-patch of $\mathcal{T}$ at x, when shifted by the vector $y - x$ coincides with the R-patch of $\mathcal{T}$ at y then $f(x) = f(y)$. If Y is a metric space then we define *(weakly) pattern equivariant* functions as functions from $\mathbb{R}^d$ to Y which are uniform limits of strongly pattern equivariant functions.

A closed subset $\mathcal{T}'$ is *locally derivable* from a closed subset $\mathcal{T}$ if there exists $R > 0$ and a continuous function $b : \mathcal{L}^R(\mathcal{T}) \to \mathcal{L}^1(\mathcal{T}')$ (a local rule) such that

$$B_1[\mathcal{T}' - x] = b(B_R[\mathcal{T} - x]) \tag{4.1}$$

for all $x \in \mathbb{R}^d$. In other words, we can construct the 1-patch of $\mathcal{T}'$ at x from the R-patch class of $\mathcal{T}$ at x. If $\mathcal{T}'$ is locally derivable from $\mathcal{T}$ and vice versa we say that the two are *mutually locally derivable*. Since (4.1) has to hold for all $x \in \mathbb{R}^d$ there is, for each $r > 0$, a unique extension $b_r : \mathcal{L}^{R+r}(\mathcal{T}) \to \mathcal{L}^{1+r}(\mathcal{T}')$ such that $B_{1+r}[\mathcal{T}' - x] = b_r(B_{R+r}[\mathcal{T} - x])$. The b_r are also continuous and so a local derivation extends to a continuous map from the hull of $\mathcal{T}$ to the hull of $\mathcal{T}'$.

Note that if S is locally derivable from $\mathcal{T}$, then the local rule b allows to locally derive a set $S(\mathcal{T}')$ for any other element of $\Omega(\mathcal{T})$. Any locally derivable uniformly discrete subset S from $\mathcal{T}$ defines a *transversal*

$$\Xi_S = \{\mathcal{T}' \in \Omega : 0 \in S(\mathcal{T}')\}.$$

Since $s \in S(\mathcal{T}')$ if and only if $0 \in S(\mathcal{T}' - s)$ we see that

$$S(\mathcal{T}') = \{x \in \mathbb{R}^d : \mathcal{T}' - x \in \Xi_S\}.$$

Clearly, the vertex set $\mathcal{V}$ of a tiling $\mathcal{T}$ is locally derivable. We call $\Xi_\mathcal{V}$ the *canonical transversal* of $\Omega(\mathcal{T})$.

Let $F : \mathbb{R}^d \to \mathbb{R}^d$ be a differentiable function whose differential dF is strongly pattern equivariant for an undecorated tiling $\mathcal{T}$. We can apply F to the vertex set $\mathcal{V}$ of $\mathcal{T}$ to obtain a new set $\mathcal{V}' = F(\mathcal{V})$ which we would like to interprete as vertex set of a new tiling $\mathcal{T}'$. If dF is close enough to the identity we can do this by (a) requiring that the new tiles are the convex hulls of their vertices, and (b) preserving the combinatorial structure, that is, the information which vertices are vertices of the same tile. One may think of F as being homotopic to the identity and the new tiling to be obtained by a continuous deformation. At least if $\mathcal{T}$ is FLC it can be shown that, if dF is close enough to the identity then F is invertible and dF^{-1} is strongly pattern equivariant for $\mathcal{T}'$. This then implies that F induces a homeomorphism $\mathcal{F} : \Omega(\mathcal{T}) \to \Omega(\mathcal{T}')$ which maps $\Xi_\mathcal{V}(\mathcal{T})$ to $\Xi_{\mathcal{V}'}(\mathcal{T}')$. Using this technique of *deformation* [10, 28] one can show the following result.

Theorem 4.1.1 ([43]) *Let $\mathcal{T}$ be a tiling of finite local complexity. There exists a tiling $\mathcal{T}'$ of finite local complexity whose vertices lie in a regular lattice and a homeomorphism between the continuous hulls of $\mathcal{T}$ and $\mathcal{T}'$ which preserves the canonical transversals.*

Up to homeomorphisms preserving the canonical transversals any tiling of FLC can therefore be understood as a Wang tiling. Indeed, if the vertices $\mathcal{V}'$ lie in a regular lattice Λ then Λ must be locally derivable from $\mathcal{T}'$ and so we may identify $\varXi_\Lambda(\mathcal{T}')$ with $\{0\} \times \varXi$ where $\varXi$ is the symbolic hull of the Wang tiling described above. It then follows that the continuous hulls of $\mathcal{T}'$ and of the Wang tiling also agree.

4.2 Operators from Tilings

We are interested in wave phenomena in a media whose structures are modeled by tilings. The waves might either be probability waves of quantum mechanics, or classical waves, like light or acoustic waves. The waves are solutions of wave equations, like the Schrödinger equation or the Helmholtz equation, which typically are differential equations. Assuming that the waves satisfy the superposition principle and the equations are linear, the stationary problem consists of solving an eigenvalue equation in the Hilbert space $L^2(\mathbb{R}^d)$, or more precisely to determine the spectrum of a linear differential operator H on $L^2(\mathbb{R}^d)$.

If no external forces are present then the locality principle requires that the differential operator H has the form

$$H = \sum_\alpha f_\alpha D^\alpha$$

where $D^\alpha \psi = \prod_{i=1}^d \left(\frac{\partial}{\partial x_i}\right)^{\alpha_i} \psi$ for $\alpha \in \mathbb{N}^d$, and the coefficients f_α are pattern equivariant functions on $\mathbb{R}^d$ for the underlying tiling, typically non-zero only for a finite number of α. External magnetic fields can be incorporated by minimal coupling, but we won't consider that case here.

H is typically an unbounded operator on $L^2(\mathbb{R}^d)$ which requires a certain amount of care, in particular as we will want to attach it to a C^*-algebra. This could be done by considering bounded functions of H, like the heat-kernel $\exp(-tH)$ or the resolvent $(H-x)^{-1}$, but alternatively by considering so-called tight binding models whose operators are from the start bounded. This is what we will do here.

4.2.1 Tight Binding Operators

Let S be a Delone set which is locally derivable from a tiling $\mathcal{T}$. For instance, the vertex set of $\mathcal{T}$ is locally derivable from $\mathcal{T}$. A tight binding operator is an operator on $\ell^2(S, \mathbb{C}^N)$ of the form

$$H\psi(x) = \sum_{y \in S} H_{xy}\psi(y) \tag{4.2}$$

where, for each pair (x, y), H_{xy} is an $N \times N$ matrix. If there is an $R > 0$ such that $H_{xy} = 0$ provided $|y - x| > R$ the model has finite range and convergence of the sum is not an issue. The locality principle now translates into the requirement that the H_{xy} are pattern equivariant in the following sense: We call *double R-patch* at $(x, y) \subset \mathbb{R}^d \times \mathbb{R}^d$ the set

$$B_R[\mathcal{T}; x, y] := B_R[\mathcal{T}; x] \times B_R[\mathcal{T}; y] \subset \mathbb{R}^d \times \mathbb{R}^d$$

and its class $[B_R[\mathcal{T}; x, y]]$ the equivalence class under the diagonal action of $\mathbb{R}^d$, $(t, t') + x = (t + x, t' + x)$. The distance $\|x - y\|$ between x and y is called the range of the double R-patch at (x, y). The set of double R-patch classes of closed subsets to $\mathbb{R}^d$ is a metric space w.r.t. the Hausdorff metric.[2] A function $\mathbb{R}^d \times \mathbb{R}^d \ni (x, y) \mapsto Y$ is then strongly pattern equivariant if, for some $R, M > 0$, there is a continuous function b defined on the set double R-patch classes with range smaller or equal to M such that $f(x, y) = b([B_R[\mathcal{T}; x, y]])$.

A simple example of an operator of the form (4.2) is the discrete Laplacian on the Penrose tiling. Here S is the vertex set of the tiling, $N = 1$, $H_{xy} = 1$ if x and y are vertices of a common edge whereas $H_{xy} = 0$ otherwise.

4.2.2   Tight Binding Operators for Wang Tilings

If the vertex set of $\mathcal{T}$ lies in a regular lattice Λ then we may take $S = \Lambda \cong \mathbb{Z}^d$ above and rewrite (4.2) as

$$H\psi(x) = \sum_{n \in \mathbb{Z}^d} H_n(x)\psi(x - n) \tag{4.3}$$

for $\psi \in \ell^2(\mathbb{Z}^d, \mathbb{C}^N)$, where $H_n(x) = H_{x\,x-n}$. H has finite rank if $H_n = 0$ once n is larger than a given size. The above notion of pattern equivariance using double R-classes simplifies and we only have to make sure that $H_n : \mathbb{Z}^d \to M_N(\mathbb{C})$ is strongly pattern equivariant for each $n \in \mathbb{Z}^d$. H depends thus only on the symbolic sequence $\xi : \mathbb{Z}^d \to \mathcal{A}$ of the Wang tiling derived from $\mathcal{T}$ and S. Using the translation operators on $\mathbb{Z}^d$, $T^n\psi(x) = \psi(x - n)$ we can write

$$H = \sum_{n \in \mathbb{Z}^d} H_n T^n.$$

By Theorem 4.1.1, any operator of the form (4.2) on a tiling of FLC can be brought into the form (4.3). This applies, for instance, also to the discrete Laplacian on the

---

[2]The Hausdorff metric between equivalence classes of compact sets is the infimum over the Hausdorff distances between representatives of the classes.

Penrose tiling, even though the version written in the form (4.3) might look a lot more complicated and does not display its fivefold symmetry. But the form (4.3) has important structural consequences which, in particular, allow us to work with crossed product algebras instead of more general groupoid algebras.

4.2.3   A Simple Class of One-dimensional Models

To illustrate the different phenomena which can occur in different type of tilings we consider four simple tight binding models defined on one-dimensional Wang tilings. The reader may consult [31] for more details.

Given a symbolic sequence $\xi \in \mathcal{A}^{\mathbb{Z}}$ over some symbol space $\mathcal{A}$, we consider the following operators on $\ell^2(\mathbb{Z}, \mathbb{C}^N)$.

1. $H = H_0 + T + T^*$ with $H_0(n) = b(\xi_n)$ for some real block sliding code $b :$ $\mathcal{A} \to \mathbb{R}$ of range 0 $(N = 1)$. H is the discrete Laplacien on $\mathbb{Z}^d$ plus an onsite potential which depends only on the symbol at site x.

2. $H = \begin{pmatrix} 0 & Q^* \\ Q & 0 \end{pmatrix}$ with $Q = \sum_{n \in \mathbb{Z}} Q_n T^n$ with $Q_n(m) = b_n(\xi_m)$ for complex block sliding codes of finite range $b_n : \ell^L(\mathcal{A}) \to \mathbb{C}$, only finitely many being non-zero $(N = 2)$. H is the Hamlitonian of a typical one dimensional tight binding model of finite range with chiral symmetry, namely H anti-commutes with the self adjoint unitary $\Gamma = \begin{pmatrix} 1 & 0 \\ 0 & -1 \end{pmatrix}$.

The sequences ξ we consider here are obtained from a cut & project scheme with dimension and codimension equal to one, namely they are constructed as follows. For a given irrational number $\theta \in [0, 1]$ we consider the Kronecker flow line

$$L := \{(t, \theta' t) | t \in \mathbb{R}\} / \mathbb{Z}^2$$

on the torus $\mathbb{R}^2/\mathbb{Z}^2$ with slope $\theta' = \frac{\theta}{1-\theta}$ (see the red sloped line in the pictures (1–3)). A *window* W is a (not necessarily connected) curve in the torus which is transversal to the flow in the sense that there exists $\epsilon > 0$ such that for every point $w \in W$ and all $0 < |t| < \epsilon$, $w + (t, \theta' t) \neq W$. We consider three types of windows. In picture (1) W is a closed continuous (blue) curve with a single slope, in (2) W a closed continuous curve with more than one slope. Finally in Picture (3) W has a single slope but is not a closed loop. It makes a jump, and we suppose that the boundary points of W belong to W and lie on a line parallel to L, but not equal to L.

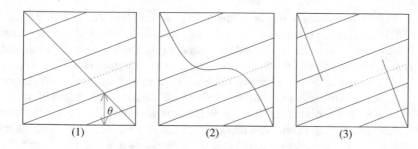

We parametrize the intersection of L with window W by a sequence of real numbers $(t_n)_{n\in\mathbb{Z}}$, that is, $(t_n, \theta' t_n) \in W + \mathbb{Z}^2$, ordered according to the standard order of the real line (there is an irrelevant choice of which element is the zeroth one). The symbolic sequence we are looking at is $\xi_n = t_{n+1} - t_n$, so $\mathcal{A}$ is a compact subset of $\mathbb{R}^+$.

If one computes the symbolic hull Ξ of ξ one finds the following. Except for an important subtlety in the third case, all elements of the hull can be obtained by moving W relative to L and repeating the same construction: take $s \in S := \{(t, 1 - t)|0 \leq t < 1\}$, parametrize the intersection of L with $s + W$, $(t_n, \theta' t_n) \in s + W + \mathbb{Z}^2$ and set $\xi_n(s) = t_{n+1} - t_n$. The choice of zeroth element coherent by demanding it to change continuously with s, which is possible except at the two points s for which $(t_0, \theta' t_0) \in s + \partial W + \mathbb{Z}^2$.

(1) If the window W is a closed continuous curve with a single slope then the sequence ξ is *periodic* and the hull consists of a single point. The models introduced above with periodic ξ have a band spectrum.

(2) If the transversal W is a closed continuous curve with more than one slope, then the sequence ξ is aperiodic and takes infinitely many values (it has infinite local complexity). Sequences of this type are *almost periodic* in the sense of Bohr. Physically they describe incommensurate structures. With an appropriate shape for W and $b(\xi_n) = \lambda \xi_n$ the operator $H = H_0 + T + T^*$ is the Harper model. The computation of its spectral properties has a long history [34]. Its spectrum is a Cantor set [3]. Its spectral type depends on the value of λ. Below a critical value for λ the operator has absolute continuous spectrum, at the critical value it has singular continuous spectrumn, and above that critical value it has pure point spectrum (for almost all θ) [24].

The map $S \ni s \mapsto \xi(s) \in \Xi$ is bijective and continuous and so yields a homeomorphism between the circle S and the symbolic hull Ξ. We may parametrize S in such a way that the shift action on Ξ corresponds to the rotation by θ.

(3) If the window W is a discontinuous curve with a single slope then the sequences ξ are so-called Sturmian sequences. They are aperiodic and have finite local complexity. They are also referred to as *quasi-periodic*, because such structures are used for one dimensional quasicrystals. With the choice of W as in the picture (3) there are only two possible values for ξ_n, let's call them a and b,

$a > b$. If the onsite potential of the above model $H = H_0 + T + T^*$ takes different values on a and b, we have the so-called Kohmoto model. Its spectrum is expected to be a Cantor set and singular continuous; this could be rigorously proven for the value of θ which corresponds to the Fibonacci sequence [14].

The jumps in W as in Picture (3) lead to an important subtlety. If the line L passes through one boundary point of $s + W$ then it passes also through the other boundary point. This situation will occur whenever $L \cap (s + \partial W) \neq \emptyset$, which happens for a countable dense set of values for $s \in S$. For these values, which we call singular, $\xi(s)$ will contain a third length c and therefore not in the symbolic hull of the original sequence ξ. However, by selecting only one of the two boundary points one obtains an element of the hull. Stated in terms of sequences, if s is singular then $\xi(s)$ contains somewhere the word bcb. If we replace this word by either ab or ba then the resulting sequence will be in the symbolic hull Ξ. As a consequence, Ξ contains the sequences $\xi(s)$ coming from the points of S which are not singular and, for each singular point of S, two sequences corresponding to the above two choices. The sequences of Ξ which come from singular points of S can be characterised by the fact that they contain a pair ab (or ba) which can be flipped to ba (or ab, respectively) so that the result lies still in Ξ. Such a flip is called a *phason flip*. It plays an important role in quasi-crystal physics.

Topologically, Ξ is the circle S disconnected along its singular points [17]. By disconnecting an interval $[0, 1]$ at some point $0 < r < 1$ we mean the following: We take r out to get two half open intervals and then add to each half open interval individually its missing boundary point. The result is the disjoint union of $[0, r]$ with $[r, 1]$ which hence has two connected components. There is an obvious map from this disjoint union back to the interval, just identify the two added boundary points.

Using inverse limits one can perform this procedure to a circle for a countable subset points. If this subset is dense then the resulting disconnected circle, which we denote S_c, is totally disconnected, the connected component of a point will contain only that point. There will then be a map from the disconnected circle S_c to the original circle S which is almost everywhere one to one, and two to one at the singular points.

The shift action on the hull can be described on S_c as follows: on the non-singular points of S the action is by rotation by θ, as in the almost periodic case, and the single orbit of singular points of S corresponds to two orbits in S_c.

(4) In the context of the above situation there is another possibility to resolve the issue of a third length c appearing in $\xi(s)$ for singular s [30]. We explain this the idea, which goes back to [16], first for the interval $[0, 1]$ which we disconnected above at some interior point r. Instead of adjoining two boundary points to the half-open intervals $[0, r)$ and $(r, 1]$ to make them closed and disconnected we may add an intervall $[0, 1]$ whose boundary points become the missing boundary points of the half-open intervals. The result is then a single (connected) interval $[0, r) \cup [r, r + 1] \cup (r + 1, 2] = [0, 2]$. When doing this for the circle with infinitely many points the result will be a again a circle,

which we call the augmented circle S_{aug}. It seems that the augmented circle has infinite circumference, but since we are only interested in its topology this is irrelevant. For the symbolic sequences ξ this has the following interpretation. If $\xi(s)$ is singular and thus contains the forbidden word bcb then, apart from the choices ab and ba, we have the possibility to replace bcb with $a_t b_t$ where $a_t = (1-t)a+tb$ and $b_t = ta+(1-t)b$, $t \in [0, 1]$. This enlarges the symbol set to $\mathcal{A} = [a, b]$. For $t \neq 0, 1$ the resulting sequences are no longer Sturmian, but the augmentation has a nice physical interpretation; it corresponds to making the phason flips a continuous motion.

This has consequences for the hull with its shift action. The *augmented* hull $\tilde{\Xi}$ is by definition the set of sequences obtained in the above way, that is, it contains Ξ together with a countable set of intervals. Topologically we may identify it with S_{aug} and then the shift action corresponds to, the rotation by θ on the points which are not singular, and a permutation of the added in intervals. Indeed, if we require that the t-variable is kept fixed there is only one possible continuous extension of the $\mathbb{Z}$-action to S_{aug}.

The effect on the spectrum of the associated Hamiltonian has been numerically analysed in [30]. If the underlying sequence contains $a_t b_t$ for $t \neq 0, 1$, the spectrum contains an additional eigenvalue in each gap.

4.2.4 Gaps and the Integrated Density of States

A gap in the spectrum of self adjoint operators is a connected component of its complement in $\mathbb{R}$. Physically a gap means that the material cannot have (electronic or wave) states at energies which lie in the gap.[3] In other words, the density of states vanishes in the gaps of the spectrum. The integrated density of states at energy E, IDS(E), is the integral over the density of states from $-\infty$ to E. It is thus a positive increasing function which is constant on gaps. Under suitable homogeneity assumptions on the material (for instance, if the tiling dynamical system is uniquely ergodic) the integrated density of states at E can be expressed as the trace per unit volume of $P_{\leq E}(H)$, the spectral projection of H to its states with energy lower than E. The trace per unit volume is a linear functional which can be evaluated on operators of $\ell^2(\mathbb{Z}^d)$ (provided they are trace-class w.r.t. it) which is invariant under conjugation with unitaries and normalised so that its value on the identity operator is 1.

We will see that, if E lies in a gap then IDS(E) is a numerical topological invariant of the operator, which takes values in a countable sub-group of $\mathbb{R}$ which depends only on the underlying tiling. The topological invariance manifests itself in the property that the number is robust against perturbations which do not close the gap.

---

[3] At low enough temperature.

4.3   Algebras for Tilings

We construct C^*-algebras for tilings for two reasons. First, by assigning in a coherent way a C^*-algebra to a tiling we can study the tiling using the tools of C^*-theory and non-commutative topology. Second, the algebra is a natural host for pattern equivariant operators for the tiling and so in particular contains operators describing the physics of the material modeled by the tiling.

The abstract definition of a C^*-algebra is as a complex $*$-Banach algebra whose norm satisfies the C^*-condition, $\|a^*a\| = \|a\|^2$. One of the fundamental theorems in C^*-theory says that any C^*-algebra can be faithfully represented on a Hilbert space. While this allows to view C^*-algebra s more concretely as the norm closed sub-algebras of the bounded operators on a Hilbert space, a point of view which is close to physics, the abstract definition has its advantages when it comes to the study of its topological invariants.

4.3.1   The Pattern Equivariant Approach

We describe the approach to tiling algebras via pattern equivariant integral kernels. When restricted to tilings of finite local complexity this is essentially the same than [26, 31].

We fix a tiling $\mathcal{T}$ which we view as the closed subset of $\mathbb{R}^d$ given by the boundary points of its tiles. We construct a C^*-algebra for $\mathcal{T}$ using double R-patch classes.

Recall that $\mathcal{L}^R(\mathcal{T})$ is the closure in the Hausdorff metric (between closed subsets of the R-ball at 0) of the set of all R-patches of $\mathcal{T}$ which are centered at 0. Deemphazising the role of the origin we may also describe $\mathcal{L}^R(\mathcal{T})$ as the closure of R-patch classes of $\mathcal{T}$ in the Hausdorff metric, defining the Hausdorff distance between two equivalence classes as the infimum of the distance taken over all possible representatives. An element of the inverse limit $\Omega(\mathcal{T}) = \lim\limits_{+\infty \leftarrow R} \mathcal{L}^R(\mathcal{T})$ can then also be described as $[T; x]$, the translational equivalence class of a tiling T together with a point x in the tiling. It corresponds to the tiling $T - x$, that is the representative of T which has x on the origin of $\mathbb{R}^d$.

Recall furthermore the definition of a double R-class of $\mathcal{T}$ as the equivalence class of a double R-patch $B_R[\mathcal{T}; x, y] = B_R[\mathcal{T}; x] \times B_R[\mathcal{T}; y]$ under the diagonal action by translation. We denote the equivalence class by $\left[B_R[\mathcal{T}; x, y] \right]$ and define, for $M \geq 0$,

$$\mathcal{L}^{R,M}(\mathcal{T}) = \overline{\{[B_R[\mathcal{T}; x, y]] : |y - x| \leq M\}}^{d_H}$$

the closure of the set of double R-classes of range bounded by M. Then $\mathcal{L}^R(\mathcal{T}) = \mathcal{L}^{R,0}(\mathcal{T})$, as we may identify $\left[B_R[\mathcal{T}; x, x] \right]$ with $B_R[\mathcal{T} - x]$, the R-patch of $\mathcal{T}$ at x shifted to 0.

If $R > R'$ we have an obvious surjection

$$\mathcal{L}^{R,M}(\mathcal{T}) \twoheadrightarrow \mathcal{L}^{R',M}(\mathcal{T}) \tag{4.4}$$

given by $[B_R[\mathcal{T}; x, y]] \mapsto [B_{R'}[\mathcal{T}; x, y]]$ and if $M < M'$ an obvious inclusion

$$\mathcal{L}^{R,M}(\mathcal{T}) \hookrightarrow \mathcal{L}^{R,M'}(\mathcal{T}). \tag{4.5}$$

Consider the continuous function $v : \mathcal{L}^{R,M}(\mathcal{T}) \to B_M(0)$

$$v([B_R[T; x, y]]) = x - y$$

and define $C_0(\mathcal{L}^{R,M}(\mathcal{T}))$ to be the space of continuous functions (sliding block codes) $b : \mathcal{L}^{R,M}(\mathcal{T}) \to \mathbb{C}$ which vanish on $v^{-1}(\partial B_M(0))$. Then the surjection (4.4) and the inclusion (4.5) induce inclusions

$$C_0(\mathcal{L}^{R',M}(\mathcal{T})) \hookrightarrow C_0(\mathcal{L}^{R,M'}(\mathcal{T}))$$

where $R' < R$ and $M < M'$, which are algebra homomorphisms w.r.t. pointwise multiplication. We define the space of strongly pattern equivariant (for $\mathcal{T}$) elements as the algebraic direct limit

$$A^{(s)}(\mathcal{T}) = \lim_{R,M \to \infty} C_0(\mathcal{L}^{R,M}(\mathcal{T}))$$

When equipped with pointwise multiplication and complex conjugation, $A^{(s)}(\mathcal{T})$ is a non-unital commutative $*$-algebra. But on $A^{(s)}(\mathcal{T})$ we have a second $*$-algebra structure, namely the non-commutative convolution product $C_0(\mathcal{L}^{R,M}(\mathcal{T})) \times C_0(\mathcal{L}^{R',M'}(\mathcal{T})) \to C_0(\mathcal{L}^{R'',M''}(\mathcal{T}), R'' = \max(R, R'), M'' = M + M'$, defined by

$$b \star b'([B_{R''}[T; x, y]]) = \int_{B_M(x) \cap B_{M'}(y)} b([B_R[T; x, z]]) b'([B_{R'}[T; z, y]]) dz$$

$$b^*([B_R[T; x, y]]) = \overline{b([B_R[T; y, x]])}$$

Every element of $\Omega(\mathcal{T})$ defines a representation π_T of $A^{(s)}(\mathcal{T})$ on $L^2(\mathbb{R}^d)$. Given $b \in C_0(\mathcal{L}^{R,M}(\mathcal{T}))$,

$$\pi_T(b)\psi(x) = \int_{B_M(x)} b([B_R[T; x, z]])\psi(z) dz.$$

This is compatible with the inclusions and one easily sees that π_T is faithful, as the translates of $\mathcal{T}$ in $\Omega(\mathcal{T})$ form a dense set. As a result $\|\pi_T(b)\| \leq \|\pi_{\mathcal{T}}(b)\|$ for all $T \in \Omega(\mathcal{T})$. The *continuous tiling algebra* $A(\mathcal{T})$ is the completion of $(A^{(s)}(\mathcal{T}), \star)$ in the norm $\|b\| := \|\pi_{\mathcal{T}}(b)\|$.

We may interpret the above in the following way: Any element $b \in A^{(s)}(\mathcal{T})$ lies in some $C_0(\mathcal{L}^{R,M}(\mathcal{T}))$ and thus defines a strongly pattern equivariant function $f_b : \mathbb{R}^d \times \mathbb{R}^d \to \mathbb{C}$ by

$$f_b(x, y) = b([B_R[\mathcal{T}; x, y]])$$

which we call its integral kernel. $A^{(s)}(\mathcal{T})$ can thus be seen as the algebra of strongly pattern equivariant integral kernels over $\mathbb{R}^d$ with their usual product and $*$-structure

$$f \star g(x, y) = \int_{\mathbb{R}^d} f(x, z)g(z, y)dz$$

$$f^*(x, y]) = \overline{f(y, x)}$$

and the representation $\pi_\mathcal{T}$ corresponds to the usual representation of integral kernels

$$\pi_\mathcal{T}(b)\psi(x) = \int_{\mathbb{R}^d} f_b(x, z)\psi(z)dz.$$

Let S be a Delone set which is locally derivable from $\mathcal{T}$. The *discrete tiling algebra* associated to S is obtained by restricting the above construction to double R-patches $B_R[\mathcal{T}; x, y]$ with $x, y \in S$. We denote $\mathcal{L}_S^{R,M}(\mathcal{T}) \subset \mathcal{L}^{R,M}(\mathcal{T})$ the closure of the double R-patch classes with $x, y \in S$ and $\|y - x\| \leq M$. The *discrete* tiling algebra $\mathfrak{A}_S(\mathcal{T})$ is the completion of the algebra $(\mathfrak{A}_S^{(s)}(\mathcal{T}), \star)$ where

$$\mathfrak{A}_S^{(s)}(\mathcal{T}) = \lim_{R,M \to \infty} C_0(\mathcal{L}_S^{R,M}(\mathcal{T}))$$

is the algebraic limit and

$$b \star b'([B_{R''}[\mathcal{T}; x, y]]) = \sum_{z \in S} b([B_R[\mathcal{T}; x, z]])b'([B_{R'}[\mathcal{T}; z, y]])$$

and the $*$-structure is $b^*([B_R[\mathcal{T}; x, y]]) = \overline{b([B_R[\mathcal{T}; y, x]])}$. Similar to the continuous case, every element of $\Xi(\mathcal{T})$ defines a representation $\pi_\mathcal{T}^S$ of $A^{(s)}(\mathcal{T})$ on $L^2(S(\mathcal{T}))$. Given $b \in C_0(\mathcal{L}^{R,M}(\mathcal{T}))$,

$$\pi_\mathcal{T}^S(b)\psi(x) = \sum_{z \in S(\mathcal{T}) \cap B_M(0)} b([B_R[\mathcal{T}; x, z]])\psi(z) \qquad (4.6)$$

and the completion is taken in the norm $\|b\| = \|\pi_\mathcal{T}^S(b)\|$. The representation $\pi_\mathcal{T}^S$ on $\ell^2(S)$ can again be written using integral kernels,

$$\pi_\mathcal{T}^S(b)\psi(x) = \sum_{z \in S} f_b(x, z)\psi(z).$$

This corresponds exactly to (4.2) and indeed, tight binding operators with local coefficients are represented by elements of the discrete tiling algebra.

One advantage of discrete tiling algebras is that they are unital. Indeed, since S is uniformly discrete, $C_0(\mathcal{L}_S^{R,M}(\mathcal{T})) = C(\mathcal{L}_S^{R,0}(\mathcal{T}))$ for some small enough $M > 0$. Therefore $\mathfrak{A}_S^{(s)}(\mathcal{T})$ contains $C(\mathcal{L}_S^{R,0}(\mathcal{T}))$ and thus a unit.

4.3.2 The Groupoid Approach

While the definition of the tiling algebras by means of pattern equivariant functions is very intuitive, as it mimicks local operators, a more abstract definition is needed for K-theory calculations. In fact, the algebras can be understood as groupoid C^*-algebra s and in the case that the groupoid is a transformation groupoid, they become crossed product algebras for which tools in K-theory have been developed.

The Gelfand spectrum of the commutative algebra $A^{(s)}(\mathcal{T})$ with pointwise multiplication is given by

$$\mathcal{G}(\mathcal{T}) = \bigcup_{M \geq 0} \lim_{\infty \leftarrow R} \mathcal{L}^{R,M}(\mathcal{T}).$$

The elements of $\mathcal{G}$ are of the form $[T; x, y]$ where $[T, x] \in \Omega(\mathcal{T})$. $\mathcal{G}$ is a the continuous tiling groupoid, it carries the groupoid product

$$[T; x, y][T'; x', y'] = [T, x, y'] \quad \text{provided } T' = T \text{ and } x' = y.$$

$C_c(\mathcal{G})$ with the standard convolution product for groupoid algebras is nothing else then $(A^{(s)}(\mathcal{T}), \star)$. $A(\mathcal{T})$ is therefore the groupoid C^*-algebra of $\mathcal{G}(\mathcal{T})$.

Under the homeomorphism $\mathcal{G}(\mathcal{T}) \ni [T; x, y] \mapsto ([T; x], x - y) \in \Omega(\mathcal{T}) \times \mathbb{R}^d$ the product becomes

$$([T; x], v)([T', x'], v') = ([T; x], v + v') \quad \text{provided } [T'; x'] = [T, x - v]$$

The product encodes thus the action $\alpha_v([T; x]) := [T; x - v]$ of $\mathbb{R}^d$ on $\Omega(\mathcal{T})$. It follows that the continuous tiling algebra is the crossed product C^*-algebra

$$A(\mathcal{T}) \cong C(\Omega(\mathcal{T})) \rtimes_\alpha \mathbb{R}^d$$

associated to the continuous dynamical system $(\Omega(\mathcal{T}), \mathbb{R}^d, \alpha)$ of the tiling.

Also $\mathfrak{A}_S(\mathcal{T})$ can be interpreted as a groupoid C^*-algebra, namely of the discrete tiling groupoid associated to S,

$$\mathfrak{G}_S(\mathcal{T}) = \bigcup_{M \geq 0} \lim_{\infty \leftarrow R} \mathcal{L}_S^{R,M}(\mathcal{T})$$

whose unit space is the discrete hull

$$\Xi_S = \lim_{\infty \leftarrow R} \mathfrak{L}_S^{R,0}(\mathcal{T}).$$

Indeed, the elements of $\mathfrak{G}_S$ are given by triples $[T; x, y]$ with $x, y \in S(T)$ and $T \in \Omega(\mathcal{T})$ and the product is the same as for $\mathcal{G}$ but x, y are restricted to $S(T)$.

4.3.3 Crossed Products with $\mathbb{Z}^d$

The case which we will describe in more detail is the one in which $\mathfrak{G}_S(\mathcal{T})$ is actually a transformation groupoid so that $\mathfrak{A}_S(\mathcal{T})$ is a crossed product by $\mathbb{Z}^d$. As some details about crossed products with $\mathbb{Z}^d$ are needed further down we recall them here, see also [15]. For that we start more generally with an action α of $\mathbb{Z}^d$ on a (unital) C^*-algebra B. Let $B_\alpha \mathbb{Z}^d$ be the linear space of Laurent polynomials $\sum_{n \in \mathbb{Z}^d} b_n u^n$ (only finitely many b_n are non-zero) with coefficients $b_n \in B$. Here $u_1, \cdots, u_d$ are d variables and $u^n = u_1^{n_1} \cdots u_d^{n_d}$. Define the associative product and $*$-structure

$$b_n u^n b_m' u^m = b_n \alpha_n (b_m') u^{n+m}, \qquad (b u^n)^* = \alpha_{-n}(b^*) u^{-n}. \qquad (4.7)$$

In particular, the $u_1, \cdots, u_d$ commute and satisfy the relations of unitaries, $u_i u_i^* = u_i^* u_i = 1$.

Any representation ρ of B on some Hilbert space $\mathcal{H}$ gives rise to a representation π_ρ of $B_\alpha \mathbb{Z}^d$ on $\ell^2(\mathbb{Z}^d, \mathcal{H})$:

$$\pi_\rho(b)\psi(n) = \rho(\alpha_{-n}(b))\psi(n), \qquad \pi_\rho(u^m)\psi(n) = \psi(n - m)$$

and if ρ is faithful then the crossed product C^*-algebra $B \rtimes_\alpha \mathbb{Z}^d$ is the completion of the above algebra $B_\alpha \mathbb{Z}^d$ w.r.t. the norm $\| \cdot \| := \|\pi_\rho(\cdot)\|$. π_ρ is called the representation induced by ρ.

We now specify to the case of discrete tiling algebras which may be written as crossed products with $\mathbb{Z}^d$. This might not always be the case and so we need to make some assumptions. We say that $\mathcal{T}$ is a *decoration of* $\mathbb{Z}^d$ if there exists an oriented graph Γ, such that

1. There is an isomorphism (of oriented graphs) ν between Γ and the Caley graph of $\mathbb{Z}^d$,
2. the set of vertices S of Γ is a Delone subset of $\mathbb{R}^d$ which is locally derivable from $\mathcal{T}$,
3. for all edges (x, y) of Γ, $\nu(x, y)$ is locally derivable from $\mathcal{T}$,
4. the combinatorial distance between $x, y \in S$ (the minimal number of edges needed to link x with y) is Lipschitz equivalent to their euclidean distance in $\mathbb{R}^d$.

Before we exploit these conditions we remark that they naturally occur in many situations. They are clearly satisfied for Wang tilings where we can take S to be the vertices and the edges of Γ to be the edges of the cubes. The orientation can be coherently fixed by a global choice of orientation on $\mathbb{R}^d$. More generally, an tiling from which we can locally derive a lattice S allows the construction of such a graph Γ in a similar way. Recall that after deformation this can always be achieved for FLC tilings. But S does not have to be a lattice, for instance, for one dimensional tilings (with a lower and an upper bound on thel tile size) we may use the tiles to define the edges of Γ, oriented according to a choice of orientation for $\mathbb{R}$.

Let v be an isomorphism between Γ and the Caley graph of $\mathbb{Z}^d$. Then we can define a map

$$\tilde{v} : \mathfrak{G}_S(\mathcal{T}) \to \mathbb{Z}^d$$

as follows. Given $(x, y) \in S \times S$ there is a path in Γ from y to x. We map this path with v to a path in the Caley graph of $\mathbb{Z}^d$. As such the path corresponds to an element of $\mathbb{Z}^d$ which we denote $v(x, y)$. While the path need not be unique, the group element is. We now define

$$\tilde{v}([\mathcal{T}; x, y]) = v(x, y)$$

By the above conditions 2. and 3., for any edge (x, y) of Γ there exists $R > 0$ such that $\tilde{v}([\mathcal{T}; x, y])$ can be derived from the double R-patch class $[B_R[\mathcal{T}; x, y]]$. The R depends on (x, y) but since, by 4., the combinatorial distance between x and y is bounded by a constant times their euclidean distance, we can, given $M > 0$, choose the R uniformly for all $(x, y) \in S \times S$ with $\|y - x\| \leq M$. This shows that $\tilde{v}$ is extends by continuity to $\mathfrak{G}_S(\mathcal{T})$. From $v(x, y) = v(x, z) + v(z, y)$ for all $z \in S$ we derive that $\tilde{v}$ is a groupoid homomorphism. Clearly, when restricted to the orbit of $[\mathcal{T}; x, x]$ in $\mathfrak{G}_S(\mathcal{T})$ (which is a group), then $\tilde{v}$ is a group isomorphism. Since the orbit of $\mathcal{T}$ is dense by construction, the restriction of $\tilde{v}$ to any orbit of $\mathfrak{G}_S(\mathcal{T})$ is an isomorphism. We therefore can define an action α on $\varXi_S$ by $\mathbb{Z}^d$,

$$\alpha_n([T; x]) = [T, y]$$

where y is the unique element of S such that $\tilde{v}([T; x, y]) = n$. Since $\tilde{v}$ is strongly pattern equivariant this action is continuous and the map

$$\mathfrak{G}_s \ni [T; x, y] \mapsto ([T; x], \tilde{v}([T; x, y])) \in \varXi_S \rtimes_\alpha \mathbb{Z}^d$$

a continuous groupoid homomorphism. Moreover, the map is bijective. Condition 4 now guarantees that its inverse is continuous. Indeed, it implies that $\tilde{v}^{-1}(e_i) \in \mathfrak{L}_S^{R,M}$ for some finite R and M, where e_i is a generator of $\mathbb{Z}^d$. Since $\mathfrak{L}_S^{R,M}$ is compact also $\tilde{v}^{-1}(e_i)$ is compact, which shows that the restriction of $\tilde{v}$ to $\tilde{v}^{-1}(e_i)$ is a homeomorphism onto $\varXi_S \times \{e_i\}$. A similar argument works for any $n \in \mathbb{Z}^d$.

It now follows from the general theory [38] that the discrete tiling algebra $\mathfrak{A}_S(\mathcal{T})$ is isomorphic to the crossed product $C(\Xi_S) \rtimes_\alpha \mathbb{Z}^d$ of $C(\Xi_S)$ with $\mathbb{Z}^d$. We find it instructive to explain this directly: We have seen that $\mathfrak{A}_S^{(s)}(\mathcal{T})$ contains $C(\mathfrak{L}_S^{R,0}(\mathcal{T}))$. It furthermore contains the indicator functions on compact open subsets of $\mathfrak{L}_S^{R,0}(\mathcal{T})$. Above we saw that $\tilde{v}^{-1}(n)$ is compact. It is also open and thus its indicator function χ_n an element of $\mathfrak{A}_S(\mathcal{T})$. Let $b \in C(\mathfrak{L}_S^{R,0}(\mathcal{T}))$. Then

$$\chi_n \star b([T; x, y]) = \chi_n([T; x, y])b([T; y, y])$$

$$b \star \chi_n([T; x, y]) = b([T; x, x])\chi_n([T; x, y]).$$

Now $\chi_n([T; x, y]) = 1$ is equivalent to $\alpha_n(b)([T; x, x]) = b([T; y, y])$ and hence we have $\chi_n \star b = \alpha_n(b) \star \chi_n$. Therefore, if $u^n = \chi_n$, the relations (4.7) are satisfied. This shows that $\mathfrak{A}_S^{(s)}(\mathcal{T})$ contains $B_\alpha \mathbb{Z}^d$ with $B = \lim_{R \to \infty} C(\mathfrak{L}_S^{R,0}(\mathcal{T}))$. Furthermore, any $b \in C_0(\mathfrak{L}_S^{R,M}(\mathcal{T}))$ can be written as a finite linear combination of elements from $C(\mathfrak{L}_S^{R,0}(\mathcal{T}))$ and χ_n. Hence $B_\alpha \mathbb{Z}^d$ is dense in $\mathfrak{A}_S(\mathcal{T})$. Since $C(\Xi_S)$ is the closure of B we have $\mathfrak{A}_S(\mathcal{T}) \cong C(\Xi_S) \rtimes_\alpha \mathbb{Z}^d$.

Given $[T; x] \in \Xi_S$ the representation π_T^S (4.6) is unitarily equivalent to a representation on $\ell^2(\mathbb{Z}^d)$ in which χ_n acts as the translation operator T^n. Indeed, the unitary intertwining the representation corresponds to the bijection $v(x, \cdot)$: $S(T - x) \to \mathbb{Z}^d$. Furthermore, the representation on $\ell^2(\mathbb{Z}^d)$ obtained in this way is the d-fold iterated induced representation of the representation $\rho_T : C(\Xi_S) \to \mathbb{C}$ given by $\rho_T(f)\psi(x) = f([T; x])\psi(x)$.

For Wang tilings, the symbolic hull Ξ plays the role of Ξ_S.

4.3.4 Extensions and Exact Sequences

One of the basic building blocks of algebraic topology are extentions. An extension of a C^*-algebra A is a C^*-algebra E together with a surjective morphism $q : E \to A$. Let $J = \ker q$. J is a closed two-sided ideal of E. Denoting by $i : J \to E$ the inclusion map we have a short exact sequence (SES)

$$0 \to J \xrightarrow{i} E \xrightarrow{q} A \to 0$$

that is, a chain of algebras with morphisms i, q such that i is injective, $\text{im}\, i = \ker q$ and q is surjective.

Of particular importance is the Toeplitz extension $\mathcal{T}(B, \beta)$ of the crossed product $A = B \rtimes_\beta \mathbb{Z}$. Its construction is close to that of $B \rtimes_\beta \mathbb{Z}$, but instead of the algebra $B_\beta \mathbb{Z}$ generated by the elements of B and a unitary u subject to the relation $ub = \beta(b)u$ we now take the algebra generated by the elements of B and a (proper) isometry $\hat{u}$ and its $*$-adjoint $\hat{u}^*$, which are subject to the relations

$$\hat{u}b = \beta(b)\hat{u}, \quad b\hat{e} = \hat{e}b$$

where $\hat{e} = 1 - \hat{u}\hat{u}^*$, and $b \in B$. Saying that $\hat{u}$ is an isometry means that $\hat{u}^*\hat{u} = 1$ and implies that $\hat{e}$ is a projection, properness requires $\hat{e}$ to be neither 0 nor 1. We denote this algebra by $B_\beta\mathbb{N}$. Elements of $B_\beta\mathbb{N}$ are thus finite sums of finite products of elements from B, $\hat{u}$ and $\hat{u}^*$. The above relation imply that also $\hat{u}^*b = \beta^{-1}(b)\hat{u}^*$.

We have a surjective $*$-algebra morphism $q : B_\beta\mathbb{N} \to B_\beta\mathbb{Z}$ which is given by

$$q(b) = b, \quad q(\hat{u}) = u$$

and hence satisfies $q(\hat{e}) = 0$. The kernel of q is the subalgebra generated by $\hat{e}$. Identifying $\hat{e}$ with the half-infinite Jacobi matrix which has as only non-zero entry a 1 at the upper left corner, and $\hat{u}$ with the Jacobi matrix which is 1 on the lower diagonal, we see that $\ker q$ can be identified with the algebra of half-infinite matrices with entries from B of which only finitely many are non-zero. This is the tensor product $B \otimes F$ where F is the algebra of half-infinite complex matrices with only finitely many are non-zero entries. We thus have a SES

$$0 \to B \otimes F \to B_\beta\mathbb{N} \xrightarrow{q} B_\beta\mathbb{Z} \to 0$$

The Toeplitz extension is the universal C^*-completion of $B_\beta\mathbb{N}$ and gives rise to the SES of C^*-algebra s

$$0 \to B \otimes \mathcal{K} \to \mathcal{T}(B, \beta) \xrightarrow{q} B \rtimes_\beta \mathbb{Z} \to 0 \tag{4.8}$$

Here $\mathcal{K}$ is the algebra of compact operators. Recall that a representation ρ of B on $\mathcal{H}$ induces a representation π_ρ of $B \rtimes_\beta \mathbb{Z}$ on $\ell^2(\mathbb{Z}, \mathcal{H})$. It also induces a representation $\hat{\pi}_\rho$ of $\mathcal{T}(B, \beta)$ on $\ell^2(\mathbb{N}, \mathcal{H})$, namely

$$\pi_\rho(b)\psi(n) = \rho(\alpha_{-n}(b))\psi(n), \quad \pi_\rho(\hat{u})\psi(n) = \psi(n - 1) \text{ with } \psi(-1) = 0.$$

4.4 K-Theoretic Invariants for Tilings

Having assigned to a tiling various C^*-algebra s, their K-groups yield invariants for the tiling. These invariants can serve to distinguish tilings, or even classify them under weaker notions of equivalence. But they also aquire physical significance, as the algebra is physically motivated.

In turns out that the discrete tiling algebras constructed above for a given tiling and a Delone set S which is locally derivable from the tiling are all strongly Morita equivalent to the continuous tiling algebra. Since K-groups do not distinguish strongly Morita equivalent algebras, the different algebras lead to the same K-groups. Having said that, K-groups can come with extra structure. The K_0-group can be ordered, for instance, and scaled (this is then called a dimension group)

[40]. Whereas the order structure is again the same for strongly Morita equivalent algebras, the scale is not, and different choices for S lead to different scales.

4.4.1   Short Definition of K-Theory

We provide a definition of the K-groups which is motivated by physics. The elements of the groups are essentially homotopy classes of Hamiltonians which have a gap in their spectrum. This approach is based on van Daele's formulation of K-theory [12] and has been adapted to topological phases in [29], though we consider only the complex case here. For the more traditional approach we refer the reader to [9, 40].

If the local arrangement of atoms in a solid are described by a tiling (in the sense of being mutually locally derivable with the tiling) then it is natural to consider the tiling algebra as the C^*-algebra of observables. But what follows can in principle be carried out with any C^*-algebra A which is motivated by the physical system one wants to consider.

The energy observable corresponds to a self-adjoint element H of A (the Hamiltonian, or a bounded function of the Hamiltonian), represented in the relevant physical representation of A. A gap in the spectrum of H indicates that the system is an insulator, provided the Fermi energy lies in that gap. The *topological phase* of the insulator is, by definition, the homotopy class of H under continuous deformation which preserves self-adjointness and the presence of a gap at the Fermi energy. Assuming that A is unital we may shift the gap and the Fermi energy to 0 and thus say that the topological phase of H is its connected component in the set of self-adjoint invertible elements $GL(A^{s.a.})$ of A. Following a standard procedure one can construct an abelian group $K_0(A)$ whose elements are generated by the homotopy classes of $GL(A^{s.a.})$. $K_1(A)$ is obtained by a similar construction, but for the set of connected components of $GL(A)$, that is, without the self-adjointness constraint. In physics, the role of an invertible element $Q \in A$ is as the *chiral* half of an insulator with chiral symmetry; the Hamiltonian of such an insulator has the form

$$H = \begin{pmatrix} 0 & Q^* \\ Q & 0 \end{pmatrix}.$$

To construct a group out of the homotopy classes of $GL(A)$ we consider homotopy classes in all $GL(M_n(A)), n \geq 1$, the invertible elements in the matrix algebras over A. To compare matrices of different sizes we use the inclusion $GL(M_n(A)) \ni x \mapsto x \oplus 1 \in GL(M_{n+1}(A))$ which induces an inclusion of homotopy classes. The direct limit of the quotient spaces by homotopy $\lim_n GL(M_n(A))/ \sim_h$ admits an addition $[x] + [y] = [x \oplus y]$ ($[x]$ denotes the homotopy class of x). It turns out that $x \oplus y$ is homotopic to $x \oplus y$ in $GL(M_2(A))$, and also to $xy \oplus 1$. The addition law defined above is therefore the abelianization of the product in A and yields an abelian group structure with neutral element $[1]$ and inverse $-[x] = [x^{-1}]$. $K_1(A)$ is this abelian group.

The construction of $K_0(A)$ follows similar lines, except one requires self-adjointness, and therefore considers the direct limit $\lim_n GL(M_n(A)^{s.a.})/ \sim_h$. Addition is still defined by $[x] + [y] = [x \oplus y]$, but since a product of two self-adjoint elements need not to be self-adjoint, it is not the abelianization of the product in A and $V(A) := \lim_n GL(M_n(A)^{s.a.})/ \sim_h$ does not contain inverse elements. Instead $V(A)$ is an abelian monoid with neutral element $[1]$. $K_0(A)$ is the associated Grothendieck group $V(A) \times V(A)/ \sim$. Here $([x_1], [y_1]) \sim ([x_2], [y_2])$ if there exist $[z] \in V(A)$ such that $[x_1] + [y_2] + [z] = [x_2] + [y_1] + [z]$ and addition is componentwise, $([x_1], [y_1]) + ([x_2], [y_2]) = ([x_1] + [x_2], [y_1] \oplus [y_2])$.

The more conventional picture of $K_0(A)$ uses homotopy classes of projections. Any $H \in GL(A^{s.a.})$ is homotopic to a self-adjoint unitary, that is, an element of the form $1 - 2p$ where p is a projection. This projection is the spectral projection onto the negative spectral part of H. A homotopy class of invertible self-adjoint elements corresponds therefore to a homotopy class of projections.

Similarily, any invertible element is homotopic to a unitary. This yields the conventional picture of $K_1(A)$ as homotopy classes of unitaries.

4.4.1.1 Basic Properties of K-Theory

We list the basic properties of K_0 and K_1.

1. **Functoriality.** Whenever we have an algebra homomorphism $\varphi : A \to B$ between two unital C^*-algebra s and φ preserves the unit, then we obtain a group homomorphism $\varphi_* : K_i(A) \to K_i(B)$ simply by setting $\varphi_*([x]) = [\varphi(x)]$ where φ is extended to $M_n(A)$ entrywise. Furthermore $(\psi \circ \varphi)_* = \psi_* \circ \varphi_*$.
2. **K_i preserves direct sums.** $K_i(A \oplus B) = K_i(A) \oplus K_i(B)$.
3. **Stability.** The map $x \mapsto x \oplus 1$ induces an isomorphism $K_i(M_n(A)) \to K_i(M_{n+1}(A))$. In particular, $K_i(\mathcal{K} \otimes A) \cong K_i(A)$, where $\mathcal{K}$ are the compact operators, and (strongly) Morita equivalent algebras have the same K-theory.
4. **Continuity.** K_i commutes with limits, $K_i(\lim_n A_n) = \lim_n K_i(A_n)$.

4.4.1.2 Definition of $K_i(A)$ for Non-unital Algebras

The definition we gave above for $K_i(A)$ needs invertibility, and hence a unit for A. If A is not unital we have to add a unit to A. This is done in the following way.

Let $A^+ = A \times \mathbb{C}$ with product $(a, \lambda)(b, \mu) = (ab + \lambda b + \mu a, \lambda \mu)$ and $*$-structure $(a, \lambda)^* = (a^*, \bar{\lambda})$. This is a unital algebra with unit $(0, 1)$. We have a surjective algebra homomorphism $\pi : A^+ \to \mathbb{C}$, $\pi(a, \lambda) = \lambda$, whose kernel is A. Since A^+ is unital we may define its K_i-groups in the above way (A^+ is a C^*-algebra in a natural way). We thus get a group homomorphism $\pi_* : K_i(A^+) \to K_i(\mathbb{C})$. By definition,

$$K_i(A) = \ker \pi_*.$$

4.4.1.3 More Involved Properties of K-Theory

The definition, in particular of $K_0(A)$ for non-unital A is not very direct, but it is crucial to obtain the following important properties.

1. **Bott periodicity.** We have not given a definition of $K_i(A)$ for $i > 1$, but the reasonable definition of $K_i(A)$ for i even, or odd, turns out to be isomorphic to $K_0(A)$, or $K_1(A)$, respectively. This is called Bott periodicity.
2. **Six term exact sequence (boundary maps).** Given a short exact sequence of C^*-algebra s

$$0 \to J \xrightarrow{i} E \xrightarrow{q} A \to 0$$

there are group homomorphisms (boundary maps) $\delta_i : K_i(A) \to K_{i-1}(J)$ making the following long sequence of K-groups exact

$$K_i(J) \xrightarrow{i_*} K_i(E) \xrightarrow{q_*} K_i(A) \xrightarrow{\delta_i} K_{i-1}(J) \cdots$$

Due to Bott periodicity, the above sequence is of period 6. We will provide formulas for the boundary maps below which can be interpreted physically.

4.4.2 Calculating K-Groups for Tiling Algebras

There are tools to calculate the K-groups for the continuous tiling algebra $C(\Omega) \rtimes \mathbb{R}^d$ and tools to calculate them in the case there is a transversal Ξ_S such that the corresponding discrete tiling algebra is a crossed product $C(\Xi_S) \rtimes_\alpha \mathbb{Z}^d$. We saw that the latter is guaranteed if the tiling has FLC. Apart from the scale, the K-groups of $C(\Omega) \rtimes \mathbb{R}^d$ and $C(\Xi_S) \rtimes_\alpha \mathbb{Z}^d$ coincide, as the algebras are strongly Morita equivalent. Both calculations need extra structure and boil down to express the K-groups as cohomology groups.

4.4.2.1 Via the Connes-Thom Isomorphism and Approximants

One method of computation of their K-groups makes use of the approximants $\mathcal{L}^R(\mathcal{T})$ of the tiling discussed above. This method is feasible if the approximates are subject to a strong hierarchical order, as, for instance, substitutional tilings [1, 31, 42].

The Connes-Thom isomorphism [11] is one of the fundamental results in K-theory which allows to relate the K-theory of a crossed product algebra with $\mathbb{R}$, $B \rtimes_\beta \mathbb{R}$, to the K-theory of B. Indeed, for any (continuous) action β of $\mathbb{R}$ on a

C^*-algebra B one has

$$K_i(B \rtimes_\beta \mathbb{R}) \cong K_{i-1}(B).$$

Starting with the continuous algebra $C(\Omega) \rtimes_\alpha \mathbb{R}^d$ we can apply iteratively the Connes-Thom isomorphism to get

$$K_i(C(\Omega) \rtimes_\alpha \mathbb{R}^d) \cong K_{i-d}(C(\Omega)).$$

As $\Omega(\mathcal{T}) = \lim_{\leftarrow} \mathcal{L}^R(\mathcal{T})$ the algebra $C(\Omega)$ is the C^*-completion of the limit $\lim_{\rightarrow} C(\mathcal{L}^R(\mathcal{T}))$ so that, using the continuity of the K-functor we get

$$K_i(C(\Omega)) = \lim_{R \to +\infty} K_i(C(\mathcal{L}^R))$$

The $\mathcal{L}^R$ are flat branched manifolds. For low enough dimension ($d \leq 3$) $K_i(C(\mathcal{L}^R))$ is isomorphic to the singular cohomology of the space $\mathcal{L}^R$ which is computable. If $\mathcal{T}$ has a hierarchical structure relating $\mathcal{L}^R$ for different R (or suitable modifications thereof) then the direct limit can be computed. For substitution tilings a computer code is available to perform these calculations [18].

4.4.2.2   Via Iteration of the Pimsner-Voiculescu Sequence and Group Cohomology

A second approach to the calculation of K-groups of tilings which are decorations of $\mathbb{Z}^d$ is based on the iteration of the Pimsner-Voiculescu sequence [35]. It is useful if one has good control on the action α of $\mathbb{Z}^d$ on the discrete hull Ξ [17, 19, 23, 25].

Consider an isomorphism $\phi : M \to M$ of an abelian group. The quotient $C_\phi M := \mathrm{coker}(\mathrm{id} - \phi)$ is called the group of coinvariants and the subgroup $I_\phi M := \ker(\mathrm{id} - \phi)$ the group of invariants of M w.r.t. to ϕ. These are the cohomology groups of the group $\mathbb{Z}$ with coefficients in M, namely $H^0(\mathbb{Z}, M) = I_\phi M$ and $H^1(\mathbb{Z}, M) = C_\phi M$. Any isomorphism β on B induces a isomorphism β_* on $K_i(B)$.

The Toeplitz extension (4.8) gives rise to a 6-term exact sequence, the Pimsner-Voiculescu sequence, which can be split into two short exact sequences ($i = 1, 2$)

$$0 \to C_{\beta_*} K_i(B) \xrightarrow{\iota_*} K_i(B \rtimes_\beta \mathbb{Z}) \xrightarrow{\delta_i} I_{\beta_*} K_{i-1}(B) \to 0. \tag{4.9}$$

where ι_* is induced by the inclusion $B \hookrightarrow B \rtimes_\beta \mathbb{Z}$ and δ_i is the boundary map from Sect. 4.4.1.3. By iteration we can express the K-group of $C(\Xi) \rtimes_\alpha \mathbb{Z}^d$ through similar short exact sequences with coinvariants and invariants of $K_i(C(\Xi))$. If $d \leq 3$ then all these sequence split and we obtain [23]

$$K_i(C(\Xi) \rtimes_\alpha \mathbb{Z}^d) \cong \bigoplus_k H^{i+2k}(\mathbb{Z}^d, C(\Xi, \mathbb{Z}))$$

The group cohomology is computable if we have a good understanding of the $\mathbb{Z}^d$-action α on Ξ. This is, for instance, the case for almost canonical cut & project tilings, and also here computer codes are available to perform the calculations [18].

4.4.3 Numerical Invariants for Pattern Equivariant Operators

We have seen that a pattern equivariant operator is an element in the representation $\pi_{\mathcal{T}}$, or $\pi_{\mathcal{T}}^S$, of the continuous or discrete algebra of some tiling $\mathcal{T}$. We denote this algebra now simply by A.

If the operator is self-adjoint and invertible it comes from a self-adjoint invertible element $h \in A$ and so defines a class in $K_0(A)$. This class is the equivalence class of $([h], [1]) \in V(A) \times V(A)$ under the relation $\sim$ defining the Grothendieck group which we discussed above, but we simply denote this class by $[h]$, as the further calculations depend essentially only on the homotopy class of h.

If the operator is only invertible thus coming from an invertible element $Q \in A$ then its homotopy class defines an element in $K_1(A)$.

The K-classes are the abstract topological invariants of the operators. They are invariants in the sense that they are stable under continuous deformation in $GL(A^{s.a.})$ or $GL(A)$, respectively.

Numerical topological invariants for pattern equivariant operators can now be obtained from additive functionals on $K_i(A)$. Such functionals can be given by index pairings with K-homology classes of A [22], or by Chern characters [11]. We will discuss here only the simplest cases, namely the functional on $K_0(A)$ defined by a trace, and the winding number on $K_1(A)$ defined by a trace and a derivation. These have physical interpretation in the examples we consider below.

4.4.3.1 Traces and Gap Labelling

Gap-labelling means the assignment of labels to the gaps in the spectrum of a self-adjoint operator. These labels are supposed to be stable under continuous deformations of the operator which do not close the gap. In Bellissard's approach to the gap-labelling by means of the K_0-group of the C^*-algebra A of observables [6, 7] the operator is an element of A and continuity means continuity in the norm of that algebra.

A (positive) trace on A is a linear functional $\mathrm{tr} : A \to \mathbb{C}$ which satisfies $\mathrm{tr}(uau^*) = \mathrm{tr}(a)$ for all $a \in A$ and unitary $u \in A^+$, and $\mathrm{tr}(a^*a) \geq 0$. It follows that $\mathrm{tr}(p) = \mathrm{tr}(p')$ for two homotopic projections. Let $P_-(h)$ be the spectral projection onto the negative energy part of h. Note that $P_-(h) \in A$ since h has a spectral gap at energy 0. Thus $h \mapsto \mathrm{tr}(P_-(h))$ is homotopy invariant in $GL(A^{s.a.})$ and

$$\mathrm{tr}_*([h]) = \mathrm{tr}(P_-(h))$$

extends to a functional on $K_0(A)$ with values in $\mathbb{R}$. Indeed, if h is in $GL(M_n(A)^{s.a.})$ we extend tr to $M_n(A)$ using the matrix trace.

We can therefore formulate the gap labelling in the following way: If h is a self-adjoint element of A and Δ a gap in the spectrum of h then, for any $E \in \Delta$ the shifted element $h - E$ belongs to $GL(A^{s.a.})$ and so defines an element of $K_0(A)$. This class $[h - E]$ is the abstract gap-label of the gap Δ of h.

If we have a trace tr on A such that, for $a \in A$, $\text{tr}(a)$ corresponds to the trace per unit volume of $\pi_{\mathcal{T}}(a)$ (or of $\pi_{\mathcal{T}}^S(a)$ in a tight binding representation), then the integrated density of states at E of the operator $\pi_{\mathcal{T}}(h)$ is given by $\text{tr}(P_-(h))$. The latter is the numerical gap-label of the gap Δ of h.

The traces on $A = C(\Xi) \rtimes_\alpha \mathbb{Z}^d$ are given by α-invariant Borel-probability measures on Ξ. Indeed, if μ is such a measure then

$$\text{tr}_\mu(a_n u^n) = \int_\Xi a_0(\xi) d\mu(\xi)$$

defines a trace on A which is normalised, $\text{tr}(1) = 1$. If μ is ergodic, then for μ-almost sure $\xi \in \Xi$ the trace per unit volume of $\pi_\xi(a)$ equals $\text{tr}(a)$. In particular, if Ξ carries a unique α-invariant ergodic probability measure μ then for all ξ the integrated density of $\pi_\xi(h)$ takes the same value $\text{IDS}(E) = \text{tr}_\mu(P_-(h - E))$ [8, 26].

4.4.3.2 Derivations and the Winding Number

A $*$-derivation on Λ is a densely defined linear map $d : A \to A$ which satisfies the Leibniz rule, $d(ab) = (da)b + a(db)$ and commutes with the $*$-operation. This implies that $d1 = 0$ and thus for any invertible Q (in the domain of d) $dQ^{-1} = -Q^{-1}dQQ^{-1}$. Moreover, if Q is unitary then $(Q^{-1}dQ)^* = -Q^{-1}dQ$.

If we have in addition a trace tr which is invariant under the derivation d, that is, $\text{tr} \circ d = 0$, then

$$\mathcal{W}_{\text{tr},d}(Q) := \frac{1}{2\pi i}\text{tr}(Q^{-1}dQ)$$

defines a real number which is stable under homotopy in $GL(A)$. It is called the winding number w.r.t. to (tr, d). $\mathcal{W}$ extends to a functional on $K_1(A)$, again by extending tr with the matrix trace if $Q \in GL(M_n(A))$ [11].

There is a little subtlety related to the fact that d is only densely defined. We need to be sure that the homotopy classes of invertibles in the domain of d are in bijection to the homotopy classes of $GL(A)$. This can be done in the above cases [11].

4.5 Bulk Boundary Correspondence

A tiling covers the whole space $\mathbb{R}^d$ and so models a material which does not have
a boundary. A real material has, of course, a boundary. The pattern equivariant
operators above describe therefore properties in the interior of the material, that is,
properties which do not depend on its boundary. Such properties are referred to as
bulk properties. The bulk-boundary correspondence principle states that certain bulk
properties correspond to *boundary* properties, i.e. properties which are confined to
the boundary of the material.

The bulk-boundary correspondence, in the form we discuss it, is of topological
nature. It is a correspondence between the topological invariants of the material
when seen without boundary and topological invariants localised near the boundary.
It manifests itself through a so-called boundary map in algebraic topology which
maps the elements of the K-group of our algebra into the K-group of an algebra
which contains the operators describing the boundary physics.

The topological nature of the bulk-boundary correspondence has physical consequences. Since topological invariants are robust against perturbations it explains the
emergence of robust boundary spectrum in topological insulators.

There is a continuous and a discrete version of the bulk boundary correspondence. We discuss the discrete version for Wang tilings which is based on the
Toeplitz extension.

4.5.1 General Philosophy

The bulk-boundary correspondence was first formulated by Halperin [20] for the
Quantum Hall Effect and by Hatsugai [21] in the context of band theory for periodic
tight binding operators. Its topological nature and universality comes to full light
through its formulation based on the K-theory of C^*-algebras. This was first worked
out for the Quantum Hall Effect in [33], and later put into a general framework [32].
Today it is quite developed and we can refer the reader to the book [37] for further
information.

Let A be the algebra containing all pattern equivariant operators of the tiling
describing the spatial structure of the material. The bulk invariants, that is the
topological invariants associated to pattern equivariant operators, are elements of
$K_*(A)$. The first goal is to find a short exact sequence

$$0 \to J \to E \xrightarrow{q} A \to 0 \tag{4.10}$$

such that the elements of the ideal J can be represented as operators localised at the
boundary. The boundary map δ from the 6-term exact sequence associated with the
above short exact sequence

$$\delta_i : K_i(A) \to K_{i+1}(J)$$

maps the bulk invariants to invariants which are expressed in terms of operators of J and therefore localised at the boundary. This is the heart (abstract form) of the bulk boundary correspondence. While there is no reason for δ_i to define a one-to-one correspondence, it will define a correspondence between some of the invariants.

Let us have a closer look at the boundary maps δ_i. Let $h \in GL(A)^{s.a.}$, the Hamiltonian of an insulator. It has a gap Δ at 0 and the inverse of the gap width $|\Delta|$ is a characteristic time $t_\Delta = \frac{2\pi}{|\Delta|}$. h defines an element of $K_0(A)$ or, if it has chiral symmetry, an element of $K_1(A)$. To construct the image of these elements under the boundary map we need a lift $\hat{h} \in E^{s.a.}$ of h, that is, $q(\hat{h}) = h$.

We consider first the situation $\delta_0 = \exp$, the so-called exponential map. It is given by

$$\exp([h]) = [U_\Delta], \qquad U_\Delta = P_\Delta(\hat{h})^\perp - e^{-it_\Delta(\hat{h}-E_0)} P_\Delta(\hat{h})$$

where $E_0 = \inf \Delta$ and $P_\Delta(\hat{h})$ is the spectral projection of $\hat{h}$ to energies in Δ and $P_\Delta(\hat{h})^\perp$ its orthocomplement. The homotopy class of U_Δ does not depend on the chosen lift $\hat{h}$. By construction $U_\Delta - 1$ belongs to the ideal J. If $\hat{h}$ is invertible then $[\hat{h}] \in K_0(E)$ is a preimage of $[h]$ and $\exp([h])$ is trivial. If however no invertible lift can be found, then $\hat{h}$ has spectrum inside the gap Δ of h, and U_Δ is the unitary of time evolution by the characteristic time t_Δ of the states of $\hat{h}$ in Δ. More generally, if $\hat{h}$ has a spectral gap overlapping with Δ and it is not protected by chiral symmetry, then it is homotopic to an invertible lift and so we see that non-triviality of $\exp([h])$ implies that any self-adjoint lift h in E must have spectrum covering Δ. It then follows that the spectrum of U_Δ covers the circle of complex numbers of absolute value 1 and no homotopy can retract that spectrum to a point.

We consider the case $\delta_1 = \mathrm{ind}$, referred to as the index map. As already mentionned $h \in GL(A)^{s.a.}$ defines an element of $K_1(A)$ if it has so-called chiral symmetry and we consider topological phases which are protected under this symmetry. We assume that the chiral symmetry is inner and thus given by a generator Γ, a self-adjoint unitary, and h anti-commutes with Γ. It gives rise to an isomorphism between A and $M_2(A_{++})$ where $A_{++} = \Pi_+ A \Pi_+$, $\Pi_\pm = \frac{1}{2}(1 \pm \Gamma)$. Written as an element of $M_2(A_{++})$, h and Γ have the form

$$h = \begin{pmatrix} 0 & Q^* \\ Q & 0 \end{pmatrix}, \qquad \Gamma = \begin{pmatrix} 1 & 0 \\ 0 & -1 \end{pmatrix}$$

and $Q \in GL(A_{++})$. The homotopy class of Q defines an element of $K_1(A_{++})$, which is isomorphic to $K_1(A)$ since A and A_{++} are strong Morita equivalent. Now the index map is given as follows

$$\mathrm{ind}[Q] = \left[P_\Delta(\hat{h})^\perp \Gamma - P_\Delta(\hat{h}) \big(\cos(t_\Delta \hat{h}) \Gamma - \sin(t_\Delta \hat{h}) \big) \right]$$

as one computes using van Daele's formula applied to the class of h [13].[4] If $\hat{h}$ is invertible and has chiral symmetry then $\mathrm{ind}([Q])$ is trivial. Conversely, non-triviality of $\mathrm{ind}([Q])$ implies that any lift $\hat{h} \in E$ with chiral symmetry must have spectrum at energy 0. Finally, if 0 is an isolated point of the spectrum of $\hat{h}$ then the spectral projection $P_0(\hat{h})$ of $\hat{h}$ onto energy 0 is an element of J and $P_\Delta(\hat{h})^\perp \Gamma - P_\Delta(\hat{h})\left(\cos(t_\Delta \hat{h})\Gamma - \sin(t_\Delta \hat{h})\right)$ homotopic to $P_0(\hat{h})^\perp - P_0(\hat{h})$, indeed, a homotopy is given by contracting $t_\Delta \hat{h}$ to $P_0(\hat{h}) + \frac{\pi}{2}\left(P_{>0}(\hat{h}) - P_{<0}(\hat{h})\right)$.

4.5.2   The Toeplitz Extension as Half Space Algebra

We now specify to the algebra $A = C(\Xi) \rtimes_\alpha \mathbb{Z}^d$ for a Wang tiling. We follow [32, 33, 37] to describe an extension of this algebra which describes the physics on half-space, the proto type of space with a boundary. We choose one direction in $\mathbb{Z}^d$, let's say along the d'th coordinate, and define the boundary to be perpendicular to it. The material with boundary is now described by the restriction of the tiling to half space, that is, we restrict the symbolic content $\xi \in \Xi$ to $\mathbb{Z}^{d-1} \times \mathbb{N}$. We can rewrite $A = B \rtimes_\beta \mathbb{Z}$ where $B = C(\Xi) \rtimes_{\alpha_\parallel} \mathbb{Z}^{d-1}$ and $\beta = \alpha^\perp$. We argue that the Toeplitz extension $\mathcal{T}(B, \beta)$ describes the physics on the half space. This is best seen by looking at its iterated induced representation. Recall that the physically motivated representation (4.6) is the iterated induced representation of $\rho_T : C(\Xi) \rightarrow \mathbb{C}$, $\rho_T(f)\psi(x) = f([T; x])\psi(x)$ on $\ell^2(\mathbb{Z}^d)$. Applying this idea to $\mathcal{T}(B, \beta)$ we iterate the induction on ρ_T $d - 1$ of times to get a representation of $B = C(\Xi) \rtimes_{\alpha_\parallel} \mathbb{Z}^{d-1}$ on $\ell^2(\mathbb{Z}^{d-1})$ and then once for the Toeplitz extension to get a representation of $\mathcal{T}(B, \beta)$ on $\ell^2(\mathbb{Z}^{d-1} \times \mathbb{N})$. The operators in this representation are precisely the pattern equivariant operators restricted to the half space. The elements of the ideal J in the short exact sequence are restrictions of pattern equivariant operators to the half space which are non-zero only near the boundary.

The exact sequence in K-theory arising from the Toeplitz extension (4.8) gives rise to the bijective correspondences

$$K_i(C(\Xi) \rtimes_\alpha \mathbb{Z}^d)/\iota_* K_i(C(\Xi) \rtimes_{\alpha_\parallel} \mathbb{Z}^{d-1}) \stackrel{\delta_i}{\cong} I_{\alpha_*^\perp} K_{i-1}(C(\Xi) \rtimes_{\alpha_\parallel} \mathbb{Z}^{d-1}). \quad (4.11)$$

This is the *K-theoretic bulk boundary correspondence*. We refer to the case with $i = 0$ as even, and that with $i = 1$ as the odd correspondence. Note that the boundary maps are explicit, as a lift $\hat{h}$ of h in the Toeplitz extension is simply given by replacing the translation operator in the dth coordinate by its truncation to $\mathbb{Z}^{d-1} \times \mathbb{N}$.

---

[4] In [37] one finds a formula for the same class with a different representative, the above is also be valid if A is a real C^*-algebra.

4.5.3   One-dimensional Examples

If the dimension is $d = 1$ then (4.11) becomes, for odd i,

$$K_1(C(\Xi) \rtimes_\alpha \mathbb{Z})/\iota_* K_1(C(\Xi))\mathrm{ind} \cong I_{\alpha_*} K_0(C(\Xi)). \tag{4.12}$$

Since Ξ is the closure of a single orbit, $I_{\alpha_*} K_0(C(\Xi))$ has a single generator, the class of -1, which corresponds to a Hamiltonian whose spectrum is purely negative. Likewise the invariants of $K_0(\mathbb{C}^k)$ are generated by the class of -1. One computes that $[u]$ is a pre-image for $[-1]$, that is, $\mathrm{ind}([u]) = [-1]$.

Let μ be an ergodic α-invariant probability measure on Ξ and recall that it defines a trace tr on $C(\Xi)$ and $C(\Xi) \rtimes_\alpha \mathbb{Z}$. Consider the $*$-derivation δ : $C(\Xi)_\alpha \mathbb{Z} \to C(\Xi)_\alpha \mathbb{Z}$ given by $\delta(bu^n) = 2\pi i n b u^n$. The trace tr defines a tracial state on $K_0(C(\Xi))$ and, together with the derivation δ a winding number $\mathcal{W}_{\mathrm{tr},\delta}$ on $K_1(C(\Xi) \rtimes_\alpha \mathbb{Z})$. One finds that $\mathcal{W}_{\mathrm{tr},\delta}(u) = \mathrm{tr}(1)$. Thus applying the winding number to the left hand side and the tracial state to the right hand side of (4.12) one obtains the odd numerical bulk-boundary correspondence

$$\mathcal{W}_{\mathrm{tr},\delta}([Q]) = \mathrm{tr}(\mathrm{ind}[Q]). \tag{4.13}$$

This has the following physical interpretation: If $h = \begin{pmatrix} 0 & Q^* \\ Q & 0 \end{pmatrix}$ is the Hamiltonian with chiral symmetry and gap Δ at energy 0 then the spectrum of any lift $\hat{h}$ of h has at most finitely many values in Δ. Moreover, these values are eigenvalues to eigenfunctions localized at the edge. Hence $\mathrm{ind}[Q] = [P_0(\hat{h})^\perp - P_0(\hat{h})]$ and thus $\mathrm{tr}(\mathrm{ind}[Q])$ is equal to the number of 0 energy eigenstates of $\hat{h}$. We thus have the following well-known result (see also [36, 37]).

Odd Bulk Boundary Correspondence *The winding number* $\mathcal{W}_{\mathrm{tr},\delta}$ *of* Q *is the number of* 0 *energy boundary states of* $\hat{h}$.

For even i (4.11) is

$$K_0(C(\Xi) \rtimes_\alpha \mathbb{Z})/\iota_* K_0(C(\Xi)) \overset{\exp}{\cong} I_{\alpha_*} K_1(C(\Xi)) \tag{4.14}$$

Other than in the case of odd i the situation depends on the nature of the hull Ξ. Again, given an ergodic α-invariant measure on Ξ we obtain a trace tr and a tracial state, now for the left hand side. To obtain a winding number for the right hand side we need a derivation on $C(\Xi)$.

We discuss these issues here for the one-dimensional systems of Sect. 4.2.3.

1. We consider a periodic sequence ξ. If the period is k the discrete hull Ξ consists of k isolated points on which $\mathbb{Z}$ acts by a cyclic permutation. Hence $C(\Xi) = \mathbb{C}^k$ and the unique invariant probability measure is the uniform probability measure

on k points. But we have $K_1(\mathbb{C}^k) = 0$ and hence the even bulk-boundary correspondence is trivial.

2. We consider an almost periodic sequence ξ. The discrete hull Ξ is homeomorphic to the circle S and α is rotation by the irrational number $\theta \in [0, 1]$. Now $K_1(C(S)) \cong \mathbb{Z}$ with generator given by the class of the function $z(s) = e^{2\pi i s}$. Hence also the coinvariants and the invariants of $K_1(C(S))$ are generated by $[z]$. Furthermore, $K_0(C(S)) \cong \mathbb{Z}$ with generator given by the class of -1, which then is also the generator of the coinvariants and the invariants of $K_0(C(S))$. $K_1(C(S) \rtimes_\alpha \mathbb{Z}) \cong \mathbb{Z}^2$ with generators $[z]$ and $[u]$. $K_0(C(S) \rtimes_\alpha \mathbb{Z}) \cong \mathbb{Z}^2$ with generators $[-1]$ and $[h_\theta]$. The second generator is the class of the function $h_\theta = 1 - 2P_\theta$ where P_θ is the famous Rieffel projection [39].

The exponential map can be computed on the generators and one finds that the even bulk-boundary correspondence is given by

$$\exp([h_\theta]) = -[z], \quad \exp([-1]) = 0.$$

Since θ is irrational, the unique ergodic invariant measure on the circle is the Lebesgues measure. Furthermore, functions of class C^1 in $C(S)$ can be derived and thus we have a densely defined derivation ∂_s. Applying the resulting tracial state to the left and the winding number $\mathcal{W}_{\mathrm{tr},\partial_s}$ to the right hand side we get $\mathrm{tr}(1) = 1$, $\mathrm{tr}(P_\theta) = \theta$ [39], $\mathcal{W}_{\mathrm{tr},\partial_s}(u) = 0$, $\mathcal{W}_{\mathrm{tr},\partial_s}(z) = 1$, so that the even numerical bulk boundary correspondence is given by

$$\mathrm{tr}([h]) + \theta \mathcal{W}_{\mathrm{tr},\partial_s}(\exp[h]) \in \mathbb{N}. \tag{4.15}$$

This has the following interpretation. The bulk invariant $\mathrm{tr}([h])$ is the IDS of h at energy 0 (Fermi energy) and it can take real values of the form $m\theta + N$ where $N, m \in \mathbb{N}$ and N is such that $m\theta + N \in [0, 1]$. Thus, except if $m = 0$, N is determined by the value of m as θ is irrational.

The boundary invariant $\mathcal{W}_{\mathrm{tr},\partial_s}(\exp[h])$ is a response coefficient related to a mechanical force. The derivation ∂_s is the infinitesimal generator of translation in the hull and translation has the effect of changing the position of the atoms (boundary points of the tiles) in the structure. This atomic motion corresponds to a variation of the potential which the electrons at the edge are subjected to, and hence they feel a gradient force. The above winding number of U_Δ corresponds to the expectation value of the gradient force on all edge states with energy inside Δ, integrated over one cycle of the circle Ξ, in units of $|\Delta|$ [30]. We summarize:

Even Bulk Boundary Correspondence *If $m\theta + N$ is the IDS of h at 0 then the work exhibited by the atomic motion through a cycle around S, on the edge states of energy inside Δ, is $-m|\Delta|$.*

3. We consider a quasi periodic sequence ξ, more precisely a Sturmian sequence. Now Ξ is the cut up circle S_c. It is totally disconnected and therefore $K_0(C(\Xi)) \cong C_\alpha C(\Xi, \mathbb{Z})$ and $K_1(C(\Xi)) \cong 0$. It follows that the even bulk-boundary correspondence is trivial, as in the periodic case. There is nevertheless a difference with the periodic case, namely $C_\alpha C(\Xi, \mathbb{Z})$ has two generators. Next to the class $[-1]$ it contains the class $[1 - 2\chi_\theta]$ where χ_θ is the characteristic function on a clopen subset of Ξ which has measure θ w.r.t. the unique ergodic invariant probability measure.

4. The reason why the even BBC is trivial in the periodic and the quasi-periodic case is, that in these cases the K_1-group of the ideal in the Toeplitz extension is trivial. This is due to the fact that the connected component of a point in the hull contains only that point. There is also no obvious definition of a winding number, as there is no notion of continuous translation in such spaces, any translation pushes a point out of its connected component. Likewise we can't define a gradient force. It is here where the augmented system comes into play.

Recall that we can make the phason flips continuous by augmenting the system, that is, adding sequences in which the phason flip is only partially carried out. Since $\tilde{\Xi}$ is a circle, $K_1(C(\tilde{\Xi}))$ is non-trivial and the even bulk boundary correspondence becomes non-trivial as well. But this means that we abandon the simple Toeplitz extension for $C(\Xi) \rtimes_\alpha \mathbb{Z}$ and replace it by a more complicated short exact sequence. We outline the construction refering the reader to [30] for more details.

The augmented hull $\tilde{\Xi}$ contains Ξ as a closed shift-invariant subset. The Toeplitz extension $\mathcal{T}(C(\tilde{\Xi}), \alpha)$ of the augmented model is therefore also an extension of $C(\Xi) \rtimes_\alpha \mathbb{Z}$. If we take this extension for (4.10) then we end up with the short exact sequence

$$0 \to C_{\alpha_*} K_0(C(\tilde{\Xi})) \xrightarrow{\tilde{i}_*} K_0(C(\Xi) \rtimes_\alpha \mathbb{Z}) \xrightarrow{\widetilde{\exp}} I_{\alpha_*} K_1(C(\tilde{\Xi})) \to 0$$

where $\tilde{i}_*$ is induced by the inclusion after restriction $C(\tilde{\Xi}) \to C(\Xi) \to C(\Xi) \rtimes_\alpha \mathbb{Z}$. Again α_* is the identity and $K_0(C(\tilde{\Xi})) \cong K_1(C(\tilde{\Xi})) \cong \mathbb{Z}$. Of the two generators of $K_0(C(\Xi) \rtimes_\alpha \mathbb{Z})$, the class $[-1]$ and the class $[1 - 2\chi_\theta]$, the first lies in the kernel of the exponential map and for the second one finds $\exp([1 - 2\chi_\theta]) = -[\tilde{z}]$ where $\tilde{z} : \tilde{\Xi} \to \mathbb{C}$ is the function with winding number -1.

There is a unique ergodic invariant probability measure on Ξ (when identifying Ξ with S_c, which is from the point of view of measure theory the same as S, the above measure is the Lebesgues measure) which we use to define the trace and tracial state on $K_0(C(\Xi) \rtimes_\alpha \mathbb{Z})$. To define the winding number on $C(\tilde{\Xi})$ we use the derivation w.r.t. the parameter t describing the continuous phason motion and the trace $\tilde{\mathrm{tr}}$ coming from the Lebesgue measure on the added intervals (see [30] for more details). We then get the non-trivial even bulk boundary correspondence: $\mathrm{tr}([\chi_\theta]) = -\theta \mathcal{W}_{\tilde{\mathrm{tr}}, \partial_t}([\tilde{z}])$ modulo $\mathbb{Z}$. As in the almost periodic case the winding number of the boundary invariant has a physical interpretation;

it is the work the phason flips exhibit on the boundary states with energy in the gap Δ, in units of $|\Delta|$ [30]. We summarize:

Even Bulk Boundary Correspondence *If $n + m\theta$ is the IDS of h at 0 then the work exhibited by a cycle of phason flips on the edge states of energy inside Δ is $-m|\Delta|$.*

This identity has been experimentally observed [2].

References

1. J.E. Anderson, I.F. Putnam, Topological invariants for substitution tilings and their associated C^*-algebras. Ergod. Theory Dyn. Syst. **18**(3), 509–537 (1998)
2. F. Baboux, E. Levy, A. Lemaitre, C. Gómez, E. Galopin, L.L. Gratiet, I. Sagnes, A Amo, J. Bloch, E. Akkermans, Measuring topological invariants from generalized edge states in polaritonic quasicrystals. Phys. Rev. B **95**, 161114(R) (2017)
3. A. Avila, S. Jitomirskaya, The ten Martini problem. Ann. Math. **170**, 303–342 (2009)
4. M. Baake, U. Grimm, *Aperiodic Order (Vol. 1)* (Cambridge University Press, Cambridge, 2013)
5. M. Baake, M. Schlottmann, P.D. Jarvis, Quasiperiodic tilings with tenfold symmetry and equivalence with respect to local derivability. J. Phys. A Math. General **24**(19), 4637 (1991)
6. J. Bellissard, K-theory of C^*-Algebras in solid state physics, in *Statistical Mechanics and Field Theory: Mathematical Aspects*, (Springer, Berlin, 1986), pp. 99–156
7. J. Bellissard, *Gap Labeling Theorems for Schrödinger Operators*, ed. by M. Waldschmidt, P. Moussa, J.-M. Luck, C. Itzykson. From Number Theory to Physics (Springer, Berlin, 1995)
8. J. Bellissard, D.J.L. Herrmann, M. Zarrouati, Hull of aperiodic solids and gap labelling theorems. Directions Math. Quasicrystals **13**, 207–258 (2000)
9. B. Blackadar, *K-theory for Operator Algebras*. Mathematical Sciences Research Institute Publications, vol. 5, 2nd edn. (Cambridge University Press, Cambridge, 1998)
10. A. Clark, L. Sadun, When shape matters: deformations of tiling spaces. Ergod. Theory Dyn. Syst. **26**(1), 69–86 (2006)
11. A. Connes, *Non-commutative Geometry* (Academic, San Diego, 1994)
12. A. van Daele, K-theory for graded Banach algebras I. Quarterly J. Math. **39**(2), 185–199 (1988)
13. A. van Daele, K-theory for graded Banach algebras II. Pacific J. Math. **135**(2), 377–392 (1988)
14. D. Damanik, M. Embree, A. Gorodetski, Spectral properties of Schrödinger operators arising in the study of quasicrystals, in *Mathematics of Aperiodic Order* (Birkhäuser, Basel, 2015), pp. 307–370
15. K. Davidson, *C^*-Algebras by Example*, vol. 6 (American Mathematical Society, Providence, 1996)
16. M.R. Herman, Sur la conjugaison différentiable des difféomorphismes du cercle à des rotations. Pub. Math. IHES **49**, 5–234 (1979)
17. A. Forrest, J. Hunton, J. Kellendonk, *Topological Invariants for Projection Method Patterns (No. 758)* (American Mathematical Society, Providence, 2002)
18. F. Gähler, Computer code, private communication
19. F. Gähler, J. Hunton, J. Kellendonk, Integral cohomology of rational projection method patterns. Algebr. Geometri. Topol. **13**(3), 1661–1708 (2013)
20. B.I. Halperin, Quantized Hall conductance, current carrying edge states, and the existence of extended states in a two-dimensional disordered potential. Phys. Rev. B **25**(4), 2185 (1982)
21. Y. Hatsugai, Chern number and edge states in the integer quantum Hall effect. Phys. Rev. Lett. **71**(22), 3697 (1993)

22. N. Higson, J. Roe, *Analytic K-homology* (OUP Oxford, Oxford, 2000)
23. J. Hunton, Spaces of projection method patterns and their cohomology, in *Mathematics of Aperiodic Order* (Birkhäuser, Basel, 2015), pp. 105–135
24. S. Jitomirskaya, Metal-insulator transition for the almost Mathieu operator. Ann. Math. **150**, 1159–1175 (1999)
25. P. Kalugin, Cohomology of quasiperiodic patterns and matching rules. J. Phys. A Math. General **38**(14), 3115 (2005)
26. J. Kellendonk, Noncommutative geometry of tilings and gap labelling. Rev. Math. Phys. **7**(07), 1133–1180 (1995)
27. J. Kellendonk, Pattern-equivariant functions and cohomology. J. Phys. A Math. General **36**(21), 5765 (2003)
28. J. Kellendonk, Pattern equivariant functions, deformations and equivalence of tiling spaces. Ergod. Theory Dynam. Syst. **28**(4), 1153–1176 (2008)
29. J. Kellendonk, On the C^*-algebraic approach to topological phases for insulators. Ann. Henri Poincaré **18**(7), 2251–2300 (2017)
30. J. Kellendonk, E. Prodan, Bulk-boundary correspondence for Sturmian Kohmoto like models. Ann. Henri Poincaré **20**(6), 2039–2070 (2019)
31. J. Kellendonk, I.F. Putnam, Tilings, C^*-algebras and K-theory, in *Directions in Mathematical Quasicrystals*, ed. by M. Baake, R.V. Moody. CRM Monograph Series, vol. 13 (2000) (American Mathematical Society, Providence, 1999), pp. 177–206
32. J. Kellendonk, S. Richard, Topological boundary maps in physics, in *Perspectives in Operator Algebras and Mathematical Physics*. Theta Series in Advanced Mathematics, vol. 8 (Theta, Bucharest, 2008), pp. 105–121
33. J. Kellendonk, T. Richter, H. Schulz-Baldes, Edge current channels and Chern numbers in the integer quantum Hall effect. Rev. Math. Phys. **14**, 87–119 (2002)
34. Y. Last, Spectral theory of Sturm-Liouville operators on infinite intervals: a review of recent developments, in *Sturm-Liouville Theory* (Birkhäuser Basel, Basel, 2005), pp. 99–120
35. M. Pimsner, D. Voiculescu. Exact sequences for K-groups of certain cross products of C^*-algebra s. J. Op. Theory **4**, 93–118 (1980)
36. E. Prodan, H. Schulz-Baldes, Non-commutative odd Chern numbers and topological phases of disordered chiral systems. J. Funct. Analy. **271**(5), 1150–1176 (2016)
37. E. Prodan, H. Schulz-Baldes, *Bulk and Boundary Invariants for Complex Topological Insulators* (Springer, Berlin, 2016)
38. J. Renault, *A groupoid Approach to C*-algebras (Vol. 793)*. (Springer, Berlin, 2006)
39. M. Rieffel, C^*-algebras associated with irrational rotations. Pacific J. Math. **93**(2), 415–429 (1981)
40. M. Rordam, F. Larsen, N.J. Laustsen, *An Introduction to K-theory of C*-Algebras* (Cambridge University Press, Cambridge, 2000)
41. L.A. Sadun, *Topology of Tiling Spaces (Vol. 46)* (American Mathematical Society, Providence, 2008)
42. L. Sadun, Cohomology of Hierarchical Tilings, in *Mathematics of Aperiodic Order* (Birkhäuser, Basel, 2015), pp. 73–104
43. L. Sadun, R.F. Williams, Tiling spaces are Cantor set fiber bundles. Ergod. Theory Dyn. Syst. **23**(1), 307–316 (2003)

Chapter 5
From Combinatorial Games
to Shape-Symmetric Morphisms

Michel Rigo

Abstract Siegel suggests in his book on combinatorial games that quite simple games provide us with challenging problems: "No general formula is known for computing arbitrary Grundy values of Wythoff's game. In general, they appear chaotic, though they exhibit a striking fractal-like pattern.". This observation is the first motivation behind this chapter. We present some of the existing connections between combinatorial game theory and combinatorics on words. In particular, multidimensional infinite words can be seen as tilings of $\mathbb{N}^d$. They naturally arise from subtraction games on d heaps of tokens. We review notions such as k-automatic, k-regular or shape-symmetric multidimensional words. The underlying general idea is to associate a finite automaton with a morphism.

5.1 Introduction

Combinatorial game theory (CGT) uses many tools from other fields: number theory, continued fractions, numeration systems, cellular automata, etc. We will limit ourselves to subtraction games played on heaps of tokens. When analyzing Sprague–Grundy values of some well-known and popular subtraction games—like Nim or Wythoff's game—notions such as k-automatic, k-regular and morphic (also called substitutive) multidimensional sequences enter naturally the picture. The aim of this chapter is to introduce these concepts and present the interplay existing between CGT and combinatorics on words (COW). The focus is put on infinite (multidimensional) words generated by iterated morphisms. The organization of this

M. Rigo (✉)
University of Liège, Department of Mathematics, Liège, Belgium
e-mail: M.Rigo@ulg.ac.be

S. Akiyama, P. Arnoux (eds.), *Substitution and Tiling Dynamics: Introduction to Self-inducing Structures*, Lecture Notes in Mathematics 2273,
https://doi.org/10.1007/978-3-030-57666-0_5

chapter faithfully reflects the course presented during the school[1] "*Tiling Dynamical System*" organized during the Jean Morlet Chair attributed to Professor Shigeki Akiyama.

After setting notation in Sect. 5.2, minimal requirements and basic results about CGT are given in Sect. 5.3. Well-known facts about k-automatic and k-regular (unidimensional) sequences are presented in Sect. 5.4. For some general references, see [8, 62, 72, 73]. Let $\mathbf{P}_W : \mathbb{N}^2 \to \{0, 1\}$ be the bidimensional infinite word defined by $\mathbf{P}_W(m, n) = 1$ if and only if (m, n) is a $\mathcal{P}$-position, i.e., a loosing position or a zero of the Sprague–Grundy function, of Wythoff's game. Section 5.5 deals with the syntactic characterization of these $\mathcal{P}$-positions in terms of representations in the Fibonacci numeration system. This result by Fraenkel is proved using the infinite Fibonacci word and a transducer computing the successor function in the Fibonacci numeration system. In particular, we make explicit the general fact that morphic words can be generated by finite automata. We look at the sequence of states reached by all the accepted words when they have been genealogically ordered. The transformation from morphisms to automata is recurrent in this chapter: uniform and non-uniform morphisms for words, uniform and non-uniform morphisms in a multidimensional setting.

The multidimensional point of view begins with Sect. 5.6. We first look at k-automatic and k-regular multidimensional words. It turns out that characterizing the $\mathcal{P}$-positions of Wythoff's game leads to a *shape-symmetric* morphism as introduced and studied by A. Maes in a logical setting. The word $\mathbf{P}_W$ depicted in Table 5.5 is the fixed point of a morphism φ_W and it has the shape-symmetric property with respect to φ_W. These notions are well fitted to a volume dedicated to tilings and are presented in Sect. 5.7. We introduce a generalization of numeration systems whose language of representations is regular: *abstract numeration systems*. Indeed, a multidimensional infinite word is $\mathcal{S}$-automatic, for some abstract numeration system $\mathcal{S}$, if and only if it is the image by a coding of a shape-symmetric word. Hence we prove that $\mathbf{P}_W$ codes the $\mathcal{P}$-positions of the Wyhtoff's game. We conclude this chapter with a short discussion about games with a finite set of rules and some bibliographic notes.

5.2 Notation and Conventions

We assume that 0 belongs to the set $\mathbb{N}$ of non-negative integers. An *alphabet* is just a finite set. A (finite) *word* over an alphabet A is a finite sequence of elements in A, i.e., a map from a finite set $\{0, \ldots, \ell - 1\}$ to A. The length of the word w is denoted by $|w|$. The set of finite words over A is denoted by A^*. Equipped with the concatenation product, it is a free monoid. A (right) *infinite word* is a map from $\mathbb{N}$ to A. It is indeed more convenient (when working with automatic sequences)

---

[1] A video is available at http://library.cirm-math.fr/.

to start indexing with 0. We will indifferently use the terminology *infinite word* or *sequence*. Compared with other chapters of this book, we will never encounter bi-infinite words. Let $d \geq 2$ be an integer. By extension, we say that a map from $\mathbb{N}^d$ to an alphabet A is a (multidimensional) *infinite word* or *sequence*. Infinite words will be denoted using a bold face letter such as $\mathbf{x}$ or $\mathbf{P}_W$.

Let $k \geq 2$ be an integer. We let $\text{rep}_k(n)$ denote the usual *base-k expansion* of $n \geq 0$. It is a word over the alphabet $\{0, \ldots, k-1\}$. We set $\text{rep}_k(0)$ to be the empty word denoted by ε. Except when stated otherwise (when dealing with a transducer computing the successor function in the proof of Theorem 5.5.20 or, when considering an automaton realizing addition in base 2 in the proof of Lemma 5.6.8), we use the most significant digit first convention (MSDF): if $n > 0$, the most significant digit of $\text{rep}_k(n)$ is written on the left—as a prefix—and is non-zero. If $w = w_\ell \cdots w_0$ is a word over an alphabet $A \subset \mathbb{Z}$, then the *k-numerical value* of w (w.r.t. A) is given by the map

$$\text{val}_k : A^* \to \mathbb{Z}, \; w \mapsto \sum_{i=0}^{\ell} w_i \, k^i \, .$$

In this chapter, we will also encounter other numeration systems, i.e., bijections from $\mathbb{N}$ to some (infinite) formal language L. We will adapt the corresponding notation to $\text{rep}_L : \mathbb{N} \to L$ and $\text{val}_L : L \to \mathbb{N}$.

We assume that the reader is familiar with basic concepts from graph theory. If $G = (V, E)$ is a directed graph and $u, v \in V$ are two of its vertices, we write $(u, v) \in E$ or $u \to v$ to denote an edge of G. Nevertheless, we recall the definition of a finite automaton. We will encounter automata in Sect. 5.4. For more, see, for instance, [79, 83, 86].

Definition 5.2.1 A *deterministic finite automaton* (DFA for short) is a 5-tuple $\mathcal{M} = (Q, q_0, A, \delta, F)$ where Q is a finite set of *states*, $q_0 \in Q$ is the *initial state*, A is the alphabet of the automaton, $\delta : Q \times A \to Q$ is the *transition function* and $F \subseteq Q$ is the set of *final states*.

For a DFA $\mathcal{M}$, the transition function is extended to $\delta : Q \times A^* \to Q$ by $\delta(q, \varepsilon) = q$ and $\delta(q, aw) = \delta(\delta(q, a), w)$ for all $q \in Q$, $a \in A$, $w \in A^*$. If the automaton is clear from the context, we simply write $q \cdot w$ instead of $\delta(q, w)$. The state reached when reading w from the initial state will also be written $\mathcal{M} \cdot w$. The language *accepted* or *recognized* by $\mathcal{M}$ is

$$L(\mathcal{M}) = \{w \in A^* \mid \mathcal{M} \cdot w \in F\} \, .$$

A language is *regular* if it is accepted by some DFA.

Definition 5.2.2 A *deterministic finite automaton with output* (DFAO for short) is given by a 5-tuple $\mathcal{M} = (Q, q_0, A, \delta, \mu)$ where the first four components are defined as for a DFA and $\mu : Q \to B$ is the output map (where B is some alphabet). A DFAO where the output map takes at most two values is a standard DFA.

Definition 5.2.3 A *non-deterministic finite automaton* (NFA for short) over an alphabet A is given by a 5-tuple $\mathcal{N} = (Q, I, A, \Delta, F)$ where Q is a finite set of states, $I \subseteq Q$ is the set of initial states, $\Delta \subset Q \times A^* \times Q$ is the (finite) transition relation and $F \subseteq Q$ is the set of final states. A word w is *accepted* by $\mathcal{N}$ if there exist an integer i, some (possibly empty) words $v_1, \ldots, v_i$ and a sequence of states $q_1, \ldots, q_{i+1}$ such that $w = v_1 \cdots v_i$ and

- $(q_1, v_1, q_2), (q_2, v_2, q_3), \ldots, (q_i, v_i, q_{i+1}) \in \Delta$
- $q_1 \in I, q_{i+1} \in F$.

Otherwise stated, there is at least one accepting path with label w from an initial state to some final state. The *language accepted* by $\mathcal{N}$ is the set of accepted words. One can assume that (q, ε, q) belongs to Δ for all $q \in Q$.

5.3 Bits of Combinatorial Game Theory

They are many kinds of games that you can think of. We deliberately restrict ourselves to one of the most simple classes: *subtraction games* (taking and breaking games) [48]. In this setting, we have two players acting in turns. They have a finite number of tokens organized in piles to start with. The tokens are given in some initial position. Typically, the players have to remove some tokens complying to prescribed rules. A player may not pass: at least one token must be removed. The set of rules is known in advance and we have the same rules for both players. There is no chance involved (no randomness) and the information is completely known for both players—no hidden information for one of the two players [1, 67]. Finally we will assume a *normal play* convention: the first player unable to move loses the game. Nim game and Wythoff's game will provide us with quite enough material.

I do not assume that the reader has any particular knowledge about CGT. Some general references arc [2, 13, 85]. I was also inspired by Ferguson's lecture notes from UCLA. This section serves as a self-contained introduction to the topic. It is not at all aimed to be exhaustive.

Example 5.3.1 (Finite Subtraction Game) Starting from a single finite pile of tokens, the players may remove either 1, 2 or 4 tokens. The player taking the last token wins the game. An example of game is given by the following sequence

$$9 \xrightarrow{-2} 7 \xrightarrow{-1} 6 \xrightarrow{-1} 5 \xrightarrow{-2} 3 \xrightarrow{-2} 1 \xrightarrow{-1} 0.$$

For instance, we can immediately see that when a player is in a position with 3 tokens left, he/she will lose the game because the only available options are to remove either 1 or 2 tokens. We will encounter finite subtraction game with several piles of tokens in Sect. 5.8.

Example 5.3.2 (Game of Nim)   Let $p \geq 1$ be the number of piles of tokens to play with. If we assume here that the piles are ordered, a *position* is coded by a p-tuple of non-negative integers. An initial position is given: a game starts with p piles containing respectively $x_1, \ldots, x_p \geq 0$ tokens. In their turn to play, the player chooses a non-empty pile and removes a positive number of tokens from it. Thus, from a position of the form $(x_1, \ldots, x_p)$ the available *options* are given by p-tuples of the form $(y_1, \ldots, y_p) \in \mathbb{N}^p$ where there exists j such that $y_j < x_j$ and $y_i = x_i$ for all $i \neq j$. In particular, the number of options available from $(x_1, \ldots, x_p)$ is just $\sum_{i=1}^{p} x_i$. The game is lost for the player left in the position $(0, \ldots, 0)$. Loosing positions will be determined in Theorem 5.3.13.

In this section, to avoid any confusion, generic positions are denoted by Greek letters: α, β, γ. Formally, if α and β are two positions of a subtraction game, we say that β is an *option* of α, if there is a move permitting the players to go from α to β. The set of options of α is denoted by $\text{Opt}(\alpha)$. We can extend Opt to a set X of positions in a natural way: $\text{Opt}(X) = \cup_{\alpha \in X} \text{Opt}(\alpha)$. The reflexive and transitive closure of Opt is denoted by Opt^*: β belongs to $\text{Opt}^*(\alpha)$ if and only if there exists a finite (possibly empty) sequence of moves from α to β. In graph-theoretic terms, one can speak of *successor* instead of option.

Example 5.3.3 (Wythoff's Game)   Consider the following modification of the game of Nim on two piles. We add extra rules to the set of Nim rules. A player can, as for Nim, either remove a positive number of tokens from one pile, or *remove simultaneously the same number of tokens from both piles*. So from a position $(x_1, x_2) \neq (0, 0)$, a player can move to either $(x_1 - i, x_2)$ with $0 < i \leq x_1$, or $(x_1, x_2 - i)$ with $0 < i \leq x_2$, or $(x_1 - i, x_2 - i)$ with $0 < i \leq \min\{x_1, x_2\}$. An example of play, starting with the position $(4, 3)$, is given below:

$$(4, 3) \xrightarrow{-(2,0)} (2, 3) \xrightarrow{-(1,1)} (1, 2) \xrightarrow{-(0,1)} (1, 1) \xrightarrow{-(1,1)} (0, 0).$$

The chosen move is indicated on the arrows. The reader may already notice this: the player in position $(1, 2)$ whatever the chosen move is, will give a position for the other player from which the game can always be won. Loosing positions will be determined by Theorem 5.3.14.

Note that it is a quite natural setting to encounter subtraction games whose rule-set extends the game of Nim.

Definition 5.3.4 (Game-Graph) A subtraction game is given (notation will not refer to the chosen game). The *game-graph* Gr_γ, for an initial position γ, is the finite directed graph whose set $V(\text{Gr}_\gamma)$ of vertices is the set of positions that can be reached from γ by a finite sequence of allowed moves, i.e., $V(\text{Gr}_\gamma) = \text{Opt}^*(\gamma)$. There is an edge from α to β if and only if β belongs to $\text{Opt}(\alpha)$. The set of edges is denoted by $E(\text{Gr}_\gamma)$.

Remark 5.3.5   In a subtraction game, consider two positions α and β. If β can be reached from α, i.e., $\beta \in \text{Opt}^*(\alpha)$, then the game-graph Gr_β associated with β is a subgraph of the game-graph Gr_α associated with α. More precisely, we have

$$E(\mathsf{Gr}_\beta) = E(\mathsf{Gr}_\alpha) \cap \left(V(\mathsf{Gr}_\beta) \times V(\mathsf{Gr}_\beta)\right)$$

and there is no edge in Gr_α from $V(\mathsf{Gr}_\beta)$ to $V(\mathsf{Gr}_\alpha) \setminus V(\mathsf{Gr}_\beta)$.

For subtraction games, the corresponding game-graph is obviously acyclic: the total number of tokens left is decreasing. Thus, there exists at least one vertex with out-degree equal to zero.[2]

Definition 5.3.6 Let $G = (V, E)$ be a directed graph. To stick to the game setting, we denote by $\mathrm{Opt}(u)$ the set of vertices v such that $(u, v) \in E$. In particular, the out-degree of u is $\# \mathrm{Opt}(u)$. A *kernel* of G is a subset K of V with the following two properties:

- K is *stable*: $\forall u \in K, \mathrm{Opt}(u) \cap K = \emptyset$;
- K is *absorbing*: $\forall v \in V \setminus K, \mathrm{Opt}(v) \cap K \neq \emptyset$.

Let $G = (V, E)$ be a finite directed acyclic graph (it is common in the literature to find the acronym DAG). For instance, DAG are the graphs with a topological sort of the vertices [74]. Let us describe an algorithm to compute a kernel of G [12]. Let $G_0 = G$. We will build a finite sequence of subgraphs $(G_i)_{i \geq 0}$.

- Let $\mathcal{P}_0$ be the set of vertices of G_0 with out-degree equal to zero (sometimes called *sink vertices*),

$$\mathcal{P}_0 = \{v \in V(G_0) \mid \mathrm{Opt}(v) = \emptyset\}.$$

- Let $\mathcal{N}_0$ be the set of vertices x of G_0 such that $\mathrm{Opt}(x) \cap \mathcal{P}_0 \neq \emptyset$.

Due to the absorbing property, the vertices of $\mathcal{P}_0$ must belong to any kernel of G. Therefore, the vertices in $\mathcal{N}_0$ do not belong to any kernel of the graph due to stability. Let G_1 be the subgraph obtained by removing the vertices in $\mathcal{P}_0 \cup \mathcal{N}_0$. This graph is again acyclic. We repeat the procedure and define $\mathcal{P}_1$ and $\mathcal{N}_1$ accordingly. If a vertex v belong to $\mathcal{P}_1$, it has no option in $V(G_1)$. In the original graph G_0, it has no option in $\mathcal{P}_0$ because otherwise we would have $v \in \mathcal{N}_0$. Thus again the vertices of $\mathcal{P}_1$ must belong to any kernel of G and the vertices in $\mathcal{N}_1$ are in no kernel of G. We define a sequence of subgraphs where G_{i+1} is obtained by removing $\mathcal{P}_i \cup \mathcal{N}_i$ from G_i. The procedure halts when we reach an empty subgraph. We directly get the next proposition.

Proposition 5.3.7 *Let $G = (V, E)$ be a finite directed acyclic graph. Applying the above algorithm, the set $\cup_i \mathcal{P}_i$ is the unique kernel of G.*

---

[2]Consider a simple path $v_0 \to v_1 \to \cdots \to v_r$ of maximal length r in a finite acyclic graph. Then v_r has out-degree zero. Proceed by contradiction and assume that there is an edge starting from v_r. Either it goes to one of the v_i's with $i < r$ and it creates a cycle. Or, it goes to some other vertex and we may build a longer simple path. Both situations lead to a contradiction.

Remark 5.3.8 If the graph is not acyclic, the situation is not so nice. Consider a simple (oriented) cycle of length n. It is easy to see that if n is odd (resp. even), then the graph has no kernel (resp. two kernels). Moreover, the reader may notice that the previous algorithm does not work anymore.

Consider a subtraction game. For a game-graph Gr_β, since it is acyclic, we have a partition of the set of vertices of Gr_β into two subsets: those belonging to the kernel and those out of it. The first ones will be called $\mathcal{P}$-positions, the other ones $\mathcal{N}$-positions. Moreover, thanks to Remark 5.3.5, we see that the status of a position is *invariant with respect to the initial position*: if α, β are two positions such that $\beta \in \mathrm{Opt}^*(\alpha)$, then the kernel of Gr_β is a subset of the kernel of Gr_α. The following definition is thus meaningful.

Definition 5.3.9 ($\mathcal{N}$- and $\mathcal{P}$-Positions) Consider a subtraction game. A position is a *losing position*, or $\mathcal{P}$-position, if all its options are $\mathcal{N}$-positions. A position is a *winning position*, or $\mathcal{N}$-position, if there exists an option which is a $\mathcal{P}$-position. The set of $\mathcal{P}$-positions (resp. $\mathcal{N}$-positions) is denoted $\mathcal{P}$ (resp. $\mathcal{N}$).

The classical meaning of $\mathcal{P}$ or $\mathcal{N}$ in CGT refers to the "previous" player or the "next" move. In a $\mathcal{P}$-position, the previous player (so not the one being in that actual position) is able to win the game. In a $\mathcal{N}$-position, the actual player may choose the correct next move to win the game. This choice is the notion of a *winning strategy*: selecting for every $\mathcal{P}$-position one available $\mathcal{N}$-position among the options.

Remark 5.3.10 In a subtraction game on p piles of tokens, if there is a move from $(x_1, \ldots, x_p)$ to $(y_1, \ldots, y_p)$, then $\sum_i x_i > \sum_i y_i$. Let $C \geq 0$ be an integer. If we know the status $\mathcal{N}$ or $\mathcal{P}$ of all positions $(x_1, \ldots, x_p)$ such that $\sum_i x_i \leq C$, then we can determine the status of any position $(x_1, \ldots, x_p)$ such that $\sum_i x_i = C + 1$.

Example 5.3.11 Consider the finite subtraction game of Example 5.3.1. Prove that n is a $\mathcal{P}$-position if and only if $n \equiv 0 \pmod 3$.

Even though the status $\mathcal{N}$ or $\mathcal{P}$ of positions can be determined by some naive methods, let us stress the fact that in terms of algorithmic complexity, computing the kernel of the game-graph or, equivalently using the above remark, has an exponential cost compared to the size of the input (size of the game position) [44]. Indeed, when playing a game like Wythoff's game, a position (x_1, x_2) is coded by a word (let us think about the base-2 expansions of x_1 and x_2 that are the inputs given to the algorithm) whose length is proportional to $\log x_1 + \log x_2 = \log(x_1.x_2)$. But the game-graph has a number of vertices equal to $(x_1 + 1).(x_2 + 1)$ which is exponentially larger than the size of the input. One can therefore ask for an algorithm determining the status of a position whose complexity is a polynomial in the length of the coding. We will see in Sect. 5.5 that deciding the status of a position of the Wythoff's game can be done in polynomial time.

Definition 5.3.12 Let m, n be two non-negative integers. Let $x = x_\ell \cdots x_0$ and $y = y_\ell \cdots y_0$ be two words over $\{0, 1\}$ such that $\mathrm{val}_2(x) = m$ and $\mathrm{val}_2(y) = n$. In other words, x and y are the base-2 expansions of m and n up to some leading zeroes ensuring that we have two words of the same length. The *Nim-sum* of m and n is the integer

$$m \oplus n = \sum_{j=0}^{\ell} \left(x_j + y_j \mod 2\right) 2^j.$$

As an example, $7 \oplus 2 \oplus 9 = 12$. It is an easy exercise to show that $(\mathbb{N}, \oplus)$ is an Abelian group.

Theorem 5.3.13 (Bouton's Theorem [15]) *For the game of Nim played on $p \geq 1$ piles of tokens, a position $(x_1, \ldots, x_p)$ is in $\mathcal{P}$ if and only if*

$$\bigoplus_{i=1}^{p} x_i = 0.$$

Proof From Definition 5.3.9, we have to prove that the set K of p-tuples with zero Nim-sum is stable and absorbing. If we consider a p-tuple whose Nim-sum is zero, playing one move of Nim will change exactly one of the piles of tokens. Thus we modify exactly one term, let us say x_j, of the Nim-sum. Therefore, at least one bit of $\mathrm{rep}_2(x_j)$ is modified and thus the Nim-sum is no more equal to zero. This shows that the set K is stable.

Now we have to prove that K is absorbing. Consider a p-tuple whose Nim-sum s is non-zero. There exists $\ell \geq 0$ such that $2^\ell \leq s < 2^{\ell+1}$. Let x_j be a component such that the base-2 expansion of x_j has a non-zero digit in the position corresponding to 2^ℓ (such an element exists), i.e., $\mathrm{rep}_2(x_j) = v1w$ where $|w| = \ell$. We can thus replace x_j with a smaller integer in such a way that the total Nim-sum is zero. $\square$

Two years after Bouton's result, Wythoff proposed a modification of the game of Nim and characterized the corresponding set of $\mathcal{P}$-position [87].

Theorem 5.3.14 (Wythoff's Theorem) *A pair (x, y) is a $\mathcal{P}$-position of the game of Wythoff if and only if it is of the form*

$$\left(\lfloor n\varphi \rfloor, \lfloor n\varphi^2 \rfloor\right) \text{ or } \left(\lfloor n\varphi^2 \rfloor, \lfloor n\varphi \rfloor\right), \quad \text{for some } n$$

where φ is the Golden ratio $(1 + \sqrt{5})/2$.

For a proof, see the original paper of Wythoff. It is enough to prove that for the game-graph associated with Wythoff's game, the above set is stable and absorbing. This can be considered as a "difficult" exercise (one has to deal with integer and fractional parts of multiple of the Golden ratio).

Definition 5.3.15 (MeX) If S is a strict subset of $\mathbb{N}$, then $\mathrm{MeX}(S) = \min(\mathbb{N} \setminus S)$. So it is the least non-negative integer not in S. It stands for "minimal excluded" value. In particular, $\mathrm{MeX}(\emptyset) = 0$.

For a subtraction game, we introduce a map associating a non-negative integer with each position. This map is defined recursively. Its use will be clear when we introduce the sum of games.

Definition 5.3.16 The *Sprague–Grundy function* $\mathcal{G}$ (of a game with an acyclic game-graph) is defined by

$$\mathcal{G}(x) = \mathrm{MeX}\left(\{\mathcal{G}(y) \mid y \in \mathrm{Opt}(x)\}\right) .$$

Example 5.3.17 (Continuing Example 5.3.1) If the players are allowed to remove 1, 2, 4 tokens from a single pile, the first values of the Sprague–Grundy function are given in Table 5.1. It is an easy exercise (left to the reader) to prove that this function is periodic of period 3: $\mathcal{G}(n) = n \mod 3$. We will reconsider games with a finite set of moves in Sect. 5.8. See Proposition 5.8.1.

Proposition 5.3.18 *Consider a subtraction game. A position α is a $\mathcal{P}$-position if and only if $\mathcal{G}(\alpha) = 0$.*

Proof One has to prove that $\{\alpha \mid \mathcal{G}(\alpha) = 0\}$ is the kernel of the game-graph. This is a direct consequence of the use of the MeX operator in the definition of $\mathcal{G}$.   □

The reader may wonder why Sprague–Grundy functions, or simply Grundy functions, could be useful. It turns out that we can consider the (disjunctive) *sum* of games. Assume that we have several games $G_1, \ldots, G_n$ in front of the two players, e.g. a Wythoff's game G_1 in position $(3, 4)$, a game of Nim G_2 in position $(1, 2, 3)$ and another Wythoff's game G_3 in position $(2, 2)$. The idea is that, at each turn, the actual player chooses the game where he/she plays a move. We use the same normal convention: when there is no more available move (in any of the n games), the game is lost.

Definition 5.3.19 (Sum of Games) Let $\mathsf{Gr}_1, \ldots, \mathsf{Gr}_n$ be the game-graphs of n games $G_1, \ldots, G_n$ respectively. The game-graph of the *sum* of $G_1, \ldots, G_n$ has the Cartesian product

$$V(\mathsf{Gr}_1) \times \cdots \times V(\mathsf{Gr}_n)$$

as set of vertices. If $(\alpha_1, \ldots, \alpha_n)$ is a vertex, i.e., α_i is a position of G_i, for all i, then there is an edge to $(\beta_1, \ldots, \beta_n)$ whenever there exists j such that $\beta_j \in \mathrm{Opt}(\alpha_j)$ (being understood that β_j is an option of α_j in the game G_j) and $\beta_i = \alpha_i$ for all $i \neq j$.

Table 5.1 First few values of $\mathcal{G}$ for a finite subtraction game

n	0	1	2	3	4	5	6	7	8	9	10	$\cdots$
$\mathcal{G}(n)$	0	1	2	0	1	2	0	1	2	0	1	$\cdots$

This theory has been independently developed by R. P. Sprague and P. M. Grundy.[3]

Theorem 5.3.20 (Sprague–Grundy Theorem) *Let G_i be impartial combinatorial games with $\mathcal{G}_i$ as respective Grundy functions, $i = 1, \ldots, n$. Then the sum of games $G_1 + \cdots + G_n$ has a Grundy function $\mathcal{G}$ given by*

$$\mathcal{G}(\alpha_1, \ldots, \alpha_n) = \mathcal{G}_1(\alpha_1) \oplus \cdots \oplus \mathcal{G}_n(\alpha_n)$$

where, for all i, α_i is a position in G_i.

Proof Let $(\alpha_1, \ldots, \alpha_n)$ be an arbitrary position in the sum of the n games. Let b be the integer $\mathcal{G}_1(\alpha_1) \oplus \cdots \oplus \mathcal{G}_n(\alpha_n)$. We will prove that

1. for all non-negative integers $a < b$, there is an option from $(\alpha_1, \ldots, \alpha_n)$ with $\mathcal{G}$-value a;
2. no option of $(\alpha_1, \ldots, \alpha_n)$ has $\mathcal{G}$-value b.

If the two items are proved, by definition of the Sprague–Grundy function, this means that $\mathcal{G}(\alpha_1, \ldots, \alpha_n) = b = \mathcal{G}_1(\alpha_1) \oplus \cdots \oplus \mathcal{G}_n(\alpha_n)$.

Let $d = a \oplus b$ and ℓ be the length of the base-2 expansion of d, i.e., $2^{\ell-1} \leq d < 2^\ell$. Since $a < b$, the base-2 expansion of b must have a 1 in position ℓ and a must have 0 in that position. Indeed, if the most significant digits of a and b are the same, then these digits will cancel each other in the Nim-sum; the position corresponding to $2^{\ell-1}$ is the most significant one where they differ:

$$\mathrm{rep}_2(a) = u0v, \quad \mathrm{rep}_2(b) = u1v', \quad |v| = |v'| = \ell - 1 .$$

But $b = \mathcal{G}_1(\alpha_1) \oplus \cdots \oplus \mathcal{G}_n(\alpha_n)$. Hence, there exists i such that the base-2 expansion of $\mathcal{G}_i(\alpha_i)$ has a 1 in position ℓ. Hence, $d \oplus \mathcal{G}_i(\alpha_i) < \mathcal{G}_i(\alpha_i)$. In the ith game, by definition of the Sprague–Grundy function $\mathcal{G}_i$, there is thus a move from α_i to some position α'_i such that $\mathcal{G}_i(\alpha'_i) = d \oplus \mathcal{G}_i(\alpha_i)$. In the sum of the games, there is a move from $(\alpha_1, \ldots, \alpha_i, \ldots, \alpha_n)$ to $(\alpha_1, \ldots, \alpha'_i, \ldots, \alpha_n)$ and

$$\mathcal{G}_1(\alpha_1) \oplus \cdots \oplus \mathcal{G}_i(\alpha'_i) \oplus \cdots \oplus \mathcal{G}_n(\alpha_n)$$

$$= d \oplus \mathcal{G}_1(\alpha_1) \oplus \cdots \oplus \mathcal{G}_i(\alpha_i) \oplus \cdots \oplus \mathcal{G}_n(\alpha_n)$$

$$= d \oplus b = a.$$

Let us consider the second item and proceed by contradiction. Assume that the position $(\alpha_1, \ldots, \alpha_i, \ldots, \alpha_n)$ has an option $(\alpha_1, \ldots, \alpha'_i, \ldots, \alpha_n)$ with the same $\mathcal{G}$-value. Hence,

$$\mathcal{G}_1(\alpha_1) \oplus \cdots \oplus \mathcal{G}_i(\alpha_i) \oplus \cdots \oplus \mathcal{G}_n(\alpha_n) = \mathcal{G}_1(\alpha_1) \oplus \cdots \oplus \mathcal{G}_i(\alpha'_i) \oplus \cdots \oplus \mathcal{G}_n(\alpha_n)$$

---

[3]The following proof is inspired by the one found in Thomas S. Ferguson's lecture notes on CGT.

and we conclude that $\mathcal{G}_i(\alpha_i) = \mathcal{G}_i(\alpha_i')$ which contradicts the fact that $\mathcal{G}_i$ is a Sprague–Grundy function and that α_i' is an option of α_i (in the ith game). $\qquad\square$

Remark 5.3.21 A game of Nim on p piles of tokens can be seen as the sum of p games of Nim on a single pile. Consequently, if we have p piles with respectively $x_1, \ldots, x_n$ tokens, then the corresponding Grundy value for this position is $\oplus_{i=1}^{p} x_i$. Also, with a single pile, the Grundy value of the position $n \geq 0$ is simply n. Thus, Bouton's theorem can retrospectively be seen as a corollary of Proposition 5.3.18 and Sprague–Grundy theorem.

In Table 5.2, we give the first few values of the Sprague–Grundy sequence $(\mathcal{G}_{NIM}(m, n))_{m,n \geq 0}$ for Nim on two piles. Our interest in this example comes from Exercises 21 and 22 found in [8, Section 16.6, p. 451].

Of course, values in this table can easily be obtained by computing a Nim-sum but we would like to know more about the structure of this table. Can we find some general pattern or recurrence occurring in it? We can also ask the same question for Wythoff's game. The following table can easily be computed using Remark 5.3.10. It will turn out that the analysis of Tables 5.2 and 5.3 are quite different.

Table 5.2 First few values of the Grundy function for Nim on two piles

	0	1	2	3	4	5	6	7	8	9
9	9	8	11	10	13	12	15	14	1	0
8	8	9	10	11	12	13	14	15	0	1
7	7	6	5	4	3	2	1	0	15	14
6	6	7	4	5	2	3	0	1	14	15
5	5	4	7	6	1	0	3	2	13	12
4	4	5	6	7	0	1	2	3	12	13
3	3	2	1	0	7	6	5	4	11	10
2	2	3	0	1	6	7	4	5	10	11
1	1	0	3	2	5	4	7	6	9	8
0	0	1	2	3	4	5	6	7	8	9

Table 5.3 First few values of the Grundy function for Wythoff's game

	0	1	2	3	4	5	6	7	8	9
9	9	10	11	12	8	7	13	14	15	16
8	8	6	7	10	1	2	5	3	4	15
7	7	8	6	9	0	1	4	5	3	14
6	6	7	8	1	9	10	3	4	5	13
5	5	3	4	0	6	8	10	1	2	7
4	4	5	3	2	7	6	9	0	1	8
3	3	4	5	6	2	0	1	9	10	12
2	2	0	1	5	3	4	8	6	7	11
1	1	2	0	4	5	3	7	8	6	10
0	0	1	2	3	4	5	6	7	8	9

It is indeed an old open problem to obtain non-trivial information about the values given in Table 5.3. Not much is known, even for the positions with $\mathcal{G}$-value equal to 1. See Sect. 5.6.2.

5.4   Automatic and Regular Sequences

The question concluding the previous section motivates us for introducing first, k-automatic sequences and then, k-regular sequences ($k \geq 2$ is an integer). The main difference between the two concepts is that the first one is limited to sequences taking values in an alphabet. We will see, in Sect. 5.6 where concepts are generalized to a multidimensional setting, that the bidimensional infinite sequence $(\mathcal{G}_{NIM}(m, n))_{m,n \geq 0}$ is 2-regular. This will provide us with a general construction scheme and recurrence relations for $(\mathcal{G}_{NIM}(m, n))_{m,n \geq 0}$.

There are many good references dealing with automatic and regular sequences. To cite just a few, we mention Allouche and Shallit's book [8], the original paper by Cobham [31] and the survey [17].

Many characterizations of these sequences do exist. A k-automatic sequence can be defined as the image under a coding, i.e., a mapping[4] on a smaller alphabet, of a fixed point of a k-uniform morphism (see Definition 5.4.5). It is the sequence of outputs of a DFA fed with base-k expansions (see Theorem 5.4.12). For every letter, the set of indices corresponding to this letter is defined by a first order formula in $\langle \mathbb{N}, +, V_k \rangle$. This is the so-called Büchi–Bruyère theorem [17]. When k is a prime power, it can be defined in terms of algebraic formal power series (this is the so-called Christol–Kamae–Mendès France–Rauzy theorem [28]). It also appears as a column of the space-time diagram of a cellular automaton [78], see Theorem 5.4.16 below. Finally, it can be defined in terms of finiteness of the its k-kernel (see Theorem 5.4.19).

5.4.1   Generalities on Sequences and Morphisms

Let us take some time to define properly some important notions. We first define a distance turning $A^{\mathbb{N}}$ into a complete ultrametric space for which the notion of convergence is usual. For general references, see, for instance, [8, 27, 62, 72].

Definition 5.4.1 Let $\mathbf{w}, \mathbf{x}$ be two distinct infinite words in $A^{\mathbb{N}}$. We let $\Lambda(\mathbf{w}, \mathbf{x})$ denote the longest common prefix of $\mathbf{w}$ and $\mathbf{x}$ and we define a map $d : A^{\mathbb{N}} \times A^{\mathbb{N}} \to [0, +\infty)$ by

$$d(\mathbf{w}, \mathbf{x}) = 2^{-|\Lambda(\mathbf{w}, \mathbf{x})|} .$$

---

[4]A *coding* is a morphism from A^* to B^* where the image of every letter has length 1.

Note that $|\Lambda(\mathbf{w}, \mathbf{x})| = \inf\{i \in \mathbb{N} \mid w_i \neq x_i\}$. Moreover, we set $d(\mathbf{w}, \mathbf{w}) = 0$. Hence, the longer prefix two words share, the closer they are. It is obvious that $d(\mathbf{w}, \mathbf{x}) > 0$ whenever $\mathbf{w} \neq \mathbf{x}$.

Definition 5.4.2 Let $(\mathbf{w}_n)_{n \geq 0}$ be a sequence of infinite words over the alphabet A. This sequence *converges* to the word $\mathbf{z} \in A^{\mathbb{N}}$ if $d(\mathbf{w}_n, \mathbf{z}) \to 0$ whenever $n \to +\infty$. Otherwise stated, for all $\ell \in \mathbb{N}$, there exists N such that, for all $n \geq N$, $\mathbf{w}_n$ and $\mathbf{z}$ share a common prefix of length at least ℓ. We say that $\mathbf{z}$ is the *limit* of the converging sequence $(\mathbf{w}_n)_{n \geq 0}$.

Let $(w_n)_{n \geq 0}$ be a sequence of finite words over A. If $\#$ is an extra symbol that does not belong to A, then any word $u \in A^*$ is in one-to-one correspondence with the infinite word $u \#^\omega \in (A \cup \{\#\})^{\mathbb{N}}$ where $\#^\omega$ means the concatenation of infinitely many copies of $\#$. We say that the sequence $(w_n)_{n \geq 0}$ *converges* to the infinite word $\mathbf{w}$ if the sequence of infinite words $(w_n \#^\omega)_{n \geq 0}$ converges to $\mathbf{w}$.

Let $f : A^* \to A^*$ be a morphism, i.e., a map such that $f(uv) = f(u)f(v)$ for all words $u, v \in A^*$. Observe that a morphism is completely defined from the images of the letters. Let a be a letter in the alphabet A. If there exists a finite word u such that $f(a) = au$, then

$$\lim_{n \to +\infty} f^n(a) = a\,u\,f(u)\,f^2(u)\,f^3(u)\cdots. \tag{5.1}$$

Note that this limit denoted by $f^\omega(a)$ is well defined in the above topology. It can possibly be equal to a finite word if, for some $k \geq 1$, we have $f^k(u) = \varepsilon$. To avoid this situation, a morphism $f : A^* \to A^*$ is said to be *prolongable* on the letter $a \in A$, not only if there exists a finite word u such that $f(a) = au$ but also, if

$$\lim_{n \to +\infty} |f^n(a)| = +\infty.$$

We will encounter a more general situation, where a second morphism is applied to the infinite word obtained by iterating a first morphism. If $g : A^* \to B^*$ is a *non-erasing* morphism, i.e., for all $a \in A$, $g(a) \neq \varepsilon$, it can be extended to a map from $A^{\mathbb{N}}$ to $B^{\mathbb{N}}$ as follows. If $\mathbf{x} = x_0 x_1 \cdots$ is an infinite word over A, then the sequence of words $(g(x_0 \cdots x_{n-1}))_{n \geq 0}$ converges to an infinite word over B. Its limit is denoted by $g(\mathbf{x}) = g(x_0)g(x_1)g(x_2)\cdots$.

Note that we can always consider non-erasing prolongable morphisms and codings.

Theorem 5.4.3 Let $f : A^* \to A^*$ be a (possibly erasing) morphism that is prolongable on a letter $a \in A$. Let $g : A^* \to B^*$ be a (possibly erasing) morphism. If the word $g(f^\omega(a))$ is infinite, there exists a non-erasing morphism $h : C^* \to C^*$ prolongable on a letter $c \in C$ and a coding $j : C^* \to B^*$ such that $g(f^\omega(a)) = j(h^\omega(c))$.

This result was already stated by Cobham in 1968 [29]. For a proof, see [8]. An alternative short proof is given in [53]. This result is also discussed in details in [21] and [23].

Definition 5.4.4 (Morphic Word) An infinite word obtained by iterating a prolongable morphism is said to be *purely substitutive* or *purely morphic*. In the literature, one also finds the term *pure morphic*. If $\mathbf{x} \in A^{\mathbb{N}}$ is pure morphic and if $g : A \rightarrow B$ is a coding, then the word $\mathbf{y} = g(\mathbf{x})$ is said to be *morphic* or *substitutive*.

5.4.2 Iterating a Constant-Length Morphism

It is now time to present k-automatic sequences. See the original paper of Cobham [30] and the comprehensive book [8].

Definition 5.4.5 (Uniform Morphism) Let $k \geq 2$. A morphism $f : A^* \rightarrow B^*$ satisfying $|f(a)| = k$, for all $a \in A$, is said to be of *constant length k* or *k-uniform*. A 1-uniform morphism is a *coding*.

Definition 5.4.6 Let A, B be two alphabets. An infinite word $\mathbf{w} \in B^{\mathbb{N}}$ is *k-automatic* if there exist a k-uniform morphism $f : A^* \rightarrow A^*$ prolongable on $a \in A$, and a 1-uniform morphism $g : A^* \rightarrow B^*$, such that $\mathbf{w} = g(f^{\omega}(a))$.

The quotient $\lfloor j/k \rfloor$ of the Euclidean division of j by k is denoted by j DIV k. So to speak, for any symbol x_j occurring in $\mathbf{x} = f^{\omega}(a)$, we can track its history, see Fig. 5.1: x_j has been produced by f from $x_{j \text{ DIV } k}$. The latter symbol appears itself in the image by f of $x_{(j \text{ DIV } k) \text{ DIV } k}$, and so on and so forth. Note that if the base-k expansion of j is $\text{rep}_k(j) = c_i \cdots c_1 c_0$, then the base-$k$ expansion of j DIV k is $c_i \cdots c_1$. This simple observation permits one to easily track the past of a given symbol x_j by considering the prefixes of $\text{rep}_k(j)$. In other words, we have the next result.

Lemma 5.4.7 *Let $f : A^* \rightarrow A^*$ be a k-uniform morphism prolongable on a and $\mathbf{x} = x_0 x_1 x_2 \cdots = f^{\omega}(a)$. Let j such that $k^m \leq j < k^{m+1}$, for some $m \geq 0$. Then $j = kq + r$ with $k^{m-1} \leq q < k^m$ and $0 \leq r < k$ and the symbol x_j is the $(r+1)$st symbol occurring in $f(x_q)$.*

The next construction will be crucial in this chapter. In particular, it explains where the term "automatic sequence" comes from. The automaton that we introduce encodes exactly the same information as the morphism f. The fact that the

Fig. 5.1 Tracking a symbol in a k-automatic word

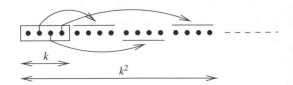

morphism has constant length implies that the DFA is *complete*, i.e., the transition function is defined for all $(b, i) \in A \times \{0, \ldots, k - 1\}$.

Definition 5.4.8 (DFA Associated with a Morphism) We associate with a k-uniform morphism $f : A^* \to A^*$ and a letter $a \in A$, a DFA $\mathcal{A}_f = (A, a, \{0, \ldots, k - 1\}, \delta_f, A)$ where $\delta_f(b, i) = w_{b,i}$ if $f(b) = w_{b,0} \cdots w_{b,k-1}$. We assume that the letter a is clear from the context, if it is not the case, one can use a notation such as $\mathcal{A}_{f,a}$. If a state b has a loop labeled by 0, then the morphism is prolongable on the letter b.

It is a bit tricky, but the alphabet A is the set of states of this automaton.

Example 5.4.9 Consider the morphism f and the associated automaton depicted in Fig. 5.2.

Proposition 5.4.10 *Let* $\mathbf{x} = f^\omega(a) = x_0 x_1 \cdots$ *with* f *a* k-*uniform morphism prolongable on the letter* a. *With the above notation, for the DFA* $\mathcal{A}_f$ *associated with* f, *we have, for all* $j \geq 0$,

$$x_j = \delta_f(a, \mathrm{rep}_k(j)) = \mathcal{A}_f \cdot \mathrm{rep}_k(j) .$$

Proof This is a direct consequence of Lemma 5.4.7. Reading an extra symbol in $\mathcal{A}_f$ corresponds to append a digit on the right (least significant digit) of a base-k expansion. The reader can also look at the proof of Theorem 5.5.6 expressed in a more general setting.   □

The converse also holds.

Proposition 5.4.11 *Let* $(A, a, \{0, \ldots, k - 1\}, \delta, A)$ *be a DFA s.t.* $\delta(a, 0) = a$. *Then the word* $\mathbf{x} = x_0 x_1 x_2 \cdots$ *defined by* $x_j = \delta(a, \mathrm{rep}_k(j))$, *for all* $j \geq 0$, *is the fixed point of a* k-*uniform morphism* f *prolongable on* a *where* $f(b) = \delta(b, 0) \cdots \delta(b, k - 1)$ *for all* $b \in A$.

Proof This is a direct consequence of Lemma 5.4.7.   □

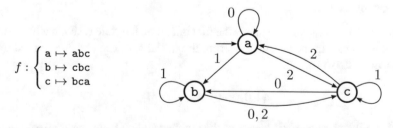

$$f : \begin{cases} a \mapsto abc \\ b \mapsto cbc \\ c \mapsto bca \end{cases}$$

Fig. 5.2 A uniform morphism and its automaton

Theorem 5.4.12 *Let* $\mathbf{w} = w_0w_1w_2\cdots$ *be an infinite word over an alphabet B. It is of the form* $g(f^{\omega}(a))$ *where* $f : A^* \to A^*$ *is a k-uniform morphism prolongable on* $a \in A$ *and* $g : A^* \to B^*$ *is a coding if and only if there exists a DFAO*

$$(A, a, \{0, \ldots, k-1\}, \delta, \mu : A \to B)$$

such that $\delta(a, 0) = a$ *and, for all* $j \geq 0$, $w_j = \mu(\delta(a, \mathrm{rep}_k(j)))$.

A proof of this result can essentially be derived from Lemma 5.4.7. One has simply to put together the previous two propositions and consider an extra coding and thus an output function for the DFA. Nevertheless, we will prove later on (see Theorems 5.5.6, 5.7.26, and Proposition 5.7.27) a general statement that includes the case discussed above.

The next notion will permit us to easily discuss recognizable series.

Definition 5.4.13 (Transition Matrices) With a DFA or a DFAO over the alphabet $A = \{0, \ldots, k-1\}$, we may associate k square matrices of $\mathbb{N}^{A \times A}$, a transition matrix M_i for each label $i < k$. The matrix M_i is defined by

$$[M_i]_{a,b} = \begin{cases} 1 \text{ , if } a \cdot i = b; \\ 0 \text{ , otherwise.} \end{cases}$$

Note that each row of these matrices contains exactly one 1. Multiplying such a matrix on the left by a row vector of the form $(0 \cdots 0\ 1\ 0 \cdots 0)$ gives a vector of the same form.

Example 5.4.14 Let us continue Example 5.4.9. The corresponding three matrices are given by

$$M_0 = \begin{pmatrix} 1 & 0 & 0 \\ 0 & 0 & 1 \\ 0 & 1 & 0 \end{pmatrix} \quad M_1 = \begin{pmatrix} 0 & 1 & 0 \\ 0 & 1 & 0 \\ 0 & 0 & 1 \end{pmatrix}, \quad M_2 = \begin{pmatrix} 0 & 0 & 1 \\ 0 & 1 & 0 \\ 1 & 0 & 0 \end{pmatrix}.$$

The next observation is a prelude to the notion of a linear representation of a recognizable series.

Remark 5.4.15 If the initial state is the first one, then the state reached when reading a word $w_1 \cdots w_\ell$ in the automaton (from the initial state) can be determined by the following matrix product:

$$\begin{pmatrix} 1 & 0 & \cdots & 0 \end{pmatrix} M_{w_1} \cdots M_{w_\ell}.$$

We obtain the characteristic vector of the reached state (all entries are 0 except one). This is easily shown by induction on the length of the input word. If the input word is empty, the matrix product is empty hence equal to the identity matrix and we get the initial characteristic vector. If $\begin{pmatrix} 1 & 0 & \cdots & 0 \end{pmatrix} M_{w_1} \cdots M_{w_i}$ is the characteristic vector of the state s reached when reading $w_1 \cdots w_i$ then the multiplication by $M_{w_{i+1}}$ is equal to $s \cdot w_{i+1}$ by definition of the matrix.

We would like to mention a connection with cellular automata.

Theorem 5.4.16 (Rowland and Yassawi [78]) *A sequence over a finite field $\mathbb{F}_q$ of characteristic p is p-automatic if and only if it occurs as a column of the space-time diagram, with eventually periodic initial conditions, of a linear cellular automaton with memory over $\mathbb{F}_q$.*

5.4.3 k-Kernel of a Sequence

The next characterization of k-automatic sequences will be extensively used in this chapter. It will lead to the notion of k-regular sequence which is based on a property of the module generated by the k-kernel of a sequence. Note that the next notion introduced by Eilenberg [39] is not related to the graph-theoretic notion introduced in Definition 5.3.6 (the standard terminology is a bit unfortunate).

Definition 5.4.17 Let $k \geq 2$. The k-*kernel* of a sequence $\mathbf{x} = (x(n))_{n \geq 0}$ is the set of subsequences:

$$\mathrm{Ker}_k(\mathbf{x}) = \left\{ (x(k^i n + s))_{n \geq 0} \mid i \geq 0,\ 0 \leq s < k^i \right\}.$$

An alternative way to define the k-kernel is first to introduce k operators of k-*decimation*, for $r \in \{0, \ldots, k-1\}$, we set

$$\partial_{k,r}((x(n))_{n \geq 0}) := (x(kn + r))_{n \geq 0}. \tag{5.2}$$

The k-kernel of $\mathbf{x}$ is thus the set of sequences of the form

$$\partial_{k,r_1} \circ \cdots \circ \partial_{k,r_m}((x(n))_{n \geq 0}), \quad r_1, \ldots, r_m \in \{0, \ldots, k-1\}. \tag{5.3}$$

Remark 5.4.18 Notice that considering a subsequence of the form $(x(k^i n + s))_{n \geq 0}$ with $\mathrm{rep}_k(s) = r_m \cdots r_1$, $m \leq i$, corresponds exactly to extracting the subsequence of indices whose base-k expansions have a suffix of length i of the form $0^{i-m} r_m \cdots r_1$. (We assume that expansions of length less than i may be preceded by leading zeroes.)

Theorem 5.4.19 (Eilenberg) *A sequence is k-automatic if and only if its k-kernel is finite.*

A proof can be found in [8, Thm. 6.6.2]. We provide an alternative proof of the fact that automaticity implies finiteness of the kernel.

Table 5.4 The 3-kernel of $f^\omega(a)$ for the morphism given in Example 5.4.9

ε	0	1	2	01	02	10	20	22	010	020	022	201	202	220	0220	2022
a	a	a	b	c	b	c	c	b	a	c	b	a	b	c	a	a
b	b	c	b	c	c	a	c	b	a	b	a	c	b	c	a	b
c	c	b	c	a	b	c	b	a	c	c	b	a	b	c	b	a

Proof We make use of Theorem 5.4.12. Let $\mathbf{x}$ be a k-automatic sequence obtained by feeding a DFAO $\mathcal{M}$ with base k-expansions, i.e., for all $n \geq 0$, $\mathbf{x}(n) = \mathcal{M} \cdot \mathrm{rep}_k(n)$. We first assume that there is no extra coding involved. Words of $\{0, \ldots, k-1\}^*$ act on the set Q of states of $\mathcal{M}$. For all words $u \in \{0, \ldots, k-1\}^*$, we have a map

$$f_u : Q \to Q, \quad q \mapsto q \cdot u.$$

The set of such maps endowed with composition is referred to as the *transition monoid* of $\mathcal{M}$. It is finite because the number of such maps is bounded by $\#Q^{\#Q}$. Let $u \in \{0, \ldots, k-1\}^*$. We observe that the subsequence $(\mathbf{x}(k^{|u|}n + \mathrm{val}_k(u)))_{n \geq 0}$ is obtained as follows

$$\mathbf{x}(k^{|u|}n + \mathrm{val}_k(u)) = \mathcal{M} \cdot (\mathrm{rep}_k(n)u) = f_u(\mathcal{M} \cdot \mathrm{rep}_k(n)).$$

Hence the number of subsequences in the k-kernel is equal to the cardinality of the transition monoid of $\mathcal{M}$ (assuming that every state of $\mathcal{M}$ can be reached from the initial state[5]). If there is an extra coding to be applied to $f^\omega(a)$, since we are considering a mapping onto a smaller alphabet, the number of distinct subsequences decreases. $\qquad \square$

How to compute the transition monoid of a DFA? Assume that the set of states Q is $\{q_1, \ldots, q_n\}$. Then starting with the n-tuple $(q_1, \ldots, q_n)$, we compute $(q_1 \cdot a, \ldots, q_n \cdot a)$ for all letters a. We list the pairwise distinct and newly created n-tuples. For each such tuple, we iterate the process and build a labeled graph where transitions are of the form

$$(r_1, \ldots, r_n) \xrightarrow{a} (r_1 \cdot a, \ldots, r_n \cdot a).$$

This algorithm explores all the n-tuples of the form $(q_1 \cdot w, \ldots, q_n \cdot w)$ where w is a word. Paths from $(q_1, \ldots, q_n)$, in this graph, correspond to the possible maps f_u. In particular, if two words u and v lead to same vertex, then $f_u = f_v$. We apply this algorithm to the automaton depicted in Fig. 5.2 and start with the 3-tuple (a, b, c) (Table 5.4). More details can, for instance, be found in [58].

---

[5]This means that every letter of the alphabet appears at least once in $\mathbf{x}$.

Example 5.4.20 (Thue–Morse Word) The *Thue–Morse word* is the fixed point $f^\omega(a)$ of the morphism $f : a \mapsto ab, b \mapsto ba$,

$$\texttt{abbabaabbaababbabaababbaabbabaab} \cdots$$

Its 2-kernel is finite and contains 2 elements. This word is ubiquitous in COW [7]. For instance, it is well-known that the Thue–Morse word is overlap-free: it does not contain any factor of the form $cvcvc$ where c is a letter and v belongs to $\{a, b\}^*$. In particular, the Thue–Morse word is not ultimately periodic.

5.4.4 Changing the Base

Example 5.3.17 shows that the sequence $(\mathcal{G}(n))_{n \geq 0}$ (when removing $1, 2$ or 4 tokens) is periodic: $(0, 1, 2)^\omega$. It is not difficult to devise a k-automatic sequence equal to $(0, 1, 2)^\omega$. For instance, assume that $k = 2$. We consider the following 2-uniform morphism and coding

$$f : a \mapsto ab, \; b \mapsto cd, \; c \mapsto ef, \; d \mapsto ab, \; e \mapsto cd, \; f \mapsto ef$$

$$g : a, d \mapsto 0, \; b, e \mapsto 1, \; c, f \mapsto 2.$$

More generally, every ultimately periodic word uv^ω, $v \neq \varepsilon$, is k-automatic for all $k \geq 2$.

What could happen when computing the 3-kernel of the Thue–Morse word? It turns out that this 3-kernel is infinite. This can be obtained from a famous result on numeration systems.

Theorem 5.4.21 (Cobham's Theorem [30]) *Let $k, \ell \geq 2$ be two multiplicatively independent integers, i.e., $\log k / \log \ell$ is irrational. If a sequence is both k-automatic and ℓ-automatic, then it is ultimately periodic.*

Proposition 5.4.22 *Let $k \geq 2$ and $n \geq 1$ be integers. A sequence is k-automatic if and only if it is k^n-automatic.*

Cobham's theorem can be considered as difficult. No immediate proof seems[6] to be known [69, 76]. On the other hand, Proposition 5.4.22 is an easy exercise. Each digit in base k^n corresponds to n digits in base k and the image by a morphism preserves the regularity of languages.

Cobham's theorem has been extended to various settings (multidimensional and logical frameworks [17], non-constant length morphism, ...). In the non-constant

---

[6] When writing this chapter, a paper by Thijmen J. P. Krebs appeared on arXiv [55].

length setting, what replaces the base k is the Perron–Frobenius eigenvalue of the matrix associated with the morphism. See the theorem of Cobham–Durand [38].

5.4.5   k-Regular Sequences

We have seen that k-automatic sequences take values in an alphabet. Nevertheless, when studying unbounded sequences (which is the case of most Grundy functions) we need to replace k-automaticity with a more general concept.

Regular sequences (again the usual terminology could be misleading) can be studied in a more general algebraic setting than the one considered here. Since, we are exclusively dealing with sequences taking integer entries, e.g. $(\mathcal{G}_{NIM}(m, n))_{m,n \geq 0}$, we will limit ourselves a bit and have a simple presentation with Definition 5.4.23. In this section, we still consider the case of infinite words. The multidimensional setting will be presented later on.

Definition 5.4.23 The sequence $\mathbf{x} \in \mathbb{Z}^{\mathbb{N}}$ is k-*regular*, if the $\mathbb{Z}$-module generated by $\mathrm{Ker}_k(\mathbf{x})$ is finitely generated, i.e., there exists $\mathbf{t}_1, \ldots, \mathbf{t}_\ell \in \mathbb{Z}^{\mathbb{N}}$ such that

$$\langle \mathrm{Ker}_k(\mathbf{x}) \rangle = \langle \mathbf{t}_1, \ldots, \mathbf{t}_\ell \rangle. \tag{5.4}$$

The ring $\mathbb{Z}$ is embedded in fields such as $\mathbb{Q}$, $\mathbb{R}$ or $\mathbb{C}$. Thus sequences we are considering can be seen as elements of $\mathbb{Q}^{\mathbb{N}}$ which is a $\mathbb{Q}$-vector space (instead of considering a $\mathbb{Z}$-module). So another way to describe k-regularity is to say that the orbit of $\mathbf{x}$ under the action of compositions of the operators (5.2) of k-decimation $\partial_{k,r}$ remains in a finite dimensional vector space.

Remark 5.4.24 The original definition given by Allouche and Shallit [6] is the following one. Let $\mathfrak{R}$ be a ring containing a commutative Noetherian ring $\mathfrak{R}'$, i.e., every ideal of $\mathfrak{R}'$ is finitely generated. A sequence $\mathbf{x}$ in $\mathfrak{R}^{\mathbb{N}}$ is $(\mathfrak{R}',k)$-*regular*, if there exists $\mathbf{t}_1, \ldots, \mathbf{t}_\ell \in \mathfrak{R}^{\mathbb{N}}$ such that every sequence in $\mathrm{Ker}_k(\mathbf{x})$ is an $\mathfrak{R}'$-linear combination of $\mathbf{t}_1, \ldots, \mathbf{t}_\ell$.

Observe that if $\mathbf{x}$ is $(\mathfrak{R}',k)$-regular, then $\langle \mathrm{Ker}_k(\mathbf{x}) \rangle$ is a submodule of a finitely generated $\mathfrak{R}'$-module. In general, this does not imply that the submodule itself is finitely generated. This means that we only have the inclusion $\langle \mathrm{Ker}_k(\mathbf{x}) \rangle \subset \langle \mathbf{t}_1, \ldots, \mathbf{t}_\ell \rangle$ instead of equality in (5.4). Nevertheless, since $\mathfrak{R}'$ is assumed to be Noetherian, one can show (using more algebra) that every submodule of a finitely generated $\mathfrak{R}'$-module is finitely generated and thus $\langle \mathrm{Ker}_k(\mathbf{x}) \rangle$ is finitely generated. From that (up to taking other generators), we can assume equality in (5.4). Dealing with integer sequences, we can take $\mathfrak{R} = \mathfrak{R}' = \mathbb{Z}$.

Similarly to recognizable formal series, with every k-regular sequence $(s(n))_{n \geq 0} \in \mathbb{Z}^{\mathbb{N}}$ is associated a *linear representation* (λ, μ, ν). There exist a positive integer r, a row vector $\lambda \in \mathbb{Z}^{1 \times r}$ and a column vector $\nu \in \mathbb{Z}^{r \times 1}$, a matrix-valued morphism $\mu : \{0, \ldots, k - 1\} \to \mathbb{Z}^{r \times r}$ such that

$$s(n) = \lambda \, \mu(c_0 \cdots c_\ell) \, v = \lambda \, \mu(c_0) \cdots \mu(c_\ell) \, v$$

for all $c_\ell, \ldots, c_0 \in \{0, \ldots, k - 1\}^*$ such that $\mathrm{val}_k(c_\ell \cdots c_0) = \sum_{i=0}^{\ell} c_i k^i = n$. The converse also holds: if there exists a linear representation associated with the canonical k-ary expansion of integers (one has to take into account the technicality of representations starting with leading zeroes), then the sequence is k-regular. See, for instance, [8, Theorem 16.2.3]. As a corollary, the nth term of a k-regular sequence can be computed with $\lfloor \log_k(n) \rfloor$ matrix multiplications.

Remark 5.4.25 These matrix multiplications are a natural extension of those encountered in Remark 5.4.15. Instead of a classical DFA, the linear representation can be seen as a *weighted automaton* with r states.

Proof *(Existence of a Linear Representation)* Let $\mathbf{s} = (s(n))_{n \geq 0} \in \mathbb{Z}^{\mathbb{N}}$ be a k-regular sequence. By definition, there exists a finite number of sequences $\mathbf{t}_1, \ldots, \mathbf{t}_\ell$ such that $\langle \mathrm{Ker}_k(\mathbf{s}) \rangle = \langle \mathbf{t}_1, \ldots, \mathbf{t}_\ell \rangle$. In particular, each $\mathbf{t}_j$ is a $\mathbb{Z}$-linear combination of elements in the k-kernel of $\mathbf{s}$. We have finitely many $\mathbf{t}_j$'s, so $\mathbf{t}_1, \ldots, \mathbf{t}_\ell$ are $\mathbb{Z}$-linear combinations of finitely many elements in $\mathrm{Ker}_k(\mathbf{s})$. Thus we can assume that $\langle \mathrm{Ker}_k(\mathbf{s}) \rangle$ is generated by finitely many elements from $\mathrm{Ker}_k(\mathbf{s})$ itself. Without loss of generality, we assume from now on that $\mathbf{t}_1, \ldots, \mathbf{t}_\ell$ belong to $\mathrm{Ker}_k(\mathbf{s})$.

From (5.3), for all $r \in \{0, \ldots, k - 1\}$ and all $i \in \{1, \ldots, \ell\}$, $\partial_{k,r}(\mathbf{t}_i)$ is a sequence in $\mathrm{Ker}_k(\mathbf{s})$ and thus, there exist coefficients $(A_r)_{1,i}, \ldots, (A_r)_{\ell,i}$ such that

$$\partial_{k,r}(\mathbf{t}_i) = \sum_{j=1}^{\ell} (A_r)_{j,i} \, \mathbf{t}_j \, .$$

Notice that A_r is an $\ell \times \ell$ matrix. Roughly, if we were in a vector space setting, this means that the matrices A_r represent the linear operators $\partial_{k,r}$ in the basis $\mathbf{t}_1, \ldots, \mathbf{t}_\ell$. Let $p \geq 0$ be an integer. Notice that if $\mathrm{rep}_k(p) = r_m \cdots r_0$, then $s(p)$ is the first term, i.e., corresponding to the index $n = 0$, of the sequence

$$(s(b^{m+1}n + p))_{n \geq 0} = \partial_{k,r_0} \circ \cdots \circ \partial_{k,r_m} \big((s(n))_{n \geq 0} \big) \, .$$

We will use the fact that $\partial_{k,r}$ is linear, i.e., if α, β are coefficients and $\mathbf{v}, \mathbf{w}$ are two sequences, then $\partial_{k,r}(\alpha \mathbf{v} + \beta \mathbf{w}) = \alpha \partial_{k,r}(\mathbf{v}) + \beta \partial_{k,r}(\mathbf{w})$. It is easy to see that

$$\partial_{k,r_0} \circ \cdots \circ \partial_{k,r_m}(\mathbf{t}_i) = \sum_{j=1}^{\ell} (A_{r_0} \cdots A_{r_m})_{j,i} \, \mathbf{t}_j \, .$$

If we have the following decomposition of $\mathbf{s}$ (in a vector space setting, we would have a unique decomposition of $\mathbf{s}$ in the basis $\mathbf{t}_1, \ldots, \mathbf{t}_\ell$)

$$\mathbf{s} = \sum_{i=1}^{\ell} \sigma_i \, \mathbf{t}_i$$

then, by linearity,

$$(s(b^{m+1}n + p))_{n \geq 0} = \sum_{i=1}^{\ell} \sigma_i \sum_{j=1}^{\ell} (A_{r_0} \cdots \cdot A_{r_m})_{j,i} \, (\mathbf{t}_j(n))_{n \geq 0} = \sum_{j=1}^{\ell} \tau_j \, (\mathbf{t}_j(n))_{n \geq 0}$$

where

$$\begin{pmatrix} \tau_1 \\ \vdots \\ \tau_\ell \end{pmatrix} = A_{r_0} \cdots \cdot A_{r_m} \begin{pmatrix} \sigma_1 \\ \vdots \\ \sigma_\ell \end{pmatrix}.$$

Consequently, $s(p)$ is obtained as

$$s(p) = \sum_{i=1}^{\ell} \tau_i \, \mathbf{t}_i(0) = \underbrace{\begin{pmatrix} t_1(0) & \cdots & t_\ell(0) \end{pmatrix}}_{=:\lambda} A_{r_0} \cdots \cdot A_{r_m} \underbrace{\begin{pmatrix} \sigma_1 \\ \vdots \\ \sigma_\ell \end{pmatrix}}_{=:\nu}$$

and $\mu(c) = A_c$. $\qquad\qquad\qquad\qquad\qquad\qquad\qquad\qquad\qquad\qquad\qquad \Box$

Example 5.4.26 For the *sum-of-digits function* given by

$$s_2(n) = \sum_{i=0}^{\ell} n_i \quad \text{whenever } \operatorname{rep}_2(n) = n_\ell \cdots n_0,$$

the sequence $\mathbf{s} = (s_2(n))_{n \geq 0}$ has a (base-2) linear representation given by

$$\lambda = \begin{pmatrix} 0 & 1 \end{pmatrix}, \quad \mu(i) = \begin{pmatrix} 1 & 0 \\ i & 1 \end{pmatrix}, \quad \nu = \begin{pmatrix} 1 \\ 0 \end{pmatrix}, \quad i = 0, 1.$$

We let $\mathbf{1}$ denote the constant sequence $(1)_{n \geq 0}$. It does not belong to the 2-kernel of $\mathbf{s}$ but it belongs to the $\mathbb{Z}$-module generated by it because it is equal to $\partial_{2,1}(\mathbf{s}) - \mathbf{s}$. Nevertheless, it is enough to see that $\partial_{2,0}(\mathbf{1}) = \partial_{2,1}(\mathbf{1}) = \mathbf{1}$ and take $\mathbf{s}$ and $\mathbf{1}$ as generators to proceed as in the proof above. From the following relations, we deduce the two columns of the matrix $\mu(0)$

$$\partial_{2,0}(\mathbf{s}) = 1 \cdot \mathbf{s} + 0 \cdot \mathbf{1}, \quad \partial_{2,0}(\mathbf{1}) = 0 \cdot \mathbf{s} + 1 \cdot \mathbf{1}$$

and, for $\mu(1)$,

$$\partial_{2,1}(\mathbf{s}) = 1 \cdot \mathbf{s} + 1 \cdot \mathbf{1}, \quad \partial_{2,1}(\mathbf{1}) = 0 \cdot \mathbf{s} + 1 \cdot \mathbf{1}.$$

The vector λ is given by $s_2(0) = 0$ and $\mathbf{1}(0) = 1$. The vector ν is obtained from $\mathbf{s} = 1 \cdot \mathbf{s} + 0 \cdot \mathbf{1}$. To compute $s_2(19)$, observe that $\mathrm{rep}_2(19) = 10011$. Thus we compute

$$\begin{pmatrix} 0 & 1 \end{pmatrix} \mu(1)\mu(1)\mu(0)\mu(0)\mu(1) \begin{pmatrix} 1 \\ 0 \end{pmatrix} = 3.$$

Remark 5.4.27 In [9, Section 6], a practical procedure to guess relations a possibly k-regular sequence will satisfy is described. Consider a sequence $(s(n))_{n \geq 0}$. The idea is to construct a matrix in which the rows represent truncated versions of elements of the k-kernel of $(s(n))_{n \geq 0}$, together with row reduction. Start with a matrix having a single row, let us say corresponding to the first m elements of the sequence. Then, in view of the first paragraph of the proof providing the existence of a linear representation, repeatedly add subsequences of the form $(s(k^\ell n + r))_{n \geq 0}$ not linearly dependent of the previous stored sequences. From this, you have candidate relations that remain to be proved.

Remark 5.4.28 There is an intermediate notion of a *k-synchronized* sequence s where a DFA, in the sense discussed in Sect. 5.6, accepts pairs of base-k expansions corresponding to $(n, s(n))$. See [18].

5.5 Characterizing $\mathcal{P}$-Position of Wythoff's Game in Polynomial Time

First let us explain what we mean by "the Fibonacci word is *coding* the $\mathcal{P}$-positions of Wythoff's game". From that result, we will introduce Zeckendorf [88] (or Fibonacci) numeration system and reobtain a well-known result by Fraenkel, Theorem 5.5.20, characterizing $\mathcal{P}$-positions of Wythoff's game from Fibonacci expansions of integers. In this section, we will deal with two copies of the Fibonacci word! One is defined as a map whose domain is $\mathbb{N}$ and the other one has domain $\mathbb{N}_{>0}$, we will try to avoid any confusion using notation $\mathbf{f}$ and $\mathbf{f}'$ respectively. The alphabet of $\mathbf{f}$ and $\mathbf{f}'$ can be either $\{a, b\}$ or, $\{1, 2\}$. It does not really matter.

Definition 5.5.1 (Fibonacci Word) The *Fibonacci word* over $\{a, b\}$

$$\mathbf{f} = \mathtt{abaababaabaababaababaaba} \cdots$$

is the unique fixed point of the morphism $\mathcal{F} : a \mapsto ab, b \mapsto a$.

A (homogeneous) *Beatty sequence* is just the sequence of integer part of positive multiples of an irrational number. In our setting, Theorem 5.3.14 provides us with a pair of *complementary* Beatty sequences because $\frac{1}{\varphi} + \frac{1}{\varphi^2} = 1$. Thus $(\lfloor n\varphi \rfloor)_{n \geq 1}$ and $(\lfloor n\varphi^2 \rfloor)_{n \geq 1}$ make a partition of $\mathbb{N}_{>0}$. This is a consequence of Beatty theorem (sometimes known as Rayleigh theorem). Regarding this result, papers of interest are [41, 68].

Theorem 5.5.2 (Beatty Theorem) *Let* $\alpha, \beta > 1$ *be irrational numbers such that* $1/\alpha + 1/\beta = 1$. *Then* $(\lfloor n\alpha \rfloor)_{n \geq 1}$ *and* $(\lfloor n\beta \rfloor)_{n \geq 1}$ *make a partition of* $\mathbb{N}_{>0}$.

There are various characterizations of Sturmian words (the Fibonacci word is Sturmian). One of those is given in terms of irrational mechanical words, i.e., the sequence of first differences of a Beatty sequence. See, for details, [63, Sec. 2.1.2]. Also see [40]. Here we start indexing infinite words with 1, because of Beatty theorem and the first non-zero $\mathcal{P}$-position of Wythoff is $(1, 2)$. In particular, using those mechanical words, one can show that the nth letter in the Fibonacci word

$$\mathbf{f'} = f_1 f_2 f_3 \cdots = 212212122122 \cdots \tag{5.5}$$

over $\{1, 2\}$ can be obtained as

$$f_n = \lfloor (n + 1)\,\varphi \rfloor - \lfloor n\,\varphi \rfloor, \quad \forall n \geq 1.$$

Wythoff's result, Theorem 5.3.14, can therefore be restated as follows.

Proposition 5.5.3 *The nth* $\mathcal{P}$-*position* $(\lfloor n\varphi \rfloor, \lfloor n\varphi^2 \rfloor)$ *of Wythoff's game is given by the indices (starting with 1) of the nth* a *and nth* b *in the Fibonacci word* $\mathbf{f'}$.

Proof Since $f_n = \lfloor (n+1)\,\varphi \rfloor - \lfloor n\,\varphi \rfloor$, if $f_n = 1$, this means $\lfloor n\,\varphi \rfloor$ and $\lfloor (n+1)\,\varphi \rfloor$ are consecutive positive integers. If $f_n = 2$, this means that there exists m such that $\lfloor n\varphi \rfloor$, $\lfloor m\varphi^2 \rfloor$ and $\lfloor (n+1)\,\varphi \rfloor$ are three consecutive integers. Indeed, because of Beatty theorem, the gap between two consecutive multiples of φ must be filled with a multiple of φ^2.

Apply the morphism $1 \mapsto$ a and $2 \mapsto$ ab to $\mathbf{f'} = 21221 \cdots$. We get exactly the Fibonacci word over $\{$a, b$\}$ because the Fibonacci word is the fixed point of the morphism $\mathcal{F}$: a $\mapsto$ ab, b $\mapsto$ a and we may replace $1, 2$ by b and a respectively. The effect of $2 \mapsto$ ab is to insert an extra symbol b between two a's (the images of 1 and 2 both start with a). Thus the multiples $\lfloor n\varphi \rfloor$ are exactly given by the indices of the a's in the Fibonacci word $\mathbf{f'}$. The remaining indices (filling the partition of $\mathbb{N}_{>0}$) corresponding to b's are thus given by the multiples $\lfloor n\varphi^2 \rfloor$. $\qquad \square$

5.5.1 From Morphic Words to Automatic Words

Since the Fibonacci morphism $\mathcal{F}$ is not of constant length (it is not uniform in the sense of Definition 5.4.5), it is time to extend Definition 5.4.8. Note that the automaton defined below is not necessarily complete: the transition function could be partial.

Definition 5.5.4 We associate with a morphism $f : A^* \to A^*$ and a letter $a \in A$, a DFA

$$\mathcal{A}_f = (A, a, \{0, \ldots, \max_{b \in A} |f(b)| - 1\}, \delta_f, A)$$

where $\delta_f(b, i) = w_{b,i}$ if $f(b) = w_{b,0} \cdots w_{b,|f(b)|-1}$. We assume that the letter a is clear from the context, if it is not the case, one can use a notation such as $\mathcal{A}_{f,a}$.

Definition 5.5.5 (Genealogical Order) Let $(A, <)$ be an ordered alphabet. We can order A^* using the *genealogical ordering* (also called *radix order*). Words are first ordered by increasing length and for words of the same length, one uses the lexicographic ordering induced by the order $<$ on A. The order is denoted by $<_{gen}$.

Compared with the lexicographic order, the genealogical order is a *well-order*, that is, every non-empty subset has a minimal element. For instance, the set $a^+b = \{a^n b \mid n > 0\}$ has no minimal element for the lexicographic order.

Here is the natural generalization of Proposition 5.4.10, base-k expansions are replaced with words accepted by $\mathcal{A}_f$. We will proceed to the proof of this result at the end of this section.

Theorem 5.5.6 *Let $(A, <)$ be an ordered alphabet. Let $\mathbf{x} \in A^{\mathbb{N}}$ be an infinite word, fixed point $f^\omega(a)$ of a morphism $f : A^* \to A^*$ prolongable on a. Consider the language L_f of words accepted by $\mathcal{A}_f$ except those starting with 0. If L_f is genealogically ordered: $L_f = \{w_0 <_{gen} w_1 <_{gen} w_2 <_{gen} \cdots\}$, then the nth symbol of $\mathbf{x}$, $n \geq 0$, is the state $\mathcal{A}_f \cdot w_n$.*

Remark 5.5.7 Adding an extra coding and considering the word $g(f^\omega(a))$ does not lead to any difficulty. The DFA is replaced with a DFAO where the output function is simply g. Thus, the nth symbol of $g(f^\omega(a))$ is given by $g(\mathcal{A}_f \cdot w_n)$. See Example 5.5.10.

Let us apply this theorem to the Fibonacci word $\mathbf{f}$. The corresponding automaton is depicted in Fig. 5.3. The language accepted by this DFA (excluding words

Fig. 5.3 DFA associated
with a $\mapsto$ ab, b $\mapsto$ a

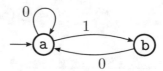

starting with 0) is $1\{0, 01\}^* \cup \{\varepsilon\}$. This is the set of words over $\{0, 1\}$ avoiding the factor 11. This is exactly the language of the Fibonacci numeration: the greedy expansions of the non-negative integers in this positional numeration system. Details on numeration systems are presented in the next section.

Remark 5.5.8 For a morphism f of constant length k, the language L_f given by Theorem 5.5.6 is $\{1, \ldots, k-1\}\{0, \ldots, k-1\}^* \cup \{\varepsilon\}$ which is exactly the language of the base-k numeration system. For the Fibonacci morphism $\mathcal{F}$, the corresponding language is associated with the so-called Fibonacci numeration system.

Nevertheless, the language L_f given by Theorem 5.5.6 is not always related to a "well-known" numeration system. The term "well-known" means associated with a positional numeration system defined in Sect. 5.5.2. In the general situation: genealogically ordering a regular language leads to abstract numeration systems discussed in Sect. 5.7.1.

Prior to the proof of Theorem 5.5.6, we need the following lemma. If $\mathbf{x} = x_0 x_1 x_2 \cdots$ is an infinite word, then the shifted word $\sigma(\mathbf{x})$ is the word $x_1 x_2 x_3 \cdots$.

Lemma 5.5.9 *Let $A = \{a_1 < \cdots < a_n\}$ be a totally ordered alphabet. Let $z \notin Q$. Let $\mathcal{A} = (Q, q_0, A, \delta_A, F)$ be a DFA where $\delta_A : Q \times A \to Q$ is (in general) a partial[7] function. Define the morphism $\psi_A : (Q \cup \{z\})^* \to (Q \cup \{z\})^*$ by $\psi_A(z) = z q_0$ and, for all $q \in Q$,*

$$\psi_A(q) = \delta_A(q, a_1) \cdots \delta_A(q, a_n).$$

In this latter expression if $\delta_A(q, a_i)$ is not defined for some i, then it is replaced by ε. Let L be the regular language accepted by (Q, q_0, A, δ, Q) where all states of $\mathcal{A}$ are final. Then the shifted word $\sigma(\psi_A^\omega(z))$ is the sequence $(x_n)_{n \in \mathbb{N}}$ of the states reached in $\mathcal{A}$ by the words of L in genealogical order, i.e., for all $n \in \mathbb{N}$,

$$x_n = \mathcal{A} \cdot w_n$$

where w_n is the $(n+1)$th word of the genealogically ordered language L.

Proof By definition of ψ, first observe that we have the following factorization, see (5.1):

$$\psi_A^\omega(z) = z x_0 x_1 x_2 \cdots = z q_0 \psi_A(q_0) \psi_A^2(q_0) \cdots$$

and $x_0 = q_0 = \delta_A(q_0, \varepsilon)$. Then by the definition of ψ_A, if $x_n = \delta_A(q_0, w_n), n \geq 0$, then the factor

$$u_n = \psi_A(x_n) = \delta_A(q_0, w_n a_1) \cdots \delta_A(q_0, w_n a_n) \tag{5.6}$$

---

[7] Compared with complete functions in Definition 5.4.8.

appears in $\psi_{\mathcal{A}}^{\omega}(z)$ with the usual convention of replacing the undefined transitions with ε. Indeed, $zx_0x_1x_2\cdots$ is a fixed point of $\psi_{\mathcal{A}}$ and each x_n produces a factor $\psi_{\mathcal{A}}(x_n) = u_n$ appearing later on in the infinite word, similar to the situation depicted in Fig. 5.1. Moreover this factor is preceded by

$$\delta_{\mathcal{A}}(q_0, w_{n-1}a_1) \cdots \delta_{\mathcal{A}}(q_0, w_{n-1}a_n)$$

and followed by

$$\delta_{\mathcal{A}}(q_0, w_{n+1}a_1) \cdots \delta_{\mathcal{A}}(q_0, w_{n+1}a_n).$$

It is therefore clear that we get all states reached from the initial state when considering the labels of all the paths in $\mathcal{A}$ in increasing genealogical order. □

Proof of Theorem 5.5.6 From Cobham's theorem on morphic words (Theorem 5.4.3), we assume that f is non-erasing. We consider the alphabet $C = \{0, \ldots, \max_{b \in A} |f(b)| - 1\}$ and the DFA $\mathcal{A}_f = (A, a, C, \delta, A)$ from Definition 5.5.4 having a as initial state.

Let $L \subseteq C^*$ be the language recognized by $\mathcal{A}_f$. Since $f(a) = au$ for some non-empty word u, it is clear that if $w \in L$ then $0w \in L$. Indeed by definition of $\mathcal{A}_f$, its initial state a has a loop labeled by 0, the first letter in C. If we apply Lemma 5.5.9 to this automaton $\mathcal{A}_f$, we obtain a morphism $\psi_{\mathcal{A}_f}$ generating the sequence of the states reached by the words of L. This morphism is defined as follows. Let $z \notin A$. We have $\psi_{\mathcal{A}_f}(z) = za$ and, for all $b \in A$, $\psi_{\mathcal{A}_f}(b) = f(b)$.

The main point leading to the conclusion is to compare $\psi_{\mathcal{A}_f}^{\omega}(z)$ and $f^{\omega}(a)$. Since $f(a) = au$, using (5.1) we have the following factorizations

$$f^{\omega}(a) = au\ f(u)\ f^2(u)\ f^3(u) \cdots$$

and

$$\psi_{\mathcal{A}_f}^{\omega}(z) = za\ a\ u\ f(a)\ f(u)\ f^2(a)\ f^2(u)\ f^3(a)\ f^3(u) \cdots.$$

If we erase the factors z, a, $f(a)$, $f^2(a)$, $\ldots$ occurring in that order in the above factorization of $\psi_{\mathcal{A}_f}^{\omega}(z)$, we recover $f^{\omega}(a)$. Recall that $\psi_{\mathcal{A}_f}^{\omega}(z)$ is, except for z, the sequence of states reached in $\mathcal{A}_f$ by considering all the possible paths in genealogical order. The second occurrence of a in $\psi_{\mathcal{A}_f}^{\omega}(z)$ is the state reached in $\mathcal{A}_f$ when reading $0 \in L$. By the property (5.6) of $\psi_{\mathcal{A}_f}$, the factor $f^n(a)$ in the above factorization corresponds to the states reached in $\mathcal{A}_f$ when reading the words in L of length $n + 1$ starting with 0. Consequently, when giving to $\mathcal{A}_f$ the words of $L_f := L \setminus 0C^*$ in increasing genealogical order, we build exactly the sequence $f^{\omega}(a) = (y_n)_{n \geq 0}$, i.e., if $w_0 <_{gen} w_1 <_{gen} w_2 <_{gen} \cdots$ are the words of L_f in genealogical order, then $y_n = \mathcal{A} \cdot w_n$. □

Example 5.5.10 Consider the alphabets $A = \{a, b, c\}$, $B = \{d, e\}$ and the morphisms

$$f : A^* \to A^*, \begin{cases} a \mapsto abc \\ b \mapsto bc \\ c \mapsto aac \end{cases} \quad \text{and} \quad g : A^* \to B^*, \begin{cases} a \mapsto d \\ b, c \mapsto e. \end{cases}$$

The corresponding automaton $\mathcal{A}_f$ is given in Fig. 5.4 and the output function is represented on the outgoing arrows.

The infinite word generated by the morphism f is $f^\omega(a) = (x_n)_{n \geq 0} = $ abcbcaacbcaacabcabcaac $\cdots$.

Corollary 5.5.11 *Let $(A, <)$ be an ordered alphabet. Let $\mathbf{x} \in A^{\mathbb{N}}$ be an infinite word, fixed point $f^\omega(a)$ of a morphism $f : A^* \to A^*$ prolongable on a. Consider the language $L_f = \{w_0 <_{gen} w_1 <_{gen} w_2 <_{gen} \cdots \}$ of words accepted by $\mathcal{A}_f$ except those starting with 0.*

Let $n \geq 0$. If $\mathcal{A}_f \cdot w_n = b$ and $|f(b)| = r_b$, then the factor $f(b)$ occurs in $\mathbf{x}$ in position corresponding to $w_n 0, \ldots, w_n(r_b - 1)$.

In Example 5.5.10, the fifth b occurring in $f^\omega(a)$ in position 17 corresponds to the word 211 in L_f. Deleting the last one gives the word 21 which corresponds to the third a occurring in $f^\omega(a)$. In particular, this symbol a gives the factor $f(a) = abc$ corresponding to the words 210, 211, 212. Finally, the first c corresponds to the word 2, so the second and third a occur in $f(c)$.

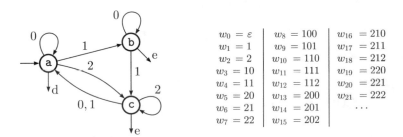

$w_0 = \varepsilon$	$w_8 = 100$	$w_{16} = 210$
$w_1 = 1$	$w_9 = 101$	$w_{17} = 211$
$w_2 = 2$	$w_{10} = 110$	$w_{18} = 212$
$w_3 = 10$	$w_{11} = 111$	$w_{19} = 220$
$w_4 = 11$	$w_{12} = 112$	$w_{20} = 221$
$w_5 = 20$	$w_{13} = 200$	$w_{21} = 222$
$w_6 = 21$	$w_{14} = 201$	$\cdots$
$w_7 = 22$	$w_{15} = 202$	

Fig. 5.4 The automaton $\mathcal{A}_f$ associated with a $\mapsto$ abc, b $\mapsto$ bc, c $\mapsto$ aac and the first few words in the associated language L_f

5.5.2 Positional Numeration Systems

A *numeration system* is a sequence $U = (U_n)_{n\geq 0}$ of integers such that U is increasing, $U_0 = 1$ and that the set $\{U_{i+1}/U_i \mid i \geq 0\}$ is bounded. This latter condition ensures that there exists an alphabet of digits used to represent integers. If $w = w_\ell \cdots w_0$ is a word over an alphabet $A \subset \mathbb{Z}$ then the numerical value of w is

$$\mathrm{val}_U(w) = \sum_{i=0}^{\ell} w_i\, U_i.$$

Using a greedy algorithm [43], every integer n has a unique *(normal) U-representation* or *U-expansion* $\mathrm{rep}_U(n) = w_\ell \cdots w_0$ which is a finite word over a minimal alphabet called the *canonical alphabet* of U and denoted by A_U. The normal U-representation satisfies

$$\mathrm{val}_U(\mathrm{rep}_U(n)) = n \text{ and for all } i \in \{0, \ldots, \ell - 1\}, \ \mathrm{val}_U(w_i \cdots w_0) < U_{i+1}.$$

Remark 5.5.12 We call these systems *positional* because the position of a digit within an expansion is relevant. A digit in ith position ($i = 0$ corresponding to the rightmost digit) is multiplied by U_i to get the numerical value of the expansion.

For some general references on numeration systems, see Frougny's chapter in [63] or [72, 73]. The greediness of the expansion has the following consequence.

Proposition 5.5.13 (Order Preserving Map) *Let m, n be two non-negative integers. We have $m < n$ if and only if $\mathrm{rep}_U(m)$ is genealogically less than $\mathrm{rep}_U(n)$.*

Definition 5.5.14 Let $B \subset \mathbb{Z}$ be an alphabet. If $w \in B^*$ is such that $\mathrm{val}_U(w) \geq 0$, then the function that maps w to $\mathrm{rep}_U(\mathrm{val}_U(w))$ is called *normalization*.

Definition 5.5.15 A numeration system U is said to be *linear* if there exist $k \in \mathbb{N} \setminus \{0\}, d_1, \ldots, d_k \in \mathbb{Z}, d_k \neq 0$, such that, for all $n \geq k$, $U_n = d_1 U_{n-1} + \cdots + d_k U_{n-k}$. The polynomial $P_U(X) = X^k - d_1 X^{k-1} - \cdots - d_{k-1}X - d_k$ is called the *characteristic polynomial* of U.

Definition 5.5.16 Recall that a *Pisot-(Vijayaraghavan) number* is an algebraic integer $\beta > 1$ whose Galois conjugates have modulus less than 1. We say that $U = (U_n)_{n\geq 0}$ is a *Pisot numeration system* if the numeration system U is linear and $P_U(X)$ is the minimal polynomial of a Pisot number β. Integer[8] base numeration systems are particular cases of Pisot systems. For instance, see [16] where it is shown that most properties related to integer base systems, can be extended to Pisot systems. For a Pisot system β, there exists some $c > 0$ such that $|U_n - c\beta^n| \to 0$, as n tends to infinity.

---

[8]Integer are the only rational numbers that are Pisot numbers.

Example 5.5.17 With the Fibonacci sequence $F = 1, 2, 3, 5, 8, 13, \ldots$, we have a Pisot numeration system associated with the Golden ratio. Thanks to the greediness of the expansions, we have $\mathrm{rep}_F(\mathbb{N}) = 1\{0, 01\}^* \cup \{\varepsilon\}$.

Thanks to the observation stated in Proposition 5.5.13, we can directly reformulate Theorem 5.5.6.

Corollary 5.5.18 *Let $(A, <)$ be an ordered alphabet. Let $\mathbf{w} \in A^{\mathbb{N}}$ be an infinite word, fixed point $f^{\omega}(a)$ of a morphism $f : A^* \to A^*$ prolongable on a. Let L_f be the language of words accepted by $\mathcal{A}_f$ except those starting with 0. If there exists a numeration system U such that $\mathrm{rep}_U(\mathbb{N}) = L_f$, then the nth symbol of $\mathbf{w}$, $n \geq 0$, is $\mathcal{A}_f \cdot \mathrm{rep}_U(n)$.*

Remark 5.5.19 The above corollary implies that if the state $a \in A$ is reached in $\mathcal{A}_f$ when reading a word w from the initial state, then the symbol with index $\mathrm{val}_U(w)$ occurring in the fixed point $\mathbf{w}$ is a. Moreover, the ordered outgoing transitions from state a with respective labels $0, \ldots, |f(a)| - 1$ are such that $f(a) = (a \cdot 0) \cdots (a \cdot (|f(a)| - 1))$.

Let us apply this result to the Fibonacci word $\mathbf{f} = f_0 f_1 f_2 \cdots = \mathtt{abaab} \cdots$ starting with index 0. It is important to note that in Theorem 5.5.6 and the above corollary, indices start with 0 (which is not the case in Proposition 5.5.3 and this difference has to be dealt with). Applying the above corollary with the DFA depicted in Fig. 5.3, we get that

- $f_j = \mathtt{a}$ if and only if $\mathrm{rep}_F(j)$ ends with 0;
- $f_j = \mathtt{b}$ if and only if $\mathrm{rep}_F(j)$ ends with 1.

As for k-automatic sequences and Lemma 5.4.7, we can still "keep track of the past": since $\mathcal{F} : \mathtt{a} \mapsto \mathtt{ab}, \mathtt{b} \mapsto \mathtt{a}$, we have

$$f_j = \mathtt{a} \text{ for some } \mathrm{rep}_F(j) = u \text{ if and only if } f_{\mathrm{val}_F(u0)} = \mathtt{a} \text{ and } f_{\mathrm{val}_F(u1)} = \mathtt{b}$$

and similarly,

$$f_j = \mathtt{b} \text{ for some } \mathrm{rep}_F(j) = v \text{ if and only if } f_{\mathrm{val}_F(v0)} = \mathtt{a} \text{ and } f_{\mathrm{val}_F(v1)} = \mathtt{a}.$$

In particular, the nth symbol $\mathtt{b}$ occurring in $\mathbf{f}$ belongs to the image by the morphism $\mathcal{F}$ of the nth $\mathtt{a}$ in $\mathbf{f}$. Otherwise stated, the nth $\mathtt{a}$ and the nth $\mathtt{b}$ in the Fibonacci word occur at indices (we recall, starting with 0) of the form:

$$\mathrm{val}_F(u0) \text{ and } \mathrm{val}_F(u01) \text{ with } u \in \mathrm{rep}_F(\mathbb{N}) . \tag{5.7}$$

If $\mathrm{rep}_F(j) = c_i \cdots c_2 c_1 c_0$, then f_j belongs to $\mathcal{F}(x_{\mathrm{val}_F(c_i \cdots c_1)})$. The symbol $x_{\mathrm{val}_F(c_i \cdots c_1)}$ appears itself in the image by the morphism $\mathcal{F}$ of the letter $x_{\mathrm{val}_F(c_i \cdots c_2)}$; x_j appears in $\mathcal{F}^2(x_{\mathrm{val}_F(c_i \cdots c_2)})$, and so on and so forth. For k-automatic sequences, we were simply using DIV k iteratively, here the result is similar: at each step, we remove the last digit of the F-representation of j.

5.5.3 Syntactic Characterization of $\mathcal{P}$-Positions of Wythoff's Game

We can now put together all the results and material of this section to obtain the following result. It provides us with a polynomial time algorithm to decide whether or not a pair is a $\mathcal{P}$-position. Note that (x, x) is a $\mathcal{N}$-position because one can directly play to $(0, 0)$.

Theorem 5.5.20 (Fraenkel [42]) *A pair (x, y), with $x < y$, is a $\mathcal{P}$-position of Wythoff's game if and only if* $\text{rep}_F(x)$ *ends with an even number of zeroes and* $\text{rep}_F(y) = \text{rep}_F(x)0$.

Proof From Proposition 5.5.3, we know that a pair (x, y), with $0 < x < y$, is a $\mathcal{P}$-position if and only if there exists n such that x is the index (starting with 1) of the nth a (resp. nth b) occurring in the Fibonacci word $\mathbf{f}'$.

We now make use of Corollary 5.5.18 and observation (5.7): (x, y) is a $\mathcal{P}$-position if and only if there exists a (valid) F-representation u such that

$$\left(\text{rep}_F(x-1), \text{rep}_F(y-1)\right) = (u0, u01). \tag{5.8}$$

We subtract 1 because Corollary 5.5.18 deals with the word $\mathbf{f}$ whose indices start with 0 and not 1 as in $\mathbf{f}'$. From (5.8) what can be said about the form of $\text{rep}_F(x)$ and $\text{rep}_F(y)$? First, we have

$$\text{rep}_F(x) = \text{rep}_F(\text{val}_F(u0) + 1) \text{ and } \text{rep}_F(y) = \text{rep}_F(\text{val}_F(u01) + 1)$$

- First case:[9] Assume that u is of the form $u'0$. We easily get valid F-representations:

$$\text{rep}_F(x) = \text{rep}_F(\text{val}_F(u'00) + 1) = u'01 \text{ ends with no zero,}$$

$$\text{rep}_F(y) = \text{rep}_F(\text{val}_F(u'001) + 1) = u'010 \text{ previous word shifted by zero.}$$

- Second case: Assume that u is of the form $u'1$ (where u' ends with 0 to have a valid F-representation). We get

$$\text{rep}_F(x) = \text{rep}_F(\text{val}_F(u'10) + 1) = \text{rep}_F(\text{val}_F(u'11)),$$

$$\text{rep}_F(y) = \text{rep}_F(\text{val}_F(u'101) + 1) = \text{rep}_F(\text{val}_F(u'110)).$$

---

[9]The first few values may be checked by hand.

Fig. 5.5 Transducer
computing the successor
function from right to left
(Fibonacci system)

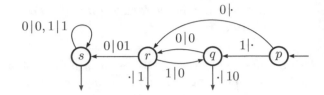

This means that we need to normalize $u'11$ and $u'110$ (in the sense of Definition 5.5.14) or equivalently, to compute the successor of $u'10$ and $u'101$. Hopefully, a transducer computing the successor function from right to left, i.e., least significant digit first, for the Fibonacci system is well-known. See [47]. This transducer is depicted in Fig. 5.5. If we feed the transducer with a valid F-representation $u'10$, it is either of the form $w100(10)^n10$ or, $10(10)^n10$, for some $n \geq 0$ (because we are considering large enough numbers). A computation for the first form is depicted below:

$$s \xleftarrow{\binom{w1}{w1}} s \xleftarrow{\binom{0}{01}} r \left[\xleftarrow{\binom{0}{0}} q \xleftarrow{\binom{1}{0}} r \right]^n \xleftarrow{\binom{0}{0}} q \xleftarrow{\binom{1}{0}} r \xleftarrow{\binom{0}{\cdot}} p \text{ :from right to left.}$$

So in the first situation, the successor of $u'10$ is

$$\underbrace{w10(01)^n0}_{u'}10 \rightarrow w101(00)^n00 \quad \text{ends with } 2n+2 \text{ zeroes, } n \geq 0$$

and for the successor of $u'101$, we get

$$\underbrace{w10(01)^n0}_{u'}101 \rightarrow w101(00)^n000 \quad \text{the previous word shifted by one zero.}$$

In the second situation, we have for the successor of $u'10$

$$\underbrace{1(01)^n0}_{u'}10 \rightarrow 100(00)^n00 \quad 2n+4 \text{ zeroes, } n \geq 0$$

and the successor of $u'101$

$$\underbrace{1(01)^n0}_{u'}101 \rightarrow 100(00)^n000 \quad \text{the previous word shifted by one zero.}$$

Putting together the different cases gives the expected result. $\square$

To conclude with this section, let us present the bidimensional word $\mathbf{P}_W$ coding the $\mathcal{P}$-positions of Wythoff's game. We have $\mathbf{P}_W(m, n) = 1$ if and only if (m, n)

Table 5.5 The bidimensional word $\mathbf{w}(m, n)$ coding $\mathcal{P}$-positions of Wythoff

10	0	0	0	0	0	0	1	0	0	0	0
9	0	0	0	0	0	0	0	0	0	0	0
8	0	0	0	0	0	0	0	0	0	0	0
7	0	0	0	0	1	0	0	0	0	0	0
6	0	0	0	0	0	0	0	0	0	0	1
5	0	0	0	1	0	0	0	0	0	0	0
4	0	0	0	0	0	0	0	1	0	0	0
3	0	0	0	0	0	1	0	0	0	0	0
2	0	1	0	0	0	0	0	0	0	0	0
1	0	0	1	0	0	0	0	0	0	0	0
0	1	0	0	0	0	0	0	0	0	0	0
	0	1	2	3	4	5	6	7	8	9	10

is a $\mathcal{P}$-position. Thanks to Proposition 5.3.18, it is a projection of $(\mathcal{G}(m, n))_{m,n \geq 0}$ given in Table 5.3.

We ask the same kind of (not yet answered) questions as before for the tables $(\mathcal{G}_{NIM}(m, n))_{m,n \geq 0}$ or $(\mathcal{G}_W(m, n))_{m,n \geq 0}$. We would like to have insight about the structure of the table, find some patterns, etc. The main difference between Table 5.5 and $(\mathcal{G}_{NIM}(m, n))_{m,n \geq 0}$ or $(\mathcal{G}_W(m, n))_{m,n \geq 0}$ is that we are with $\mathbf{P}_W$ over an alphabet $\{0, 1\}$.

In Sect. 5.6.1, we prove that $(\mathcal{G}_{NIM}(m, n))_{m,n \geq 0}$ is 2-regular. It is well-known that the mapping on a finite alphabet of a 2-regular word is 2-automatic [6]. In Sect. 5.7, we prove that $(\mathbf{P}_W(m, n))_{m,n \geq 0}$ is a morphic shape-symmetric word.

5.6   Extension to a Multidimensional Setting

This extension is pretty straightforward. The only technical part is that one has to pad shorter expansions to deal with words of the same length. In the first sections of this chapter, we have considered DFA over alphabets such as $\{0, \ldots, k-1\}$. But there is no objection to consider other finite sets as alphabets such as

$$\{0, 1\}^2 = \{0, 1\} \times \{0, 1\} = \left\{ \begin{pmatrix} 0 \\ 0 \end{pmatrix}, \begin{pmatrix} 0 \\ 1 \end{pmatrix}, \begin{pmatrix} 1 \\ 0 \end{pmatrix}, \begin{pmatrix} 1 \\ 1 \end{pmatrix} \right\}. \tag{5.9}$$

With this alphabet, the corresponding languages are subsets of $(\{0, 1\}^2)^*$. We make no distinction between pairs written horizontally or vertically, but it seems more natural to write them as column vectors because that is what the machine should read simultaneously. Examples of DFA over $(\{0, 1\}^2)^*$ are given in Figs. 5.8 and 5.20. There is no objection to take d-tuples instead of pairs and also, we can have alphabets with more than two symbols. A DFA over $(\{0, 1\}^3)^*$ is given in Fig. 5.9.

Definition 5.6.1 Let $d \geq 1, k \geq 2$ be integers. The base-k expansion of a d-tuple of non-negative integers is defined by

$$\mathrm{rep}_k : \mathbb{N}^d \to \left(\{0, \ldots, k-1\}^d \right)^*, \quad (n_1, \ldots, n_d) \mapsto \begin{pmatrix} 0^{\ell - |\mathrm{rep}_k(n_1)|} \, \mathrm{rep}_k(n_1) \\ \vdots \\ 0^{\ell - |\mathrm{rep}_k(n_d)|} \, \mathrm{rep}_k(n_d) \end{pmatrix}$$

where $\ell = \max\{|\mathrm{rep}_k(n_1)|, \ldots, |\mathrm{rep}_k(n_d)|\}$.

Let $d \geq 1, k \geq 2$ be integers. A map $f : A \to A^{k^d}$ where the image of a letter in A is a d-dimensional cube of size k, can be extended to a k-uniform morphism over the set of d-dimensional cubes. The image of a cube of size ℓ is a cube of size $k \cdot \ell$. The morphism is *prolongable* on the letter $a \in A$ if the "lower-left" corner of $f(a)$ is equal to a.

Example 5.6.2 (Pascal Triangle mod 2) Consider the morphism

$$\psi : 0 \mapsto \begin{array}{|c|c|} \hline 0 & 0 \\ \hline 0 & 0 \\ \hline \end{array}, \quad 1 \mapsto \begin{array}{|c|c|} \hline 1 & 0 \\ \hline 1 & 1 \\ \hline \end{array}.$$

Iterating n times the morphism ψ from 1 gives (explanations will follow) the first 2^n rows of the binomial coefficients modulo 2 (up to a rotation of $3\pi/4$). The binomial coefficient $\binom{m}{n}$ (mod 2) corresponds to the point of coordinates $(m-n, n)$ in the picture on the left of Fig. 5.6. It is important to keep in mind this "change of variables". Note that the sequence of normalized sets with ratio $1/2^n$ where 1's (resp. 0's) are full (resp. empty) unit squares, is a sequence of compact sets converging, for the Hausdorff distance, to the Sierpiński gasket [3, 11]. On fractal patterns of the Pascal triangle and generalizations, see, e.g., [4, 5, 50, 61].

Lemma 5.4.7 and thus Theorem 5.4.12 are translated verbatim.

Lemma 5.6.3 *Let f be a k-uniform d-dimensional morphism prolongable on a and $\mathbf{x} = f^\omega(a) = (x(n_1, \ldots, n_d))_{n_1, \ldots, n_d \geq 0}$. Let $(j_1, \ldots, j_d)$ such that $k^{m_i} \leq j_i < k^{m_i+1}$, for some $m_i \geq 0$, for all i. Then $j_i = kq_i + r_i$ with $k^{m_i-1} \leq q_i < k^{m_i}$ and $0 \leq r_i < k$ and the symbol $x_{(j_1, \ldots, j_d)}$ occurs in the d-cube of size k $f(x_{(q_1, \ldots, q_d)})$ in position $(r_1, \ldots, r_d)$.*

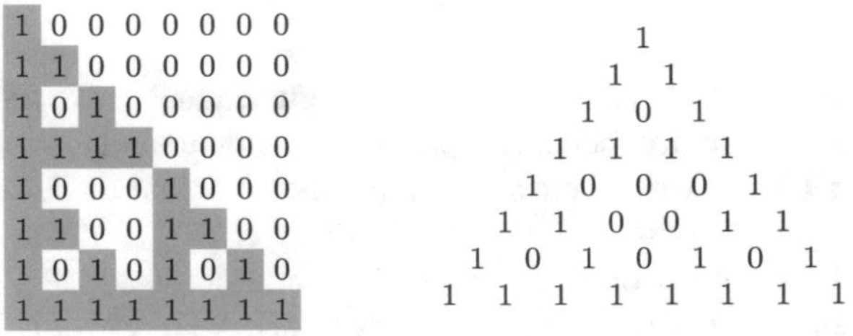

Fig. 5.6 Third iterate of ψ and the corresponding Pascal triangle modulo 2

Fig. 5.7 Tracking the past of a symbol in a bidimensional 2-automatic word

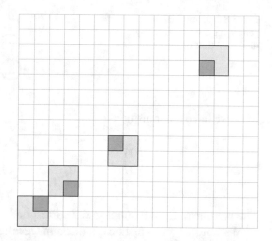

Example 5.6.4 With notation of the previous lemma, $k = d = 2$, consider Fig. 5.7. The symbol in position $(12, 10)$, where $\mathrm{rep}_2(12) = 1100$ and $\mathrm{rep}_2(10) = 1010$, is in the image of the symbol which is in position $(\mathrm{val}_2(110), \mathrm{val}_2(101)) = (6, 5)$. Moreover, since the last letter of 1100 and 1010 are both 0, $x(12, 10)$ is the lower-left corner of $f(x(6, 5))$. Now, $x(6, 5)$ appears in the image of $x((\mathrm{val}_2(11), \mathrm{val}_2(10)) = x(3, 2)$. Since the last digits in the base-2 expansion of 6 and 5 are respectively 0 and 1, $x(6, 5)$ is the upper-left corner of $f(x(3, 2))$. We can continue this way until we reach $x(1, 1)$.

Theorem 5.6.5 *[81, 82] Let $\mathbf{w} : \mathbb{N}^d \to B$ be an infinite d-dimensional word over an alphabet B. It is of the form $g(f^\omega(a))$ where $f : A \to A^{k^d}$ is a k-uniform morphism, prolongable on $a \in A$ and $g : A \to B$ is a coding if and only if there exists a DFAO*

$$(A, a, \{0, \ldots, k-1\}^d, \delta, \mu : A \to B)$$

such that $\delta(a, (0, \ldots, 0)) = a$ and, for all $(j_1, \ldots, j_d) \in \mathbb{N}^d$,

$$\mathbf{w}(j_1, \ldots, j_d) = \mu(\delta(a, \mathrm{rep}_k(j_1, \ldots, j_d)))$$

with the base-k expansion given in Definition 5.6.1.

The next example should suffice to explain how to derive a suitable DFAO from a morphism. If images of letters are d-dimensional cubes of size k, then transitions are labeled by d-tuples of digits in $\{0, \ldots, k-1\}$.

Example 5.6.6 (Pascal Triangle Modulo 2) We associate with the morphism ψ an automaton with input alphabet (5.9). Let $r, s, t, u, v \in \{0, 1\}$. If

$$\psi(r) = \begin{array}{|c|c|} \hline u & v \\ \hline s & t \\ \hline \end{array}$$

Fig. 5.8 An automaton
generating the Pascal triangle
modulo 2 (left of Fig. 5.6)

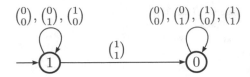

then, we have the transitions

$$r \xrightarrow{\binom{0}{0}} s, \quad r \xrightarrow{\binom{1}{0}} t, \quad r \xrightarrow{\binom{0}{1}} u, \quad r \xrightarrow{\binom{1}{1}} v.$$

The morphism from Example 5.6.2 corresponds to the DFA depicted in Fig. 5.8.

We can make use of Lucas's theorem (recalled below) for $p = 2$. From (5.10), the binomial coefficient $\binom{m}{n}$, with $m \geq n$, is even if and only if $\mathrm{rep}_2(m, n)$ contains the pair of digits $\binom{0}{1}$ (because the other three pairs give binomial coefficients equal to 1). To prove that the morphism ψ generates the Pascal triangle modulo 2, recall that we have some change of variables. It suffices to observe that $\mathrm{rep}_2(m, n)$ contains the pair of digits $\binom{0}{1}$ if and only if $\mathrm{rep}_2(m - n, n)$ contains the pair of digits $\binom{1}{1}$. The reader can probably carry on a proof of this necessary and sufficient condition.[10] We take the opportunity of Lemma 5.6.8 to give a proof involving automata. The latter condition is recognized by the DFA depicted in Fig. 5.8 and associated with the morphism ψ.

Theorem 5.6.7 (Lucas's Theorem) *Let p be a prime. Let m and n be two non-negative integers. If*

$$\mathrm{rep}_p(m, n) = \begin{pmatrix} m_k & m_{k-1} & \cdots & m_1 & m_0 \\ n_k & n_{k-1} & \cdots & n_1 & n_0 \end{pmatrix}$$

then the following congruence relation holds

$$\binom{m}{n} \equiv \prod_{i=0}^{k} \binom{m_i}{n_i} \mod p, \tag{5.10}$$

using the following convention: $\binom{m}{n} = 0$ *if* $m < n$.

Lemma 5.6.8 *Let $m \geq n$. The pair $\mathrm{rep}_2(m, n)$ contains the pair of digits $\binom{0}{1}$ if and only if $\mathrm{rep}_2(m - n, n)$ contains the pair of digits $\binom{1}{1}$.*

---

[10]One can relate this result to a theorem of Kummer. The p-adic valuation of $\binom{m}{n}$ is the number of carries when adding n to $m - n$ in base p. See, e.g., [77] and the references therein.

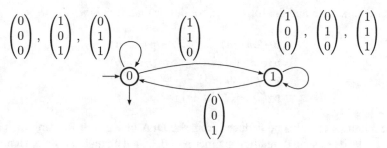

Fig. 5.9 A DFA accepting $\mathrm{rep}_2(x, y, x + y) \subset \{0, 1\}^3$

Proof The DFA depicted in Fig. 5.9 is mimicking base-2 addition. It recognizes 3-tuples of words (u, v, w) such that u, v and w have the same length and $\mathrm{val}_2(u) + \mathrm{val}_2(v) = \mathrm{val}_2(w)$. Successful paths are those starting and ending in state 0. The DFA reads least significant digits first. State 1 corresponds to the situation where there is a carry to take into account. The undefined transitions lead to some sink state.

We can make use of this DFA to recognize $\mathrm{rep}_2(m - n, n, m)$. Notice that whenever $\mathrm{rep}_2(n, m)$ contains a pair $\binom{1}{0}$ (the last two of the three components), in a successful path, we must use the unique transition from state 0 to state 1. This means that $\mathrm{rep}_2(m - n, n)$ contains the pair $\binom{1}{1}$ (the first two of the three components) and conversely. □

Definition 5.6.9 For the sake of simplicity, consider the case $d = 2$. Consider a bidimensional sequence $\mathbf{x} = (x(m, n))_{m,n \geq 0}$. The *k-kernel* of $\mathbf{x}$ is the set of bidimensional subsequences

$$\mathrm{Ker}_k(\mathbf{x}) = \left\{ (x(k^i m + r, k^i n + s))_{m,n \geq 0} \mid i \geq 0, 0 \leq r, s < k^i \right\}.$$

Note that we have the same multiplicative factor k^i for both components. One element of the k-kernel corresponds to selecting two suffixes

$$(0^{i-p} r_p \cdots r_1, 0^{i-q} s_q \cdots s_1)$$

where $\mathrm{rep}_k(r) = r_p \cdots r_1$ and $\mathrm{rep}_k(s) = s_q \cdots s_1$. Theorem 5.4.19 can be extended to this setting: a d-dimensional word satisfies the conditions of Theorem 5.6.5 if and only if its k-kernel is finite. See, for instance, [8, 81, 82].

Example 5.6.10 (Pascal Triangle Modulo 2) Let $(\mathbf{p}_2(m, n))_{m,n \geq 0}$ be the fixed point of the morphism ψ given in Example 5.6.2. We compute its 2-kernel as follows.

From Lemma 5.6.3 and the form of ψ, we derive that, for all $m, n \geq 0$,

$$
\begin{aligned}
\mathbf{p}_2(2m, 2n) &= \mathbf{p}_2(m, n), \\
\mathbf{p}_2(2m + 1, 2n) &= \mathbf{p}_2(m, n), \\
\mathbf{p}_2(2m, 2n + 1) &= \mathbf{p}_2(m, n), \\
\mathbf{p}_2(2m + 1, 2n + 1) &= 0.
\end{aligned}
$$

These relations can also be deduced from the DFA in Fig. 5.8: Reading the pair of digits $\binom{1}{1}$ leads to state 0, reading another pair does not change the state. Hence, the 2-kernel only contains the sequence itself and the null sequence. An alternative is to proceed as in the Proof of Theorem 5.4.19 and compute the transition monoid of the DFA in Fig. 5.8.

5.6.1 2-Regularity for Grundy Values of the Game of Nim

Compared with Example 5.6.10 where the 2-kernel of the Pascal triangle modulo 2 is finite, we can define a multidimensional k-regular sequence: the $\mathbb{Z}$-module generated by $\mathrm{Ker}_k(\mathbf{x})$ is finitely generated. It is now time to reconsider Table 5.2.

Proposition 5.6.11 (Exercises 21 and 22, Section 16.6, p. 451 [8]) *If we consider the game of Nim on two token piles, the bidimensional sequence $(\mathcal{G}_N(m, n))_{m,n \geq 0}$ is 2-regular.*

Proof From Bouton's theorem and simple base-2 manipulations, we get

$$
\begin{aligned}
\mathcal{G}_N(2m, 2n) &= 2m \oplus 2n &&= 2\mathcal{G}_N(m, n), \\
\mathcal{G}_N(2m + 1, 2n) &= (2m + 1) \oplus 2n &&= 2\mathcal{G}_N(m, n) + 1, \\
\mathcal{G}_N(2m, 2n + 1) &= 2m \oplus (2n + 1) &&= 2\mathcal{G}_N(m, n) + 1, \\
\mathcal{G}_N(2m + 1, 2n + 1) &= (2m + 1) \oplus (2n + 1) &&= 2\mathcal{G}_N(m, n).
\end{aligned}
$$

Hence the 2-kernel is generated by $(\mathcal{G}_N(m, n))_{m,n \geq 0}$ and the constant sequence $(1)_{m,n \geq 0}$. $\qquad\square$

Remark 5.6.12 To prove that a sequence $(s(n))_{n \geq 0}$ is k-regular, it is enough to find some $j \geq 1$ and k^j linear relations expressing, for all $r < k^j$, $s(k^j n + r)$ in terms of elements of the form $s(k^i n + t)$ with $i < j$ and $t < k^i$. This observation is similar in a multidimensional setting.

From the relations given in the proof of Proposition 5.6.11, we can express any element $(\mathcal{G}_N(2^j m + r, 2^j n + t))_{m,n \geq 0}$ of the 2-kernel as a linear combination of $(\mathcal{G}_N(m, n))_{m,n \geq 0}$ and $(1)_{m,n \geq 0}$. An example should be enough to understand the reasoning. Can $(\mathcal{G}_N(8m + 5, 8n + 2))_{m,n \geq 0}$ be expressed as a $\mathbb{Z}$-linear combination of these two sequences?

$$\mathcal{G}_N(8m + 5, 8n + 2) = \mathcal{G}_N\left(2(4m + 2) + 1, 2(4n + 1)\right)$$
$$= 2\,\mathcal{G}_N(4m + 2, 4n + 1) + 1$$
$$= 2\,\mathcal{G}_N(2(2m + 1), 2.2n + 1) + 1$$
$$= 2\,[2\,\mathcal{G}_N(2m + 1, 2n) + 1] + 1$$
$$= 4\,\mathcal{G}_N(2m + 1, 2n) + 3$$
$$= 4\,[2\,\mathcal{G}_N(m, n) + 1] + 3$$
$$= 8\,\mathcal{G}_N(m, n) + 7.$$

We can discuss a bit further the meaning of the relations given in the proof of Proposition 5.6.11. The situation is similar to the one of Lemma 5.6.3 and Fig. 5.7. The only difference is that we have a function whose domain is not necessarily bounded. A 2×2 block in the bidimensional word $(\mathcal{G}_N(i, j))_{i,j \geq 0}$ whose lower-left corner has coordinates $(2m, 2n)$ is completely determined from the value of $\mathcal{G}_N(m, n)$:

$$\mathcal{G}_N(m, n) \mapsto \begin{array}{|c|c|} \hline 2\,\mathcal{G}_N(m, n) + 1 & 2\,\mathcal{G}_N(m, n) \\ \hline 2\,\mathcal{G}_N(m, n) & 2\,\mathcal{G}_N(m, n) \mid 1 \\ \hline \end{array} \tag{5.11}$$

The situation is depicted within Table 5.6. The value of $\mathcal{G}_N(2, 1)$ determines the 2×2 block with lower-left corner of coordinates $(4, 2)$. The four elements of that block give the 4×4 block whose lower-left corner has coordinates $(8, 4)$ and so on and so forth.

Table 5.6 First few values of $\mathcal{G}_N(m, n)$ highlighting the action of the map (5.11)

9	8	11	10	13	12	15	14	1	0
8	9	10	11	12	13	14	15	0	1
7	6	5	4	3	2	1	0	15	14	13	12
6	7	4	5	2	3	0	1	14	15	12	13
5	4	7	6	1	0	3	2	13	12	15	14
4	5	6	7	0	1	2	3	12	13	14	15
3	2	1	0	7	6	5	4	11	10
2	3	0	1	6	7	4	5	10	11
1	0	3	2	5	4	7	6	9	8
0	1	2	3	4	5	6	7	8	9

Fig. 5.10 Graph of
$\mathcal{G}_W(5, n) - n$ for
$50 \leq n \leq 150$

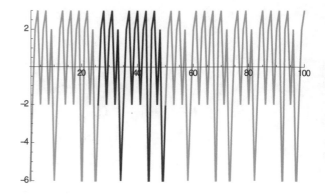

5.6.2   Grundy Values of the Game of Wythoff

Even though it looks quite similar to the situation encountered with the game of
Nim, Table 5.3 is challenging! We quote the book [85, p. 200]: "*No general formula
is known for computing arbitrary $\mathcal{G}$-values of Wythoff. In general, they appear
chaotic, though they exhibit a striking fractal-like pattern. . . Despite this apparent
chaos, the $\mathcal{G}$-values nonetheless have a high degree of geometric regularity.*"

We collect some results from Blass and Fraenkel about $\mathcal{G}_W(m, n)$ [14]:

- On every parallel to the main diagonal, $(\mathcal{G}_W(n, n + j))_{n \geq 0}$ takes every possible
  value.
- Points with Grundy value 1 are "close" to those with value 0.
- Recursive algorithms to determine the points with Grundy value 1 are provided.

Definition 5.6.13 A sequence $(a_j)_{j \geq 0}$ is *additively periodic* if

$$\exists p, q, \ \forall j \geq q : a_{j+p} = a_j + p.$$

Note that $(a_j)_{j \geq 0}$ is additively periodic if and only if $(a_j - j)_{j \geq 0}$ is ultimately
periodic.

Example 5.6.14 As an example, the row $\mathcal{G}_W(5, n)$ is such that for all $n \geq 27$,
$\mathcal{G}_W(5, n + 24) = \mathcal{G}_W(5, n) + 24$ (Fig. 5.10).

Every row and column of $\mathcal{G}_W(m, n)$ is additively periodic, see [32, 56].

5.7   Shape-Symmetry

Table 5.5 will permit us to introduce the notion of shape-symmetry. Our aim in
this section is to prove Theorem 5.7.14 stating that Table 5.5 is the fixed point
of a morphism with the shape-symmetry property. A picture is the analogue in a

multidimensional setting of a finite word. It is a bit more intricate to define the concatenation of pictures. The results and material presented in this section come from [64] and [22]. We first start with formal definitions, Example 5.7.4 is presented afterwards.

Definition 5.7.1 (Picture) Let A be an alphabet. Let $s_1, \ldots, s_d \geq 1$ be in $\mathbb{N}_{>0} \cup \{\infty\}$. A d-*dimensional picture* over A is a map

$$x : \{0, \ldots, s_1 - 1\} \times \cdots \times \{0, \ldots, s_d - 1\} \to A$$

and $(s_1, \ldots, s_d)$ is the *shape* of x. It is denoted by $|x|$ and $|x|_i = s_i$. If $s_i < \infty$, for all i, x is *bounded*. The set of bounded pictures over A is denoted by $\mathcal{B}_d(A)$.

Definition 5.7.2 (Factor) Let $x \in \mathcal{B}_d(A)$ be a bounded picture of shape $(s_1, \ldots, s_d)$. Let $i_1, j_1, \ldots, i_d, j_d$ be integers such that $0 \leq i_k \leq j_k < s_k$ for all $k \in \{1, \ldots, d\}$. We let

$$x[(i_1, \ldots, i_d), (j_1, \ldots, j_d)]$$

denote the picture y of shape $(j_1 - i_1 + 1, \ldots, j_d - i_d + 1)$ defined by

$$y(n_1, \ldots, n_d) = x(i_1 + n_1, \ldots, i_d + n_d)$$

for all $n_1 < j_1 - i_1, \ldots, n_d < j_d - i_d$. In 2 dimensions, this just means that we specify the lower-left and upper-right corner of a sub-picture.

We let $\mathbf{0}$ (resp. $\mathbf{1}$) denote the row vector (of convenient dimension) whose entries are all equal to zero (resp. one). Let $i \in \{1, \ldots, s\}$. If z is a d-tuple, we let $z_{\hat{i}}$ denote the $(d - 1)$-tuple where the ith coordinate has been removed.

Definition 5.7.3 (Concatenation) Let $x, y \in \mathcal{B}_d(A)$. If we have for some $i \in \{1, \ldots, d\}$, $|x|_{\hat{i}} = |y|_{\hat{i}} = (s_1, \ldots, s_{i-1}, s_{i+1}, \ldots, s_d)$, then we define the *concatenation* of x and y *along the ith direction* to be the d-dimensional picture $x \odot^i y$ of shape

$$(s_1, \ldots, s_{i-1}, |x|_i + |y|_i, s_{i+1}, \ldots, s_d).$$

satisfying

1. $x = (x \odot^i y)[\mathbf{0}, |x| - \mathbf{1}]$ and
2. $y = (x \odot^i y)[(0, \ldots, 0, |x|_i, 0, \ldots, 0), (0, \ldots, 0, |x|_i, 0, \ldots, 0) + |y| - \mathbf{1}]$.

In our examples, taking the usual convention for matrices, we first count the number of rows, then the number of columns. Concatenation along the direction 1 (resp. 2) follows the vertical (resp. horizontal) axis.

Example 5.7.4 Consider the two pictures

$$x = \begin{array}{|c|c|} \hline a & b \\ \hline c & d \\ \hline \end{array} \quad \text{and} \quad y = \begin{array}{|c|c|c|} \hline a & a & b \\ \hline b & c & d \\ \hline \end{array}$$

of respective shape $|x| = (2, 2)$ and $|y| = (2, 3)$. Since $|x|_{\widehat{2}} = |y|_{\widehat{2}} = 2$, we get

$$x \odot^2 y = \begin{array}{|c|c|c|c|c|} \hline a & b & a & a & b \\ \hline c & d & b & c & d \\ \hline \end{array} \quad \text{and} \quad y \odot^2 x = \begin{array}{|c|c|c|c|c|} \hline a & a & b & a & b \\ \hline b & c & d & c & d \\ \hline \end{array}.$$

However $x \odot^1 y$ is not defined because $2 = |x|_{\widehat{1}} \neq |y|_{\widehat{1}} = 3$. The pictures $x \odot^2 y$ and $y \odot^2 x$ both have shape $(2, 5)$. Thus we can, for instance, define

$$z = (x \odot^2 y) \odot^1 (y \odot^2 x) = \begin{array}{|c|c|c|c|c|} \hline a & a & b & a & b \\ \hline b & c & d & c & d \\ \hline a & b & a & a & b \\ \hline c & d & b & c & d \\ \hline \end{array}.$$

Remark 5.7.5 A map $\gamma : A \rightarrow \mathcal{B}_d(A)$ cannot necessarily be extended to a morphism $\gamma : \mathcal{B}_d(A) \rightarrow \mathcal{B}_d(A)$. As an example, consider the map defined by

$$\gamma : a \mapsto \begin{array}{|c|c|} \hline b & d \\ \hline a & a \\ \hline \end{array}, \quad b \mapsto \begin{array}{|c|} \hline b \\ \hline c \\ \hline \end{array}, \quad c \mapsto \begin{array}{|c|c|} \hline a & a \\ \hline \end{array}, \quad d \mapsto \begin{array}{|c|} \hline d \\ \hline \end{array}.$$

Take the following bounded picture

$$x = \begin{array}{|c|c|} \hline c & d \\ \hline a & b \\ \hline \end{array}.$$

Considering the two rows of x, we may apply $\odot^2$ because $|\gamma(a)|_{\widehat{2}} = |\gamma(b)|_{\widehat{2}} = 2$ and $|\gamma(c)|_{\widehat{2}} = |\gamma(d)|_{\widehat{2}} = 1$. Considering the two columns of x, we may apply $\odot^1$ because $|\gamma(a)|_{\widehat{1}} = |\gamma(c)|_{\widehat{1}} = 2$ and $|\gamma(b)|_{\widehat{1}} = |\gamma(d)|_{\widehat{1}} = 1$. Hence the image of x by γ is well-defined:

$$\gamma(x) = \begin{array}{|c|c|c|} \hline a & a & d \\ \hline b & d & b \\ \hline a & a & c \\ \hline \end{array}$$

but $\gamma^2(x)$ is not well-defined! Indeed, see Fig. 5.11, if we try to put together the images by γ of the different letters, we do not get a picture: Conditions to have a well-defined morphism are given in (5.12) and Theorem 5.7.13.

Fig. 5.11 A map cannot necessarily be extended to a morphism

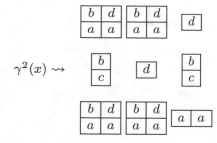

$$\gamma^2(x) \rightsquigarrow$$

Consider the section of a picture by a hyperplane. In this discrete setting, all the hyperplanes that we consider have equation of the form $x_i = k$ for some i.

Definition 5.7.6 (Section) Let x be a d-dimensional picture of shape $|x| = (s_1, \ldots, s_d)$. For all $i \in \{1, \ldots, d\}$ and $k < s_i$, we let $x_{|i,k}$ denote the $(d-1)$-dimensional picture of shape

$$|x|_{\hat{i}} = (s_1, \ldots, s_{i-1}, s_{i+1}, \ldots, s_d)$$

defined by setting the ith coordinate equal to k in x, that is,

$$x_{|i,k}(n_1, \ldots, n_{i-1}, n_{i+1}, \ldots, n_d) = x(n_1, \ldots, n_{i-1}, k, n_{i+1}, \ldots, n_d)$$

for all $n_j < s_j$ with $j \in \{1, \ldots, d\} \setminus \{i\}$.

What is the exact meaning for a morphism to be well-defined on a given picture? Let $\gamma : A \to \mathcal{B}_d(A)$ be a map and x be a bounded d-dimensional picture such that

$$\forall i \in \{1, \ldots, d\}, \forall k < |x|_i, \forall a, b \in \mathrm{Alph}(x_{|i,k}) : |\gamma(a)|_i = |\gamma(b)|_i. \tag{5.12}$$

We let $\mathrm{Alph}(x_{|i,k})$ denote the set of letters occurring in $x_{|i,k}$. Then the *image* of x by γ is the d-dimensional picture defined as

$$\gamma(x) = \odot^1_{0 \le n_1 < |x|_1} \left(\cdots \left(\odot^d_{0 \le n_d < |x|_d} \gamma(x(n_1, \ldots, n_d)) \right) \cdots \right).$$

Note that the ordering of the products in the different directions is unimportant. If a bounded picture x does not satisfy (5.12), then $\gamma(x)$ is *undefined*. This means that the map γ can be extended to a subset of $\mathcal{B}_d(A)$. It is a quite restrictive requirement.

In two dimensions, condition (5.12) simply means that the images by γ of all the elements belonging to the same column (resp. rows) are pictures having the same number of columns (resp. row), see Fig. 5.12. With d dimensions, (5.12) means that, for all sections by a hyperplane $x_i = k$, the images of all letters in this section have a shape with the same component along the direction orthogonal to that hyperplane. This ensures that building $\gamma(x)$ by concatenating the images of letters, we will not obtain "holes" or "overlaps".

Fig. 5.12 Illustration of the
condition (5.12)

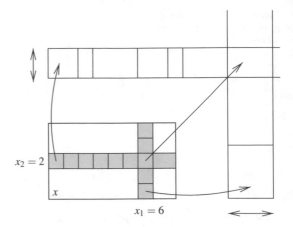

$x_2 = 2$

x

$x_1 = 6$

Definition 5.7.7 (Multidimensional Morphism) If for all $a \in A$ and all $n \geq 1$, $\gamma^n(a)$ is well-defined from $\gamma^{n-1}(a)$, then γ is said to be a *d-dimensional morphism*. If there exists a letter b such that $\gamma(b)_{0,\dots,0} = b$ and, for all i, $|\gamma^n(b)|_i \to +\infty$ when $n \to +\infty$, then γ is said to be *prolongable* on b. We assume that the picture grows in every direction. It is not a strong assumption. If this is not the case and one of the direction remains constant, then we have a finite union of hyperplanes and words of lower dimension.

Remark 5.7.8 Let x, y be two bounded pictures such that $|x|_{\hat{i}} = |y|_{\hat{i}}$ and a morphism γ. Assume that $\gamma(x)$ and $\gamma(y)$ are well-defined pictures. If $\gamma(x \odot^i y)$ is well-defined, then it is equal to $\gamma(x) \odot^i \gamma(y)$ which is thus well-defined. On the other hand, if $\gamma(x) \odot^i \gamma(y)$ is defined, there is no reason for $\gamma(x \odot^i y)$ to be well-defined. As an example, take

$$\gamma : a \mapsto \begin{array}{|c|c|} \hline b & b \\ \hline a & a \\ \hline \end{array}, \quad b \mapsto \begin{array}{|c|c|} \hline b & b \\ \hline \end{array} \quad \text{and} \quad x = \begin{array}{|c|} \hline a \\ \hline b \\ \hline \end{array} \quad y = \begin{array}{|c|} \hline b \\ \hline a \\ \hline \end{array}$$

We have

$$x \odot^2 y = \begin{array}{|c|c|} \hline a & b \\ \hline b & a \\ \hline \end{array}, \quad \gamma(x) \odot^2 \gamma(y) = \begin{array}{|c|c|c|c|} \hline b & b & b & b \\ \hline a & a & b & b \\ \hline b & b & a & a \\ \hline \end{array}$$

but $\gamma(x \odot^2 y)$ is undefined because (5.12) is not satisfied. The images by γ of letters on each row of $x \odot^2 y$ have different number of rows.

Definition 5.7.9 (Projection) Let $\gamma : A \to \mathcal{B}_d(A)$ be a d-dimensional morphism prolongable on a. The ith *projection* of γ is a unidimensional morphism γ_i defined as follows. For all b, $\gamma_i(b)$ is the intersection of the picture $\gamma(b)$ with the hyperplane $x_i = 0$.

Definition 5.7.10 (Shape-Symmetry) Let $\gamma: \mathcal{B}_d(A) \rightarrow \mathcal{B}_d(A)$ be a d-dimensional morphism having the d-dimensional infinite word **x** as a fixed point. This word is *shape-symmetric with respect to* γ if, for all permutations v of $\{1, \ldots, d\}$, we have, for all $n_1, \ldots, n_d \geq 0$,

$$|\gamma(\mathbf{x}(n_1, \ldots, n_d))| = (s_1, \ldots, s_d) \Rightarrow |\gamma(\mathbf{x}(n_{v(1)}, \ldots, n_{v(d)}))| = (s_{v(1)}, \ldots, s_{v(d)}).$$

Note that, on the diagonal, $\gamma(\mathbf{x}(j, \ldots, j))$ must be a cube.

Example 5.7.11 The multidimensional k-automatic words are trivially shape-symmetric with respect to a morphism that sends letters to hypercubes of size k.

Example 5.7.12 The following morphism will be extensively used

$$\varphi_W : a \mapsto \begin{array}{|c|c|} \hline c & d \\ \hline a & b \\ \hline \end{array} \quad b \mapsto \begin{array}{|c|} \hline e \\ \hline i \\ \hline \end{array} \quad c \mapsto \begin{array}{|c|c|} \hline i & j \\ \hline \end{array} \quad d \mapsto \begin{array}{|c|} \hline i \\ \hline \end{array} \quad e \mapsto \begin{array}{|c|c|} \hline f & b \\ \hline \end{array}$$

$$f \mapsto \begin{array}{|c|c|} \hline h & d \\ \hline g & b \\ \hline \end{array} \quad g \mapsto \begin{array}{|c|c|} \hline h & d \\ \hline f & b \\ \hline \end{array} \quad h \mapsto \begin{array}{|c|c|} \hline i & m \\ \hline \end{array} \quad i \mapsto \begin{array}{|c|c|} \hline h & d \\ \hline i & m \\ \hline \end{array}$$

$$j \mapsto \begin{array}{|c|} \hline c \\ \hline k \\ \hline \end{array} \quad k \mapsto \begin{array}{|c|c|} \hline c & d \\ \hline l & m \\ \hline \end{array} \quad l \mapsto \begin{array}{|c|c|} \hline c & d \\ \hline k & m \\ \hline \end{array} \quad m \mapsto \begin{array}{|c|} \hline h \\ \hline i \\ \hline \end{array}$$

with the coding

$$\mu_W : a, e, g, j, l \mapsto 1, \quad b, c, d, f, h, i, k, m \mapsto 0.$$

The fourth iterate of $\varphi_W^4(a)$ is given in Table 5.7.

It can be shown that φ_W is a two-dimensional morphism, see the next theorem. The size of the nth iterate from a is given by the nth Fibonacci number. The infinite word with a in position $(0, \ldots, 0)$ and which is a fixed point of φ_W, is shape-

Table 5.7 The fourth iterate of $\varphi_W^4(a)$

i	m	i	i	**j**	i	m	i
h	d	h	c	d	h	d	h
i	m	i	**l**	m	i	m	i
h	d	c	h	d	h	d	**e**
i	m	k	i	m	**g**	b	i
i	**j**	i	f	b	i	m	i
c	d	**e**	h	d	h	d	h
a	b	i	i	m	i	m	i

symmetric with respect to φ_W. For instance, the image of e (second row, third column) has shape $(1, 2)$ and the image of j (second column, third row) has a transposed shape $(2, 1)$.

The two projections of φ_W are given by

$$[\varphi_W]_1 : a \mapsto ac, \ c \mapsto i, \ i \mapsto ih, \ h \mapsto i \qquad (5.13)$$

and

$$[\varphi_W]_2 : a \mapsto ab, \ b \mapsto i, \ i \mapsto im, \ m \mapsto i . \qquad (5.14)$$

Up to some renaming of letters, these two morphisms are equal.

The following result can be found in Maes's thesis [64]. In this thesis, Maes provides two different proofs. We only present the one using automata.

Theorem 5.7.13 (A. Maes) *Let $d \geq 1$. Let A be an alphabet.*

- *Determining whether or not a map $\mu \colon A \to \mathcal{B}_d(A)$ can be extended to a d-dimensional morphism, prolongable on a letter a, is a decidable problem.*
- *If μ is prolongable on the letter a, then it is decidable whether or not the fixed point $\mu^\omega(a)$ is shape-symmetric with respect to μ.*

Sketch of the Proof The reader should be used to the construction. We associate a DFA $\mathcal{A}_\mu$, just as in Example 5.6.6, with the map $\mu \colon A \to \mathcal{B}_d(A)$ and a specified initial symbol a. The set of states is A and transitions are labeled by d-tuples of digits: if $\mu(b)$ has shape $|\mu(b)| = (s_1, \ldots, s_d)$, we have a transition $(c_1, \ldots, c_d)$ for $0 \leq c_i < s_i$ and for all $i \in \{1, \ldots, d\}$. So the number of outgoing transitions from $b \in A$ is $\prod_i |\mu(b)|_i$. The DFA associated with the map γ of Remark 5.7.5 is depicted in Fig. 5.13. From this DFA, we build d new NFAs. Let $i \in \{1, \ldots, d\}$. We define a NFA $\mathcal{N}_i$ where, for every transition, we only keep the ith component of the label (we proceed to a projection that explains the non-determinism). Moreover, the set I_i of initial states is made of all the states that can be reached from a when reading a word in 0^* (Fig. 5.14).

Fig. 5.13 The DFA associated with the map γ of Remark 5.7.5

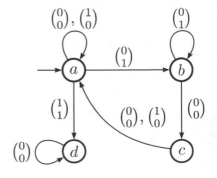

Fig. 5.14 The NFAs $\mathcal{N}_1$ and $\mathcal{N}_2$ derived from $\mathcal{A}_\gamma$

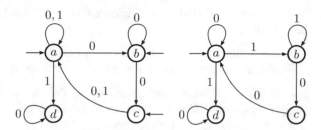

We apply the subset construction (Rabin–Scott theorem) to these NFAs. The corresponding DFAs are denoted by $\mathcal{D}_i$, $i = 1, \ldots, d$. Recall that states of $\mathcal{D}_i$ are subsets of states of $\mathcal{N}_i$ and thus of $\mathcal{A}_\mu$, i.e., a state of $\mathcal{D}_i$ is a subset of the alphabet A. Observe that the initial state I_i of $\mathcal{D}_i$ is made of the states that can be reached in $\mathcal{A}_\mu$ by a word in $(\mathbb{N}^d)^*$ whose ith component is in 0^*. If μ is a morphism prolongable on a, then I_i is exactly the set of letters occurring in the intersection of $\mu^\omega(a)$ with the hyperplane $x_i = 0$. Therefore, in view of (5.12), for all $b \in I_i$, the quantities $|\mu(b)|_i$ must all be the same. Let Q be a state of $\mathcal{D}_i$. There exists a word $w \in \mathbb{N}^*$ such that, for all b in Q, there exists a word in $(\mathbb{N}^d)^*$ which is the label of a path from a to b in $\mathcal{A}_\mu$ and whose ith component is in 0^*w. If μ is a morphism prolongable on a, then w can be chosen in L_{μ_i} where μ_i is the ith projection of μ (see Definition 5.7.9) and L_{μ_i} is the language given by Theorem 5.5.6 where words do not start with 0. If w is the $(n+1)$st word, $n \geq 0$, in the genealogically ordered language L_{μ_i}, then Q is exactly the set of letters occurring in the intersection of $\mu^\omega(a)$ with the hyperplane $x_i = n$. In view of (5.12), for all $b \in Q$, the quantities $|\mu(b)|_i$ must all be the same. We conclude that, if μ is a morphism prolongable on a, then:

- For all $i \in \{1, \ldots, d\}$, for any two letters b, c belonging to the same state of $\mathcal{D}_i$, we have $|\mu(b)|_i = |\mu(c)|_i$.

Conversely, if the above condition holds, then proceed by induction on the iterate $j \geq 0$. Assume that $\mu^j(a)$ exists. Let $n \geq 0$. Then the set of letters occurring in the intersection of $\mu^j(a)$ with $x_i = n$ is a subset of some state in $\mathcal{D}_i$. Therefore, the above condition and (5.12) permit us to define $\mu^{j+1}(a)$. In Table 5.8, we have applied the subset construction to $\mathcal{N}_1$. This shows that, starting with a, the map γ cannot be extended to a morphism. The last column contains the first component of the shape of $\gamma(e)$ for the letters e in the corresponding subset.

Table 5.8 Subset construction applied to $\mathcal{N}_1$

	State of $\mathcal{D}_1$	$|\gamma(\cdot)|_1$
$\mathcal{D}_1 \cdot \varepsilon = I_1$	$\{a, b, c\}$	$2, 2, 1$
$\mathcal{D}_1 \cdot 1$	$\{a, d\}$	$2, 1$
$\mathcal{D}_1 \cdot 10$	$\{a, b, d\}$	$2, 2, 1$
$\mathcal{D}_1 \cdot 100$	$\{a, b, c, d\}$	$2, 2, 1, 1$

Let us now turn to the second decision problem. From Definition 5.7.10, it should be clear that:

- Given a morphism μ prolongable on a letter a, the fixed point $\mu^\omega(a)$ is shape-symmetric with respect to μ if and only if the languages $L_{\mu_1}, \ldots, L_{\mu_d}$ are equal.

It is well-known that testing equality of regular languages is decidable (see, e.g., [86]). For instance, from (5.13) and (5.14), we directly see that if $\varphi_W^\omega(a)$ exists then it is shape-symmetric with respect to φ_W.   □

In Fig. 5.15, we have depicted the first iteration (two viewpoints) of what a shape-symmetric morphism looks like in three dimensions.

We can already state the main result related to Wythoff's game because it is related to shape-symmetric morphisms.

Theorem 5.7.14 *The morphism φ_W defined in Example 5.7.12 and the coding μ_W give the two-dimensional infinite word coding the $\mathcal{P}$-positions of Wythoff.*

The proof will be given in Sect. 5.7.3. In particular, the DFA associated with φ_W is depicted in Fig. 5.20. If we apply the procedure described in the proof of Theorem 5.7.13, when determinizing the two NFAs we get the following Table 5.9 showing that φ_W is a prolongable morphism.

Example 5.7.15 (Product of Substitutions) Let $\mu : A^* \to A^*$ and $\nu : B^* \to B^*$ be two morphisms prolongable respectively on $a \in A$ and $b \in B$. We define the *product* (called *direct product* in Priebe Frank's chapter (this volume)) $\mu \times \nu : A \times B \to \mathcal{B}_d(A \times B)$ where $(\mu \times \nu)(c, d)$ is the picture obtained as the cross product of the two finite words $\mu(c)$ and $\nu(d)$. It is easy to check that the map $\mu \times \nu$ is a morphism prolongable on (a, b). The corresponding fixed point is the cross product of $\mu^\omega(a)$ and $\nu^\omega(b)$. Take the Thue–Morse morphism $f : 0 \mapsto 01, 1 \mapsto 10$

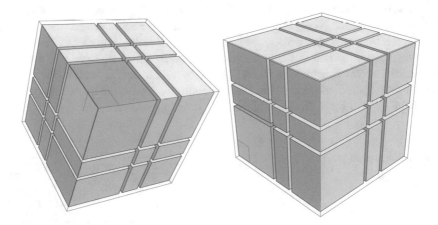

Fig. 5.15 Initial blocks of some two-dimensional shape-symmetric morphism

Table 5.9 Applying Maes's procedure to φ_W

	State of $\mathcal{D}_1$	$|\gamma(\cdot)|_1$
$\mathcal{D}_1 \cdot \varepsilon = I_1$	$\{a, b, i, m\}$	$2, 2, 2, 2$
$\mathcal{D}_1 \cdot 1$	$\{c, d, e, h\}$	$1, 1, 1, 1$
$\mathcal{D}_1 \cdot 10$	$\{b, f, i, j, m\}$	$2, 2, 2, 2, 2$
$\mathcal{D}_1 \cdot 11$	$\{\}$	
$\mathcal{D}_1 \cdot 100$	$\{b, g, i, k, m\}$	$2, 2, 2, 2, 2$
$\mathcal{D}_1 \cdot 1000$	$\{b, f, i, l, m\}$	$2, 2, 2, 2, 2$

	State of $\mathcal{D}_2$	$|\gamma(\cdot)|_2$
$\mathcal{D}_2 \cdot \varepsilon = I_2$	$\{a, c, h, i\}$	$2, 2, 2, 2$
$\mathcal{D}_2 \cdot 1$	$\{b, d, j, m\}$	$1, 1, 1, 1$
$\mathcal{D}_2 \cdot 10$	$\{c, e, h, i, k\}$	$2, 2, 2, 2, 2$
$\mathcal{D}_2 \cdot 11$	$\{\}$	
$\mathcal{D}_2 \cdot 100$	$\{c, f, h, i, l\}$	$2, 2, 2, 2, 2$
$\mathcal{D}_2 \cdot 1000$	$\{c, g, h, i, k\}$	$2, 2, 2, 2, 2$

of Example 5.4.20 and the Fibonacci morphism $\mathcal{F} :$ a $\mapsto$ ab, b $\mapsto$ a of Definition 5.5.1.

$$f \times \mathcal{F} : (0, a) \mapsto \frac{(0, b)\ (1, b)}{(0, a)\ (1, a)}, \quad (0, b) \mapsto \boxed{(0, a)\ (1, a)},$$

$$(1, a) \mapsto \frac{(1, b)\ (0, b)}{(1, a)\ (0, a)}, \quad (1, b) \mapsto \boxed{(1, a)\ (0, a)}.$$

Note that we get a word shape-symmetric with respect to $\mu \times \nu$ if and only if L_μ and L_ν are equal (where the languages are given by Theorem 5.5.6). In particular, the product of a morphism by itself gives shape-symmetry. See also [80].

5.7.1 Abstract Numeration Systems

Abstract numeration systems are based on an infinite (regular) language over a totally ordered alphabet. They are natural generalizations of classical systems such as integer base systems or Pisot systems [16]. Recall that the genealogical ordering was introduced in Definition 5.5.5. For a survey chapter introducing abstract numeration systems, see [59].

Definition 5.7.16 An *abstract numeration system* (or *ANS* for short) is a triple $S = (L, A, <)$ where L is an infinite regular[11] language over a totally ordered alphabet $(A, <)$. The map $\mathrm{rep}_S : \mathbb{N} \to L$ is the one-to-one correspondence mapping $n \in \mathbb{N}$ to the $(n+1)$st word in the genealogically ordered language L, which is called the *S-representation* of n. The S-representation of 0 is the first word in L. The inverse map is denoted by $\mathrm{val}_S : L \to \mathbb{N}$. If w is a word in L, $\mathrm{val}_S(w)$ is its *S-numerical value*.

ANS were introduced in [60]. Note that $\mathrm{val}_S(w)$ is sometimes called the *rank* of w. See, for instance, [26].

Remark 5.7.17 A motivation for studying abstract numeration systems is that it is quite convenient to have a regular language of admissible representations. Given a finite word, one can decide in linear time with respect to the length of the entry, using a DFA, whether or not this word is a valid representation.

Another motivation comes from Cobham's theorem about base dependence. See Theorem 5.4.21 for its statement in terms of k-recognizable[12] sets of integers. In view of this theorem of Cobham, if a set of integers is recognizable within two "sufficiently different" systems, then this set is ultimately periodic. Moreover, every ultimately periodic set is always k-recognizable for every integer base $k \geq 2$. Therefore, if one thinks about a possible generalization of this theorem of Cobham, then a minimal requirement is that ultimately periodic sets—in particular $\mathbb{N}$—should have a set of S-representations which is a regular language.

Example 5.7.18 (Integer Base System) Let $k \geq 2$ be an integer. Consider the language

$$L = \{\varepsilon\} \cup \{1, \ldots, k-1\}\{0, \ldots, k-1\}^*.$$

The ANS built on L using the natural ordering of the digits in $\{0, \ldots, k-1\}$ is the usual base-k numeration system. Note that we do *not* allow leading zeroes in representations. Indeed, adding leading zeroes would change the length of the word and therefore the ordering (and thus the value) of this word. Recall that, in the genealogical ordering, words are first ordered with respect to their length.

Example 5.7.19 (Unambiguous Integer Base System) Let $k \geq 2$ be an integer. Consider the language

$$U = \{1, \ldots, k\}^*.$$

---

[11]One could relax the assumption about regularity of the language on which the numeration system is built to encompass a larger framework. Nevertheless, most of the nice properties that we shall present (in particular, the equivalence with morphic words) do not hold without the regularity assumption.

[12]A set $X \subseteq \mathbb{N}$ is *k-recognizable* if $\mathrm{rep}_k(X) \subseteq \{0, \ldots, k-1\}^*$ is recognized by a DFA or, equivalently, if the characteristic sequence of X is k-automatic.

As the reader may observe, in this system, the digit set is $\{1, \ldots, k\}$ instead of $\{0, \ldots, k-1\}$. Therefore, we avoid any discussion about possible leading zeroes. For $k = 2$, the first few words in the ordered language U, using the natural ordering of $\{1, \ldots, k\}$, are

$$\varepsilon <_{gen} 1 <_{gen} 2 <_{gen} 11 <_{gen} 12 <_{gen} 21 <_{gen} 22$$

$$<_{gen} 111 <_{gen} 112 <_{gen} 121 <_{gen} 122 <_{gen} \cdots .$$

Let $\mathcal{U}$ be the ANS built on U. Note that if $c_\ell \cdots c_0$ is a word over $\{1, \ldots, k\}$, then

$$\mathrm{val}_{\mathcal{U}}(c_\ell \cdots c_0) = \sum_{i=0}^{\ell} c_i \, k^i .$$

Let $n \geq 0$. In particular, note that $\mathrm{val}_{\mathcal{U}}(k^n) = k\frac{k^n-1}{k-1}$ and the next word in the genealogical ordering, i.e., the first word of the next length, gives $\mathrm{val}_{\mathcal{U}}(1^{n+1}) = \frac{k^{n+1}-1}{k-1}$. For more about unambiguous systems, see [51, 52].

Example 5.7.20 We can reconsider Example 5.5.17. We have $\mathrm{rep}_F(\mathbb{N}) = 1\{0, 01\}^* \cup \{\varepsilon\}$. Hence, ordering the words of this regular language gives an abstract numeration system. Because of Proposition 5.5.13, this remark can be applied to every positional numeration system whose language of U-representations is regular.

Example 5.7.21 Consider $L = \mathrm{a}^*\mathrm{b}^*$ with $\mathrm{a} < \mathrm{b}$ and the ANS $\mathcal{S} = (L, \{\mathrm{a}, \mathrm{b}\}, <)$. The first few words in L in increasing genealogical order are

$$\varepsilon <_{gen} \mathrm{a} <_{gen} \mathrm{b} <_{gen} \mathrm{aa} <_{gen} \mathrm{ab} <_{gen} \mathrm{bb} <_{gen}$$

$$\mathrm{aaa} <_{gen} \mathrm{aab} <_{gen} \mathrm{abb} <_{gen} \mathrm{bbb} <_{gen} \cdots .$$

For example, $\mathrm{val}_{\mathcal{S}}(\mathrm{abb}) = 8$ and $\mathrm{rep}_{\mathcal{S}}(3) = \mathrm{aa}$. If we consider the bijection from L to $\mathbb{N}^2$ mapping the word $\mathrm{a}^i\mathrm{b}^j$ to the pair (i, j), $i, j \geq 0$, it is not difficult to see that the genealogical ordering of L corresponds to the primitive recursive *Peano enumeration* of $\mathbb{N}^2$, that is

$$\mathrm{val}_{\mathcal{S}}(\mathrm{a}^i\mathrm{b}^j) = \frac{1}{2}(i + j)(i + j + 1) + j = \binom{i + j + 1}{2} + \binom{j}{1}. \tag{5.15}$$

Many papers are dedicated to numeration systems satisfying conditions of the form (5.15). For more on these combinatorial numeration systems, see [25] and the references therein, in particular [54].

Example 5.7.22 (Prefix-Closed Language) Consider the prefix-closed language

$$\{\mathrm{a}, \mathrm{ba}\}^*\{\varepsilon, \mathrm{b}\}.$$

Fig. 5.16 A trie for words of
length ≤ 3 in $\{a, ba\}^* \{\varepsilon, b\}$

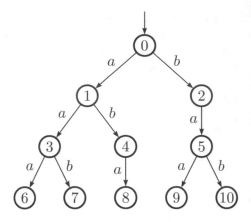

When considering such a language ordered by genealogical order, the nth level of
the trie contains all words of L of length n in lexicographic order from left to right
assuming that the children of a node are also ordered with respect to the ordering of
the alphabet. To enumerate the words in the language: proceed one level at a time,
from left to right. In Fig. 5.16 we represent the first four levels of the corresponding
trie, i.e., a rooted tree where the edges are labeled by letters from A, and the nodes
are labeled by prefixes of words in the considered language L. Let $u \in A^*$, $a \in A$.
If ua is (a prefix of) a word in L, then there is an edge between u and ua. Note that
for a prefix-closed language L, all prefixes of words in L belong to L. In the nodes,
we have written the S-numerical value of the corresponding words in L. The root is
associated with ϵ. See also [65, 66].

In Sect. 5.4 and in particular, with Theorem 5.4.12, we have seen that k-automatic
sequences can be obtained by feeding a DFAO with base-k expansions of integers.
Now that we have generalized numeration systems, we can feed a DFAO with S-
representations of integers.

Definition 5.7.23 Let $S = (L, A, <)$ be an ANS. We say that an infinite word $\mathbf{x} =
x_0 x_1 x_2 \cdots \in B^{\mathbb{N}}$ is S-*automatic*, if there exists a DFAO $(Q, q_0, A, \delta, \mu : Q \to B)$
such that $x_n = \mu(\delta(q_0, \mathrm{rep}_S(n)))$ for all $n \geq 0$.

This notion was introduced in [70, 75]. For an intermediate notion, see [84].

Example 5.7.24 Let $k \geq 2$. Every k-automatic sequence is S-automatic for the
ANS introduced in Example 5.7.18.

Example 5.7.25 We consider the alphabets $A = \{a, b\}$, $B = \{0, 1, 2, 3\}$, the ANS
$S = (a^*b^*, A, a < b)$ of Example 5.7.21 and the DFAO depicted in Fig. 5.17. We
obtain the first few terms of the corresponding S-automatic sequence

$$\mathbf{x} = 0102303120023101012302303120312023100231012 \cdots .$$

Fig. 5.17 A DFAO with output alphabet $\{0, 1, 2, 3\}$

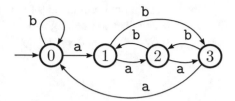

Notice that taking another ANS such as $\mathcal{R} = (\{a, ba\}^*\{\varepsilon, b\}, \{a, b\}, a < b)$, we obtain with the same DFAO another infinite word $\mathbf{y} = 01023\underline{1}31\underline{0}23\cdots$ which is $\mathcal{R}$-automatic (underlined letters indicate the differences between $\mathbf{x}$ and $\mathbf{y}$). This stresses the fact that a S-automatic sequence really depends on two ingredients: an ANS and a DFAO.

Theorem 5.5.6 can directly be restated in terms of ANS: morphic words are S-automatic for some ANS S. This is just a question of terminology. We state it in the case of a pure morphic word. It is not difficult to add an extra coding taking into account in the output function of the corresponding DFAO.

Theorem 5.7.26 *Let* $(A, <)$ *be an ordered alphabet. Let* $\mathbf{w} \in A^{\mathbb{N}}$ *be an infinite word, fixed point* $f^\omega(a)$ *of a morphism* $f : A^* \to A^*$ *prolongable on* a. *Consider the language* L_f *of words accepted by the automaton* $\mathcal{A}_f$ *associated with* f, *except those starting with* 0. *Then the word* $\mathbf{w}$ *is* S-*automatic for the ANS* $S = (L_f, A, <)$ *and the DFAO* $\mathcal{A}_F$.

Now we turn to the converse of Theorem 5.7.26.

Proposition 5.7.27 *Let* S *be an ANS. Every* S-*automatic sequence is a morphic word.*

Proof Let $S = (L, A, <)$ be an ANS. Let $\mathcal{A} = (Q, q_0, A, \delta_\mathcal{A}, F)$ be a complete DFA accepting L. Let $\mathcal{B} = (R, r_0, A, \delta_B, \mu : R \to B)$ be a DFAO generating an S-automatic sequence $\mathbf{x} = (x_n)_{n \geq 0}$ over B, i.e., for all $n \geq 0$, $x_n = \mu(\delta_B(r_0, \text{rep}_S(n)))$.

Consider the Cartesian product automaton $\mathcal{P} = \mathcal{A} \times \mathcal{B}$ defined as follows. The set of states of $\mathcal{P}$ is $Q \times R$. The initial state is (q_0, r_0) and the alphabet is A. For any word $w \in A^*$, the transition function $\Delta : (Q \times R) \times A^* \to Q \times R$ is given by

$$\Delta((q, r), w) = (\delta_\mathcal{A}(q, w), \delta_B(r, w)).$$

This means that the product automaton mimics the behaviors of both $\mathcal{A}$ and $\mathcal{B}$ in a single automaton. In particular, after reading w in $\mathcal{P}$, $\Delta((q_0, r_0), w)$ belongs to $F \times R$ if and only if w belongs to L. Moreover if $\text{rep}_S(n) = w$ and $\Delta((q_0, r_0), w) = (q, r)$, then $x_n = \mu(r)$.

Now we can apply Lemma 5.5.9 to $\mathcal{P}$ and define a morphism $\psi_{\mathcal{P}}$ prolongable on a letter z which does not belong to $Q \times R$. In view of the previous paragraph, we define $\nu : ((Q \times R) \cup \{z\})^* \to B^*$ by

$$\nu(q, r) = \begin{cases} \mu(r), & \text{if } q \in F; \\ \varepsilon, & \text{otherwise;} \end{cases}$$

and $\nu(z) = \varepsilon$. As Lemma 5.5.9 can be used to describe the sequence of reached states, $\nu(\psi_{\mathcal{P}}(z))$ is exactly the sequence $(x_n)_{n \geq 0}$. This proves that $\mathbf{x}$ is morphic. $\square$

Note that the morphisms obtained at the end of this proof are erasing. This is not a problem thanks to Theorem 5.4.3.

5.7.2 Multidimensional S-Automatic Sequences

We have something similar to Definition 5.6.1 but since the digit 0 modifies the length of a word (see Examples 5.7.18 and 5.7.19), we use an extra padding symbol not in the original alphabet.

Definition 5.7.28 If $w_1, \ldots, w_d$ are finite words over the alphabet A, the *padding map* $(\cdot)^{\#} : (A^*)^d \to ((A \cup \{\#\})^d)^*$ is defined as

$$(w_1, \ldots, w_d)^{\#} := (\#^{m-|w_1|} w_1, \ldots, \#^{m-|w_d|} w_d)$$

where $m = \max\{|w_1|, \ldots, |w_d|\}$. For all $n_1, \ldots, n_d \geq 0$, we set

$$\text{rep}_{\mathcal{S}}(n_1, \ldots, n_d) := (\text{rep}_{\mathcal{S}}(n_1), \ldots, \text{rep}_{\mathcal{S}}(n_d))^{\#} .$$

As an example, $(ab, bbaa)^{\#} = (\#\#ab, bbaa)$.

Definition 5.7.29 An infinite d-dimensional word $\mathbf{x} \in B^{\mathbb{N}}$ is *S-automatic* for an abstract numeration system $\mathcal{S} = (L, A, <)$, if there exists a DFAO $\mathcal{A} = (Q, q_0, (A \cup \#)^d, \delta, \tau : Q \to B)$ such that, for all $n_1, \ldots, n_d \geq 0$,

$$\tau(\delta(q_0, \text{rep}_{\mathcal{S}}(n_1, \ldots, n_d))) = \mathbf{x}_{n_1, \ldots, n_d}.$$

In this case, we say that the DFAO $\mathcal{A}$ generates the infinite word $\mathbf{x}$.

Example 5.7.30 Consider the ANS $\mathcal{S} = (\{a, ba\}^* \{\varepsilon, b\}, \{a, b\}, a < b)$ and the DFAO depicted in Fig. 5.18. Since this automaton is fed with entries of the form $(\text{rep}_{\mathcal{S}}(n_1), \text{rep}_{\mathcal{S}}(n_2))^{\#}$, we do not consider the transitions on input $(\#, \#)$. If the outputs of the DFAO are considered to be the states themselves, then the DFAO generates the bidimensional infinite $\mathcal{S}$-automatic word given in Fig. 5.19.

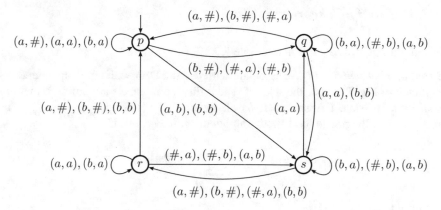

Fig. 5.18 A deterministic finite automaton with output

	ε	a	b	aa	ab	ba	aaa	aab	
aab	q	p	s	p	s	s	p	s	
aaa	p	p	s	p	s	q	p	s	
ba	p	s	q	p	s	q	s	q	
ab	q	p	s	p	s	s	s	r	
aa	p	p	s	p	s	q	q	s	
b	q	p	s	q	s	q	p	s	
a	p	p	s	s	q	s	p	s	
ε	p	q	q	p	q	p	q	q	...

Fig. 5.19 A bidimensional infinite $\mathcal{S}$-automatic word

The main goal of [22] is to prove the following result, an extension of Salon's theorem.

Theorem 5.7.31 *Let $d \geq 1$. The d-dimensional infinite word $\mathbf{x}$ is $\mathcal{S}$-automatic for some abstract numeration system $\mathcal{S} = (L, A, <)$ where $\varepsilon \in L$ if and only if $\mathbf{x}$ is the image by a coding of a shape-symmetric infinite d-dimensional word.*

This result is not surprising: we have seen in the proof of Theorem 5.7.13 that an infinite word is shape-symmetric with respect to a morphism μ if and only if the languages L_{μ_i} are all equal. But to define a $\mathcal{S}$-automatic word, we use the same ANS for the representation of every component.

5.7.3   Proof of Theorem 5.7.14

Recall that φ_W was defined in Example 5.7.12.

Proof of Theorem 5.7.14 We associate with φ_W the DFA with input alphabet (5.9) depicted in Fig. 5.20. For the sake of readability, some labels are omitted. Ingoing transitions to state i (resp. m, d, h) all have label $\binom{0}{0}$ (resp. $\binom{1}{0}$, $\binom{1}{1}$, $\binom{0}{1}$). For r, s, t, u, v belonging to the 13-letter alphabet $\{a, b, \ldots, m\}$, if

$$\varphi_W(r) = \begin{array}{|c|c|} \hline u & v \\ \hline s & t \\ \hline \end{array}, \quad \begin{array}{|c|c|} \hline s & t \\ \hline \end{array}, \quad \begin{array}{|c|} \hline u \\ \hline s \\ \hline \end{array} \quad \text{or} \quad \begin{array}{|c|} \hline s \\ \hline \end{array}$$

we have transitions like

$$r \xrightarrow{\binom{0}{0}} s, \quad r \xrightarrow{\binom{1}{0}} t, \quad r \xrightarrow{\binom{0}{1}} u, \quad r \xrightarrow{\binom{1}{1}} v.$$

From Table 5.9 and Theorem 5.7.13, we already know that φ_W is a morphism prolongable on a.

First observe that, if all states are assumed to be final, this automaton accepts the words

$$\binom{u}{v}$$

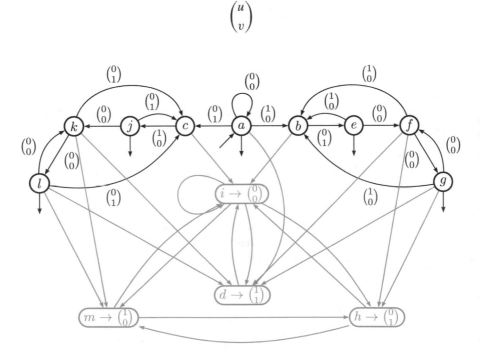

Fig. 5.20 The DFA associated with the morphism φ_W

where $|u| = |v|$ and u, v are both valid F-representation (possibly padded with leading zeroes to get two words of the same length).

Second, if we restrict to the "black" part (still assuming $a, b, c, e, f, g, j, k, l$ to be final), the automaton accepts exactly the words

$$\begin{pmatrix} 0w_1 \cdots w_\ell \\ w_1 \cdots w_\ell 0 \end{pmatrix} \text{ and } \begin{pmatrix} w_1 \cdots w_\ell 0 \\ 0w_1 \cdots w_\ell \end{pmatrix}$$

where $w_1 \cdots w_\ell$ is a valid F-representation.

Finally, taking into account the coding given in Example 5.7.12, the set of final states is $\{a, e, g, j, l\}$. We have the extra acceptance condition that $w_1 \cdots w_\ell$ ends with an even number of zeroes. With our previous characterization of $\mathcal{P}$-positions given by Theorem 5.5.20, this concludes the proof.   □

As a concluding remark, we try to answer the following question. The reader may wonder how we got the morphism φ_W having only access to Table 5.5. We considered some kind of "reverse engineering" strategy. We first conjectured that the Fibonacci word is playing a role (clear from Theorem 5.5.20). Hence, we cut the bidimensional word using, on both axis, the directive sequence $2, 1, 2, 2, 1, 2, 1, 2, 2, 1, \ldots$ deduced from the Fibonacci word $\mathbf{f}'$ (5.5), see Table 5.10. It produces pictures of shape $(2, 2)$, $(2, 1)$, $(1, 2)$ and $(1, 1)$. If a morphism φ_W exists, then every time we see a given symbol, its image must have a constant shape (same observation for iterates). Thus by looking at the future of a symbol, we may distinguish several types of 0's and 1's. With this heuristic, we get a finite number of candidates for symbols and images providing the morphism φ_W. For instance the first three 1's in positions $(0, 0)$, $(1, 2)$ and $(2, 1)$ must correspond to different symbols. Indeed, they should give rise to images of respective shape $(2, 2)$, $(2, 1)$ and $(1, 2)$.

Table 5.10 The bidimensional word $\mathbf{w}(m, n)$ coding $\mathcal{P}$-positions of Wythoff

10	0	0	0	0	0	0	1	0	0	0	0
9	0	0	0	0	0	0	0	0	0	0	0
8	0	0	0	0	0	0	0	0	0	0	0
7	0	0	0	0	1	0	0	0	0	0	0
6	0	0	0	0	0	0	0	0	0	0	1
5	0	0	0	1	0	0	0	0	0	0	0
4	0	0	0	0	0	0	0	1	0	0	0
3	0	0	0	0	0	1	0	0	0	0	0
2	0	1	0	0	0	0	0	0	0	0	0
1	0	0	1	0	0	0	0	0	0	0	0
0	1	0	0	0	0	0	0	0	0	0	0
	0	1	2	3	4	5	6	7	8	9	10

5.8 Games with a Finite Set of Moves

In the very first Example 5.3.1, we were considering a single pile of tokens and subtraction games where only finitely many moves are available. Note that Nim does not belong to this category, one can remove an unbounded number of tokens (whenever available). In this short section, we introduce finite subtraction games on several piles. For one pile, the situation is completely understood [85].

Proposition 5.8.1 *Every finite subtraction game on one pile, i.e., the set $I \subset \mathbb{N}$ of moves is finite, has an ultimately periodic Sprague–Grundy function.*

Proof This is a classical pigeonhole principle argument. Let $m = \#I$ be the maximal number of options for any position. Hence $\mathcal{G}(n) \leq m$ for all n. Let $k = \max I$. Hence, from position $n \geq k$, the options are in $\{n-k, \ldots, n-1\}$. There are $(m+1)^k$ possible k-tuples taking values in $\{0, \ldots, m\}$. Since $\mathcal{G}(n)$ depends only on $\mathcal{G}(n - \ell)$ for $1 \leq \ell \leq k$, hence, by pigeonhole principle, there exist $i < j$ such that

$$\mathcal{G}(i + n) = \mathcal{G}(j + n) \text{ for all } n \in \{0, \ldots, k - 1\}.$$

Thus $j - i$ is a period of $\mathcal{G}$ with preperiod i. □

Another similar result is given in [85, p. 188].

Proposition 5.8.2 *Consider a finite subtraction game on one pile with $I \subset \mathbb{N}$ as set of moves. If there exist $N \geq 0$ and $p \geq 1$ such that*

$$\mathcal{G}(n + p) = \mathcal{G}(n), \quad \forall n \in \{N, \ldots, N - 1 + \max I\}$$

then $\mathcal{G}(n + p) = \mathcal{G}(n)$ for all $n \geq N$.

If we may optionally split a pile, the situation is more intricate. It gives rise to the notion of an *octal game* which is played with tokens divided into piles. Two players take turns moving until no moves are possible. The name came from the fact that the rules are coded by words over $\{0, \ldots, 7\}$.

Definition 5.8.3 (Octal Game) Every move consists of selecting just one of the piles, and either

- removing all of the tokens in the pile, leaving no pile,
- removing some but not all of the tokens, leaving one smaller pile, or
- removing some of the tokens and dividing the remaining tokens into two nonempty piles.

Piles other than the selected pile remain unchanged. The last player to move wins in a normal play convention.

The coding of an octal game (known as *Conway code*) is an infinite word

$$d_0 \bullet d_1 d_2 d_3 \cdots \quad d_i \in \{0, \ldots, 7\}$$

where d_i written in base 2 has a fixed 3-digit length, $e_2^{(i)} e_1^{(i)} e_0^{(i)} \in \{0, 1\}^3$. It gives the conditions under which i tokens may be removed.

- if $e_0^{(i)} = 1$, then a (full) pile with i tokens can be suppressed;
- if $e_1^{(i)} = 1$, then a pile with $n > i$ tokens can be replaced with a pile with $n - i$ tokens left;
- if $e_2^{(i)} = 1$, then a pile with $n > i + 1$ tokens can be replaced with two piles containing respectively a and b tokens, $a, b \geq 1, a + b = n - i$.

Example 5.8.4 The game of Nim on an arbitrary number of piles is coded by $0 \bullet 3333 \cdots$. Indeed, with a leading zero $\text{rep}_2(3) = 011$. In general, classical subtraction games where a pile cannot be split into two piles is coded by a word over $\{0, 1, 2, 3\}$.

A subtraction game is finite if and only if it is coded by a finite word (i.e., an infinite word with only finitely many non-zero digits), e.g. with a set of moves $I = \{3, 5, 6\}$, the game is coded by $0 \bullet 003033$. The game of Example 5.3.1 is coded by $0 \bullet 3303$.

Theorem 5.8.5 (Octal Game Periodicity [85]) *Consider a finite octal game coded by $d_0 \bullet d_1 d_2 \cdots d_k$ with $d_k \neq 0$. If there exist $N \geq 0$ and $p \geq 1$ such that*

$$\mathcal{G}(n + p) = \mathcal{G}(n), \quad \forall n \text{ with } N \leq n < 2N + p + \max I$$

then $\mathcal{G}(n + p) = \mathcal{G}(n)$ for all $n \geq N$.

A general open problem is to determine whether all finite octal games have an ultimately periodic Grundy function. For instance, $0 \bullet 07$ has period 34 and preperiod 53; $0 \bullet 165$ has period 1550 and preperiod 5181; $0 \bullet 106$ has period $\simeq 3.10^{11}$ and preperiod $\simeq 4.10^{11}$. Up to our knowledge, $0 \bullet 007$ has no known periodicity. See [10, 49, 85].

In view of Proposition 5.8.1, one can conjecture that for a finite subtraction game on two (or more) piles of tokens, the Sprague–Grundy function should be definable in the Presburger arithmetic $\langle \mathbb{N}, + \rangle$ or, equivalently, each value of $\mathcal{G}$ should correspond to a semilinear set in $\mathbb{N}^2$. Indeed, this is exactly the situation encountered when considering generalizations to $\mathbb{N}^d$ of Cobham's theorem (Theorem 5.4.21) to Cobham–Semenov theorem [17]. See the latter survey for precise definitions.

Definition 5.8.6 A set X of $\mathbb{N}^n$ is *linear* if there exist $v_0, v_1, \ldots, v_k \in \mathbb{N}^n$ such that

$$X = v_0 + \mathbb{N} v_1 + \cdots + \mathbb{N} v_k.$$

The vectors $v_1, \ldots, v_k$ are usually called the *periods* of X. A set X of $\mathbb{N}^n$ is *semilinear* if it is a finite union of linear sets.

The next facts follow from a work in progress with X. Badin De Montjoye, V. Gledel, V. Marsault and A. Massuir.

Proposition 5.8.7 *Every finite subtraction game on two piles with at most two moves, has a Sprague–Grundy function definable in* $\langle \mathbb{N}, + \rangle$.

In Fig. 5.21, we have represented $\mathcal{G}$-value 0 in white. Darker points correspond to higher Grundy values. In this figure, we have considered two piles of tokens with the moves $\alpha = (2, 1)$ and $\beta = (3, 5)$. The reader may notice that $\alpha + \beta$ is a vector of period. On the right of Fig. 5.21, we have depicted an example with 3 moves $(1, 3)$, $(3, 1)$ and $(4, 4)$.

Proof If there is a single move, the result is clear. Assume that we have the moves $\alpha = (\alpha_1, \alpha_2)$ and $\beta = (\beta_1, \beta_2)$. The first $2 \max(\alpha_1, \beta_1)$ columns of $(\mathcal{G}(m, n))_{m,n \geq 0}$ are ultimately periodic. Similarly, the first $2 \max(\alpha_2, \beta_2)$ rows are ultimately periodic. We show that $\alpha + \beta$ is a vector of period. Let $x \in \mathbb{N}^2$ be a position such that $x - \alpha$ and $x - \beta$ belong to $\mathbb{N}^2$. We will prove that

$$\mathcal{G}(x) = \mathcal{G}[x + \alpha + \beta].$$

Assume that $\mathcal{G}(x) = 0$ and $\mathcal{G}[x + \alpha + \beta] = 2$. From the position $x + \alpha + \beta$, we can either subtract α or β. By definition of the Grundy function, $\{\mathcal{G}[x + \alpha], \mathcal{G}[x + \beta]\} = \{0, 1\}$. But this contradicts the fact that $\mathcal{G}(x) = 0$; we would have two consecutive positions x and either $x + \alpha$ or $x + \beta$ with the same $\mathcal{G}$-value. The reasoning is similar if we assume that $\mathcal{G}[x + \alpha + \beta] = 1$.

Assume that $\mathcal{G}(x) = 1$ and $\mathcal{G}[x + \alpha + \beta] = 2$. From the last equality, we have $\{\mathcal{G}[x + \alpha], \mathcal{G}[x + \beta]\} = \{0, 1\}$. One of these two positions leads directly to x. This is a contradiction.

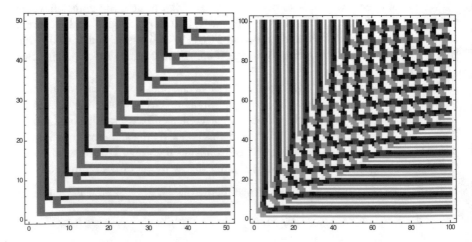

Fig. 5.21 First values of $\mathcal{G}$ with $I = \{(2, 1), (3, 5)\}$ (left) and $I = \{(1, 3), (3, 1), (4, 4)\}$ (right)

Assume that $\mathcal{G}(x) = 1$ and $\mathcal{G}[x + \alpha + \beta] = 0$. These equalities imply that $\mathcal{G}[x + \alpha] = \mathcal{G}[x + \beta] = 2$. Then, considering the options of these last two positions, we deduce that $\mathcal{G}[x + \alpha - \beta] = \mathcal{G}[x + \beta - \alpha] = 0$. Since, $\mathcal{G}(x) = 1$, either $\mathcal{G}[x - \beta] = 0$ or, $\mathcal{G}[x - \alpha] = 0$ contradicting the previous equality.

Assume that $\mathcal{G}(x) = 2$ and $\mathcal{G}[x + \alpha + \beta] = 0$. This implies that $\mathcal{G}[x + \alpha] = \mathcal{G}[x + \beta] = 1$. Then, considering the options of these two positions, we deduce that $\mathcal{G}[x + \alpha - \beta] = \mathcal{G}[x + \beta - \alpha] = 0$. Since, $\mathcal{G}(x) = 2$, then $\{\mathcal{G}[x - \beta], \mathcal{G}[x - \alpha]\} = \{0, 1\}$ contradicting the previous equality. The reasoning is similar if we assume that $\mathcal{G}[x + \alpha + \beta] = 1$. $\square$

We conjecture that with two piles of tokens and the set of moves

$$I = \{(1, 2), (2, 1), (3, 5), (5, 3), (2, 2)\},$$

the corresponding word $(\mathcal{G}(m, n))_{m,n \geq 0}$ is not definable in $\langle \mathbb{N}, + \rangle$. We can therefore wonder how Proposition 5.8.1 could be generalized to two or more piles of tokens.

5.9   Bibliographic Notes

The link between morphisms and automata which is a cornerstone of this chapter can already be found in the fundamental work of Cobham [31]. In this chapter, we did not present the Dumont–Thomas approach relating substitutions to numeration systems. See [37]. The papers [65, 66] also develop similar constructions. Several surveys are of interest, see [17] for integer base systems, [71] where connections with games are also mentioned.

A few papers are dealing with combinatorial games linked with morphic words. See [34–36, 46, 57]. In particular, Theorem 5.7.14 was proved in [33]. For instance, we have also considered alterations (adding/removing moves) of the set of moves in order to keep the same set of $\mathcal{P}$-positions as the original game. We may characterize moves that can be adjoined without changing the $\mathcal{P}$-positions of Wythoff's game. No move is redundant. The notion of *invariant game* is introduced in [36]. About Wythoff's game, see [42] and [45]. For connections between games and Beatty sequences, see [36] and then [20, 57].

We did not discuss much about the logical characterization of k-automatic sequences (and generalizations to Pisot systems). See again [17, 24] and also the last chapter of [73] for a comprehensive introduction. Maes's motivations were primarily set on decidability of arithmetic theories: which expansions of $\langle \mathbb{N}, < \rangle$ by morphic predicates or automata are decidable? See also [19] where it is shown that for a morphic predicate P the associated monadic second-order theory $MTh\langle \mathbb{N}, <, P \rangle$ is decidable. A trace of abstract numeration system (not mentioned in these terms at that time) can already be found in Maes's thesis [64, Rem. 6.9, p. 134]: *"The set of codes of $\mathbb{N}$ given by the above automaton is of course a regular language... The language read by A is 0^*L. However, the above coding is not a numeration system* *in the sense of [16]. Indeed, the representation of a natural number is not obtained using a 'Euclidean division' algorithm.".* In some sense, Maes was conjecturing Theorem 5.7.26 and Proposition 5.7.27: the set of morphic words is equal to the set of S-automatic words, for some ANS S.

Acknowledgments Even though it was some hard work preparing the lectures and this chapter, I am quite happy with the final result (with this kind of exercise, you always selfishly learn a lot and ask yourself new questions). I therefore warmly thank Professors Shigeki Akiyama and Pierre Arnoux for their invitation to contribute to this school. I would also like to thank Eric Duchêne for his constant help when collaborating on game related problems. I also had several colleagues reading drafts of this chapter: first M. Stipulanti and then, E. Duchêne, J. Leroy and A. Parreau. I thank them all for their feedback. Finally, I thank the anonymous reviewer for his/her careful reading and many suggestions.

References

1. M.H. Albert, R.J. Nowakowski (eds.), Games of no chance 3, in *Mathematical Sciences Research Institute Publications*, vol. 56 (Cambridge University Press, Cambridge, 2009). Papers from the Combinatorial Game Theory Workshop held in Banff, AB, June 2005
2. M.H. Albert, R.J. Nowakowski, D. Wolfe, *Lessons in Play: An Introduction to Combinatorial Game Theory* (A K Peters/CRC Press, 2007)
3. J.-P. Allouche, V. Berthé, Triangle de Pascal, complexité et automates. Bull. Belg. Math. Soc. Simon Stevin **4**(1), 1–23 (1997). Journées Montoises (Mons, 1994)
4. J.-P. Allouche, F. von Haeseler, H.-O. Peitgen, A. Petersen, G. Skordev, Automaticity of double sequences generated by one-dimensional linear cellular automata. Theoret. Comput. Sci. **188**(1–2), 195–209 (1997)
5. J.-P. Allouche, F. von Haeseler, Peitgen, H.-O., G. Skordev, Linear cellular automata, finite automata and Pascal's triangle. Discrete Appl. Math. **66**(1), 1–22 (1996)
6. J.-P. Allouche, J. Shallit, The ring of k-regular sequences. Theoret. Comput. Sci. **98**(2), 163–197 (1992)
7. J.-P. Allouche, J. Shallit, The ubiquitous Prouhet–Thue–Morse sequence, in *Sequences and Their Applications (Singapore, 1998)*. Springer Series of Discrete Mathematics and Theoretical Computer Science (Springer, London, 1999), pp. 1–16
8. J.-P. Allouche, J. Shallit, *Automatic Sequences, Theory, Applications, Generalizations* (Cambridge University Press, Cambridge, 2003)
9. J.-P. Allouche, J. Shallit, The ring of k-regular sequences. II. Theoret. Comput. Sci. **307**(1), 3–29 (2003)
10. R.B. Austin, *Impartial and Partisan Games*. Master's thesis, Univ. of Calgary (1976)
11. A. Barbé, F. von Haeseler, Limit sets of automatic sequences. Adv. Math. **175**(2), 169–196 (2003)
12. C. Berge, *The Theory of Graphs* (Dover Publications, Mineola, 2001)
13. E.R. Berlekamp, J.H. Conway, R.K. Guy, *Winning Ways for Your Mathematical Plays*, vol. 1, 2nd edn. (A K Peters, Natick, 2001)
14. U. Blass, A.S. Fraenkel, The Sprague–Grundy function for Wythoff's game. Theoret. Comput. Sci. **75**(3), 311–333 (1990)
15. C.L. Bouton, Nim, a game with a complete mathematical theory. Ann. Math. **3**, 35–39 (1905)
16. V. Bruyère, G. Hansel, Bertrand numeration systems and recognizability. Theoret. Comput. Sci. **181**(1), 17–43 (1997)
17. V. Bruyère, G. Hansel, C. Michaux, R. Villemaire, Logic and p-recognizable sets of integers. Bull. Belg. Math. Soc. Simon Stevin **1**(2), 191–238 (1994)

18. A. Carpi, C. Maggi, On synchronized sequences and their separators. Theor. Inform. Appl. **35**(6), 513–524 (2002) (2001). A tribute to Aldo de Luca
19. O. Carton, W. Thomas, The monadic theory of morphic infinite words and generalizations. Inf. Comput. **176**, 51–65 (2002)
20. J. Cassaigne, E. Duchêne, M. Rigo, Non-homogeneous Beatty sequences leading to invariant games. SIAM J. Discret. Math. **30**, 1798–1829 (2016)
21. J. Cassaigne, F. Nicolas, Factor complexity, in *Combinatorics, Automata and Number Theory*. Encyclopedia of Mathematics and Its Applications, vol. 135 (Cambridge University Press, Cambridge, 2010), pp. 163–247
22. É. Charlier, T. Kärki, M. Rigo, Multidimensional generalized automatic sequences and shape-symmetric morphic words. Discret. Math. **310**(6–7), 1238–1252 (2010)
23. É. Charlier, J. Leroy, M. Rigo, Asymptotic properties of free monoid morphisms. Linear Algebra Appl. **500**, 119–148 (2016)
24. É. Charlier, N. Rampersad, J. Shallit, Enumeration and decidable properties of automatic sequences. Internat. J. Found. Comput. Sci. **23**(5), 1035–1066 (2012)
25. É. Charlier, M. Rigo, W. Steiner, Abstract numeration systems on bounded languages and multiplication by a constant. Integers **8**, #A35 (2008)
26. Ch. Choffrut, W. Goldwurm, Rational transductions and complexity of counting problems. Math. Syst. Theory **28**(5), 437–450 (1995)
27. Ch. Choffrut, J. Karhumäki, Combinatorics of words, in *Handbook of Formal Languages*, vol. 1 (Springer, Berlin, 1997), pp. 329–438
28. G. Christol, T. Kamae, M. Mendès France, G. Rauzy, Suites algébriques, automates et substitutions. Bull. Soc. Math. France **108**(4), 401–419 (1980)
29. A. Cobham, On the Hartmanis-Stearns problem for a class of tag machines, in *IEEE Conference Record of 1968 Ninth Annual Symposium on Switching and Automata Theory* (1968), pp. 51–60. Also appeared as IBM Research Technical Report RC-2178, August 23 1968
30. A. Cobham, On the base-dependence of sets of numbers recognizable by finite automata. Math. Syst. Theory **3**, 186–192 (1969)
31. A. Cobham, Uniform tag sequences. Math. Syst. Theory **6**, 164–192 (1972)
32. A. Dress, A. Flammenkamp, N. Pink, Additive periodicity of the Sprague–Grundy function of certain Nim games. Adv. Appl. Math. **22**, 249–270 (1999)
33. E. Duchêne, A.S. Fraenkel, R.J. Nowakowski, M. Rigo, Extensions and restrictions of Wythoff's game preserving its *P* positions. J. Combin. Theory Ser. A **117**(5), 545–567 (2010)
34. E. Duchêne, M. Rigo, Cubic Pisot unit combinatorial games. Monatsh. Math. **155**(3–4), 217–249 (2008)
35. E. Duchêne, M. Rigo, A morphic approach to combinatorial games: the Tribonacci case. Theor. Inform. Appl. **42**(2), 375–393 (2008)
36. E. Duchêne, M. Rigo, Invariant games. Theor. Comput. Sci. **411**(34–36), 3169–3180 (2010)
37. J.M. Dumont, A. Thomas, Systèmes de numération et fonctions fractales relatifs aux substitutions. Theoret. Comput. Sci. **65** (1989)
38. F. Durand, Cobham's theorem for substitutions. J. Eur. Math. Soc. **13**(6), 1799–1814 (2011)
39. S. Eilenberg, Automata, Languages, and Machines, vol. A (Academic Press [A subsidiary of Harcourt Brace Jovanovich, Publishers], New York, 1974). Pure and Applied Mathematics, vol. 58
40. N.P. Fogg, in *Substitutions in Dynamics, Arithmetics and Combinatorics*, ed. by V. Berthé, S. Ferenczi, C. Mauduit A. Siegel. Lecture Notes in Mathematics, vol. 1794 (Springer, Berlin, 2002)
41. A.S. Fraenkel, The bracket function and complementary sets of integers. Can. J. Math. **21**, 6–27 (1969)
42. A.S. Fraenkel, How to beat your Wythoff games' opponent on three fronts. Am. Math. Mon. **89**(6), 353–361 (1982)

43. A.S. Fraenkel, The use and usefulness of numeration systems, in *Combinatorial Algorithms on Words (Maratea, 1984)*. NATO Adv. Sci. Inst. Ser. F Comput. Systems Sci., vol. 12 (Springer, Berlin, 1985), pp. 187–203
44. A.S. Fraenkel, Complexity, appeal and challenges of combinatorial games. Theoret. Comput. Sci. **313**(3), 393–415 (2004). Algorithmic combinatorial game theory
45. A.S. Fraenkel, Euclid and Wythoff games. Discret. Math. **304**, 65–68 (2005)
46. A.S. Fraenkel, Complementary iterated floor words and the Flora game. SIAM J. Discret. Math. **24**(2), 570–588 (2010)
47. Ch. Frougny, On the sequentiality of the successor function. Inf. Comput. **139**(1), 17–38 (1997)
48. S.W. Golomb, A mathematical investigation of games of "take-away". J. Combin. Theory **1**, 443–458 (1966)
49. R.K. Guy, C.A. Smith, The *G*-values of various games. Proc. Camb. Philos. Soc. **52**, 514–526 (1956)
50. F. von Haeseler, H.-O. Peitgen, G. Skordev, Self-similar structure of rescaled evolution sets of cellular automata. I. Internat. J. Bifur. Chaos Appl. Sci. Eng. **11**(4), 913–926 (2001)
51. J. Honkala, Bases and ambiguity of number systems. Theoret. Comput. Sci. **31**(1–2), 61–71 (1984)
52. J. Honkala, On unambiguous number systems with a prime power base. Acta Cybernet. **10**(3), 155–163 (1992)
53. J. Honkala, On the simplification of infinite morphic words. Theoret. Comput. Sci. **410**(8–10), 997–1000 (2009)
54. G. Katona, A theorem of finite sets, in *Theory of Graphs (Proceedings of the Colloquium Tihany, 1966)* (Academic Press, New York, 1968), pp. 187–207
55. Krebs, T.J.P.: A more reasonable proof of Cobham's theorem. CoRR abs/1801.06704 (2018). http://arxiv.org/abs/1801.06704
56. H.A. Landman, More games of no chance, in *Chap. A Simple FSM-Based Proof of the Additive Periodicity of the Sprague–Grundy Function of Wythoff's Game* (Cambridge University Press, Cambridge, 2002)
57. U. Larsson, P. Hegarty, A.S. Fraenkel, Invariant and dual subtraction games resolving the Duchêne–Rigo conjecture. Theoret. Comput. Sci. **412**(8–10), 729–735 (2011)
58. M.V. Lawson, *Finite Automata* (Chapman & Hall/CRC, Boca Raton, 2004)
59. P. Lecomte, M. Rigo, Abstract numeration systems, in *Combinatorics, Automata and Number Theory*. Encyclopedia of Mathematics and its Applications, vol. 135 (Cambridge University Press, Cambridge, 2010), pp. 108–162
60. P.B.A. Lecomte, M. Rigo, Numeration systems on a regular language. Theory Comput. Syst. **34**(1), 27–44 (2001)
61. J. Leroy, M. Rigo, M. Stipulanti, Generalized Pascal triangle for binomial coefficients of words. Adv. in Appl. Math. **80**, 24–47 (2016)
62. M. Lothaire, in *Combinatorics on Words*. Cambridge Mathematical Library (Cambridge University Press, Cambridge, 1997)
63. M. Lothaire, in *Algebraic Combinatorics on Words*. Encyclopedia of Mathematics and its Applications, vol. 90 (Cambridge University Press, Cambridge, 2002)
64. A. Maes, Morphic predicates and applications to the decidability of arithmetic theories. Ph.D. thesis, UMH Univ. Mons-Hainaut (1999)
65. V. Marsault, J. Sakarovitch, The signature of rational languages. Theoret. Comput. Sci. **658**, 216–234 (2017)
66. V. Marsault, J. Sakarovitch, Trees and languages with periodic signature. Indag. Math. **28**(1), 221–246 (2017)
67. R.J. Nowakowski (ed.), Games of no chance 4, in *Mathematical Sciences Research Institute Publications*, vol. 63 (Cambridge University Press, New York, 2015). Selected papers from the Banff International Research Station (BIRS) Workshop on Combinatorial Games held in Banff, AB, 2008
68. K. O'Bryant, Fraenkel's partition and brown's decomposition. Integers **3** (2003)

69. D. Perrin, Finite automata, in *Handbook of Theoretical Computer Science*, vol. B (Elsevier, Amsterdam, 1990), pp. 1–57
70. M. Rigo, Generalization of automatic sequences for numeration systems on a regular language. Theoret. Comput. Sci. **244**(1–2), 271–281 (2000)
71. M. Rigo, Numeration systems: a link between number theory and formal language theory, in *Developments in Language Theory*. Lecture Notes in Computer Science, vol. 6224 (Springer, Berlin, 2010), pp. 33–53
72. M. Rigo, *Formal Languages, Automata and Numeration Systems 1*. (ISTE/ Wiley, London/Hoboken, 2014). Introduction to Combinatorics on Words, With a foreword by Valérie Bethé
73. M. Rigo, *Formal Languages, Automata and Numeration Systems 2*. Networks and Telecommunications Series (ISTE/ Wiley, London/Hoboken, 2014). Applications to recognizability and decidability, With a foreword by Valérie Bethé
74. M. Rigo, Advanced graph theory and combinatorics, in *Networks and Telecommunications Series*. ISTE, London (Wiley, Hoboken, 2016)
75. M. Rigo, A. Maes, More on generalized automatic sequences. J. Autom. Lang. Comb. **7**(3), 351–376 (2002)
76. M. Rigo, L. Waxweiler, A note on syndeticity, recognizable sets and Cobham's theorem. Bull. Eur. Assoc. Theor. Comput. Sci. EATCS **88**, 169–173 (2006)
77. E. Rowland, Binomial coefficients, valuations, and words, in *Developments in Language Theory*. Lecture Notes in Computer Science, vol. 10396 (Springer, Berlin, 2017), pp. 68–74
78. E. Rowland, R. Yassawi, A characterization of p-automatic sequences as columns of linear cellular automata. Adv. in Appl. Math. **63**, 68–89 (2015)
79. J. Sakarovitch, *Elements of Automata Theory* (Cambridge University Press, Cambridge, 2009)
80. P.V. Salimov, On uniform recurrence of a direct product. Discret. Math. Theoret. Comput. Sci. **12**, 1–8 (2010)
81. O. Salon, Suites automatiques à multi-indices et algébricité. C. R. Acad. Sci. Paris Sér. I Math. **305**(12), 501–504 (1987)
82. O. Salon, Suites automatiques à multi-indices, in *Séminaire de Théorie des Nombres de Bordeaux* (1986/1987), pp. 4.01–4.27. Followed by an appendix by J. Shallit
83. J. Shallit, *A Second Course in Formal Languages and Automata Theory* (Cambridge University Press, Cambridge, 2008)
84. J.O. Shallit, A generalization of automatic sequences. Theoret. Comput. Sci. **61**, 1–16 (1988)
85. A.N. Siegel, in *Combinatorial Game Theory*. Graduate Studies in Mathematics, vol. 146 (American Mathematical Society, Providence, 2013)
86. T.A. Sudkamp, *Languages and Machines: An Introduction to the Theory of Computer Science* (Addison-Wesley, Reading, 2006)
87. W.A. Wythoff, A modification of the game of nim. Nieuw Arch. Wiskd. **7**, 199–202 (1907)
88. É. Zeckendorf, Représentation des nombres naturels par une somme de nombres de Fibonacci ou de nombres de Lucas. Bull. Soc. Roy. Sci. Liège **41**, 179–182 (1972)

Chapter 6
The Undecidability of the Domino Problem

Emmanuel Jeandel and Pascal Vanier

Abstract One of the most fundamental problems in tiling theory is to decide, given a surface, a set of tiles and a tiling rule, whether there exists a way to tile the surface using the set of tiles and following the rules. As proven by Berger (The Undecidability of the Domino Problem. Ph.D. thesis, Harvard University, 1964) in the 1960s, this problem is undecidable in general. When formulated in terms of tilings of the discrete plane $\mathbb{Z}^2$ by unit tiles with colored constraints, this is called the Domino Problem and was introduced by Wang (Bell Syst Tech J 40:1–41, 1961) in an effort to solve satisfaction problems for $\forall\exists\forall$ formulas by translating the problem into a geometric problem. There exist a few different proofs of this result. The most well-known proof is probably the proof by Robinson (Invent Math 12(3):177–209, 1971) which is a variation on the proof of Berger. A relatively new proof by Kari (Machines, computations, and universality (MCU). vol. 4664, 2007, pp. 72–79) has some nice ramifications for tilings of surfaces and groups. In terms of ingredients, one can divide the proofs in 4 categories. The remaining two categories are given by the proof of Aanderaa and Lewis (J Symb Log 39(3):519–548, 1974) and the fixed point method of Durand et al. (J Comput Syst Sci 78(3):731–764 , 2012). In this course, we will give a brief description of the problem and to the meaning of the word "undecidable", and then give the four different proofs. As we will explain, the undecidability of the Domino Problem has as a consequence the existence of an aperiodic tileset. All four sections will be organized in such a way that the interested reader can first extract from the proof the aperiodic tileset into consideration, before we go into more details to actually prove the undecidability of the problem.

E. Jeandel (✉)
Université de Lorraine, CNRS, Inria, LORIA, Nancy, France
e-mail: emmanuel.jeandel@loria.fr

P. Vanier
Laboratoire d'Algorithmique, Complexité et Logique Université de Paris-Est, LACL, UPEC, Créteil, France
e-mail: pascal.vanier@lacl.fr

© The Editor(s) (if applicable) and The Author(s), under exclusive license
to Springer Nature Switzerland AG 2020
S. Akiyama, P. Arnoux (eds.), *Substitution and Tiling Dynamics: Introduction to Self-inducing Structures*, Lecture Notes in Mathematics 2273,
https://doi.org/10.1007/978-3-030-57666-0_6

293

6.1 Statement and Consequences

6.1.1 Definitions and Statement

Definition 6.1.1 (Wang [51]) Let C be a finite set, called the set of colors. A Wang tile t over C is a map from $\{N, S, E, W\}$ to C.

A tileset τ is a finite set of Wang tiles.

A first example is given in Fig. 6.1. C can be thought as a set of colors, patterns, symbols or integers.

Definition 6.1.2 Let $P \subseteq \mathbb{Z}^2$, a coloring x of support P assigns to each element of P a tile of τ. When $P = \mathbb{Z}^2$, x is a *configuration*.

A tiling of P by τ assigns to each element of P a tile from τ such that colors of adjacents tiles agree on their common border. Formally, a tiling is a map $x : P \to \tau$ such that:

- If $(i, j) \in P$ and $(i + 1, j) \in P$ then $x(i, j)(E) = x(i + 1, j)(W)$
- If $(i, j) \in P$ and $(i, j + 1) \in P$ then $x(i, j)(N) = x(i, j + 1)(S)$

When $P = \mathbb{Z}^2$, it is a *tiling of the plane*. We say that τ *tiles* P if there exists a tiling of P by τ.

An example of a tiling of a finite set by the tileset τ of Fig. 6.1 is given in Fig. 6.2.

Although this particular tileset, as evidenced by the figure, can be used to tile a large rectangle, it turns out that it doesn't tile the entire plane. In fact, it cannot tile a square of size 15. In general, knowing if a tileset τ tiles the plane is a hard problem:

Theorem 6.1.3 (Berger [7]) *There is no algorithm that, given a set of Wang tiles τ, decides if τ tiles the plane.*

In the rest of these notes, we will explain what this theorem means, and how to prove it.

6.1.2 Algorithmic Consequences

In this section, we want to investigate what the theorem of Berger actually means in practice when one studies tilings.

Fig. 6.1 A tileset τ_1 composed of 7 tiles

Fig. 6.2   A tiling of a finite rectangle by τ_1

Let τ be a set of Wang tiles. τ falls in three different cases:

- τ does not tile the plane
- τ tiles the plane, and a periodic tiling by τ exists
- τ tiles the plan, but no periodic tiling exists. We will say that τ is *aperiodic*.

There are various possible definitions of what a periodic tiling is, but they are all equivalent in our case. We will say that a tiling x is periodic if there exists p s.t $x(i, j) = x(i \mod p, j \mod p)$.

From the algorithmic point of view, it is easy to see if τ does not tile the plane. Indeed

Proposition 6.1.4 (Compactness (Folklore)) τ *tiles the plane iff for all n, the tileset τ tiles a square of size $n \times n$.*

This proposition gives us an algorithm to prove that τ does not tile the plane: For all n, find a way to tile with τ an $n \times n$ square. If it is not possible for one value of n, then accept.

The second case of the trichotomy is also easy to test algorithmically: For all n, try to find a way to tile a $n \times n$ square in a periodic manner. If this succeeds for some n, then accept.

As the Domino Problem is undecidable, this means there is no algorithm to test for the third case: Indeed, combining the three algorithms together would solve the Domino Problem. Intuitively, this means that there exists aperiodic tilesets τ for which we cannot prove that they are aperiodic. This intuition is made a bit more precise if we look at what Berger proved exactly:

Theorem 6.1.5 (Berger [7]) *We can transform effectively any program p into a tileset τ s.t.*

- *If p halts, then τ does not tile the plane.*
- *If p does not halt, then τ tiles the plane aperiodically.*

It is well known that no algorithm can decide if a program halts (see Theorem 6.2.4 for a proof), and therefore this theorem implies the undecidability of the Domino problem. It also means in terms of provability that, starting from a program that tries to prove some mathematical statement P, we could build a tileset τ s.t. proving that τ is aperiodic is equivalent to proving the property P.

Berger's theorem tells us that we cannot decide whether a set τ tiles the plane. However, it doesn't say anything about what the tilings, if they exist, look like.

Recall that Berger's proof, starting from a program p, builds a set of tiles τ s.t. τ tiles the plane iff p does not halt. It turns out that the tilings by τ, if they exist, are easy to build: There exists a tiling x by τ and a program that on input (n, m) outputs in finite time the tile of x at position (n, m). This program doesn't even need to know if there exists a tiling by τ: If no such tiling exists, it will output "error" for some value of n and m.

The situation as described by Berger is therefore not as hopeless as it could be. Later results show however that there exist tilesets that are much more complicated that the ones Berger built.

Theorem 6.1.6 (Hanf–Myers [21, 41]) *There exists a tileset τ s.t. tilings by τ exist but no tilings by τ can be produced by a program.*

We also mention the following curiosity:

Theorem 6.1.7 (Levin [35]) *There exists a tileset τ s.t.:*

- *Tilings by τ exist*
- *No tilings by τ can be produced by a program.*
- *There exists a probabilistic program (i.e. with a random coin) that will produce a tiling by τ with probability at least $1/2$*

The constant $1/2$ can be replaced by $1 - \epsilon$ for any $\epsilon > 0$. This result is stated in [35] but no formal proof is given. One could obtain the result using e.g. the results of Simpson [49] and some basic notions of algorithmic randomness theory.

6.1.3   Algorithms to Prove A Tileset Does Not Tile the Plane

While the theorem of Berger prevents any algorithm to succeed in deciding that a tileset tiles the plane in the general case, it is still interesting to find algorithms that can do it in particular cases. As evidenced by the previous discussion, the easy part is to prove that a tileset does not tile the plane: it is sufficient to find an integer N s.t. no tilings of a square of size N exists.

From the point of view of complexity, better algorithms exist and we will investigate them in this section. Of course, none of these algorithms work: Some of them will not always terminate, some of them may answer "I don't know".

6.1.3.1   The First Algorithm

An easy algorithm is therefore to test, for all N, if there exists a tiling of a square of size $N \times N$. While this is the easiest algorithm to describe, it is not easy to implement in a reasonable manner. It is important for this algorithm to not only test if there exists a tiling of a square of size $N \times N$, but to enumerate all of them: This information will indeed be useful when trying to tile squares of size $(N+1) \times (N+1)$: it suffices to take all squares of size $N \times N$ and to find all ways of completing them into squares of bigger sizes.

Of course there might be a large number of possible tilings: If we denote by $|\tau|$ the number of tiles of τ, we see that we can have at most $|\tau|^{N \times N}$ tilings of a $N \times N$ square. This means generating all of them might be costly, and that our algorithm will be of complexity $2^{O(N^2)})$. We can do a bit better by realizing that we do not need to know what the tilings of the square look like, but only what they look like on the *border*. Indeed, the only information we need to complete a tiling of a $N \times N$ square into a tiling of a $(N+1) \times (N+1)$ square is the colors on the border of the square. Doing this reduces the complexity of the whole operation to $2^{O(N)}$. See Fig. 6.3 for an example.

For the tileset τ_1 of Fig. 6.2, it can be proven that there is no tiling of a 15×15 square, so that the tileset τ_1 indeed does not tile the plane.

Fig. 6.3 As the two colorings have the same border, they can be considered to be equal when listing all tilings of a 2×2 square

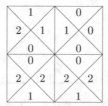
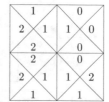

6.1.3.2 Graphs and Automata

It is possible for a tileset τ to tile a square of size $N \times N$ but for it to not even tile an entire row. We might expect therefore better results by looking at tilings of N consecutive rows rather than tilings of squares. Before explaining how this can be done, we remark the following:

Proposition 6.1.8 *Let τ be a tileset consisting of $q = |\tau|$ Wang tiles.*
If τ tiles a horizontal strip of height N, τ tiles a square of size $N \times N$.
If τ tiles a square of size $N \times N$, it tiles a horizontal strip of height $\lfloor \log_q N \rfloor$.

Proof The first statement is obvious. For the second statement, suppose that τ tiles a square of size $N \times N$. In particular, τ tiles a rectangle of height $\lfloor \log_q N \rfloor$ and length N.

We look at each column of colors we see inside this rectangle. As the rectangle is of length N, there are $N + 1$ different positions. As the rectangle is of height $\lfloor \log_q N \rfloor$, at most $q^{\lfloor \log_q N \rfloor} < N + 1$ different associations of color can appear.

By the pigeonhole principle, some column of colors appear at least twice. We can therefore obtain a tiling of an horizontal strip by repeating periodically a subrectangle. See Fig. 6.4.

The bound we obtain is not tight.

It remains to explain how to test efficiently if there is a tiling of a horizontal strip of height N.

The key is to represent the tileset as a finite automaton (more accurately finite transducers). The representation is as follows: The states of our finite automaton are the colors C. For each tile $t \in \tau$, there is an edge from $t[W]$ to $t[E]$ labelled with $(t[S], t[N])$.

Figure 6.5 represents τ_1 as an automaton. In this representation, it is easy to see if τ_1 tiles a infinite strip of height 1. Indeed, an infinite strip of height 1 represents one (biinfinite) run of the automaton.

Proposition 6.1.9 τ_1 *tiles an infinite strip of height 1 iff the automaton that represents τ_1 has a directed cycle.*

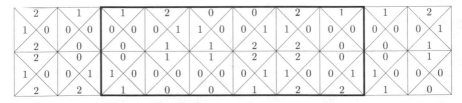

Fig. 6.4 This rectangle of large width contains (by pigeonhole) a smaller rectangle (in bold) with the same colors on the east side and the west side. This rectangle can be used in a periodic fashion to obtain a tiling of a horizontal strip of height 2

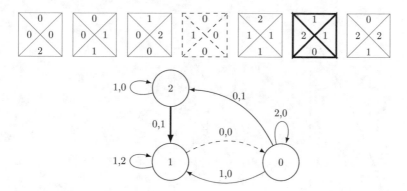

Fig. 6.5 The tileset τ_1 and its representation as an automaton. The bold (resp. dashed) transition correspond to the bold (resp. dashed) tile

A strip of height 2 represents therefore two infinite runs of the automaton, s.t. the north labels of the automaton at the bottom agree with the south labels of the automaton on top. This suggest the following definition:

Proposition 6.1.10 *Let $\mathcal{A}$ and $\mathcal{B}$ be two automata, with states Q_A and Q_B. The product of $\mathcal{A}$ and $\mathcal{B}$ is the automaton with states $Q_A \times Q_B$ where there is an edge from (q_a, q_b) to (q'_a, q'_b) labeled (s_a, n_b) iff there is a color c and an edge from q_a to q'_a labeled (s_a, c) in $\mathcal{A}$ and an edge from q_b to q'_b labeled (c, n_b) in $\mathcal{B}$.*

Intuitively, $\mathcal{A}$ is the automaton on the bottom and $\mathcal{B}$ the automaton on the top.

An example is given in Fig. 6.6. Let A be the automaton corresponding to the tileset τ and A^n its n-times composition. Intuitively, an edge in A^n corresponds to a well-tiled pattern of size $1 \times n$, as shown in the figure for $n = 2$. As a consequence, a path in A^n exists iff τ tiles an infinite strip of height n.

We therefore obtain again an algorithm to semi-test if a tileset tiles the plane by computing A^n for all values of n.

In terms of efficiency, it is easy to see that we obtain a complexity similar to the previous algorithm, although we are actually testing for a stronger property. Of course, one can do this both in the vertical and the horizontal direction. Doing this we can prove that τ_1 does not tile a vertical strip of width 8, although it can tile a 14×14 square.

The main interest however is in the fact that translating the problem into the theory of automata means we can use all the tools available in the theory of automata. Indeed, the translation of the fact that τ tiles the plane is about the iteration of the automaton A, which makes it clear that it doesn't really depend on τ, but on the language of infinite words encoded by τ. In particular, we can *minimize* in some way the automaton A (and actually all the automata A^n) under consideration, without changing the problem. We leave [26] as an additional reference about finite automata. These techniques have been used successfully to prove that the smallest aperiodic tileset has 11 tiles [27].

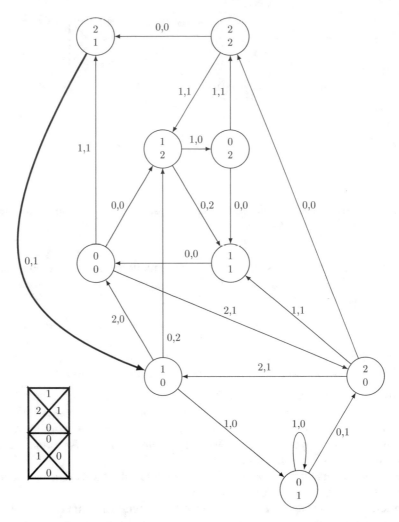

Fig. 6.6 The composition of the automaton corresponding to τ_1 with itself. The bold edge correspond to the bold and dashed edges in the previous picture, and to the 2×1 pattern on the bottom left

6.1.3.3 A Semi-Algorithm Based on the Anderson–Putnam Complex

We will briefly present here a method that can prove, in some cases, that a tileset τ does not tile the plane without even trying to tile a square with τ.

This method is based on the Anderson-Putnam complex and its second singular homology group, and can be transformed into an "iff" condition [10]. However, we will use here a down to earth approach.

To understand the idea, first suppose that τ has a periodic tiling, say, τ tiles a square of size $p \times p$.

Let $c \in C$ be a color that can appear horizontally. If we look at a square that is tiled periodically by τ, it is obvious that in this square the color c should appear exactly the same number of times on the east side as on the west side. If we denote by x_i the number of times the i-th tile of τ appear, this gives us an equation relating all tiles that contain the color c:

Definition 6.1.11 Let $\tau = \{t_1 \ldots t_n\}$ be a tileset. Let $x_1 \ldots x_n$ be n real variables, where x_i is associated to the tile t_i of τ.

Let c be a color.

The horizontal equation for c is the equation:

$$\sum_{t_i \in E(c)} x_i = \sum_{t_i \in W(c)} x_i$$

where $E(c)$ is the set of all tiles that have the color c in their east side, and $W(c)$ the set of all tiles that have the color c in their west side.

We define similarly the vertical equation for c.

Basically the horizontal equation for c states that the number of times the color c appear somewhere on the east side in the periodic tiling should be equal to the number of times c appear on the west side.

Rather than considering the number of times the tiles appear, we will consider their density. We therefore have an additional equation, stating that the sum of all densities should be one:

Definition 6.1.12 The set of equations for τ is the combination of all horizontal equations, all vertical equations, and the unit equation:

$$\sum_i x_i = 1$$

Proposition 6.1.13 *If τ tiles a square periodically, then the set of equations for τ has a nonnegative solution.*

Proof Take x_i to be the density of the tile t_i.

What is interesting is that this is actually true more generally:

Proposition 6.1.14 *If τ tiles the plane, then the set of equations for τ has a nonnegative solution.*

This is true even if τ has no periodic tiling.

Proof Suppose that τ tiles the plane. Let $n \in \mathbb{N}$. By hypothesis, τ tiles an $n \times n$ square. Let $p_{i,n}$ be the number of times the tile t_i appear in this square and $x_{i,n} = p_{i,n}/n^2$ the density of the tile t_i.

The unit equation is obviously satisfied by the $x_{i,n}$.

Let c be a color. It is easy to see that

$$\sum_{t_i \in E(c)} p_{i,n} - \sum_{t_i \in W(c)} p_{i,n} = O(n)$$

Indeed the difference between these quantities is bounded by the number of times the color c appear in the border of the square.

We therefore get

$$\sum_{t_i \in E(c)} x_{i,n} - \sum_{t_i \in W(c)} x_{i,n} = O\left(\frac{1}{n}\right)$$

We now extract from the sequence $(x_{i,n})_{n \in \mathbb{N}} \in [0, 1]^{|\tau|}$ a converging subsequence to get the result.

In our proof we see that the solution x_i we obtain to the system of equations is the limit of densities. Using more complex results from ergodic theory, we can prove that we can take x_i to actually be the density of the tile t_i in a specific tiling of the plane.

This is of course only a necessary condition for τ to tile the plane. Let's look for example at the tileset τ_1 from Fig. 6.2.

We obtain the following set of equations:

$$x_1 + x_2 + x_3 = x_1 + x_4 \tag{6.1}$$

$$x_4 + x_5 = x_2 + x_5 + x_6 \tag{6.2}$$

$$x_6 + x_7 = x_3 + x_7 \tag{6.3}$$

$$x_1 + x_2 + x_4 + x_7 = x_3 + x_4 + x_6 \tag{6.4}$$

$$x_3 + x_6 = x_2 + x_5 + x_7 \tag{6.5}$$

$$x_5 = x_1 \tag{6.6}$$

$$x_1 + x_2 + x_3 + x_4 + x_5 + x_6 + x_7 = 1 \tag{6.7}$$

The first three equations are the horizontal equations, the following three equations correspond to the vertical constraints.

Even though τ_1 does not tile the plane, the system admits solutions. Among them, there is for example: $x_3 = x_4 = x_6 = 1/5$ and $x_7 = 2/5$, the other variables being 0.

This solution corresponds to the tiling of the pattern in Fig. 6.7. Each color appear the same number of times in the east and west side of this figure, but this pattern cannot be used to tile the plane periodically. However, it can be used to tile *some* surface: Just glue the north side and the south side of the pattern together and do the same in the east and west side. One can prove that solutions of the system of equations can always be used to tile some surface but that the genre of the surface

Fig. 6.7 A tiling of a finite pattern by τ_1 where colors appear the same number of times in the east (resp. north) side and the west (resp. south) side

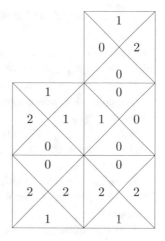

will have to grow linearly in the number of tiles if the tileset τ does not have a tiling. This is in essence the result of [10].

6.2 Main Ingredients in the Proofs

There are basically four vastly different proofs of the undecidability of the Domino Problem: The proof by Berger [7, 8] later simplified by Robinson [47], the proof by Aanderaa and Lewis [1], best presented in the book of Lewis [36], the proof by Kari [32], and the proof by Durand, Romashchenko and Shen [14]. All other proofs of the result the authors are aware of can be roughly characterized as variants of the construction of Berger.

It is interesting to note that Kari's proof relies on the proof by Hooper [25] of the undecidability of the immortality problem of Turing machines (more on this later). As a consequence, three of the different proofs were obtained by, or use results of, PhD students of Hao Wang (Robert Berger, Philip Hooper and Stål Aanderaa), working on the undecidability of the ∀∃∀ fragment of first order logic.

We will focus in this section on the common elements in all the proofs. First, there is a need to define precisely what it is meant by "undecidable", for which we need a formal definition of an algorithm (or a program). Then we will prove the undecidability of the fixed domino problem, which is the first brick in most of the proofs. We will then explain why the previous proof cannot be adapted directly to obtain the main result, and what is needed to accomplish it.

6.2.1   Turing Machines

The theorem by Berger states that no algorithm can decide, starting from a set of Wang tiles τ, if τ tiles the plane.

To prove such a theorem, one needs a formal definition of a program. Since 1930, various equivalent definitions, due to Church, Turing or Herbrand–Gdel [11, 34, 50] have been given.

For our purpose, the definition by Turing will be the most suitable, although some results have been obtained [33] using the concept of 2-counter machines introduced by Minsky [39].

We refer the reader to [37, 39, 44] for more information about Turing machines. The book [26] offers in particular an introduction to both Turing machines and finite automata.

A Turing machine is a formal definition of a computation device. This device takes as input a word in some given alphabet, and outputs, after several steps, a word in some other alphabet.

The machine uses the concept of a *tape*, which is an infinite or biinfinite array of cells. Each position of the cell contains a symbol, and one specific position of the tape is marked by the *head*. The Turing machine makes a decision at each step depending on the symbol it can read on the tape at the position of the head, and its internal *state*. Based on these informations, the machine can shift the position of the head, change the symbol of the tape, and change its internal state.

Definition 6.2.1   A Turing machine is given by:

- A finite set of states Q. One distinguishes in Q a specific state $q_0 \in Q$, called the initial state, and two special states q_a, q_r, the accepting and refusing state.[1]
- An input alphabet Σ
- A work alphabet $\Gamma \supset \Sigma$, it contains a special symbol B (not present in Σ), called the blank symbol.
- A transition function: $\delta : \Gamma \times Q \to \Gamma \times Q \times \{-1, 1\}$

Intuitively $\delta(a, q) = (b, q', d)$ means that, if the Turing machine reads the symbol a on its head and is currently in state q, then it will write the symbol b instead of a, change its internal state to q', and move its head in direction d.

The transition function is best depicted with a graph. Vertices represent states of the machine, and there is an edge from q to q' labeled with $a/b, d$ if $\delta(a, q) = (b, q', d)$. It is also customary to write $+1, -1$ as $\to$ and $\leftarrow$ as they denote movement of the head.

An example of a Turing machine is given in Fig. 6.8. The machine works as follows: we first write the input on the tape (see the first row of Fig. 6.9). The Turing

---

[1] The set of states is called Q rather than S for some forgotten historical reason. Alan M. Turing uses the vocabulary "configuration" rather than "state", but the word "state" has become more common nowadays.

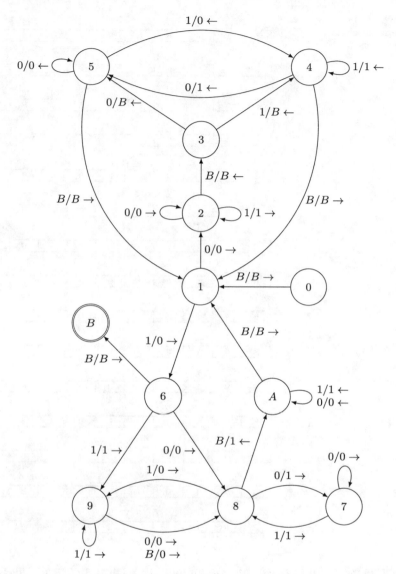

Fig. 6.8 An example of a Turing machine. The initial state is $q_0 = 0$. The accepting state is state $q_a = B$

machine starts with initial state q_0. At each step, the Turing machine looks at its internal state and the symbol at the position of the head. The transition function then tells us what symbol should be written at the position of the head, and what is the new state and the new position of the head. Figure 6.9 contains the first steps of the execution of a Turing Machine.

Fig. 6.9 The first 15 steps of the Turing machine of Fig. 6.8 on input 11

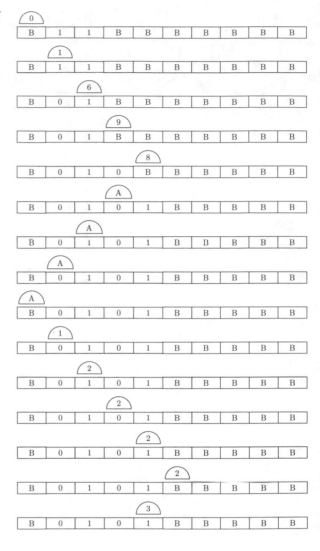

Definition 6.2.2 An *instantaneous description* (also called a *configuration*) is a tuple $w = (c, h, q) \in \Gamma^{\mathbb{Z}} \times \mathbb{Z} \times Q$. The content of the tape is represented by c, while h contains the position of the head on the tape, and q the internal state of the machine.

For $w = (c, h, q)$ an ID, the successor of w is the ID $w' = (c', h', q')$ defined by:

- Let $\delta(c_h, q) = (b, q', d)$
- $c'_j = c_j$ for $j \neq h$ and $c'_h = b$.
- $h' = h + d$

We write $M(w)$ for the successor of w and $M^n(w)$ is defined in the usual way.

Definition 6.2.3 Let M be a Turing machine.
Let u be a word. The initial ID is $w_u = (c, h, q)$ where

- $c_{i+1} = u_i$ when it make sense and $c_i = B$ otherwise (the word u is represented on the tape, starting at position 1.
- $h = 0$ (The head is initially at position 0)
- $q = q_0$ (The initial state is q_0).

We say that u is accepted if there exists n s.t. $M^n(w_u) = (c', h', q_a)$. We say that a Turing machine accepts a language L if on every input the Turing machine eventually reaches either the accepting state q_a or the rejecting state q_r, and L is exactly the set of accepted words.

Notice that there are three possible behaviours of a Turing machine on input u: Either it eventually reaches the accepting or the rejecting state, or it keeps running forever.

If the Turing machine is defined with an infinite array of cells rather than an biinfinite array (i.e. configurations are in $\Gamma^\mathbb{N} \times \mathbb{N} \times Q$), a fourth possibility arises: the Turing machine may try to go left when the head is in the leftmost cell, in which case it crashes. It is easy to impose safeguards so that the last case never happens.

If we look at what is written on the tape of the Turing machine when it reaches an accepting state, we can define a *function* computed by a Turing machine. We leave the details to the reader.

A first example of a Turing machine is depicted on Fig. 6.8, with the first 15 steps of the machine represented in Fig. 6.9.

This machine has not been chosed arbitrarily, and we can explain what it does. The input of the machine has to be understood as an integer coded in binary, written from least to most significant bit, i.e. the integer 6 is written 011. The state 1 has two different outgoing transitions, depending on whether the first symbol is a 0 or a 1. This means we have two different behaviours depending if the integer is even or odd.

We will first focus on the case where the integer is even. When in state 2, the head keeps going right, until it reaches the end of the word (state 3) and then we go to state 4 or 5 depending on the last symbol. States 4 and 5 essentially shift the input to the left. This means replacing each symbol with the symbol previously seen; the state 4 (resp. 5) means the previously seen symbol was a 0 (resp. 1). When we reach the begining of the word, we go again to state 1. We have said that, if an integer is even, we will delete its first bit, meaning that states 2 to 5 essentially divide an integer by 2.

States 7, 8, 9 basically multiply an integer by 3. Multiplying an integer by 3 in binary may produce a carry of value 0, 1 and 2, which correspond respectively to the three states 7, 8, 9. As an example, suppose the carry is currently 1, and the integer to multiply by 3 begins with a 1. As $3 \times 1 + 1 = 0 + 2 \times 2$, this means the first bit of the output should be a 0 and our new carry is equal to 2. This corresponds to the transition from state 8 to state 9. The state 6 is essentially there to kickstart the process, and detect if the output is the integer 1. The state A rewinds the head so that it is again at the initial position. In other words, states 6 to A, starting from an integer n, produces the integer $3n + 1$, except if $n = 1$ in which case the Turing machine stops.

As a whole, our Turing Machine is essentially mimicking the Collatz sequence: Starting from an integer n written in binary, we do the operation $n \mapsto n/2$ or $n \mapsto 3n + 1$ until n is equal to 1. Knowing if this particular Turing machine halts on all inputs that ends with the symbol 1 is therefore equivalent to the Collatz Conjecture. (For simplicity, the Turing machine we designed does not work if the input ends with a symbol 0, i.e. if it codes the integer 0, or if an integer is coded with additional symbols 0 at the end).

This somewhat easy example should convince the reader that Turing machines are indeed able to represent arbitrary computations. We have somehow coded traditional constructions from programming languages: We have a `while` loop that ends when the integer n is equal to 1 (this is done by states 1 and 6), and we have used *subroutines* that divides by 2 or multiply by 3 respectively (states 3/4 and states 7/8/9).

One can prove that any program in your favorite programming language may be translated into an equivalent Turing machine. The *Church-Turing* thesis essentially states that all reasonable models of computation we can devise are equivalent, in the sense that they define exactly the same computable functions. In fact, almost all models also use, up to a polynomial, the same ressources in time and memory, this is the extent of the extended Church-Turing thesis.

With the vocabulary of Turing machines, we can give the first formal example of an undecidable problem:

Theorem 6.2.4 *There is no algorithm that, given a Turing machine M and a word u, decides if M halts on input u.*

This problem is called the "Halting problem".

Proof To prove that such an algorithm does not exist, we need to formalize exactly its input. So we need an encoding of descriptions of Turing machines into words. We write machine(u) for the Turing machine whose encoding is u.

Suppose such an algorithm exists. As algorithms can be turned into Turing machines, we therefore have the existence of a Turing machine N s.t., on input u and v, works as follows:

- If machine(u) halts on input v, N accepts
- If machine(u) does not halt on input v, N rejects.

We obtain from N a Turing machine N' that works in the following way: On input v:

- If machine(v) halts on input v, N' does not halt
- If machine(v) of code v does not halt on input v, N' halts.

We then obtain a contradiction by looking at the behaviour of N' on input w s.t. machine$(w) = N'$.

Corollary 6.2.5 *There is no algorithm that, given a Turing machine M, decides if M halts on the empty input.*

Proof *(Sketch)* Given a Turing machine M and an input u, we can easily build a Turing machine N s.t. N on the empty input simulates M on input u: N proceeds by first writing u on its tape, and then "calls" the Turing machine M. Therefore an algorithm that solves the halting problem on the empty input can solve it on any input.

6.2.2   Encoding of Turing Machines in Tilings: The Fixed Domino Problem

Figure 6.9 shows that the execution of a Turing machine can be written essentially as a two-dimensional object, with each row representing a step of the Turing machine. It is therefore not surprising that this model is quite easy to encode into Wang tiles. There exists multiple equivalent encodings in the litterature.

In the case of Turing machines with an infinite (rather than biinfinite) tape, such an encoding is proposed in Fig. 6.10, and a quarter of plane using these tiles is given in Fig. 6.11.

If we write the input on the bottom left of a quarter of a plane, then the only way to fill up the quarter of a plane is to simulate the execution of the Turing machine. If furthermore, we delete all tiles involving the accepting state q_a, then a tiling is possible iff the Turing machine does not halt.

Using the three tiles at the bottom of Fig. 6.10, we can initialize the computation so that it starts with the empty input and state q_0. In fact, any tiling of the quarter plane that contains at the origin the tile that is at the bottom-left of Fig. 6.10 simulates the execution of the Turing machine with empty input. This is in fact a tiling of the entire plane, as the rest can be filled up with the blank tiles. Therefore, we have a tiling of the entire plane, with a specific tile at the origin, iff the Turing machine does not halt on empty input.

We have proven:

Theorem 6.2.6 (Undecidability of the Fixed Domino Problem, see Wang [52], Büchi [9], and Kahr et al. [29]) *There is no algorithm that decides, given a set of Wang tiles τ and a predescribed tile $t \in \tau$ whether there exists a tiling of the plane by τ that contains t.*

If we look at the construction, we can obtain

Corollary 6.2.7 *There is no algorithm that decides, given a set of Wang tiles τ and a predescribed tile $t \in \tau$ whether there exists a tiling of the quarter of the plane by τ with t at the corner.*

Fig. 6.10 Coding of a Turing machine by Wang Tiles

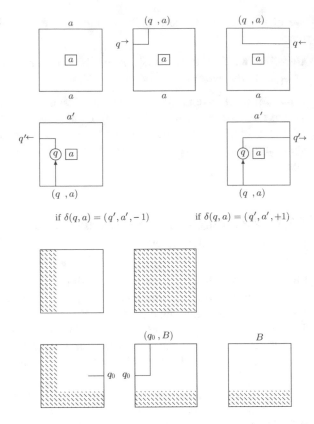

if $\delta(q,a) = (q',a',-1)$ if $\delta(q,a) = (q',a',+1)$

It is possible to adjust slightly the set of tiles so that we can, starting from an input u, produce a finite pattern P s.t. there is a tiling by τ that contains the pattern P iff the Turing machine does not halt with input u. Doing this we obtain

Theorem 6.2.8 (Robinson [47]) *There exists a set of Wang tiles τ s.t. there is no algorithm that decides, given a finite pattern P whether there exists a tiling of the plane by τ that contains P.*

Sketch of Proof Let M be the Turing machine that on input u simulates machine(u) on the empty input. This machine is indeed implementable, and it is easy to see that there is no algorithm that decides, given an input u, if M halts on input u. We now code this Turing machine as in Fig. 6.10.

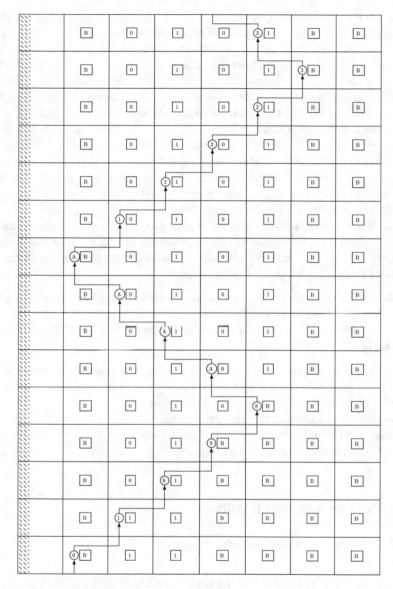

Fig. 6.11 A tiling of a quarter of plane by the set of Wang tiles of Fig. 6.10. The link with Fig. 6.9 should be obvious

6.2.3   Towards the Domino Problem

With one notable exception, the coding we presented before is central in all the proofs of the undecidability of the Domino Problem. However this is far from sufficient.

Indeed, the construction gives rise to a lot of tilings if the primordial tile is not used. First, we can have tilings with no head per row, using e.g. only the first tile on the top left of Fig. 6.10, or even using only the blank tile. Second, we can have tilings with more than one head per row. There are easy tricks to stop this problem from happening, e.g. by having different background colors before and after the head. The third problem is that there can be tilings that correspond to uninitialized computations, i.e. biinfinite computations rather than a computation starting from a finite input and the initial state.

There are no easy solutions to the remaining problems. Three of the methods outlined built a set of Wang tiles in which a quarter of plane appears, in such a way that the corner of the quarter of the plane can be locally identified. It is then a routine exercise to enforce that a specific tile appears in that corner, so that we can use the previous construction inside this quarter of plane.

The last method uses a very different encoding technique, which will enforce that every row codes a configuration of a Turing machine in such a way that the head is always present. The remaining problem is therefore uninitialized computations, which is solved using the following theorem:

Theorem 6.2.9 (Hooper [25]) *There is no algorithm that, given a Turing machine M, decides if M halts on all configurations.*

This is still true if we restrict ourselves to Turing machines with a working alphabet of size 2.

This last proof of the undecidability of the Domino problem is tremendously simpler, but one could say that part of the complexity of the proof is hidden inside the proof of Theorem 6.2.9, which is far from trivial.

It is interesting to remark that Hooper was, like Robert Berger, a PhD Student of Hao Wang. In fact, this theorem was obtained for the same purpose as Berger's, i.e. proving the undecidability of the $\forall\exists\forall$ fragment of first order logic.

6.3 The Substitutive Method 1/2

6.3.1 Preliminary Discussion

We now start with the first proof of the Undecidability of the Domino Problem. The proof we present here follows closely Berger [7] and Robinson [47] although we will use some recent results and concepts from Ollinger [43].

As we saw in the previous section, it is easy to use Turing machines to prove the undecidability of the Fixed Domino Problem, i.e. whether there exists a tiling by τ that contains some specific tile t. The problem of the construction is, of course, that τ always has some trivial tilings that do not contain t. We need therefore some way to force the tile t to appear.

This suggests a nave approach to prove the undecidability of the (General) Domino problem: First build a specific tileset τ_s with the property that every tiling of the plane with τ_s contains exactly one copy of a specific tile t_s. Now, given a tileset τ and a tile t, superimpose τ_s with τ by forcing the tile t to appear exactly on top of t_s. Let's call $\tau \times \tau_s$ this new tileset. Then it is clear that $\tau \times \tau_s$ tiles the plane iff there exists a tiling of the plane by τ that contains t. Therefore we cannot decide the General Domino Problem as it would allow us to decide the Fixed Domino Problem, case closed.

However, such a tileset τ_s cannot exist:

Proposition 6.3.1 *Let τ_s be a tileset. If every tiling by τ_s contains the tile t_s, then there exists some integer n s.t. every tiling of a square of size n contains the tile t_s.*

Proof If it were not the case, we would have tilings of arbitrary large squares without the tile t_s. By compactness (Proposition 6.1.4), this would imply that there is a tiling without t_s.

Therefore the tile t_s, on top of which we want to kickstart the computation process of τ appears in every sufficiently large square. This means that our construction would have to contain infinitely many instances of the computation process. How to manage all these different computations in such a way that they do not overlap is not an easy task.

The solution is provided by building tileset τ_s with some specific properties. The general idea is to first build an aperiodic tileset, and then try to embed computation in it. While in theory, the method can be used starting from any aperiodic tileset, the only tilesets that have been used successfully for this method are substitutive tilesets, that we now define.

6.3.2   Substitutions

Definition 6.3.2 Let A be a finite alphabet. A (square) *substitution* is a map ϕ from A to $A^{n \times n}$. If w is a (two-dimensional) word of size $m \times m$, then the image by ϕ of w, named $\phi(w)$ is the word of size $nm \times nm$ defined by

$$\phi(w)_{(ni_1+i_2, nj_1+j_2)} = \phi(w_{i_1,j_1})_{i_2,j_2}$$

We write $\phi^k(w)$ for the k-th iterate of ϕ. The map ϕ is extended naturally to $w \in A^{\mathbb{Z} \times \mathbb{Z}}$ using the same formula.

The *substitutive subshift* associated to ϕ is the set S_ϕ of all configurations of $A^{\mathbb{Z} \times \mathbb{Z}}$ s.t. $x \in S_\phi$ iff there exists an infinite sequence $(w_i)_{i<0}$ s.t. $\phi(w_i) = w_{i+1}$ and $w_0 = x$.

An example of a substitution is given in Fig. 6.12. This substitution is quite random and has no relevance in the rest of the paper.

Fig. 6.12 A substitution ϕ on $A = \{0, 1\}$ and $\phi^k(1)$ for $k = 0, 1, 2, 3$

$$0 \mapsto \begin{matrix} 0 & 1 \\ 1 & 0 \end{matrix} \qquad\qquad 1 \mapsto \begin{matrix} 1 & 1 \\ 0 & 1 \end{matrix}$$

```
                                    1 1 1 1 1 1 1 1
                                    0 1 0 1 0 1 0 1
                        1 1 1 1     0 1 1 1 0 1 1 1
              1 1       0 1 0 1     1 0 0 1 1 0 0 1
    1         0 1       0 1 1 1     0 1 1 1 1 1 1 1
                        1 0 0 1     1 0 0 1 0 1 0 1
                                    1 1 0 1 0 1 1 1
                                    0 1 1 0 1 0 0 1
```

Intuitively the substitutive subshift S_ϕ looks like $\phi^k(w)$ at the infinity. The term "subshift" comes from the symbolic dynamics school. We refer to other articles in this volume for details on subshifts.

Definition 6.3.3 (Informal Definition) A tileset τ is substitutive if the set of tilings by τ is, up to a recoloring, a substitutive subshift.

There are a lot of examples of substitutive tilesets [7, 47]. In fact, one can show that, for any ϕ, S_ϕ can be realized as a tileset τ [17, 20, 40].

Most constructions of substitutive tilesets are hindered by the fact that the tilings are not directly substitutive, but this is only true of their recolorings.

We therefore strive to find a tileset that is intrinsically substitutive:

Definition 6.3.4 ([5, 13, 14, 43]) Let τ be a tileset. We say that τ is intrinsically substitutive with factor 2 if there exists a one-to-one substitution $\phi : \tau \to \tau^{2\times2}$ s.t.:

- For any finite pattern w, $\phi(w)$ is a valid tiling precisely when w is a valid tiling
- For any tiling of the plane x by τ, there exist y s.t. $\phi(y) = x$ upto shift.

The second point ensure that ϕ somehow exhausts all possible patterns that can appear in a tiling of the plane. Notice that, in the second property y is automatically a valid tiling (as every pattern of y is valid by the first property).

Theorem 6.3.5 ([13, 43]) *There exists an intrinsically substitutive aperiodic tileset.*

The tileset obtained by Ollinger is depicted in Fig. 6.13. Although the tiles are decorated for cosmetic reasons, they are really Wang tiles: two tiles can be put together if they match on their common edge. Each tile is composed of three layers, as seen in Fig. 6.14. An example of a large square tiled by τ is given in Fig. 6.15.

The first layer ensures that, in each tiling of the plane, the tiles are grouped in squares of size 2×2 to form a small red square.

The substitution that corresponds to τ_s is given in Fig. 6.16.

Proposition 6.3.6 *The tileset τ_s depicted in Fig. 6.13 is intrinsically substitutive and aperiodic.*

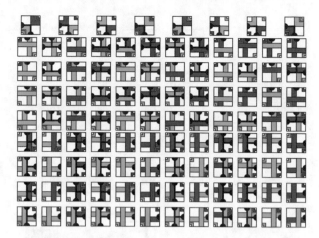

Fig. 6.13 The tileset τ_s

Fig. 6.14 The three layers inside a tile

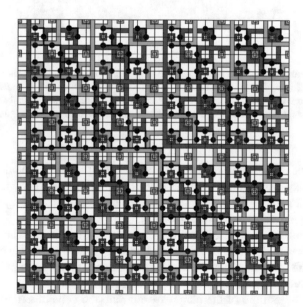

Fig. 6.15 A portion of a tiling by τ_s

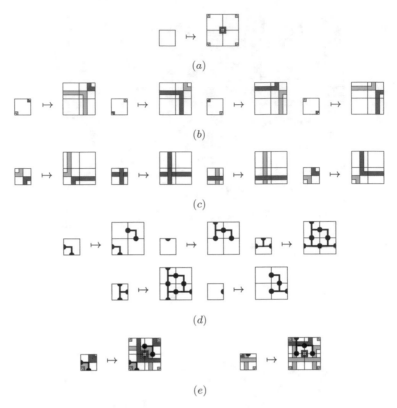

Fig. 6.16 The substitution. The layer 1 of $\phi(t)$ is the same for all tiles t, as shown in subfigure (**a**). The second layer of $\phi(t)$ is cut in half. First, The layer 1 of the tile t is stretched (subfigure (**b**)). Second, the wires in the layer 2 of t are extended horizontally and vertically (subfigure (**c**)). Finally subfigure (**d**) explains how the layer 3 of $\phi(t)$ is obtained from t. Subfigure (**e**) gives two examples

Proof Let ϕ be the substitution defined in Fig. 6.16. It is easy to see that ϕ is one to one, and an easy inspection of ϕ shows that $\phi(w)$ is valid precisely when w is valid.

We say that a tiling of a 2×2 square is well behaved if its first layer consists of a small red square (i.e. similar to Fig. 6.16a).

We now look at every possible tiling of a 4×4 square by τ where the central 2×2 square is well behaved. By an exhaustive analysis, we see that every such 2×2 square is of the form $\phi(w)$ for some w. (This is easily seen by writing and running a small program. A detailed proof is also given in [5].)

Now let x be a tiling of the plane. We regroup the cells of x into 2×2 well-behaved squares $(x_{i,j})_{i,j \in \mathbb{Z}} \in (\tau^{2 \times 2})^{\mathbb{Z}^2}$. By the previous argument, each of them is of the form $\phi(w_{i,j})$, which means that $x = \phi(w)$ upto shift.

Therefore τ is intrinsically substitutive.

τ should be aperiodic. Indeed, suppose that there exists a tiling x by τ of period p, and choose p as small as possible. As x can be divided into 2×2 squares that are not periodic, p should be even. Now $x = \phi(w)$ for some tiling of the plane w. As x is periodic of period p, $\phi(w)$ should be periodic of period $p/2$, a contradiction.

The reader familiar with the work of Robinson [2, 19, 28, 47] might recognize something familiar in Fig. 6.15. The tileset of Robinson can indeed be obtained from the tileset τ_s by some identification of colors, see the bibliographic notes below.

6.3.3   Grids Inside Tilings

We now explain how to use the tileset we built previously to obtain the undecidability of the Domino Problem. The proof highly depends on properties of this tileset, and it is unknown whether a similar reasoning could be used starting from any substitutive tileset.

First, to understand the tilings, we need to simplify the pictures. We will forget about the first and third layer, and focus only on the wires of color red in the second layer. It is easy to see that they form arbitrarily large squares. We obtain Fig. 6.17

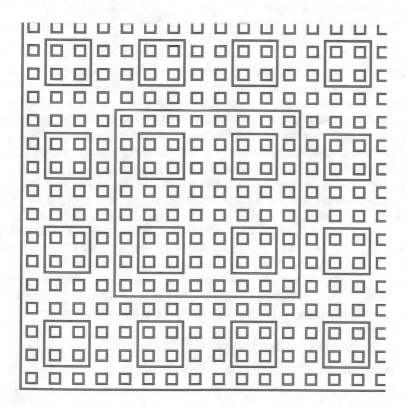

Fig. 6.17 Squares inside the tilings by τ_s

by considering only half of the squares. This is done in practice by changing τ_s to have two different shades of red (say, light red and dark red) and forcing light red wires to cross only dark red wires. Figure 6.17 then represents only, say, the light red squares.

We now see that tilings by τ_s are formed of squares that do not cross.

Remember that by Corollary 6.2.7 there is no algorithm to decide, given a tileset τ and a tile t, whether there exists a tiling of a quarter of plane by τ with t at the corner. By compactness, it means there is no algorithm to decide, given a tileset τ and a tile t whether one can tile arbitrary large squares by τ with t at the bottom-left corner.

Our goal now is clear: starting with a tileset τ and a tile t, we will fill up the squares of τ_s by tilings by τ in such a way that t appear in the bottom-left corner. One difficulty remains: While the squares in τ_s do not cross, they are included one in each other, which means that we cannot use all the space in one square, as some of it is already occupied by other squares. We will therefore first extract in each square the largest "free" subsquare.

Let S be a square, that we see as a subset of positions of the form $C \times D$ for some segments C and D. Let U (for "used") be the set of positions inside S that are occupied by a smallest square.

By a free subsquare of S, we denote a set of horizontal positions $H \subseteq C$ and vertical positions $V \subseteq D$ s.t. $H \times D$ and $C \times V$ do not intersect U.

To obtain the largest free subsquare of S, we just have to cross out every horizontal (resp. vertical) coordinate on which a smallest square lies. We will call these coordinates "obstructed". In our particular case this can be done easily by local rules: The idea is that each square will emit horizontal (resp. vertical) signals on their exterior borders that propagate horizontally (resp. vertically) and that die out when they encounter another border. This is explained nicely in [47], see Fig. 6.18. To be exact, notice that the parts that are not crossed do not correspond to whole tiles, but only quarters of them. This has no bearing in the following.

As can be seen in the figure, larger squares of the tiling have larger free subsquares formed by joining the unobstructed space. It remains to put the tileset τ inside the free subsquare.

The idea is that the position (i, j) of the supposed tiling by τ will be contained in the position (h_i, v_h) of the square S, where h_k (resp. v_k) is the k-th element of H (resp. V). As no element of $H \times D$ is occupied by a smallest square, we can use cells of $H \times (D \setminus V)$ to propage the vertical constraints of τ inside the grey area of the tiling of Fig. 6.18. We can also force the tile at the bottom-left quarter of the square easily, by enforcing that the only tile that can appear on top of a corner tile of τ_s is the tile t.

Putting everything together we have the desired proof.

Theorem 6.3.7 *The Domino Problem is undecidable.*

Proof Let τ be a tileset and $t \in \tau$. We use the construction above to obtain a new tileset τ_r combining τ and τ_s.

Fig. 6.18 The squares with the obstructed rows and columns grayed out. Inside a square, the remaining unobstructed space, in white, forms a square whose size is a quarter of its container square

Suppose this new tileset τ_r admits a tiling. This tiling contains arbitary large squares, and therefore arbitary large free subsquares. But these subsquares form a tiling by τ with the tile t at the corner. Therefore there exist arbitary large squares tiled by τ with t at the corner.

Conversely, if there exist arbitary large squares tiled by τ with t at the corner, we can obtain a tiling of the plane by τ_r by putting in each free subsquare of each square a valid tiling.

We have therefore proven that there exists arbitary large squares tiled by τ with t at the corner iff there exists a tiling by τ_r.

Therefore no algorithm can solve the Domino Problem: If such an algorithm existed, it could be used, via the given transformation, to decide, given a tileset τ

and a tile t, whether there exists tilings of arbitrary large squares by τ with t at the corner, which is undecidable by Corollary 6.2.7.

6.3.4 Bibliographic Notes

There exist a lot of different proofs of the Undecidability of the Domino Problem based on building first an aperiodic substitutive tileset and then encoding tilings with a fixed tile t inside the squares it produces. Notice that in theory, this proof doesn't need a substitutive tileset, but only a tileset from which we can extract arbitrary large squares. However all known variants of this proof uses substitutive (or more accurately near substitutive) tilesets.

The first proof of the kind is the proof by Berger [7]. Technically the construction of Berger is not a division into squares which overlap, but a division into infinite vertical strips of arbitrary width which can overlap, but the principle stays the same.

The proof we presented here is essentially from Robinson [47], except we used the tileset of Ollinger [43] as a basis instead of the tileset of Robinson. The tileset used in Robinson's article is very similar to Ollinger; it uses the same tiles as Fig. 6.13 except that the green wires are now *centered*, i.e. the two colors "green on top" and "green on bottom" of the east/west side of the second layer are identified, and the same goes for the colors "green on left" and "green on right" on the north/south side. Robinson also speaks in his note [46] of a set of 52 tiles, that was later published in an article of Poizat[45]. This set is the same as Fig. 6.13 without the first layer. The fact that these two tilesets of Robinson look very close to Ollinger's tileset and the fact that Robinson himself hints at a set of 104 tiles in his note [46] suggests that the set discovered by Ollinger was already known to Robinson.

There exist a lot of literature explaining and reexplaining Robinson's tileset [2, 28, 48]. The tileset of Ollinger is explained in [43]. Further details that prove the tileset is substitutive and minimal can be found in the thesis of Ballier [5].

The simplifications made by Robinson lead to a smaller tileset but also have a major drawback: the tileset is not intrinsically substitutive, which makes the proof of its aperiodicity (and a complete description of the possible tilings) more difficult. The tileset of Ollinger is not the first intrinsically substitutive tileset that was discovered. In fact, Berger set of 103 tiles is intrinsically substitutive, although this was never proven. This property is also used very often to prove aperiodicity of tilesets in the plane $\mathbb{R}^2$, where it is sometimes called "the unique composition property", see e.g. [3, p. 2].

To finish, remark that the whole construction is based on the fact that an aperiodic substitutive tileset with some specific property exist, but the tileset we use (and similar tilesets used by Berger, Robinson, or others) appears somehow by miracle. Sect. 6.6 of this document explains how to prove using tools from computability theory and programming languages that such a tileset exist.

6.4   The Construction of Aanderaa and Lewis

The construction by Aanderaa and Lewis we present here is not well known. As with all early constructions of an aperiodic tilesets, this tileset was built in the context of the undecidability of the $\forall\exists\forall$ fragment of first order logic, and is actually the first published example of an aperiodic *deterministic* tileset, i.e. a tileset s.t. any row determines uniquely all rows in the upper half plane above this row. This also gives a proof of the undecidability of the nilpotency of one-dimensional cellular automata.

The first proof of the idea was published by Aanderaa and Lewis in 1974 [1]; a better version may be found in the book by Lewis [36]. Our exposition here follows somewhat Lewis but the construction is actually quite different, in particular with the introduction of sofic shifts and p-adic numbers.

This construction is quite interesting as it uses only elementary number theory, and is actually building at first a one dimensional object rather than a two-dimensional tiling.

While not stricly necessary, it is however recommended to have a good understanding of the basics of symbolic dynamics. We refer the reader to the book [38] and to other articles of this book.

6.4.1   Tiling Problems on One-dimensional Objects

6.4.1.1   Symbolic Dynamics

We briefly recall in this section classical notions from symbolic dynamics.

The concept of Wang tiles is not intrinsically two-dimensional, and one can define Wang tiles for tilings of the discrete line $\mathbb{Z}$ rather than the discrete plane $\mathbb{Z}^2$. This can also be generalized obviously to higher dimensions, or even to tilings of finitely generated groups:

Definition 6.4.1 (One-dimensional Wang Tiles) Let C be a finite set, called the set of colors. A *Wang tilet* over C is a map from $\{E, W\}$ to C. A *tileset* τ is a finite set of Wang tiles.

Definition 6.4.2 A *tiling* of $\mathbb{Z}$ by τ assigns to each element of $\mathbb{Z}$ a tile from τ s.t. colors of adjacents tiles agree on their common border.

Formally, a tiling is a map $x : \mathbb{Z} \to \tau$ s.t.

- $\forall i, x(i)(E) = x(i+1)(W)$

We say that τ *tiles* $\mathbb{Z}$ if there exists a tiling of $\mathbb{Z}$ by τ.

Proposition 6.4.3 *The Domino Problem is decidable on $\mathbb{Z}$: there exists an algorithm, that on input a tileset τ, decides if τ tiles the discrete line $\mathbb{Z}$.*

The idea is very similar to the construction presented in Fig. 6.5.

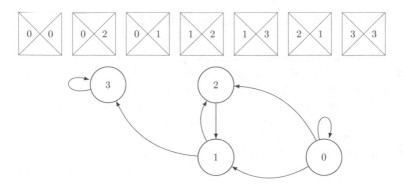

Fig. 6.19 The tileset τ_2 and its representation as a graph

Proof Let τ be a one-dimensional tileset. Consider the (multi) graph whose vertices are the colors C, and there is an edge from u to v if there exists a Wang tile with u on the west side and v on the east side. Then it is obvious that there exists a tiling by τ if and only if the graph G contains a cycle. See Fig. 6.19 for an example.

In the following, we will be interested in sofic shifts, which correspond to Wang tiles with an additional symbol. We defined them only for one-dimensional objects, although we will also need their two-dimensional variant:

Definition 6.4.4 (Decorated one-dimensional Wang Tiles) Let C be a finite set, called the set of colors, and A a finite set, called an alphabet. A *decorated Wang tile* t over C is as 2-tuple (t', a) where t' is a Wang tile and $a \in A$ is the decoration. Given a decorated Wang tile $t = (t', a)$, we note $\pi(t) = a$. A *decorated tileset* τ is a finite set of decorated Wang tiles.

Tilings by decorated Wang tiles are defined exactly as tilings by Wang tiles.

Definition 6.4.5 If x is a tiling of $\mathbb{Z}$ by a decorated tileset τ, we write $\pi(x)$ the extension of π to words over the alphabet A. It is defined by $\pi(x)_i = \pi(x(i))$.

The sofic shift defined by τ, denoted X_τ is the set of all words $\pi(x)$, for x a tiling by τ.

An example of a decorated tileset τ, a tiling by τ, and the corresponding infinite word are given in Fig. 6.20.

The term "shift" corresponds to the fact that the tilings are invariant by shift:

Definition 6.4.6 Let x be a bi-infinite word over an alphabet A. The shift of x, $\sigma(x)$, is the bi-infinite word defined by $\sigma(x)_i = x_{i+1}$

We define $\sigma^n(x)$ for $n \in \mathbb{Z}$ similarly.

Definition 6.4.7 A set of infinite words $X \subseteq A^{\mathbb{Z}}$ is a subshift if it is closed under shift ($x \in X \iff \sigma(x) \in X$) and topologically closed (for the product topology on $A^{\mathbb{Z}}$)

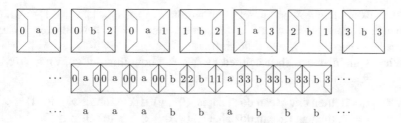

Fig. 6.20 The tileset τ_3, a tiling x by τ_3, and the corresponding bi-infinite word $\pi(x)$

The subshift *generated* by a set T is the smallest subshift that contains T, i.e. the topological closure of $\{\sigma^n(x), x \in T, n \in \mathbb{Z}\}$

We refer to other articles in this volume for more reference about symbolic dynamics.

Proposition 6.4.8 X_τ *is a subshift.*

If we decorate the edges from the graph obtained in Fig. 6.19 with the labels from Fig. 6.20, we obtain a finite automaton such that the projections $\pi(x)$ correspond exactly to the biinfinite paths in this automaton.

In a sense, this means that the behaviour of one-dimensional sets of Wang tiles are entirely understood once we understand the theory of finite automata. In particular:

Proposition 6.4.9 *Let L be the set of finite words that can occur in some infinite word in X_τ. Then L is regular.*

Conversely, let L be a regular language. The set of infinite words, all factors of which are in L is a sofic shift.

See [38, Chapter 3] for more details.

6.4.1.2 Distance Shifts

Now that we have the concept of a sofic shift, we can introduce the main construction of Aanderaa and Lewis. The core of the proof is based on the following definition:

Definition 6.4.10 Given a sofic shift $X \subseteq (A \times A)^{\mathbb{Z}}$, the *distance shift* X^Δ corresponding to X is the set of all pairs $(x, y) \in A^{\mathbb{Z}} \times A^{\mathbb{Z}}$ s.t. for all $i \in \mathbb{Z}$, $(x, \sigma^i(y)) \in X$.

The definition is of course symmetric in x and y: a pair (x, y) is in the distance shift iff for all $n, m \in \mathbb{Z}$, $(\sigma^n(x), \sigma^m(y)) \in X$.

We call this shift a distance shift, as it can compare patterns in x and in y that are at any distance from each other, just by shifting x or y to put them in the same position.

As an example, let X be the set of all words $(x, y) \in (\{0, 1\} \times \{0, 1\})^{\mathbb{Z}}$ with the following property: If $x_i = y_i = 1$ for some i, then $x = y$. It is an exercise left to the reader to show that X is a sofic shift.

What is the distance shift defined by X? To know this, let $(x, y) \in (\{0, 1\} \times \{0, 1\})^{\mathbb{Z}}$ such that for all $n, m \in \mathbb{N}$, $(\sigma^n(x), \sigma^m(y)) \in X$.

- If $\forall i, x_i = 0$, then any y can do. That is, $(0^{\mathbb{Z}}, y) \in X^{\Delta}$ for all $y \in \{0, 1\}^{\mathbb{Z}}$.
- If $\forall i, y_i = 0$, then any x can do. That is, $(x, 0^{\mathbb{Z}}) \in X^{\Delta}$ for all $x \in \{0, 1\}^{\mathbb{Z}}$.
- Suppose that x contains only one occurrence of the letter 1, say in position i. Then it is easy to see that y should contain at most one occurrence of 1: shift y so that one of the occurences of 1 in y coincides with the only occurence in x: As $\sigma^n(y)_i = x_i = 1$ we get $\sigma^n(y) = x$.
- Similarly, if y contains only one occurence of the letter 1, then x contains at most one occurrence of the letter 1.
- Otherwise x and y contains at least two occurences of the letter 1. Then it is easy to see that x and y are periodic words. Indeed suppose that $x_i = x_j = 1$ and that $y_k = 1$. Then $\sigma^i(x)_0 = \sigma^k(y)_0 = 1$ and therefore $\sigma^k(y) = \sigma^i(x)$. Similary, $\sigma^j(x) = \sigma^k(y)$ and therefore $\sigma^i(x) = \sigma^j(x)$. With a bit more work, we can show that x and y are (up to shift) of the form $(0^p 1)^{\mathbb{Z}}$ for some p.

This easy example shows that the structure of X^{Δ} can already be quite complicated. In particular, if X is sofic, X^{Δ} might be far from sofic.

The result of Aanderaa and Lewis is the following:

Theorem 6.4.11 ([1, 36]) *There is no algorithm that decides, given a sofic shift X whether X^{Δ} is empty.*

The proof of this theorem will be given over the next sections.

6.4.1.3 Application to the Domino Problem

It is not obvious from the theorem how it can be applied for the undecidability of the Domino Problem. The key idea is the following: Starting from a sofic shift $X \subseteq (A \times A)^{\mathbb{Z}}$, we build a two-dimensional object X_2 s.t. for each configuration $(x, y) \in X$, the n-th row of the corresponding configuration in X_2 is the pair $(x, \sigma^n(y))$. The result is depicted on Fig. 6.21. It is easy to see that X_2 is indeed a sofic shift:

Proposition 6.4.12 *Let X be a one-dimensional sofic shift over an alphabet $A \times A$. Consider the following two-dimensional shift X_2:*

- *Every cell of the subshift is composed of two layers.*
- *The first layer contains a word $w \in A^{\mathbb{Z}^2}$ that is identical on all rows $w_{i,j} = x_i$ for some $x \in A^{\mathbb{Z}}$.*
- *The second layer contains a word $z \in A^{\mathbb{Z}^2}$ that is shifted on consecutive rows: $z_{i,j+1} = z_{i+1,j}$. In other words $z_{i,j} = y_{i+j}$ for some $y \in A^{\mathbb{Z}}$*
- *The word in each row is in X.*

x_0, y_9	x_1, y_{10}	x_2, y_{11}	x_3, y_{12}	x_4, y_{13}	x_5, y_{14}	x_6, y_{15}	x_7, y_{16}	x_8, y_{17}	x_9, y_{18}
x_0, y_8	x_1, y_9	x_2, y_{10}	x_3, y_{11}	x_4, y_{12}	x_5, y_{13}	x_6, y_{14}	x_7, y_{15}	x_8, y_{16}	x_9, y_{17}
x_0, y_7	x_1, y_8	x_2, y_9	x_3, y_{10}	x_4, y_{11}	x_5, y_{12}	x_6, y_{13}	x_7, y_{14}	x_8, y_{15}	x_9, y_{16}
x_0, y_6	x_1, y_7	x_2, y_8	x_3, y_9	x_4, y_{10}	x_5, y_{11}	x_6, y_{12}	x_7, y_{13}	x_8, y_{14}	x_9, y_{15}
x_0, y_5	x_1, y_6	x_2, y_7	x_3, y_8	x_4, y_9	x_5, y_{10}	x_6, y_{11}	x_7, y_{12}	x_8, y_{13}	x_9, y_{14}
x_0, y_4	x_1, y_5	x_2, y_6	x_3, y_7	x_4, y_8	x_5, y_9	x_6, y_{10}	x_7, y_{11}	x_8, y_{12}	x_9, y_{13}
x_0, y_3	x_1, y_4	x_2, y_5	x_3, y_6	x_4, y_7	x_5, y_8	x_6, y_9	x_7, y_{10}	x_8, y_{11}	x_9, y_{12}
x_0, y_2	x_1, y_3	x_2, y_4	x_3, y_5	x_4, y_6	x_5, y_7	x_6, y_8	x_7, y_9	x_8, y_{10}	x_9, y_{11}
x_0, y_1	x_1, y_2	x_2, y_3	x_3, y_4	x_4, y_5	x_5, y_6	x_6, y_7	x_7, y_8	x_8, y_9	x_9, y_{10}
x_0, y_0	x_1, y_1	x_2, y_2	x_3, y_3	x_4, y_4	x_5, y_5	x_6, y_6	x_7, y_7	x_8, y_8	x_9, y_9

Fig. 6.21 A portion of the shift X_2

Then X_2 is *sofic and each row of X_2 is composed of pairs (x, y) s.t. for all j,* $(x, \sigma^j(y)) \in X$.

Furthermore, there exists an algorithm that can obtain a set of decorated Wang tiles for X_2 from a set of decorated Wang tiles for X.

In fact the j-th row of X_2 proves that $(x, \sigma^j(y)) \in X$. Therefore X_2 is empty iff X^Δ is empty.

Corollary 6.4.13 *The Domino problem is undecidable.*

Notice that, even taking into account the projection π that deleted the colors, the tilings we obtain using the Aanderaa-Lewis method are quite different from the tilings of the previous section.

In fact, the construction used above can give us a stronger result if done correctly.

Recall that each row should be part of a sofic shift X, which means they correspond to decorations of Wang tiles. We represent in Fig. 6.22 the shift X_2 with the Wang tiles these decorations are part from.

Now it is well known that each sofic shift can be obtained from a *right-resolving* set of Wang tiles, that is a set of Wang tiles where the colors on the east side is entirely determined from the colors on the west side and the label of the tile (This is obtained by the usual process of determinization of a finite automaton).

If this is the case, the tiling we obtain has a deterministic property: the half plane at the right of the line is entirely determined by the line. In fact, the symbols at the black position are entirely determined by the three gray symbols that are immediately above and at its left: x_4 is obtained from the tile on the top, y_8 from the tile at the top left, z_4^4 by the tile at the left, and z_4^5 is uniquely determined knowing the three others.

This means that we can convert the picture and the shift X_2 into a north-west deterministic set of Wang tiles. (A set of Wang tiles is north-west deterministic if

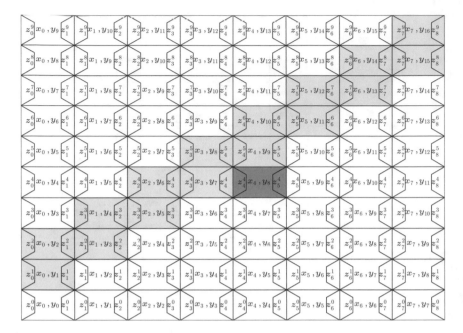

Fig. 6.22 A tiling from X_2 with the (one-dimensional) Wang tiles drawn

the colors on the east and south part of the Wang tile are uniquely determined by the two other colors.) It is then easy to see that Theorem 6.4.11 implies the following statements:

Theorem 6.4.14 *The Domino problem for north-west deterministic set of Wang tiles is undecidable*

Corollary 6.4.15 *The nilpotency problem for one-dimensional cellular automata is undecidable.*

The classical proof of this result is attributed to Kari [30] but in fact this result is already mentioned in the original article of Aanderaa and Lewis [1].

The rest of the section is devoted to the proof of Theorem 6.4.11. As the proof is quite long and difficult, we will focus on proving that there exists a nonempty distance shift with no periodic points, which implies that there exists a two-dimensional aperiodic tileset. The modification to obtain the undecidability result will be mentioned briefly at the end.

The general idea is to find a set S s.t. if $x, y \in S$, then the subword of x that corresponds to the position where it differs from y (assuming this set of positions is biinfinite) is itself in S. By comparing in particular x with $\sigma^n(x)$, this implies in particular some kind of hierarchical structure in x.

It is easy, with a sofic shift or equivalently with a finite automaton, to isolate the positions where x and y differs. Therefore there is some hope to be able to define such a set S as a distance shift (more precisely, $S \times S$ will be our distance shift).

The set S that will work is the subshift S_p we build below that corresponds to a Toeplitz subshift based on p-adic numbers. There will be an additional difficulty in the proof: if x, $y \in S_p$, then the subword of x that corresponds to the position where it differs from y (assuming this set of positions is biinfinite) is "almost" in S_p. What this "almost" part means is quite technical and will be describe later.

We will first describe this set S_p and how it related to p-adic numbers. We will then describe the hierarchical properties it satisfies. Finally, we will explain why the difference between two words of S_p is almost a word in S_p and how to use this to build our distance shift. This distance shift, when passed through the two-dimensional machinery explained in the few previous pages, will gives us a two-dimensional aperiodic tileset.

We will finish by briefly indicating what remains to be done to obtain the undecidability of the emptiness of a distance shift, and therefore the undecidability of the Domino problem

6.4.2   The Subshift S_p and p-Adic Numbers

Let p be an integer Let $a_p(n)$ be the first non-zero digit in the expansion of n in base p. That is, $a_p(n) = r$ if $n = q \times p^{k+1} + p^k r$ for some q and $r \neq 0 \mod p$. For example $a_{10}(1664) = 4$, $a_{10}(71,500) = 5$ and $a_{10}(71,050) = 5$.

Note that $a_p(0)$ is not well defined. We define it as taking all values in the alphabet $\{0, 1, \ldots, p - 1\}$ (this makes a_p a multivalued function).

We will be interested in the hierarchical structure of the word u defined by $u_n = a_p(n)$, and of all other words similar to u.

For reference, here are the 80 digits of u around 0 when $p = 3$ with the arbitrary choice $a_3(0) = 2$.

2212212112212212112212121121121122122121121212.2.12112212112112212212112212112112212112212112211

The main part of the construction of Aanderaa and Lewis is to build the subshift generated by this point, for large values of p. To understand how to do it, we first have to understand this subshift.

For this we need the notion of p-adic integers. p-adic integers are formal sums

$$m = \sum_{k \geq 0} m_k p^k$$

with $m_k \in \{0, \ldots, p - 1\}$. p-adic integers form a group under addition, where addition is done componentwise with propagation of carry, and even a ring with the multiplication defined the usual way. The integers $\mathbb{Z}$ can be seen as a subset of this group: Nonnegative integers correspond to formal sums where all but finitely many terms are 0. Negative integers correspond to formal sums where all but finitely many terms are equal to $p-1$. The ring of p-adic integers will be denoted $\mathbb{Z}_p$. It has the natural, discrete topology: $d(n,m) = p^{-\min\{i, n_i \neq m_i\}}$ which makes it a compact space, in which $\mathbb{Z}$ is dense.

The function a can be extended to $\mathbb{Z}_p$ in the natural way: For $m \in \mathbb{Z}_p$, let i be the minimal position s.t. $m_i \neq 0$. Then $a_p(m) = m_i$. It is easy to see that a_p is a continuous function.[2]

Proposition 6.4.16 *Let p be an integer and S_p the subshift generated by the sequence u with $u_i = a_p(i)$. S_p contains exactly all points $i \to a_p(m+i)$ for $m \in \mathbb{Z}_p$.*

Proof For $m \in \mathbb{Z}_p$, let u^m be the sequence defined by $(u^m)_i = a_p(m+i)$ and let $X = \{u^m, m \in \mathbb{Z}_p\}$. By continuity of a_p, we get that if $m_k \xrightarrow{k \to \infty} m$ then $u^{m_k} \xrightarrow{k \to \infty} u^m$, so X is closed. In particular X is a subshift and thus $S_p \subseteq X$.

For $n \in \mathbb{Z}, u^n = \sigma^n(u)$ and therefore $u^n \in S_p$ as S_p is shift-invariant. If $m \in \mathbb{Z}_p$ then $m = \lim m_k$ for $m_k \in \mathbb{Z}$ and therefore $u^m = \lim u^{m_k} \in S_p$ as S_p is topologically closed. Therefore $X \subseteq S_p$, and thus $X = S_p$.

Notice that all the points $i \to a_p(m+i)$ are distinct. This gives a *continuous* map:

$$f : S_p \mapsto \mathbb{Z}_p$$
$$x \to f(x)$$

s.t. $f(\sigma(x)) = f(x) + 1$. This makes our dynamical system an almost 1-1 extension of the odometer (the dynamical system $(\mathbb{Z}_p, T)$ on the odometer defined by $T(x) = x + 1$). The extension is almost 1-1 due to the fact that all points of $\mathbb{Z}_p$ except the integers have only one preimage.

Here are a few useful properties of a_p:

- $a_p(pm) = a_p(m)$ for all $m \in \mathbb{Z}_p$.
- $a_p(m+p) = a_p(m)$ if $m \neq 0 \mod p$
- $a_p(m+p^k) = a_p(m)$ if $m \neq 0 \mod p^k$

Notice that $m = 0 \mod p$ makes sense not only for $m \in \mathbb{Z}$ but also for $m \in \mathbb{Z}_p$.

The following proposition will be useful:

---

[2]This statement is somewhat wrong, as a_p is a multivalued function at 0. (An exact statement would be that a_p is upper hemicontinuous: If the sequence y^n converges to y, then $a_p(y)$ contains (but may contain more) all limit points of $a_p(y^n)$). The fact that a_p is multivalued is a burden that is not worth the hassle, and we urge the reader to consider a_p as a normal, single valued function, even though it is not.

Proposition 6.4.17 *Let x be a word s.t. for all k, there exists $n_k \in \mathbb{Z}$ s.t. $x_i = a_p(n_k + i)$ if $n_k + i \neq 0 \mod p^k$. Then there exist $m \in \mathbb{Z}_p$ s.t. $x_i = a_p(m + i)$ for all i. In particular $x \in S_p$.*

Proof Let $j \in] - p^k/2, p^k/2]$ s.t. $j = -n_k \mod p^k$. Let $w = x_j$ be the character at position j. By replacing n_k by $n_k + p^k r$ for some suitable r we can assume wlog that $j + n_k = p^{k+1}w$. In particular $a_p(j + n_k) = w = x_j$.

Therefore, $x_i = a_p(i + n_k)$ for $i \in] - p^k/2, p^k/2]$.

Let m be a limit point of the sequence $(n_k)_{k \in \mathbb{N}}$. Then by construction $x_i = a_p(m + i)$ for all i. $\blacksquare$

6.4.3 S_p as a Toeplitz Subshift

S_p has also the structure of a Toeplitz subshift (as all symbolic almost 1-1 extensions of odometers). Let us recall what the point $u \in S_p$ looks like when $p = 5$:

1234112342123431234412341123411234212343123441234212341123421234312344123431234112342123431234412

By definition of a_p, $a_p(n + p) = a_p(n)$ if $n \neq 0 \mod p$. So if we delete one symbol every p symbols in u, we obtain a periodic point.

This can be formalized in the following way:

Definition 6.4.18 If x is a sequence on the alphabet A, the p-skeleton of x denoted $\mathrm{skel}_p(x)$ is the sequence over the alphabet $A \cup \{\star\}$ defined by:

$$\mathrm{skel}_p(x)_i = \begin{cases} x_i & \text{if } \forall n, x_{i+np} = x_i \\ \star & \text{otherwise} \end{cases}$$

Intuitively, and in what follows, the $\star$ symbol represents a symbol the value of which is unknown. The 5-skeleton of u for $p = 5$ is therefore:

..1234$\star$1234$\star$1234$\star$1234$\star$1234$\star$1234$\star$1234$\star$1234$\star$1234$\star$1234$\star$1234$\star$1234$\star$1234$\star$1234$\star$1234$\star$1234$\star$1234..

It is easy to see that any word of S_p actually has the p-skeleton:

...12...$p'\star$12...$p'\star$12...$p'\star$12...$p'\star$12...$p'\star$12...$p'\star$12...$p'\star$12...$p'\star$12...$p'\star$12...$p'\star$12...$p'\star$12...$p'\star$12...

with $p' = p - 1$. Furthermore, any word of S_p has the following p^2-skeleton (the differences with the p-skeleton have been highlighted)

...12...p'112...p'21............$p'\mathbf{p'}$12...$p'\star$12...p'112...p'21.........$p'\mathbf{p'}$12...$p'\star$12...p'112...p'21...$p'\mathbf{p'}$12...

As the p-skeleton contains only the periodic part of a word, we will be interested mainly in the complement of the p-skeleton:

Definition 6.4.19 If x is a sequence on the alphabet A, the p-coskeleton of x denoted $coskel_p(x)$ is the sequence over $A \cup \{\_\}$ defined by:

The *reduced* p-coskeleton is the image of $coskel_p(x)_i$ under the morphism h defined by $h(a) = a$ for $a \in A$ and $h(\_) = \epsilon$

It is a bit painful to define exactly how to apply a morphism on a biinfinite word. Note however that applying an erasing morphism (like h) to a biinfinite word might produce in some cases a finite word, or a word that is finite in one direction and infinite in the other direction.

We go back to the running example:

1234112342123431234412341123411234212343123441234212341123421234312344123412341123421234312344123

Its 5-coskeleton is:

..._____1_____2_____3_____4_____1_____1_____2_____3_____4_____2_____1_____2_____3...

and its reduced 5-coskeleton is:

1234112342123431234412341123411234212343123441234212341123421234312344123412341123421234312344123

In fact, the following proposition is an easy consequence of the fact that $a_p(pn) = a_p(n)$, and the proof is left to the reader:

Proposition 6.4.20 *Let $x \in S_p$. Then the reduced p-coskeleton of x is also in S_p.*

This gives us an easy way to show that a word $x \in S_p$:

Proposition 6.4.21 *Recall that $p' = p - 1$ and let S be a set of configurations s.t.:*

1. *Every $x \in S$ has the form $(12 \ldots p'\star)^{\mathbb{Z}}$. That is, there exists $n \in \mathbb{Z}$ s.t. $x_{i+n} = a_p(i)$ if $i \neq 0 \mod p$.*
2. *The word x' defined by $x'_i = x_{pi+n}$ is in S with n as defined in (1).*

Then $S \subseteq S_p$.

Notice that x' is exactly the reduced p-coskeleton of x.

Proof We first prove by induction on k that if $x \in S$, then there exists some $n \in [0, p^k - 1]$ s.t. $x_{i+n} = a_p(i)$ for $i \neq 0 \mod p^k$. This is true for $k = 1$ by definition. Now suppose the result is true for k and let $x \in S$ and x', n as in the definition. We can suppose wlog that $n \in [0, p - 1]$. By induction hypothesis, there exists $n' \in [0, p^k - 1]$ s.t. $x'_{i+n'} = a_p(i)$ for $i \neq 0 \mod p^k$. Let $N = n + pn' \in [0, p^{k+1} - 1]$. Then $x_{i+N} = a_p(i)$ for all $i \neq 0 \mod p^{k+1}$ which ends the induction.

We now conclude by Proposition 6.4.17.

6.4.4 $S_p \times S_p$ as a Distance Shift

Our goal now is to prove that $S_p \times S_p$ is a distance shift, which means that we can somehow decide if $(x, y) \in S_p \times S_p$ by local rules involving x and all possible shifts of y.

The previous proposition is a first step toward a solution. The problem is that there is no easy way to extract the p^k-coskeleton of a configuration easily. There is however a workaround. It turns out that, if we compare two points $x, y \in S_p$, then the set of positions where they differ looks like the p^k-coskeleton for some k.

Let's look for example at the point x defined by $x_n = a_5(n)$ and the point y defined by $y_n = a_5(n + 10)$. The third row shows x when it differs from y.

The only difference between the two last sequences is that some symbols are missing (i.e. replaced by _) in the third sequence. The same is true more generally and is illustrated in Fig. 6.23.

Definition 6.4.22 If x and y are two configurations on alphabet A, the difference of x and y, denoted by $\mathrm{diff}(x, y)$ is the configuration on alphabet $A \cup \{\_\}$ defined by

$$\mathrm{diff}(x, y)_i = \begin{cases} \_ & \text{if } x_i = y_i \\ x_i & \text{otherwise} \end{cases}$$

It is of course obvious that most symbols should be the same in x and y. If $x_i = a_p(i)$ and $y_i = a_p(i + p^k)$ then all positions of valuation less than k will be identical in x and y.

Proposition 6.4.23 Let $x_i = a_p(i)$ and $y_i = a_i(i + p^k)$.
Then $\mathrm{diff}(x, y) \subseteq coskel_{p^k}(x)$, in the sense that $\mathrm{diff}(x, y)_i = coskel_{p^k}(x)$ whenever $\mathrm{diff}(x, y)_i \neq \_$.

```
12341123421234312344123411234112342123431234412342123411234212343123441234312341123421234312344123
23412342123431234123411234212343123421234212341234212343123412343123412342123431234123412341234312342
41123412343124412341123411234123443123412441123412343124412341123412343124412341123412341123412
2341234212343123412341234212343123412342123412342123431234123412343123412342123431234123412341234312342
3411234212341234412341123411234213412344123421341123421341234412341234112342134123444123441234112342
234123421234312341234123421234312341234212341234212343123412343123412342123431234123412341234312342
3411234213412344123411234112342134123441234213411234213412344123412341123421341234412344123411234112342
2342234312341234112341234223431234123412341123422343123412343123411234223431234123412341123422343
```

Fig. 6.23 In the first row, the sequence a_5. In the second row, the values of $a_5(1 + n)$ when it differs from $a_5(n)$ (values that are equal are not shown) In the third row, the values of $a_5(3 + n)$ when it differs from $a_5(n)$ The other rows correspond to the choices 10, 25, 30, and 50

The above examples seem to suggest that there are only few positions where the reverse does not hold. The following lemma explains, in the case of k of valuation 0, that at most 4 symbols of the 1-skeleton in every block of size p^2 will be replaced by the symbol $\_$ in $\text{diff}(x, y)$.

Lemma 6.4.24 *Let $k \neq 0 \mod p$. If $a_p(i + k) = a_p(i)$ then*

- $i = 0 \mod p^2$ *or*
- $i = -k \mod p^2$ *or*
- $i = pk \mod p^2$ *or*
- $i = -kp - k \mod p^2$.

In particular, given any $k \neq 0 \mod p$ there are (at most) four values of $i \mod p^2$ s.t $a_p(i + k) = a_p(i)$ and no three of them are consecutive.

Proof

- If $i + k \neq 0 \mod p$ and $i \neq 0 \mod p$, then $a_p(i) = i \mod p$ and $a_p(i + k) = i + k \mod p$ and therefore $a_p(i) \neq a_p(i + k)$.
- If $i + k = 0 \mod p$ but $i + k \neq 0 \mod p^2$. Then $a_p(i + k) = (i + k)/p \mod p$ and $a_p(i) = i \mod p$. Therefore we obtain $i = (i + k)/p \mod p$ and $i = -k(p + 1) \mod p^2$.
- If $i = 0 \mod p$ but $i \neq 0 \mod p^2$ we get similarly $i + k = i/p \mod p$ and $i = pk \mod p^2$

Corollary 6.4.25 *Let $m, n \in \mathbb{Z}_p$ s.t $m \neq n$ but m and n coincide exactly on their first r digits, i.e. $m - n = kp^r$ with $k \neq 0$.*
If $a_p(m + i) = a_p(n + i)$ then

- $i \neq -m \mod p^r$ *or*
- $i = -m \mod p^{r+2}$ *or*
- $i = -n \mod p^{r+2}$ *or*
- $i = -n + p(m - n) \mod p^{r+2}$ *or*
- $i = -m - p(m - n) \mod p^{r+2}$.

The proof is an easy generalization of the lemma and is left as an exercise. The corollary states the following: For any x, y, $\text{diff}(x, y)$ looks almost like the p^r-coskeleton of x, except $\text{diff}(x, y)$ might contain at most 4 additional symbols replaced with a symbol $\_$ compared to the coskeleton.

Now we have seen that, if we delete the symbol $\_$ from the p^r-coskeleton of x, we obtain an element of S_p, and in particular an infinite concatenation of words of the form:

$$12 \ldots p' 1 1 2 \ldots p' 2 1 \ldots p' p' 1 2 \ldots p' \star$$

where $\star$ denotes any symbol.

As a consequence, if we delete the $\_$ symbol from $\text{diff}(x, y)$, we also obtain infinite concatenations of words of the same form, except that 4 symbols might be missing in each block.

This is formalized in the next proposition.

Proposition 6.4.26 *Let W be the collection of blocks of size p^2 of the form*

$$4 \ldots p' \, 1 \, 1 \, 2 \ldots p' \, 2 \, 1 \ldots p' \, p' \, 1 \, 2 \ldots p' \star 1 \, 2 \, 3$$

where $\star$ denotes any symbol. Let $\hat{W}$ be the set of subwords of words of W of length at least $p^2 - 4$.

Let $x, y \in S_p$ and suppose that x and y differ in more than one position.

Then $h(\mathrm{diff}(x, y)) \in (\hat{W})^{\mathbb{Z}}$ where h is the erasing morphism: $h(\_) = \epsilon$ and $h(a) = a$ otherwise.

In particular, let $x \neq y \in S_p$. Then the image by h of every factor of $\mathrm{diff}(x, y)$ is a factor of $(\hat{W})^{\mathbb{Z}}$

The words x and y may differ in only one position when $x_i = y_i = a_p(n+i)$ with $n \in \mathbb{Z}$ rather than $n \in \mathbb{Z}_p$.

It would seem more natural to change the words in W so that the factor 123 appears at the beginning rather than at the end. The version we use is essentially cosmetic but greatly simplifies the proof.

What is interesting and nontrivial is that this proposition is a characterization of words in S_p.

Definition 6.4.27 We say that a pair of words (x, y) satisfies property $\mathcal{P}$ if every factor of $h(\mathrm{diff}(x, y))$ (resp. $h(\mathrm{diff}(y, x)))$ is a factor of $(\hat{W})^{\mathbb{Z}}$.

Notice that "every factor of $h(\mathrm{diff}(x, y))$ is a factor of $(\hat{W})^{\mathbb{Z}}$" is the same as "$h(\mathrm{diff}(x, y)) \in (\hat{W})^{\mathbb{Z}}$" when $h(\mathrm{diff}(x, y))$ is a biinfinite word, however it also makes sense even if $\mathrm{diff}(x, y)$ is a finite word, or is simply infinite.

We are now in position to state the main theorem:

Theorem 6.4.28 *Let $p > 6$.*

Let x, y be two words over the alphabet $\{1, \ldots, p-1\}$ s.t.

- $x \in W^{\mathbb{Z}}$
- $y \in W^{\mathbb{Z}}$
- *For all k, the words x and $\sigma^k(y)$ have property $\mathcal{P}$*

Then $x \in S_p$ and $y \in S_p$.

The theorem is an easy consequence of the following proposition:

Proposition 6.4.29 *Suppose that x and y are composed of blocks of size p of the form*

$$1 \, 2 \ldots p' \star$$

and that for all k, the words x and $\sigma^k(y)$ have property $\mathcal{P}$. Then $x \in W^{\mathbb{Z}}$ and $y \in W^{\mathbb{Z}}$

Proof of the Theorem Let $\mathcal{S}$ be the set of all configurations x s.t. there exists y s.t.:

- $x \in (12 \ldots p' \star)^{\mathbb{Z}}$
- $y \in (12 \ldots p' \star)^{\mathbb{Z}}$
- For all k, the words x and $\sigma^k(y)$ have property $\mathcal{P}$

Let $x \in \mathcal{S}$ and y that witnesses it. Upto shift, we may suppose that $y_i = x_i = i$ mod p for $i \neq p$. Let $x_i' = x_{pi}$ and $y_i' = y_{pi}$.

By the previous proposition, $x \in W^{\mathbb{Z}}$ which implies that $x' \in (12 \ldots p' \star)^{\mathbb{Z}}$, and the same is true for y'.

Furthermore, we can prove that x' and y' satisfy property $\mathcal{P}$. Indeed we have $h(\text{diff}(x', o^k(y'))) - h(\text{diff}(x, \sigma^{pk}(y)))$ as x and y have the same p-skeleton and therefore disagree only on the positions that are inside x' and y'.

Therefore we have proven that $x' \in \mathcal{S}$. We conclude by Proposition 6.4.21.

Proof of the Proposition The proof of the proposition is quite technical and should be skipped at first read.

The various depictions here suppose that $p = 10$, that is $p - 1 = 9$. The general pictures can be obtained by replacing all occurrences of 9 with "$p - 1$" and all occurrences of 8 with "$p - 2$".

By taking shifts of x and y we can suppose wlog that $x_i = i \mod p$ for $i \neq 0$ mod p, and similarly for y. We will prove by comparing x and $\sigma^2(y)$ that x and y almost have the predescribed structure. Considering x and $\sigma^{-2}(y)$ will finish the proof.

First we compare x and $\sigma^2(y)$:

$i \mod p$	4	$\ldots$	8	9	0	1	2	3
x	4	$\ldots$	8	9	$\star$	1	2	3
$\sigma^2(y)$	6	$\ldots$	$\star$	1	2	3	4	5
$z = \text{diff}(x, \sigma^2(y))$	4	$\ldots$	?	9	?	1	2	3

The $\star$ symbol corresponds to positions in x and $\sigma^2(y)$ where the value is currently unknown. The only symbols in $z = \text{diff}(x, \sigma^2(y))$ that could be equal to the symbol $\_$ are the question mark symbols. By assumption, we know that $h(z) \in (\hat{W})^{\mathbb{Z}}$. This gives us a decomposition of z into blocks: write $z = \ldots w_{-1} w_0 w_1 \ldots$ s.t. $h(w_i) \in \hat{W}$.

It should be obvious that the only possibility is for each w_i to be of size exactly p^2, and to be of the form $45 \ldots p' \star 45 \ldots \ldots p' \star 123$. Indeed, z contains the factor "34" periodically with period p, and this factor appear at most $p - 1$ times inside a word of $\hat{W}$ (which is of length between $p^2 - 4$ and p^2), so at least one out of every p occurrences of the factor "34" should correspond to the last letter of a word in $\hat{W}$ followed by the first letter of a word in $\hat{W}$, which proves the result.

We now look at each word w_i independently. We will look at the situation from the point of view of editing distance:

w_i	4	5	...	?	9	?	1	2	...	?	9	?	1	...	...	?	9	?	1	2	...	9	?	1	2	3
t	4	5	...	8	9	1	1	2	...	8	9	2	1	...	...	8	9	9	1	2	...	9	★	1	2	3

The question marks inside w_i represents symbols that are either equal to the symbol of x, or are equal to the symbol _. The word t is a word of W. We know that w_i, when deleting the _ symbols, is equal to a word of $\hat{W}$, i.e. equal to the word t with at most 4 symbols removed. In other words, we should delete at most 4 symbols from w_i, and that can only be done in the places where the _ symbol appears, and add at most 4 symbols to obtain the word t depicted.

This is only possible if every symbol of w_i which is not a _ symbol has the same value that the symbol at the same position in t.

We now look at the p^2 symbols of x that correspond to w_i. We have just said that any symbol of x that was not changed into a symbol _, that is, every position in x that contains a different value that $\sigma^2(y)$, has the same symbol as t. Now looking back at the table, the only possibility for a ★ symbol in x (i.e. a symbol for which we did not know previously the value) to become a _ symbol, if if this symbol was actually a symbol 2.

We have therefore proven that the p^2 symbols of x that correspond to w_i are of the form

$$4\,5\,\ldots\,8\,9\,\underline{1}\,1\,2\,\ldots\,8\,9\,\underline{2}\,1\,\ldots\ldots\,9\,\underline{9}\,1\,2\,\ldots\,9\,\star\,1\,2\,3$$

where at most 4 of the underlined symbols might actually have been replaced by the symbol 2. Thus x is a concatenation of blocks of W where at most four symbols might have been replaced with the symbol 2.

We now start the reasoning again with x and $\sigma^{-2}(y)$, which proves that x is a concatenation of blocks of W, where at most four symbols might have been replaced with the symbol $p - 2$.

Both situations cannot happen simultaneously unless no symbols were actually replaced, i.e. $x \in W^{\mathbb{Z}}$.

Corollary 6.4.30 *There exists a sofic shift X s.t. the distance shift X^{Δ} corresponding to X is exactly $S_p \times S_p$. In particular the distance shift has no periodic point and therefore gives rise to an aperiodic 2-dimensional SFT.*

Proof The sofic shift is the set of all x, y s.t.

- $x \in W^{\mathbb{Z}}$
- $y \in W^{\mathbb{Z}}$
- (x, y) satisfy property $\mathcal{P}$.

The only nontrivial part of this statement might be that the set of all (x, y) that satisfy property $\mathcal{P}$ is a sofic shift.

Indeed, let $\mathcal{W}$ be the language of $\hat{W}^{\mathbb{Z}}$, i.e. the set of all factors of words in $\hat{W}^{\mathbb{Z}}$. $\mathcal{W}$ is easily seen to be a regular language as $\hat{W}$ is finite. Now let $\mathcal{L}$ be the set of all words (u, v) over the alphabet $(A \times A)$ s.t. $h(\text{diff}(u, v)) \in \mathcal{W}$. Then $\mathcal{L}$ is a regular language as the inverse image by a morphism of a regular language. The set of (x, y) all factors of which are in $\mathcal{L}$ is therefore a sofic shift.

6.4.5 Undecidability for Distance Shifts

In this section we will briefly describe what should be changed to prove that there is no algorithm that decide if a distance shift is empty.

The subshift X we build in the previous corollary codes the distance shift $S_p \times S_p$. Now suppose that the letters are colored with two colors, gray and light gray. We do the same construction, but taking for W all words where color alternates between the levels. That is, W contains all words of the form

$$4 \cdots p' \; 1 \; 1 \; 2 \cdots p' \; 2 \; 1 \; 2 \cdots\cdots\cdots p' \; p' \; 1 \; 2 \cdots p' \; \star \; 1 \; 2 \; 3$$

and all words of the form

$$4 \cdots p' \; 1 \; 1 \; 2 \cdots p' \; 2 \; 1 \; 2 \cdots\cdots\cdots p' \; p' \; 1 \; 2 \cdots p' \; \star \; 1 \; 2 \; 3$$

and we take as a sofic subshift the set X of all pairs (x, y) s.t.:

- x and y are of the form $\left(\begin{array}{cccc} 1 & 2 \cdots p' & \star \end{array} \right)^{\mathbb{Z}}$
- (x, y) satisfy property $\mathcal{P}$ (for this new set W)

The reader should be convinced that the previous reasoning go through and that the distance shift defined by X is exactly the same as $S_p \times S_p$ except that symbols are colored depending on their valuation (remember that the valuation of an integer n, or a p-adic number n, is k if p^k is the greatest integer that divides n). More precisely if $x_i = a_p(i + n)$ for $n \in \mathbb{Z}_p$, then x_i is light gray if $i + n$ is of valuation k for k even, and dark gray otherwise.

We can go a bit further. Let B be a finite alphabet. Let $P \subseteq B \times B$. We now look at words over the alphabet $\{1, 2, \ldots, p - 1\} \times B$. Symbols in $\{1, 2, \ldots, p - 1\}$ and symbols and B will be written in two different layers to simplify the exposition.

Consider the following set of words W:

$$4 \ldots p' \; 1\,1\,2 \ldots p' \; 2\,1 \ldots p' \; p' \; 1\,2 \ldots p' \star 1\,2\,3$$
$$a \ldots a \; b\,a\,a \ldots a \; b\,a \ldots a \; b \; a\,a \ldots a \star a\,a\,a$$

for all pairs $(a, b) \in P$.

And we take as a sofic subshift the set X of all pairs (x, y) s.t.:

- x and y are of the form

$$
\begin{array}{l}
1 \ \ 2 \ \ldots \ p' \star 1 \ \ 2 \ \ldots \ p' \star 1 \ \ 2 \ \ldots \ p' \star \ldots \\
a_0 \ a_0 \ldots a_0 \star a_0 \ a_0 \ldots a_0 \star a_0 \ a_0 \ldots a_0 \star \ldots
\end{array}
$$

- (x, y) satisfy property $\mathcal{P}$ (for this new set W)

Now the distance shift defined by this new set X as the following form: Its elements are pairs $(x, y) \in S_p \times S_p$. Furthermore, if $x_i = a_p(i + n)$ then:

- x_i has the symbol a_0 if $i + n$ is of valuation 0
- All positions j s.t. $j + n$ is of valuation $k \in \mathbb{N}$ contain the same symbol a_k
- For all k, $(a_k, a_{k+1}) \in P$

Now suppose that B are one-dimensional Wang tiles and that P codes the adjacency relation: $(t, t') \in P$ if t' can be placed on the right of t.

Then it is is easy to see that this distance shift codes somehow a tiling of $\mathbb{N}$ by this set of Wang tiles.

It seems a strange idea to code a tiling in this complicated way, but we have gained something here: we were able to fix the tile that should be in position 0 (namely the symbol a_0). This is fundamental: As we saw earlier, this is precisely the difficulty in proving the undecidability of the domino problem: find a way to fix the symbol at position $(0, 0)$.

To obtain the undecidability of the domino problem, we therefore have to find a way to do the same construction but for two-dimensional Wang tiles.

The idea is to construct a distance shift that code $S_p \times S_q$ for two different (relatively prime) values p and q. Using the same idea as explained before, we will be able to guarantee that all positions of valuation k for p and k' for q have the same symbol $a_{k,k'}$, that the symbol $a_{0,0}$ is the predescribed symbol, and that the pairs $(a_{k,k'}, a_{k+1,k'})$ and $(a_{k,k'}, a_{k,k'+1})$ satisfy the horizontal and vertical rules.

6.4.6   Bibliographic Notes

Our presentation differs greatly from the previous presentations by Aanderaa-Lewis [1] and Lewis [36]. In fact, the concept of p-adic numbers, of subshifts and sofic shifts are not present in the original articles. Aanderaa and Lewis introduce in particular the notion of perfect tilings, normal tilings and admissible tilings; The admissible tilings they define in a awkward way are just the elements of the subshift generated by the normal tilings.

There are important differences between the construction of Aanderaa-Lewis and the construction we gave here. In fact, the concept of distance shift that they use are called *sampling systems*, and are *apriori* weaker than distance shifts. As a consequence, they are not able to obtain a construction for something as simple as $S_p \times S_p$.

The construction they obtain correspond more or less to $T_p \times T_p$ where T_p is the subshift generated by the word u defined by $u_n = b_p(n)$ where $b_p(n)$ are the *last two* nonzero coefficients in the expansion of n in base p, i.e. $b_{10}(13{,}370) = 37$ and $b_{10}(13{,}070) = 07$. The subshift they build is actually bigger than $T_p \times T_p$: Each cell n contains in theory the last two nonzero coefficients in the expansion of n in base p, but the second-to-last coefficient (the "3" in $b_{10}(13{,}370) = 37$) can actually differ from his normal value, but in a controlled way.

6.5   The Construction of Kari

We now present the construction by Kari [31] for which the technique is very different from the ones of Berger and Anderaa-Lewis.

We will first present a construction of an aperiodic tileset that was initially designed to have a very small number of tiles, and then we will explain what to change to obtain the undecidability of the Domino Problem. The presentation here is inspired by the two groundbreaking articles of Kari [31, 32], see also [16].

While the Anderaa-Lewis construction can be explained through p-adic numbers, the core of Kari's construction is to represent numbers in a bi-infinite way so that each row of the tileset represents a different number. For the small aperiodic tileset, this will be enforced by ensuring that a row on top of another one can only be obtained by multiplying the previous number by a factor 2 or a factor $\frac{2}{3}$. The fact that it is not possible to obtain 1 by any multiplication of 2's and $\frac{2}{3}$'s will imply aperiodicity.

Let us first talk about the bi-infinite representations of numbers that will be used.

6.5.1   Balanced Representations of Reals

Definition 6.5.1 Let α be a positive real number, $a \in \{\lfloor \alpha \rfloor, (\lfloor \alpha \rfloor + 1)\}^{\mathbb{Z}}$ is a balanced representation of α iff for any subword $a_{[i,i+n-1]}$ we have:

$$n\alpha - 1 \leq \sum_{k=i}^{i+n-1} a_k \leq n\alpha + 1$$

In other words, a balanced representation of a real number is a sequence composed only of its two nearest integers where the average over finite subwords of length n is within $\frac{1}{n}$ of α. This means in particular that if we take longer and longer subwords, their average converges to α.

Fig. 6.24 The Beatty
sequence and balanced
representation of $\phi = \frac{1+\sqrt{5}}{2}$.
The ones and the twos of the
balanced representation
correspond respectively to
cutting a vertical and a
horizontal integer line

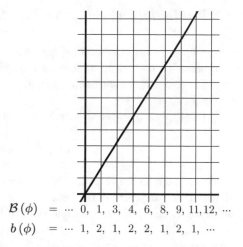

$$\mathcal{B}(\phi) \;=\; \cdots \;\; 0,\; 1,\; 3,\; 4,\; 6,\; 8,\; 9,\; 11, 12,\; \cdots$$
$$b(\phi) \;=\; \cdots \;\; 1,\; 2,\; 1,\; 2,\; 2,\; 1,\; 2,\; 1,\; \cdots$$

Such a balanced representation for a finite word is easy to construct by means of
Beatty sequences [6]. Given a real number α we denote $\mathcal{B}(\alpha)$ its Beatty sequence,
with $\mathcal{B}(\alpha)_k = \lfloor k\alpha \rfloor$. Informally, the Beatty sequence of a number α corresponds to
how many horizontal lines of the $\mathbb{Z}^2$ grid the line $y = \alpha x$ has crossed at coordinate
x, see Fig. 6.24.

Now if we take the sequence of first differences of the Beatty sequence of α

$$b(\alpha)_k = \mathcal{B}(\alpha)_{k+1} - \mathcal{B}(\alpha)_k,$$

it is a symbolic sequence on alphabet $\{\lfloor \alpha \rfloor, \lfloor \alpha + 1 \rfloor\}$. And it is a balanced
representation of α as for any i:

$$\sum_{k=i}^{i+n-1} b(\alpha)_k = \lfloor (i+n)\alpha \rfloor - \lfloor i\alpha \rfloor$$

and

$$n\alpha - 1 \le \lfloor (i+n)\alpha \rfloor - \lfloor i\alpha \rfloor < n\alpha + 1.$$

So the sequence $b(\alpha)$ is a balanced representation of α it is also sometimes called
the β-sequence for α, see [42], or a rotation sequence [18]. It is interesting to note
that when α is irrational this sequence corresponds to the sturmian word for the
cutting line $y = \alpha x$.

6.5.2 Multiplying Balanced Representations

Given a positive rational number $q = \frac{n}{m}$, it turns out that by multiplying a balanced representation of $(a_i)_{i \in \mathbb{Z}}$ of some real number α by q one still obtains a balanced representation. We however have to define what we mean by multiplying a balanced representation.

To multiply a by q we will multiply each digit individually while using a carry:

$$c_i + qa_i = b_i + c_{i+1}$$

where c_i and c_{i+1} are respectively the carry inherited from the previous multiplication and the carry sent to the next. With such a definition, when multiplying some subword $a_{[i,i+n-1]}$ by q, one obtains a word $b_{[i,i+n-1]}$ with

$$c_i + q \sum_{k=i}^{i+n-1} a_k = \sum_{k=i}^{i+n-1} b_k + c_{i+n}.$$

See Fig. 6.25 for an example of multiplication of a balanced representation.

Lemma 6.5.2 *If a is a balanced representation of some α then there exists a balanced representation b of $q\alpha$ that can be obtained by multiplying a by q. Furthermore, the set of carries being used is finite and only depends on q.*

Proof Take b any balanced representation of $q\alpha$. Since

$$n\alpha - 1 \le \sum_{k=i}^{i+n} a_i \le n\alpha + 1$$

and

$$nq\alpha - 1 \le \sum_{k=i}^{i+n} b_i \le nq\alpha + 1$$

we have

$$-(1+q) \le \sum_{k=i}^{i+n} qa_i - b_i \le 1 + q$$

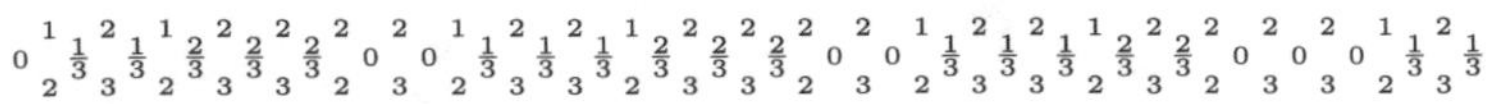

$$0 \; \tfrac{1}{2} \; \tfrac{2}{3} \; \tfrac{1}{2} \; \tfrac{2}{3} \; \tfrac{2}{3} \; \tfrac{2}{3} \; \tfrac{2}{3} \; \tfrac{2}{3} \; 0 \; 0 \; \tfrac{1}{2} \; \tfrac{2}{3} \; \tfrac{1}{2} \; \tfrac{2}{3} \; \tfrac{1}{2} \; \tfrac{2}{3} \; \tfrac{2}{3} \; \tfrac{2}{3} \; \tfrac{2}{3} \; 0 \; 0 \; \tfrac{1}{2} \; \tfrac{2}{3} \; \tfrac{1}{2} \; \tfrac{2}{3} \; \tfrac{1}{2} \; \tfrac{2}{3} \; \tfrac{2}{3} \; \tfrac{2}{3} \; \tfrac{2}{3} \; 0 \; 0 \; 0 \; \tfrac{1}{2} \; \tfrac{2}{3} \; \tfrac{1}{3}$$

Fig. 6.25 An example of multiplication of a balanced representation. On the bottom row, a balanced representation of $\phi + 1$, on the top a balanced representation of $\frac{2}{3}(\phi + 1)$. The carries are shown in the middle

Thus, it suffices to take carries that are in $[-(1+q), (1+q)]$, and because the a_i's and b_i's are integers and $q = \frac{n}{m}$ is a rational, the carries are the multiples of $\frac{1}{m}$ in this interval.

We now know that this set of carries is finite, which would be sufficient to construct a tileset as will be seen in the next section. However Kari noticed that it is possible to restrict oneself to a smaller set of carries by looking directly at how the Beatty sequences for α and $q\alpha$ are related.

Lemma 6.5.3 *The balanced sequence $b(q\alpha)$ can be obtained by multiplying $b(\alpha)$ using carries belonging only to*

$$S = \left\{ -\frac{n-1}{m}, -\frac{n-2}{m}, \ldots, \frac{m-1}{m} \right\}$$

Proof Since

$$q\lfloor \alpha \rfloor - 1 < \lfloor q\alpha \rfloor < q(\lfloor \alpha \rfloor + 1)$$

we have that

$$-q < q\lfloor \alpha \rfloor - \lfloor q\alpha \rfloor < 1$$

which means in particular that for any α:

$$(q\lfloor \alpha \rfloor - \lfloor q\alpha \rfloor) \in \left\{ -\frac{n-1}{m}, -\frac{n-2}{m}, \ldots, \frac{m-1}{m} \right\}$$

Now suppose we have incarry $c_i \in S$ and $c_i = qB(\alpha)_{i-1} - B(q\alpha)_{i-1}$:

$$c_i + qb(\alpha)_i - b(q\alpha)_i = qB(\alpha)_{i-1} - B(q\alpha)_{i-1}$$
$$+ qB(\alpha)_i - qB(\alpha)_{i-1}$$
$$- B(q\alpha)_i + B(q\alpha)_{i-1}$$
$$= (qB(\alpha)_i - B(q\alpha)_i)$$
$$= (q\lfloor i\alpha \rfloor - \lfloor qi\alpha \rfloor) \in S$$

It is thus possible to obtain $b(q\alpha)$ by multiplying $b(\alpha)$ by q using only carries from S.

So we now know that multiplying balanced representations by some positive rational can yield balanced representations. Let us now see in the next subsection how this translates to aperiodic tilings.

6.5.3 An Aperiodic Tileset

The idea now is to use the fact that our multiplication uses only a finite number of rational carries (see Lemma 6.5.2) to implement multiplication by using non-deterministic transducers.

For a multiplication by $q = \frac{n}{m}$, the states will correspond to the possible carries:

$$S = \left\{ -\frac{n-1}{m}, -\frac{n-2}{m}, \ldots, \frac{m-1}{m} \right\}$$

The idea is to take balanced representations of numbers that belong only to the interval $[1, 3]$, avoiding any possible representation of 0. In particular, numbers in $[1, 1.5]$ will be multiplied by 2 and numbers in $[1.5, 3]$ by $\frac{2}{3}$. The alphabet on which the transducers will act will hence be $\Sigma = \{1, 2, 3\}$. The transition table will contain all transitions

$$s \xrightarrow{a|b} s'$$

verifying $aq + s = b + s'$.

We can now construct a tileset corresponding to our transducer: for each transition we get one tile where the bottom and top represent the input and the output respectively and the left and right edges represent the inward carry and the outward carry respectively.

So in order to apply multiplications by 2 and $\frac{2}{3}$ we need two different transducers and corresponding sets of tiles. We can assume their set of states is disjoint, up to renaming the states. Limiting the output of the transducer multiplying by 2 to $\{2, 3\}$ de facto limits the input interval to $[1, 1.5]$.

Figure 6.26 shows the tiles for 2 and Figs. 6.27 and 6.28 show the tiles for $\frac{2}{3}$.

Lemma 6.5.4 *The tileset T tiles the plane.*

Proof Take some irrational real α in $[1, 3]$, one can tile $\mathbb{Z} \times \mathbb{N}$, by applying iteratively a multiplication by 2 to $b(\alpha)$ when $\alpha \in [1, 1.5]$ and a multiplication

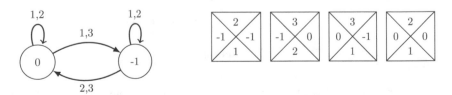

Fig. 6.26 On the left the transducer corresponding to multiplication by 2 and on the right the corresponding tiles. Notice how the input must belong to $\{1, 2\}$ and the outputs to $\{2, 3\}$, which translates the fact that the input belongs to $[1, 2]$ and the output to $[2, 3]$

Fig. 6.27 The transducer corresponding to multiplication by $\frac{2}{3}$ for inputs in [1, 2] and outputs in [1, 3]

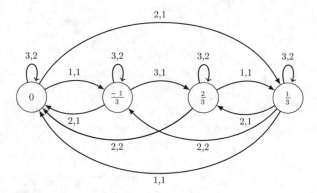

Fig. 6.28 The set of tiles for the multiplication by $\frac{2}{3}$. We assume the 0 of this part of the tileset is different from the one of the multiplication by 2

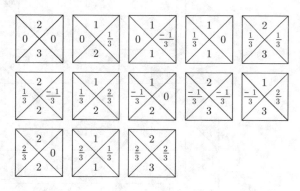

by $\frac{2}{3}$ when $\alpha \in [1.5, 3]$: the result always stays in [1, 3]. By compactness the tileset then tiles the whole plane.

We now know that we can tile the plane aperiodically however, lines of a valid tiling do not necessarily correspond to a balanced representation, we therefore still have to prove that the tileset always tiles aperiodically.

Lemma 6.5.5 *The tileset T tiles only aperiodically.*

Proof Suppose T tiles periodically, then there exists a tiling $c : \mathbb{Z}^2 \to T$ periodic in both directions. Let a and b be its horizontal and vertical periods respectively and let x_i be the finite sum of the values represented on top of $c(1, i)$ to $c(a, i)$. Since all the values are positive, x_i cannot be zero. And since the carry on the left is equal to the carry on the right of the period, $x_i = q_i x_{(i-1)}$ with q_i equal to 2 or $\frac{2}{3}$.

Therefore, $x_0 = q_n \cdots q_1 x_0$ which is imposible since no nonempty product of 2's and $\frac{2}{3}$'s can equal 1.

See Fig. 6.29 for an example of tiling by this tileset. The tileset obtained here is not Kari's tileset but a tileset built using the same techniques. Kari's original tileset used only numbers in the interval $[\frac{1}{2}, 2]$ but still used multiplications by 2 and $\frac{2}{3}$ and used a technical trick to avoid 0.

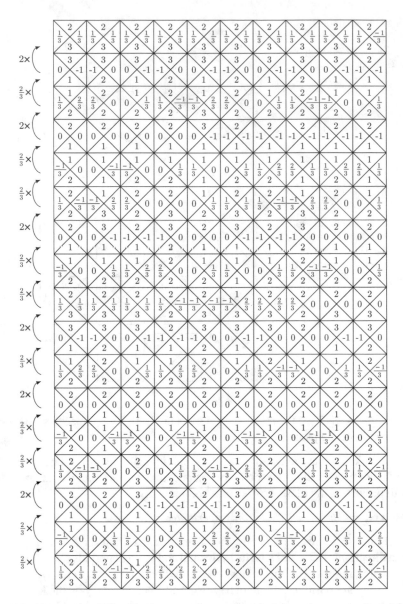

Fig. 6.29 An example of tiling, the bottom row starts from the balanced representation of $\phi + 1$, and the transducers are then applied according to which interval the resulting multiplied integer belongs to

6.5.4 Undecidability

What is interesting about this construction of aperiodic tilesets is that it allows a different encoding of computation: we will use piecewise affine maps in order to encode computations. While in the usual encoding of Turing machines in tilesets, the head and the current state are explicitly encoded in some tile, using piecewise affine functions will allow to encode the state implicitly in two real numbers representing the full instantaneous description of the Turing machine at some timestep.

6.5.4.1 Computing with Piecewise Affine Maps

Let us first explain how Turing machine computations may be encoded by piecewise affine maps. Let M be a Turing machine working on alphabet $\Gamma = \{0, 1\}$ and with state set $Q = \{0, \ldots, n\}$.

Let $(c, h, q) \in \Gamma^{\mathbb{Z}} \times \mathbb{Z} \times Q$ be an ID of said Turing machine. We will encode this ID inside two real numbers (l, r) so that the fractional part of l and r will encode the left and right side of the infinite tape respectively and $\lfloor l \rfloor$ will encode the symbol currently under the head and $\lfloor r \rfloor$ the current state of the machine. This encoding will be done in a way reminiscent of Cantor middle thirds: if the symbol currently under the head is 1, then $l \in [2, 3]$ and if it is 0 then $l \in [0, 1]$.

Knowing the state of the machine is thus equivalent to determining to which $[k, k + 1[$ interval r belongs to, moving to a new state corresponds to adding/-substracting some integer to r and moving the head corresponds to shifting (multiplying) the tape and writing the new symbol (adding a constant).

The transition function hence corresponds to applying a rational piecewise affine map depending only on $\lfloor r \rfloor$ and $\lfloor l \rfloor$. This map will not be defined on $[1, 2]$ reflecting the fact that $[1, 2]$ does not code for any valid symbol of the tape.

Let's see an example, with the Turing machine in state q and the following tape:

$$\ldots 0000101001001.1.0101010101010101010 \ldots$$

(l, r) can be as follows:

$$l = 2.2002002020000 \ldots$$

$$r = q.0202020202020202020 \ldots$$

If the machine goes from state q to state q', writes s and shifts the tape to the right, it is equivalent to do the following operations on l and r:

$$l := (l - \lfloor l \rfloor) \cdot 10$$

$$r := ((r - \lfloor r \rfloor) + 2 \cdot s)/10 + q'$$

which leads to the following new values of l and r:

$$l = 2.002002020000\ldots$$

$$r = q'.(2s)02020202020202020202\ldots$$

Thus our Turing machine has become a set of affine maps whose application depend only on which rational intervals r and l belong to. Let us now formalize this.

Definition 6.5.6 (l, r) is a valid coding of ID (c, h, q) iff

- for any $i \in \mathbb{N}$, $10(10^i \cdot l - \lfloor 10^i \cdot l \rfloor)$ is in $[0, 1]$ when $c_{h-i} = 0$ and is in $[2, 3]$ when $c_{h-i} = 1$,
- for any $i \in \mathbb{N}^*$, $10(10^i \cdot r - \lfloor 10^i \cdot r \rfloor)$ is in $[0, 1]$ when $c_{h+i} = 0$ and is in $[2, 3]$ when $c_{h+i} = 1$,
- and $\lfloor r \rfloor \in Q$.

Theorem 6.5.7 *Let M be a Turing machine and f a piecewise affine map constructed as stated before:*

1. *If c is an ID of M, and (l, r) a valid coding of c, then $f(l, r)$ is a valid coding of c' the ID obtained after one more step of M.*
2. *If (l, r) is such that $f^n(l, r)$ is defined for all $n \in \mathbb{N}$, then there exists an ID c and a valid coding (l', r') of c such that $f^n(l', r')$ is defined for all $n \in \mathbb{N}$.*

Proof (1) is straightforward. For (2), let (l, r) be such reals, suppose (l, r) is not a valid coding, and let $l = \sum_{i=0}^{\infty} l_i 10^{-i}$ and $r = \sum_{i=0}^{\infty} r_i 10^{-i}$, since (l, r) is not valid, this means that some l_i and r_i are not zeroes or twos, but since $f^n(l, r)$ is always defined, this also means that these never appear as l_0, so the computation to which they correspond is a computation that never reaches them. It then suffices to replace them by zeroes to obtain a valid coding, c then corresponds to this valid coding. See [12] for a more precise proof.

Using two disjoint sets like $[0, 1]$ and $[2, 3]$ allows us to avoid some decoding problems that would necessarily happen if the tape was directly coded with its representation in binary.[3]

6.5.4.2 Undecidability of the Domino Problem

The piecewise affine map that we have constructed can easily be transformed into a set of tiles: it has rational coefficients, adding or substracting a rational number can be done in a similar way to what was done for multiplication and the fact that we

---

[3]The representations $0.100\ldots$ and $0.011\ldots$ code the same real but should not encode the same tape.

are now considering two reals and not only one does not constitute an obstacle, the transducers can as easily be constructed.

The outputs of any transducer coding for a transition will belong either to $\{0, 1\}$ or to $\{2, 3\}$, thus ensuring that numbers written on one row either belong to $[0, 1]$ or to $[2, 3]$ and that invalid codings can never appear as a result of applying piecewise affine map.

In order to obtain the undecidability of the domino problem, we will use the immortality problem:

Theorem 6.2.9 (Hooper [25]) *There is no algorithm that, given a Turing machine M, decides if M halts on all configurations.*

This is still true if we restrict ourselves to Turing machines with a working alphabet of size 2.

By not having any representation for any halting state in the tileset corresponding to the piecewise affine map, any tiling by this tileset will represent an infinite computation of our Turing machine. And thus the tileset will only be able to tile a half plane if the Turing machine is immortal, by compactness, if the tileset tiles the half plane, it tiles the whole plane.

6.6   The Substitutive Method 2/2

The construction of an aperiodic tileset, and the undecidability result, that we describe in this section comes from the work of Durand, Romashchenko and Shen [14] and is inspired by the work of Gács in the 70's.

We saw in Sect. 6.3 the concept of an intrinsically substitutive tileset, and we gave an example of such a tileset. However, we gave no indication how this tileset was obtained, or how to prove in general that such a tileset should exist. The core of the construction of Durand, Romashchenko and Shen is to use recursion theory to prove this.

We will first redefine more generally the notion of substitutive tiling, and of a simulation

Definition 6.6.1 ([5, 13, 14, 43]) Let σ, τ be two tilesets.

We say that σ simulates τ with zoom factor N if there exists a one-to-one function $\phi : \tau \to \sigma^{N \times N}$ s.t.

- For any finite pattern w, $\phi(w)$ is a valid tiling for σ precisely when w is a valid tiling for τ.
- For any tiling of the plane x by σ, there exist w s.t. $\phi(w) = x$ upto shift. More precisely there exists a *unique* pair $(i', j') \in [0, N-1] \times [0, N-1]$ and a unique w s.t. $x_{i+i', j+j'} = \phi(w)_{i, j}$.

The uniqueness property means that there is only one way to shift x to obtain an image of ϕ.

Theorem 6.6.2 ([14]) *There exists a tileset that simulates itself.*

We will explain in this section how this theorem is proven, and then how to apply it to obtain the undecidabilty of the domino problem.

The idea is that, given a tileset τ, it is easy to build a tileset σ that simulates τ with a given zoom factor N.

We now somehow want to have $\sigma = \tau$ in the previous construction.

How to do this is similar to the fixed point theorem of recursion theorem, which allows in particular to prove that functions that call themselves are (partial) computable.

We will first illustrate this fixed point theorem in a particular example.

Suppose we want to compute recursively the factorial function, in a language where recursion is not allowed. Here is what we would write in Python:

```python
def f(n):
    if n == 0:
        return 1
    else:
        return n*f(n-1)
```

But in our fictive language, recursion is not allowed, so we cannot write call f.

One solution is to have f call another function g, and then somehow set g to be equal to f. A first version would be:

```python
def f(n):
    if n == 0:
        return 1
    else:
        return n*g(n-1)
g = f
```

In this example, g is a global variable that was somehow defined before the definition of f and then changed value afterwards. This is bad programming practice, so we prefer to give it as an argument to f:

```python
def f(g,n):
    if n == 0:
        return 1
    else:
        return n*g(g,n-1)
```

we can then call f with input (f, n) to obtain the result. This version would however make type theorists jump out of their skin,[4] as it not easy to find what type is f.

---

[4]It however works in python.

We will then proceed differently. Suppose that our programming language has a way to convert the code of a program (which is just a string of characters) into a function. Let call this procedure eval. Then we can write:

```
def f(x,n):
    if n == 0:
        return 1
    else:
        return n*eval(x)(x,n-1)
```

Now f is just a function that takes an integer and a string and outputs an integer. Now if we feed f with n and its own code, then we obtain the factorial function.[5]
The tileset we will construct mimics exactly this idea.

6.6.1   The Fixed Point Theorem of Computability Theory

The generalization of what preceeds is called Kleene's fixed point theorem and is one of the fundamental theorems of computability theory. Usually in computability theory functions, strings, integers are all represented by finite binary sequences that may be seen as integers, and thus all theorems are on computable functions on integers.

Theorem 6.6.3 (Kleene's Fixed Point)   *If* $f : \mathbb{N} \times \mathbb{N} \to \mathbb{N}$ *is a computable function, then there exists a computable function g such that* $f(g, n) = g(n)$.

Proof   Let U be a universal program: given $p, n_1, \ldots, n_k$ as arguments, U computes p with arguments $n_1, \ldots, n_k$, that is to say:

$$U(p, n_1, \ldots, n_k) = p(n_1, \ldots, n_k).$$

This is essentially what the eval function from before does.
Consider $s(p) = U(p, p)$, the function that applies p to its own code and denote $h(i, n) = f(s(i), n)$. Now set $g(n) = h(h, n)$ we have:

$$g(n) = h(h, n) = f(s(h), n) = f(h(h), n) = f(g, n)$$

---

[5]Up to a few syntactic changes, the reasoning we took actually works in python, as shown by the following two lines:
```
x = "lambda n,x: 1 if n == 0 else n*eval(x)(n-1,x)"
f = lambda n: eval(x)(n,x).
```

One way to interpret this theorem is that there exists a function g that "mimics" an infinite application of f on itself: $g(n) = f(g, n) = f(f(g, n)) = f(f(f(\ldots), n), n)$.

The universal Turing machine of the proof induces some overhead to the time $t(n)$ that the program p would take by itself on an input of size n. Using several tapes, one can design a universal TM taking an overhead at most polynomial in $t(n)$, see for instance [4, 22].

The aim is to construct a tileset using Kleene's fixed point theorem to obtain a self simulating tileset.

6.6.2   Simulating a Tileset with a Turing Machine

A tile can be summed up by the colors $c = \langle n, w, s, e \rangle$ of its four borders:

And thus a tileset can be seen as a computable function $g(c)$ that accepts when c represents a tile belonging to the tileset and rejects otherwise:

Definition 6.6.4 A computable function g *codes a tileset τ with k colors* if:

- g interprets its argument as a quadruplet of colors: g rejects if its input is not of size $4 \log k$.
- g accepts only when c represents a tile of τ.

Now if g is a function defining a tileset τ, we are going to design a computable function $f(N, k, g, c)$ that verifies if the tile c is in a tileset τ' simulating τ with zoom factor N. That is to say, f builds the tileset τ' from g and then checks that its input c belongs to it.

The tileset constructed by f will divide the plane into *macrotiles*, virtual tiles formed by an $N \times N$ square of smaller tiles. Each macrotile then runs the function g on input $C = \langle n, w, s, e \rangle$, the 4 colors coded in the border of the macrotile. The program g is only allowed to accept, so that only valid macrotiles may be formed. Let us now see the details.

The tileset constructed by f first starts by separating the grid into $N \times N$ squares:

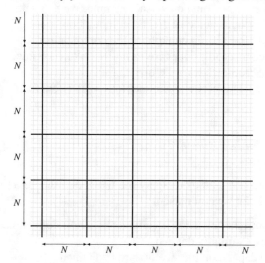

This can easily be done by having tiles that count modulo N, the borders of the macrotile appearing on the transition between 0 and $N - 1$:

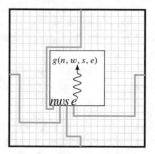

The center of each macrotile then contains a zone on each side that represents the color of the side and wires that route them towards a computation zone which applies g to c:

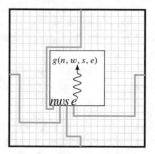

We make the computation zone of size $N/2 \times N/2$ inside each macrotile.

If N is big enough, this computation zone has enough time and space to run g:

Theorem 6.6.5 *If g codes τ with k colors, there exists N such that for any $N' > N$, $f(N', k, g, c)$ codes a tileset simulating τ with zoom factor N'.*

Proof Let t be an upper bound of the time taken by g on any accepting input of size $4 \log k$, when g is executed on a universal machine. We now take $N = 2t$.

Let $N' > N$. By our choice of N, the computation zone inside each macrotile defined by $f(N', k, g, c)$ is big enough for the computation of g to fit.

In this case, each macrotile codes colors n, s, e, w on its border s.t. $g(\langle n, s, e, w \rangle)$ accepts, i.e. the macrotile codes a tile of τ.

Therefore the whole tileset τ' is simulating τ with zoom factor N'.

6.6.3   A Fixedpoint Based Tileset

Wishful thinking would lead us to apply the fixedpoint theorem to f however this is not possible as is for several reasons:

- First, g does not take the same arguments as f. As defined before, g supposes that N and k are externally defined.
- N and k would depend on the size of f to be defined.

The solution to this is straightforward and is to make k and N become arguments of g. The only modification needed in the previous construction for is to hardcode N and k as inputs of g inside the macrotiles that are constructed by f. As the number of tiles generated by f is $O(N^2)$, this means that taking $k = N^2$ colors is sufficient to code them all. Furthermore, there is always room in the macrotiles to carry N^2 colors, as this only takes $2 \log N$ bits.

So we now have a function $f(g, N, c)$ which takes a function $g(N, c)$ as an argument.

It now suffices to apply Kleene's fixed point theorem to f. We obtain a new program ρ such that $f(\rho, N, c) = \rho(N, c)$.

Theorem 6.6.6 *There exists N such that for any $n > N$, $\rho(N, c)$ codes a tileset τ simulating itself with zoom factor N.*

Proof As f is polynomial in $\log N$, $|g|$ and $|c|$, this means that ρ's runtime is polynomial in $\log N$. So taking $N \gg poly(\log N)$ suffices to have enough runtime inside the macrotiles.

ρ defines a tileset τ which produces macrotiles of size $N \times N$ that themselves check if they belong to τ by running ρ in their computation zone. Thus ρ simulates itself with zoom factor N.

We can now show that τ indeed tiles the plane by inductively defining macrotiles of level k, this will also give us a picture of what tiling by τ look like:

- The macrotiles of level 0 are the macrotiles formed by $N \times N$ squares of tiles of τ.
- The macrotiles formed $N \times N$ squares of macrotiles of level k are macrotiles of level $k + 1$, see Fig. 6.30.

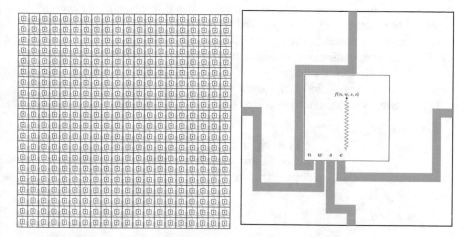

Fig. 6.30 An $N \times N$ square of macrotiles of level k forms a macrotile of level $k+1$

τ does indeed tile the plane since for any k one can tile an $N^{2^k} \times N^{2^k}$ square just by taking a macrotile of level k, which implies that there exists a tiling of the plane, by compactness.

One may also show that τ is aperiodic:

Lemma 6.6.7 *A tileset that simulates itself is aperiodic.*

Proof Let T be a tileset that simulates itself with zoom factor N and suppose it is periodic. Since any tiling is uniquely subdivided in $N \times N$ zones, the period must be divisible by N. However, since the tileset simulates itself, any tiling can also be uniquely subdivided in $N^2 \times N^2$ zones, so N^2 must also divide the period. By the same argument this must be true for any N^{2^k}, which is impossible.

6.6.4 Undecidability

The tileset τ we constructed in the previous subsection, while aperiodic, does not prove the undecidability of the Domino Problem. In order to do this, given a Turing machine M we will construct τ_M in a similar fashion, but we will embed computations of M into the macrotiles which will prevent the tileability in case M halts.

We cannot use exactly the construction from before, as a zoom factor of N for each level of macrotiles will mean that the macrotiles of different levels have the same computation space and thus all computations would be bounded by the same N. The goal being that macrotiles of level k simultaneously generate macrotiles of level $k+1$ and simulate a computation of M for more and more time steps.

Thus, if M halts, some macrotile of level k would uncover it and prevent the formation of macrotiles of level $k + 1$ and the tileset to tile the plane.

To achieve this, instead of making f hardcode N as an input to g, f hardcodes $2N$. Thus the computation zone is doubled with each simulation. So now a macrotile of level $k + 1$ divides the plane in $2^k N \times 2^k N$ macrotiles of level k. And the computation zone inside the macrotiles of level $k + 1$ now is $2^k N/2$.

The program run by a macrotile can be summed up by the following pseudocode:

```
def f(g, N, c):
    constuct Tₖ: a tileset dividing the plane in N × N
                 macrotiles which contain a
                 computation of g(2N, c').
    if c ∉ Tₖ:
        reject
    else:
        launch N/2 timesteps of the simulation of M
        if the simulation halts reject
```

Instead of a constant zoom factor between the different levels, each level k now has zoom factor $N_k = 2^k N$ which gives more and more computational room. So each level of macrotiles now ensures more simulation steps of M and thus if M halts, it will happen at some finite level which will prevent the tileability.

6.6.5   Bibliographic Notes

The fixed point tileset construction originally appeared in [15]. Its versatility allowed Durand, Romashchenko and Shen to prove new and original results on tilings as well as reprove ancient results [14]. For instance, among the results that can be proven with almost no modifications of the construction is a result by Hanf and Myers [21, 41] stating that there exists tilesets producing only non computable tilings of the plane. Other results that can be proved using variations of this technique include:

- Substitutive tilesets defined by a rectangular substitution are sofic, as was proved by Mozes [40].
- d-dimensional effective subshifts are subactions of $d + 1$ dimensional sofic subshifts, a result originally discovered only with $d + 2$-dimensional sofic shifts by Hochman [23].
- A characterization of the entropy of tilings by right recursively enumerable numbers (Hochman and Meyerovitch [24]).
- That there exists tilesets robust to errors.
- A characterization of the set of non-expansive directions of tilings [54].
- The subshift whose configurations are constituted only of squares whose sizes are chosen from a co-recursively enumerable set [53].

References

1. S.O. Aanderaa, H.R. Lewis, Linear sampling and the ∀∃∀ case of the decision problem. J. Symb. Log. **39**(3), 519–548 (1974). See also [H.R. Lewis, *Unsolvable Classes of Quantificational Formulas* (Addison-Wesley, Reading, 1979)]
2. C. Allauzen, B. Durand, Tiling problems. in *The Classical Decision Problem, Perspectives in Mathematical Logic*, chap. A (Springer, Berlin, 2001), pp. 407–420
3. R. Ammann, B. Grunbaum, G. Shephard, Aperiodic tiles. Discret. Comput. Geom. **8**(1), 1–25 (1992)
4. S. Arora, B. Barak, *Computational Complexity: A Modern Approach*, 1st edn (Cambridge University Press, New York, 2009)
5. A. Ballier, Propriétés structurelles, combinatoires et logiques des pavages. Ph.D. thesis, Aix-Marseille Université (2009)
6. S. Beatty, Problem 3173. Am. Math. Mon. **33**, 159 (1926)
7. R. Berger, The Undecidability of the Domino Problem. Ph.D. thesis, Harvard University (1964)
8. R. Berger, The Undecidability of the Domino Problem. No. 66 in Memoirs of the American Mathematical Society (The American Mathematical Society, Providence, 1966)
9. J.R. Büchi, Turing-Machines and the Entscheidungsproblem. Math. Ann. **148**(3), 201–213 (1962). https://doi.org/10.1007/BF01470748
10. J.R. Chazottes, J.M. Gambaudo, F. Gautero, Tilings of the plane and Thurston semi-norm. Geom. Dedicata **173**(1), 129–142 (2014). https://doi.org/10.1007/s10711-013-9932-4
11. A. Church, An unsolvable problem of elementary number theory. Am. J. Math. **58**(2), 345–363 (1936)
12. P. Collins, J.H. van Schuppen, Observability of hybrid systems and turing machines, in *43rd IEEE Conference on Decision and Control* (2004), pp. 7–12. https://doi.org/0.1109/CDC.2004.1428598
13. B. Durand, L.A. Levin, A. Shen, Local rules and global order, or aperiodic tilings. Math. Intell. **27**(1), 64–68 (2004)
14. B. Durand, A. Romashchenko, A. Shen, Fixed-point tile sets and their applications. J. Comput. Syst. Sci. **78**(3), 731–764 (2012). https://doi.org/10.1016/j.jcss.2011.11.001
15. B. Durand, A. Shen, A. Romashchenko, Fixed Point and Aperiodic Tilings. Tech. Rep. TR08-030, ECCC (2008)
16. S. Eigen, J. Navarro, V.S. Prasad, An aperiodic tiling using a dynamical system and Beatty sequences, in *Recent Progress in Dynamics*. MSRI Publications, vol. 54 (Cambridge University Press, Cambridge, 2007)
17. T. Fernique, N. Ollinger, Combinatorial substitutions and Sofic Tilings, in *Journées Automates Cellulaires (JAC)*, TUCS (2010), pp. 100–110
18. N.P. Fogg, in *Substitutions in Dynamics, Arithmetics and Combinatorics, chap. Sturmian Sequences*. Lecture Notes in Mathematics (Springer, Berlin, 2002)
19. F. Gähler, A. Julien, I. Savinien, *Combinatorics and Topology of the Robinson Tiling*. Comptes Rendus de l'Académie des Sciences de Paris, Série I—Mathématiques (2012), pp. 627–631. https://doi.org/10.1016/j.crma.2012.06.007
20. C. Goodman-Strauss, Matching rules and substitution tilings. Ann. Math. **147**(1), 181–223 (1998)
21. W. Hanf, Non recursive tilings of the plane I. J. Symbolic Log. **39**(2), 283–285 (1974)
22. F.C. Hennie, R.E. Stearns, Two-tape simulation of multitape turing machines. J. ACM **13**(4), 533–546 (1966). https://doi.org/10.1145/321356.321362
23. M. Hochman, On the dynamics and recursive properties of multidimensional symbolic systems. Invent. Math. **176**(1), 2009 (2009)
24. M. Hochman, T. Meyerovitch, A characterization of the entropies of multidimensional shifts of finite type. Ann. Math. **171**(3), 2011–2038 (2010). https://doi.org/10.4007/annals.2010.171.2011

25. P.K. Hooper, The undecidability of the turing machine immortality problem. J. Symbolic Log. **31**(2), 219–234 (1966)
26. J.E. Hopcroft, J.D. Ullman, *Introduction to Automata Theory, Languages and Computation* (Addison-Wesley, Reading, 1979)
27. E. Jeandel, M. Rao, An aperiodic set of 11 Wang tiles. Preprint
28. A. Johnson, K. Madden, Putting the pieces together: understanding Robinson's nonperiodic tilings. Coll. Math. J. **28**(3), 172–181 (1997)
29. A. Kahr, E.F. Moore, H. Wang, Entscheidungsproblem reduced to the ∀∃∀ case. Proc. Natl. Acad. Sci. U. S. A. **48**(3), 365–377 (1962)
30. J. Kari, The nilpotency problem of one-dimensional cellular automata. SIAM J. Comput. **21**(3), 571–586 (1992)
31. J. Kari, A small aperiodic set of Wang tiles. Discret. Math. **160**, 259–264 (1996)
32. J. Kari, The tiling problem revisited, in *Machines, Computations, and Universality (MCU)*. Lecture Notes in Computer Science, vol. 4664 (2007), pp. 72–79
33. J. Kari, N. Ollinger, Periodicity and immortality in reversible computing, in *MFCS 2008*. Lecture Notes in Computer Science, vol. 5162 (2008), pp. 419–430
34. S. Kleene, Two papers on the predicate calculus., chap., in *Finite Axiomatizability of Theories in the Predicate Calculus Using Additional Predicate Symbols*. No. 10 in Memoirs of the American Mathematical Society (American Mathematical Society, Providence, 1952)
35. L.A. Levin, Forbidden information. J. ACM **60**(2), 9:1–9 (2013). https://doi.org/10.1145/2450142.2450145
36. H.R. Lewis, *Unsolvable Classes of Quantificational Formulas* (Addison-Wesley, Reading, 1979)
37. H.R. Lewis, C.H. Papadimitriou, *Elements of the Theory of Computation* (Prentice-Hall, Englewood Cliffs, 1998)
38. D.A. Lind, B. Marcus, *An Introduction to Symbolic Dynamics and Coding* (Cambridge University Press, New York, 1995)
39. M.L. Minsky, *Computation: Finite and Infinite Machines* (Prentice-Hall, Englewood Cliffs, 1967)
40. S. Mozes, Tilings, substitutions systems and dynamical systems generated by them. J. Anal. Math. **53**, 139–186 (1989)
41. D. Myers, Non recursive tilings of the plane II. J. Symbolic Log. **39**(2), 286–294 (1974)
42. R. Nillsen, K. Tognetti, G. Winley, *Bernoulli (Beta) and Integer Part Sequences* (Australian Mathematical Society, Canberra, 1999)
43. N. Ollinger, Two-by-two substitution systems and the undecidability of the domino problem, in *CiE 2008*. Lecture Notes in Computer Science, vol. 5028 (2008), pp. 476–485
44. C.H. Papadimitriou, *Computational Complexity* (Addison Wesley, Reading, 1995)
45. B. Poizat, Une théorie finiement axiomatisable et superstable. Groupe d'études de théories stables **3**, 1–9 (1980)
46. R.M. Robinson, Seven polygons which permit only nonperiodic tilings of the plane. Not. Am. Math. Soc. **14**, 835 (1967)
47. R.M. Robinson, Undecidability and nonperiodicity for tilings of the plane. Invent. Math. **12**(3), 177–209 (1971). https://doi.org/10.1007/BF01418780
48. O. Salon, Quelles tuiles! (Pavages apériodiques du plan et automates bidimensionnels). J. Théor. Nombres Bordeaux **1**(1), 1–26 (1989)
49. S.G. Simpson, Medvedev degrees of two-dimensional subshifts of finite type. Ergodic Theory Dynam. Syst. **34**, 679–688 (2014). https://doi.org/10.1017/etds.2012.152
50. A.M. Turing, On computable numbers, with an application to the entscheidungsproblem. Proc. Lond. Math. Soc. **s2-42**(1), 230–265 (1937)
51. H. Wang, Proving theorems by pattern recognition II. Bell Syst. Tech. J. **40**, 1–41 (1961)

52. H. Wang, Dominoes and the ∀∃∀ case of the decision problem, in *Mathematical Theory of Automata* (1963), pp. 23–55
53. L.B. Westrick, Seas of squares with sizes from a Π_1^0 set. Isr. J. Math. **222**(1), 431–462 (2017). https://doi.org/10.1007/s11856-017-1596-6
54. C. Zinoviadis, Hierarchy and expansiveness in 2d subshifts of finite type, in Proceedings of the Ninth International Conference on Language and Automata Theory and Applications, LATA 2015, Nice, March 2–6, 2015 (2015), pp. 365–377

Chapter 7
Renormalisation of Pair Correlations and Their Fourier Transforms for Primitive Block Substitutions

Michael Baake and Uwe Grimm

Abstract For point sets and tilings that can be constructed with the projection method, one has a good understanding of the correlation structure, and also of the corresponding spectra, both in the dynamical and in the diffraction sense. For systems defined by substitution or inflation rules, the situation is less favourable, in particular beyond the much-studied class of Pisot substitutions. In this contribution, the geometric inflation rule is employed to access the pair correlation measures of self-similar and self-affine inflation tilings and their Fourier transforms by means of exact renormalisation relations. In particular, we look into sufficient criteria for the absence of absolutely continuous spectral contributions, and illustrate this with examples from the class of block substitutions. We also discuss the Frank–Robinson tiling, as a planar example with infinite local complexity and singular continuous spectrum.

7.1 Introduction

The theory of model sets via the projection method, see [7, 56] and references therein for background, has led to a reasonably good understanding of mathematical models for perfect quasicrystals. This is particularly true of systems with pure point spectrum, and applies to spectra both in the diffraction and in the dynamical sense; see [7, 13, 14, 38, 40] and references therein for more, in particular on equivalence results for the different types of spectra.

M. Baake
Fakultät für Mathematik, Universität Bielefeld, Bielefeld, Germany
e-mail: mbaake@math.uni-bielefeld.de

U. Grimm (✉)
School of Mathematics & Statistics, The Open University, Milton Keynes, UK
e-mail: uwe.grimm@open.ac.uk

359

S. Akiyama, P. Arnoux (eds.), *Substitution and Tiling Dynamics: Introduction to Self-inducing Structures*, Lecture Notes in Mathematics 2273,
https://doi.org/10.1007/978-3-030-57666-0_7

Another intensely-studied approach starts from a substitution on a finite alphabet, or considers an inflation rule for a finite set of prototiles; see [7, 29, 55] and references therein for more. If the inflation multiplier happens to be a Pisot–Vijayaraghavan (PV) number, one meets an interesting overlap with the projection method via systems that can both be described by inflation and as a regular model set; see [7, Ch. 7] for some classic examples. However, the still open Pisot substitution conjecture, compare [1, 51], shows that important parts of the picture are still missing.

Considerably less is known for more general substitution or inflation schemes, be it beyond the PV case, in higher dimensions, or both. In particular, the study of non-PV substitutions is only at its beginning. Some recent progress [4, 6] in one dimension was possible by realising that such systems admit an exact renormalisation approach to their pair correlation measures; see [18, 19] for related results on the spectral measures for these systems.

The purpose of this contribution is to show how to extend such an exact renormalisation approach to higher dimensions, and also beyond the case of inflation tilings of finite local complexity (FLC). To be able to discuss some interesting classes of examples, we will build on several results from [6]. One of our goals is to formulate an effective sufficient criterion for the absence of absolutely continuous (ac) diffraction, which then implies that the diffraction measure is a singular measure, with the analogous result on the spectral measure of maximal type where possible at present. This clearly is expected to be the typical situation for inflation systems with vanishing topological entropy, but no general classification is known so far.

To formulate a criterion for the absence of ac components, it will be instrumental to identify a natural cocycle attached to the inflation rule together with an appropriate Lyapunov exponent. Implicitly, this amounts to an asymptotic analysis of infinite matrix products of Riesz product type. They have shown up in various ways in the spectral theory of inflation systems [4, 10, 18, 19, 48]. It should not be surprising to meet them again, in a slightly different fashion. In fact, they provide perhaps the most natural point of entry for a renormalisation type analysis of inflation systems.

This contribution is both a summary of known results, including those from [3, 5, 6, 44], and their extension to some new territory, in particular in higher dimensions (as announced in [45] and discussed in [6]). For the latter purpose, we proceed in an example-oriented manner via the class of block substitutions (not necessarily of constant size), which is still sufficiently simple to see the underlying ideas, yet rich enough to illustrate some new phenomena. In particular, in view of recent general interest [28, 31–33, 41, 55], we include some examples of infinite local complexity as well. Various general results that we employ are discussed and proved in [6], for which we only give a brief account here.

The material presented below is organised as follows. In Sect. 7.2, we set the scene by recalling some basic material, including some proofs for convenience, in particular where we are not aware of a good reference. Section 7.3 continues this account, covering some important aspects of uniform distribution and averages, which will be instrumental in most of our later calculations. Then, in Sect. 7.4, we discuss inflation systems in one dimension, from the viewpoint of exact renormalisation of the pair correlation measures and their Fourier transforms, with one concrete example of recent interest being discussed in Sect. 7.5. For further fully worked-out examples, we refer to [4, 5, 10, 44].

Starting with Sect. 7.6, we develop the entire theory for higher-dimensional inflation tilings with finitely many prototiles up to translations, which is then applied to various examples. In particular, we treat binary block substitutions of constant size (Sect. 7.7) and a rather versatile family of block substitutions with squares (Sect. 7.8), which comprises tilings with infinite local complexity. This is also a feature of the Frank–Robinson tiling (Sect. 7.9), which is shown to have singular continuous diffraction beyond the trivial Bragg peak at the origin. Some concluding remarks and open problems follow in Sect. 7.10.

7.2 Preliminaries

Our general references for concepts, notation and background are [7, 9]. Here, we collect further methods and results, where we begin with a simple property of Hermitian matrices,

Fact 7.2.1 *Let $H = (h_{ij})_{1 \leqslant i,j \leqslant d} \in \mathrm{Mat}(d, \mathbb{C})$ be Hermitian and positive semi-definite, with rank m. Then, all diagonal elements of H are non-negative. If $h_{ii} = 0$ for some i, one has $h_{ij} = h_{ji} = 0$ for all $1 \leqslant j \leqslant d$. In particular, $H = 0$ iff $m = 0$.*

Whenever $H \neq 0$, there are m Hermitian, positive semi-definite matrices $H_1, \ldots, H_m$ of rank 1 such that $H = \sum_{r=1}^{m} H_r$ together with $H_r H_s = 0$ for $r \neq s$.

Proof By Sylvester's criterion, H positive semi-definite means that all principal minors are non-negative, hence in particular all diagonal elements of H. Assume $h_{ii} = 0$ for some i, and select any $j \in \{1, \ldots, d\}$. By semi-definiteness in conjunction with Hermiticity, one finds

$$0 = h_{ii} h_{jj} \geqslant h_{ij} h_{ji} = |h_{ij}|^2 \geqslant 0,$$

which implies the second claim. The equivalence of $H = 0$ with $m = 0$ is clear.

Employing Dirac's notation, the spectral theorem for Hermitian matrices asserts that one has $H = \sum_{i=1}^{d} |v_i\rangle \lambda_i \langle v_i|$, where the eigenvectors $|v_i\rangle$ can be chosen to form an orthonormal basis (so $\langle v_i | v_j \rangle = \delta_{i,j}$ and $|v_i\rangle\langle v_i|$ is a projector of rank 1), while all eigenvalues are non-negative due to positive semi-definiteness. The rank of H is the number of positive eigenvalues, counted with multiplicities. Ordering the eigenvalues as $\lambda_1 \geqslant \lambda_2 \geqslant \cdots \geqslant \lambda_d \geqslant 0$, one can choose $H_r = |v_r\rangle \lambda_r \langle v_r|$ for $1 \leqslant r \leqslant m$, and the claim is obvious. $\square$

7.2.1 Logarithmic Integrals and Mahler Measures

The logarithmic Mahler measure of a polynomial $p \in \mathbb{C}[x]$ is defined as

$$\mathfrak{m}(p) := \int_0^1 \log\left|p\left(e^{2\pi i t}\right)\right| dt. \tag{7.1}$$

It was originally introduced by Mahler as a measure of the complexity of p; compare [25]. If $p(x) = a \prod_{i=1}^s (x - \alpha_i)$, it follows from Jensen's formula [49, Prop. 16.1] that

$$\mathfrak{m}(p) = \log|a| + \sum_{i=1}^s \log\left(\max\{1, |\alpha_i|\}\right). \tag{7.2}$$

This has the following immediate consequence.

Fact 7.2.2 *If p is a monic polynomial that has no roots outside the unit disk, one has $\mathfrak{m}(p) = 0$. In particular, this holds if p is a cyclotomic polynomial,[1] or a product of such a polynomial with a monomial.*   □

Clearly, for polynomials p and q, one has $\mathfrak{m}(pq) = \mathfrak{m}(p) + \mathfrak{m}(q)$. If $p \in \mathbb{Z}[x]$, one can say more about the possible values of $\mathfrak{m}(p)$. They are of interest both in number theory and in dynamical systems; see [3, 25] and references therein.

Mahler measures of multivariate (or multi-variable) polynomials are defined by an integration over the corresponding torus. Concretely, for any $p \in \mathbb{C}[x_1, \ldots, x_d]$, one has

$$\mathfrak{m}(p) := \int_{\mathbb{T}^d} \log\left|p\left(e^{2\pi i t_1}, \ldots, e^{2\pi i t_d}\right)\right| dt_1 \cdots dt_d, \tag{7.3}$$

where $\mathbb{T}^d = \mathbb{R}^d/\mathbb{Z}^d$ denotes the d-torus. Unfortunately, in contrast to the one-dimensional situation, there is no simple general way to calculate such integrals. If we need to single out a variable, we do so by a subscript. For instance, $\mathfrak{m}_x(1+x+xy)$ denotes the logarithmic Mahler measure of $1 + x + xy$, viewed as a polynomial in x, with y being a coefficient. We refer to [25] for general background and examples.

7.2.2 Radon Measures

Let μ denote a (generally complex) Radon measure on $\mathbb{R}^d$, which we primarily view as a linear functional over the space $C_c(\mathbb{R}^d)$ of compactly supported continuous

---

[1]A non-constant polynomial $p \in \mathbb{Z}[x]$ is called *cyclotomic* if $p(x)$ divides $x^n - 1$ for some $n \in \mathbb{N}$.

functions. The 'flipped-over' version $\widetilde{\mu}$ is defined by $\widetilde{\mu}(g) = \overline{\mu(\widetilde{g})}$, where $\widetilde{g}(x) := \overline{g(-x)}$. A measure μ is called *positive* when $\mu(g) \geqslant 0$ for all $g \geqslant 0$, and *positive definite* when $\mu(g * \widetilde{g}) \geqslant 0$ for all $g \in C_c(\mathbb{R}^d)$. Here, $g * h$ refers to the convolution of two integrable functions, as defined by $(g * h)(x) = \int_{\mathbb{R}^d} g(x - y)h(y)\,dy$. By $|\mu|$, we denote the *total variation measure* of μ. If $|\mu|(\mathbb{R}^d) < \infty$, the measure is *bounded* or *finite*, while we call it *translation bounded* when $\sup_{t \in \mathbb{R}^d} |\mu|(t + K) < \infty$ holds for some compact set $K \subset \mathbb{R}^d$ with non-empty interior.

If $f \colon \mathbb{R}^d \longrightarrow \mathbb{R}^d$ is an invertible mapping, we define the *pushforward* $f.\mu$ of a measure μ by $(f.\mu)(g) = \mu(g \circ f)$, where g is an arbitrary test function. Viewing μ as a regular Borel measure via the general Riesz–Markov representation theorem, compare [50], the matching relation for a bounded Borel set $\mathcal{E}$ is

$$(f.\mu)(\mathcal{E}) = (f.\mu)(1_{\mathcal{E}}) = \mu(1_{\mathcal{E}} \circ f) = \mu(1_{f^{-1}(\mathcal{E})}) = \mu(f^{-1}(\mathcal{E})).$$

Of particular importance is the *Dirac measure* at x, denoted by δ_x, which is defined by $\delta_x(g) = g(x)$ for test functions. For Borel sets, the matching relation is

$$\delta_x(\mathcal{E}) = \begin{cases} 1, & \text{if } x \in \mathcal{E}, \\ 0, & \text{otherwise,} \end{cases}$$

which is often used in the form $\delta_x(\mathcal{E}) = \delta_x(1_{\mathcal{E}})$. For a point set $S \subset \mathbb{R}^d$, which is at most countable in our setting [7], one defines the corresponding *Dirac comb* as $\delta_S = \sum_{x \in S} \delta_x$.

When ν is absolutely continuous relative to μ, denoted by $\nu \ll \mu$, with Radon–Nikodym density h, we write $\nu = h\mu$, so that $(h\mu)(g) = \mu(hg)$. For the pushforward, this leads to

$$f.(h\mu) = (h \circ f^{-1}) \cdot (f.\mu), \tag{7.4}$$

as follows from a simple calculation with a test function. Now, when we have $f(x) = Ax$ with $A \in \mathrm{GL}(d, \mathbb{R})$ and μ is Lebesgue measure, it is sometimes more convenient to rewrite this relation as

$$A.h := f.h = \frac{h \circ f^{-1}}{|\det(A)|} = \frac{h \circ A^{-1}}{|\det(A)|}, \tag{7.5}$$

to be understood as a relation between absolutely continuous measures. The *convolution* $\mu * \nu$ of two finite measures is defined by

$$(\mu * \nu)(g) = \int_{\mathbb{R}^d} \int_{\mathbb{R}^d} g(x + y)\,d\mu(x)\,d\nu(y),$$

which can be extended in various ways, in particular to the case where one measure is finite and the other is translation bounded [16, Prop. 1.13].

Lemma 7.2.3 *If μ and ν are two convolvable measures on $\mathbb{R}^d$ and if the mapping $f: \mathbb{R}^d \longrightarrow \mathbb{R}^d$ is both invertible and linear, the pushforward operation satisfies the relation $f.(\mu * \nu) = (f.\mu) * (f.\nu)$.*

Proof Let g be a general test function and define g_a by $g_a(x) = g(a + x)$. Then, one has

$$
\begin{aligned}
\left(f.(\mu * \nu)\right)(g) &= \int_{\mathbb{R}^d} \int_{\mathbb{R}^d} g\big(f(x + y)\big) \, d\mu(x) \, d\nu(y) \\
&= \int_{\mathbb{R}^d} \int_{\mathbb{R}^d} g\big(f(x) + f(y)\big) \, d\mu(x) \, d\nu(y) \\
&= \int_{\mathbb{R}^d} \mu\big(g_{f(y)} \circ f\big) \, d\nu(y) = \int_{\mathbb{R}^d} (f.\mu)\big(g_{f(y)}\big) \, d\nu(y) \\
&= \int_{\mathbb{R}^d} \int_{\mathbb{R}^d} g\big(x + f(y)\big) \, d\big(f.\mu\big)(x) \, d\nu(y) \\
&= \int_{\mathbb{R}^d} \int_{\mathbb{R}^d} g_x\big(f(y)\big) \, d\nu(y) \, d\big(f.\mu\big)(x) \\
&= \int_{\mathbb{R}^d} \nu(g_x \circ f) \, d\big(f.\mu\big)(x) = \int_{\mathbb{R}^d} (f.\nu)(g_x) \, d\big(f.\mu\big)(x) \\
&= \int_{\mathbb{R}^d} \int_{\mathbb{R}^d} g(x + y) \, d\big(f.\nu\big)(y) \, d\big(f.\mu\big)(x) \\
&= \big((f.\nu) * (f.\mu)\big)(g) = \big((f.\mu) * (f.\nu)\big)(g),
\end{aligned}
$$

where the step from the fourth to the fifth line, as well as the last step, rely on Fubini's theorem. $\qquad \square$

A linear map f on $\mathbb{R}^d$ is *expansive* if there exists a constant $\alpha > 1$ such that $\|f(x)\| \geqslant \alpha \|x\|$ for all $x \in \mathbb{R}^d$. This implies that all eigenvalues satisfy $|\lambda| \geqslant \alpha$ and that f is invertible.

Lemma 7.2.4 *Let μ be a Radon measure on $\mathbb{R}^d$ such that $\mu|_U = \mu(\{0\}) \delta_0$ holds for some open neighbourhood U of 0. Then, if f is an expansive linear map on $\mathbb{R}^d$, one has*

$$
\lim_{n \to \infty} f^n.\mu = \mu(\{0\}) \delta_0.
$$

Proof Let $\mathcal{E} \subset \mathbb{R}^d$ be a fixed, bounded Borel set. Viewing μ as a regular Borel measure, one has $(f^n.\mu)(\mathcal{E}) = \mu(f^{-n}(\mathcal{E}))$. Since f is expansive, with expansion constant $\alpha > 1$, it is invertible, and f^{-1} is contractive, with $\|f^{-1}(x)\| \leqslant \frac{1}{\alpha}\|x\|$ for all $x \in \mathbb{R}^d$. Consequently, $f^{-n}(\mathcal{E}) \subset U$ for n sufficiently large.

Now, the set $\mathcal{E}$ contains 0 if and only if $f^{-n}(\mathcal{E})$ does, so

$$(f^n.\mu)(\mathcal{E}) = \mu(\{0\})\, \delta_0(\mathcal{E})$$

for n large enough. Since $\mathcal{E}$ was bounded but otherwise arbitrary, our claim on the measure μ follows. $\qquad\square$

The Fourier transform of measures will play an important role in many of our arguments. We follow the classical approach as outlined in [16, Ch. 1], see also [7, Ch. 8] as well as [46], where the Fourier transform of an integrable function f is given by

$$\widehat{f}(k) = \int_{\mathbb{R}^d} \mathrm{e}^{-2\pi \mathrm{i}\langle k|x\rangle}\, f(x)\, \mathrm{d}x$$

as usual, where $\langle \cdot|\cdot \rangle$ denotes the standard inner product of $\mathbb{R}^d$. If μ is a finite measure, its Fourier transform is a continuous function, written as

$$\widehat{\mu}(k) = \int_{\mathbb{R}^d} \mathrm{e}^{-2\pi \mathrm{i}\langle k|x\rangle}\, \mathrm{d}\mu(x).$$

For translation-bounded measures, we shall also employ standard notions and techniques from the theory of tempered distributions; compare [50, Sec. 6.2].

Below, we will make frequent use of a relation that tracks the consequence of an invertible linear map under Fourier transform.

Lemma 7.2.5 *Let μ be a Fourier-transformable measure on $\mathbb{R}^d$, and consider a matrix $A \in \mathrm{GL}(d, \mathbb{R})$. Then, with $A^* := (A^T)^{-1}$ denoting the dual matrix, one has*

$$\widehat{A.\mu} = \frac{A^*.\widehat{\mu}}{|\det(A)|}.$$

Moreover, when $\widehat{\mu}$ is absolutely continuous relative to Lebesgue measure, hence represented by a locally integrable function, the relation simplifies to

$$\widehat{A.\mu} = \widehat{\mu} \circ A^T = A^T.\widehat{\mu}.$$

Proof If g is an arbitrary test function, one has

$$\widehat{A.\mu}\,(g) = (A.\mu)(\widehat{g}) = \mu(\widehat{g} \circ A) = \int_{\mathbb{R}^d} \int_{\mathbb{R}^d} \mathrm{e}^{-2\pi \mathrm{i}\langle x|At\rangle} g(x)\, \mathrm{d}x\, \mathrm{d}\mu(t).$$

Observing $\langle x|At\rangle = \langle t|A^T x\rangle$ and setting $x = A^* y$, hence $\mathrm{d}x = |\det(A^*)|\,\mathrm{d}y$, one finds

$$\widehat{A.\mu}(g) = \int_{\mathbb{R}^d}\int_{\mathbb{R}^d} \mathrm{e}^{-2\pi\mathrm{i}\langle t|y\rangle} g(A^* y)\,\frac{\mathrm{d}y\,\mathrm{d}\mu(t)}{|\det(A)|} = \int_{\mathbb{R}^d}(\widehat{g\circ A^*})(t)\frac{\mathrm{d}\mu(t)}{|\det(A)|}$$

$$= \frac{\mu(\widehat{g\circ A^*})}{|\det(A)|} = \frac{\widehat{\mu}(g\circ A^*)}{|\det(A)|} = \frac{(A^*.\widehat{\mu})(g)}{|\det(A)|},$$

which implies the first claim. The second claim is a consequence of Eq. (7.5).   □

7.2.3   Riesz Products

Of particular interest in the context of singular measures are measures that have a representation as infinite Riesz products. Let us recall one paradigmatic example of pure point type, and then generalise it. Here, an expression of the form $\prod_{m\geqslant 0} f_m(k)$ with continuous functions f_m is a short-hand for the measure that is defined as the vague limit of a sequence of absolutely continuous measures, the latter being given by the Radon–Nikodym densities $\prod_{m=0}^{n} f_m(k)$ with $n \geqslant 0$.

Lemma 7.2.6 *As a relation between translation-bounded measures on $\mathbb{R}$, one has*

$$\prod_{m\geqslant 0}\bigl(1 + \cos(2\pi\,2^m k)\bigr) = \delta_{\mathbb{Z}},$$

where $k \in \mathbb{R}$ and convergence is in the vague topology.

Proof We employ a method that is well known from the theory of Bernoulli convolutions; compare [47]. Define $\mu = \delta_0 + \delta_1$ and consider the measure

$$\nu = \frac{1}{2}\,\mu * \widetilde{\mu} = \delta_0 + \frac{1}{2}(\delta_1 + \delta_{-1}),$$

which means that

$$\widehat{\nu}(k) = 1 + \cos(2\pi k).$$

With $f(x) = 2x$, one has $f.\nu = \frac{1}{2}\,(f.\mu)*(f.\widetilde{\mu})$ by Lemma 7.2.3, where $f.\widetilde{\mu} = \widetilde{f.\mu}$. Moreover, one has $\mu*(f.\mu)*\ldots*(f^{n-1}.\mu) = \sum_{\ell=0}^{2^n-1}\delta_\ell$. Now, for $n \geqslant 1$, a simple convolution calculation gives

$$\underset{m=0}{\overset{n-1}{\Large *}} f^m.\nu = \sum_{\ell=1-2^n}^{2^n-1} \frac{2^n - |\ell|}{2^n} \delta_\ell \xrightarrow{n\to\infty} \delta_{\mathbb{Z}},$$

with convergence in the vague topology.

With $\widehat{f^m.\nu} = \widehat{\nu} \circ f^m$, which follows from Lemma 7.2.5, an application of the convolution theorem in conjunction with the continuity of the Fourier transform leads to

$$\underset{m=0}{\overset{n-1}{\widehat{\Large *\; f^m.\nu}}} = \prod_{m=0}^{n-1} (\widehat{\nu} \circ f^m) \xrightarrow{n\to\infty} \widehat{\delta_{\mathbb{Z}}} = \delta_{\mathbb{Z}}, \tag{7.6}$$

where the last step is the Poisson summation formula (PSF); see the general version in [7, Prop. 9.4]. Our claim now follows via the observation that $(\widehat{\nu} \circ f^m)(k) = 1 + \cos(2\pi 2^m k)$. □

More generally, let $2 \leqslant M \in \mathbb{N}$ be fixed and consider

$$\mu = \sum_{\ell=0}^{M-1} \delta_\ell \quad \text{and} \quad \nu = \frac{\mu * \widetilde{\mu}}{M} = \delta_0 + \sum_{\ell=1}^{M-1} \frac{M-\ell}{M}(\delta_\ell + \delta_{-\ell}).$$

With $f(x) = Mx$, one has $\mu * (f.\mu) * \ldots * (f^{n-1}.\mu) = \sum_{\ell=0}^{M^n-1} \delta_\ell$ and

$$\underset{m=0}{\overset{n-1}{\Large *}} f^m.\nu = \sum_{\ell=1-M^n}^{M^n-1} \frac{M^n - |\ell|}{M^n} \delta_\ell \xrightarrow{n\to\infty} \delta_{\mathbb{Z}}$$

in the vague topology, so that Eq. (7.6) holds here as well. Observe that

$$\widehat{\nu}(k) = 1 + 2 \sum_{\ell=1}^{M-1} \frac{M-\ell}{M} \cos(2\pi \ell k),$$

which satisfies $\widehat{\nu}(k) \geqslant 0$ and $\int_0^1 \widehat{\nu}(k)\,dk = 1$. Moreover, $\prod_{m=0}^{n-1}(\widehat{\nu} \circ f^m)$ defines a probability density on $[0, 1]$ for each $n \in \mathbb{N}$. Now, the generalisation of Lemma 7.2.6 reads as follows.

Proposition 7.2.7 *For any $2 \leqslant M \in \mathbb{N}$, one has*

$$\prod_{m \geqslant 0} \Big(1 + 2 \sum_{\ell=1}^{M-1} \frac{M-\ell}{M} \cos(2\pi \ell M^m k)\Big) = \delta_{\mathbb{Z}},$$

where $k \in \mathbb{R}$ and convergence is in the vague topology. □

It is clear how to extend this to more than one dimension, the details of which are left to the interested reader.

7.3   Uniform Distribution and Averages

While uniform distribution results are usually stated for one dimension, many of them have natural, though less well-known, generalisations to higher dimensions. We shall need some of them to calculate limits of various Birkhoff sums in our examples. To formulate the results, we represent $\mathbb{T}^d = \mathbb{R}^d / \mathbb{Z}^d$ as the half-open unit cube $[0, 1)^d$ with (coordinate-wise) addition modulo 1. As before, we use $\langle x \,|\, y \rangle = \sum_{i=1}^d x_i y_i$ for the standard inner product in $\mathbb{R}^d$. Let us recall some useful properties of non-singular linear forms.

Fact 7.3.1 *Consider a non-singular linear form $f : \mathbb{R}^d \longrightarrow \mathbb{R}$, which can thus be written as $f(x) = \langle a \,|\, x \rangle$ with $0 \neq a \in \mathbb{R}^d$. Then, if $\mathcal{E}$ is a Lebesgue null set in $\mathbb{R}$, its preimage $f^{-1}(\mathcal{E})$ is a Lebesgue null set in $\mathbb{R}^d$.*

Proof Let μ_{L} and ν_{L} denote Lebesgue measure in $\mathbb{R}^d$ and $\mathbb{R}$, respectively. Clearly, the linear mapping f is differentiable, with $\nabla f(x) = a \neq 0$ for all $x \in \mathbb{R}^d$, hence certainly measurable and surjective. Now, the pushforward $f.\mu_{\mathrm{L}}$ defines a regular Borel measure on $\mathbb{R}$, with $\big(f.\mu_{\mathrm{L}}\big)(\mathcal{E}) = \mu_{\mathrm{L}}\big(f^{-1}(\mathcal{E})\big)$ for *any* Borel set $\mathcal{E} \subseteq \mathbb{R}$; compare [36, Thm. 39.C]. Due to the linearity of f, for any $t \in \mathbb{R}$, we have the relation $f^{-1}(t + \mathcal{E}) = z_t + f^{-1}(\mathcal{E})$ for some $z_t \in \mathbb{R}^d$ with $f(z_t) = t$, which covers the empty set via the standard convention $x + \varnothing = \varnothing$.

This property implies $\big(f.\mu_{\mathrm{L}}\big)(t + \mathcal{E}) = \big(f.\mu_{\mathrm{L}}\big)(\mathcal{E})$ for all $t \in \mathbb{R}$ and all Borel sets $\mathcal{E}$, which means that $f.\mu_{\mathrm{L}}$ is translation invariant and thus a multiple of Haar measure on $\mathbb{R}$. Consequently, we have $f.\mu_{\mathrm{L}} = c \nu_{\mathrm{L}}$, where $c > 0$ follows from $a \neq 0$. This means that $f.\mu_{\mathrm{L}}$ and ν_{L} are equivalent as measures, and our claim on the Lebesgue null sets follows.   □

Fact 7.3.2 *Let $\alpha \in \mathbb{R}$ with $|\alpha| > 1$ be given and let f be the linear form from Fact 7.3.1. Then, for Lebesgue-a.e. $x \in \mathbb{R}^d$, the sequence $\big(f(\alpha^n x)\big)_{n \in \mathbb{N}}$ is uniformly distributed modulo 1.*

Proof One has $f(\alpha^n x) = \alpha^n t$ with $t = \langle a \,|\, x \rangle$ and $a \neq 0$. Clearly, $(\alpha^n t)_{n \in \mathbb{N}}$ is uniformly distributed modulo 1 for a.e. $t \in \mathbb{R}$ by standard results from uniform distribution theory; compare [39, Thm. 4.3 and Exc. 4.3]. If $\mathcal{E}$ is the corresponding null set of exceptional points, uniform distribution modulo 1 of $\big(f(\alpha^n x)\big)_{n \in \mathbb{N}}$ fails precisely for all $x \in f^{-1}(\mathcal{E}) \subset \mathbb{R}^d$. By Fact 7.3.1, $f^{-1}(\mathcal{E})$ is a null set in $\mathbb{R}^d$, which implies the claim.   □

Lemma 7.3.3 *Let $\alpha \in \mathbb{R}$ with $|\alpha| > 1$ be fixed. Then, for Lebesgue-a.e. $x \in \mathbb{R}^d$, the sequence $(\alpha^n x)_{n \in \mathbb{N}}$ taken modulo 1 is uniformly distributed in $\mathbb{T}^d$.*

Proof For $d = 1$, this is a well-known result from metric equidistribution theory [39, Ch. 4], as mentioned earlier. For $d > 1$ and any given $x \in \mathbb{R}^d$, it is convenient to employ Weyl's criterion [39, Thm. 6.2] and consider the convergence behaviour of character sums. In fact, this implies that uniform distribution of $(\alpha^n x)_{n \in \mathbb{N}}$ modulo 1 is equivalent to uniform distribution modulo 1 of the sequences $(\alpha^n \langle k | x \rangle)_{n \in \mathbb{N}}$ for all $k \in \mathbb{Z}^d \setminus \{0\}$; compare [39, Thm. 6.3]. For each such k, let $\mathcal{E}_k$ be the exceptional set of points $x \in \mathbb{R}^d$ where uniform distribution fails, which is a null set by Fact 7.3.2. Since $\mathbb{Z}^d \setminus \{0\}$ is countable, the set $\bigcup_{k \in \mathbb{Z}^d \setminus \{0\}} \mathcal{E}_k$ is still a null set in $\mathbb{R}^d$, and the claim follows. □

Next, we need to understand averages of various types of periodic and almost periodic functions, in particular along exponential sequences of the above type.

Lemma 7.3.4 *Let* $\alpha \in \mathbb{R}$ *with* $|\alpha| > 1$ *be fixed. For any* $a \in \mathbb{R}^d$ *and then a.e.* $x \in \mathbb{R}^d$, *one has*

$$\lim_{N \to \infty} \frac{1}{N} \sum_{n=0}^{N-1} e^{2\pi i \alpha^n \langle a | x \rangle} = \delta_{a,0}.$$

Proof When $a = 0$, the limit is 1 for *all* $x \in \mathbb{R}^d$, so let $a \neq 0$. Then, by Fact 7.3.2, $(\alpha^n \langle a | x \rangle)_{n \in \mathbb{N}}$ is uniformly distributed modulo 1 for a.e. $x \in \mathbb{R}^d$, where the null set $\mathcal{E}_a$ of exceptions depends on a. So, for any given a and then every $x \in \mathbb{R}^d \setminus \mathcal{E}_a$, we get

$$\lim_{N \to \infty} \frac{1}{N} \sum_{n=0}^{N-1} e^{2\pi i \alpha^n \langle a | x \rangle} = \int_0^1 e^{2\pi i t} \, dt = 0$$

by Weyl's lemma. □

The next step is an extension to (complex) trigonometric polynomials, as given by

$$P_m(x) = c_0 + \sum_{\ell=1}^{m} c_\ell \, e^{2\pi i \langle k_\ell | x \rangle}$$

with $m \in \mathbb{N}_0$ and coefficients $c_\ell \in \mathbb{C}$. When $m \geqslant 1$, the frequency vectors $k_1, \ldots, k_m$ are assumed to be non-zero and distinct. Clearly, under the conditions of Lemma 7.3.4, one obtains

$$\lim_{N \to \infty} \frac{1}{N} \sum_{n=0}^{N-1} P_m(\alpha^n x) = c_0 = \mathbb{M}(P_m) \tag{7.7}$$

for a.e. $x \in \mathbb{R}^d$. Here, $\mathbb{M}(f)$ is the *mean* of a bounded function,

$$\mathbb{M}(f) := \lim_{n \to \infty} \frac{1}{\mathrm{vol}(A_n)} \int_{A_n} f(x) \, \mathrm{d}x, \tag{7.8}$$

where $\mathcal{A} = (A_n)_{n \in \mathbb{N}}$ is a fixed sequence of growing sets for the averaging process. The sets A_n are supposed to be sufficiently 'nice', which means that one assumes a property of Følner or van Hove type. To be concrete, we can think of A_n as the closed cube of sidelength n centred at 0. It is clear that the limit in (7.8) exists for trigonometric polynomials. More generally, it exists for all functions that are *uniformly almost periodic*, which are often also called Bohr almost periodic. They are the continuous functions that can uniformly be approximated by trigonometric polynomials. In other words, the space of uniformly almost periodic functions is the $\|.\|_\infty$-closure of the space of trigonometric polynomials; see [23] for general results.

Proposition 7.3.5 *Let* $f : \mathbb{R}^d \longrightarrow \mathbb{C}$ *be a uniformly (or Bohr) almost periodic function, and let* $\alpha \in \mathbb{R}$ *with* $|\alpha| > 1$ *be given. Then, for a.e.* $x \in \mathbb{R}^d$, *one has*

$$\lim_{N \to \infty} \frac{1}{N} \sum_{n=0}^{N-1} f(\alpha^n x) = \mathbb{M}(f).$$

In particular, this applies to functions of the form $f = \log(g)$ *with* g *a non-negative, uniformly almost periodic function that is bounded away from 0.*

Proof The first claim for $d = 1$ is [11, Thm. 6.4.4]. A close inspection of its proof reveals that the same chain of arguments also applies to the case $d > 1$, which is all we need here.

The second claim follows from the first because $g(x) \geqslant \delta > 0$ for all $x \in \mathbb{R}^d$ implies that $\log(g)$ is again uniformly almost periodic [4, Fact 6.14]. $\qquad \square$

In the attempt to generalise Proposition 7.3.5 beyond uniformly almost periodic functions, one difficulty emerges when f is no longer locally Riemann-integrable. Let us first look at periodic functions, where we begin by recalling a classic result.

Fact 7.3.6 ([11, Lemma 6.3.3]) *Let* $q \in \mathbb{Z}$ *with* $|q| \geqslant 2$ *be fixed, and consider a function* $f \in L^1_{\mathrm{loc}}(\mathbb{R})$ *that is* 1-*periodic. Then,*

$$\frac{1}{N} \sum_{n=0}^{N-1} f(q^n x) \xrightarrow{N \to \infty} \int_0^1 f(y) \, \mathrm{d}y = \mathbb{M}(f)$$

holds for a.e. $x \in \mathbb{R}$. $\qquad \square$

The key ingredient to Fact 7.3.6 is the ergodicity of Lebesgue measure on $\mathbb{T}$ for the dynamical system defined by $x \mapsto qx$ modulo 1, which permits to use Birkhoff's ergodic theorem instead of Weyl's lemma and uniform distribution of $(q^n x)_{n \in \mathbb{N}}$ for a.e. $x \in \mathbb{R}$. The natural counterpart on $\mathbb{T}^d$ can be stated as follows.

Lemma 7.3.7 *Let Q be a non-singular endomorphism of $\mathbb{T}^d$ such that no eigenvalue is a root of unity, and consider a $\mathbb{Z}^d$-periodic function $f \in L^1_{\text{loc}}(\mathbb{R}^d)$. Then, for a.e. $x \in \mathbb{R}^d$, one has*

$$\frac{1}{N} \sum_{n=0}^{N-1} f(Q^n x) \xrightarrow{N \to \infty} \int_{\mathbb{T}^d} f(y)\, dy = \mathbb{M}(f).$$

In particular, this result applies to every toral endomorphism that is expansive.

Proof Under our assumptions, Lebesgue measure is an invariant and ergodic measure for the dynamical system defined by Q on $\mathbb{T}^d$; see [24, Cor. 2.20]. The main statement now follows from Birkhoff's ergodic theorem. Since all eigenvalues of an expansive $Q \in \text{End}(\mathbb{T}^d)$ satisfy $|\lambda| > 1$, the last claim is clear. $\quad\square$

Beyond Fact 7.3.6 and Lemma 7.3.7, we will need the following result, which can be viewed as a variant of Sobol's theorem [52]; see also [11, 37].

Lemma 7.3.8 *Let $p \geqslant 0$ be a trigonometric polynomial in d variables, and let $\alpha \in \mathbb{R}$ with $|\alpha| > 1$ be fixed. Let us further assume that, for some sufficiently small $\delta > 0$, the critical points of p with value in $[0, \delta]$ are isolated. Then, for Lebesgue-a.e. $x \in \mathbb{R}^d$, one has*

$$\lim_{N \to \infty} \frac{1}{N} \sum_{n=0}^{N-1} \log(p(\alpha^n x)) = \mathbb{M}(\log(p)).$$

Proof *(Sketch)* Since the case $p(k) \geqslant \delta > 0$ for all $k \in \mathbb{R}^d$ is covered by Proposition 7.3.5, we assume $\inf_{k \in \mathbb{R}^d} p(k) = 0$, hence $\inf_{k \in \mathbb{R}^d} \log(p(k)) = -\infty$, which is the origin of the complication. Note, however, that all singularities of $\log(p)$ are of logarithmic type and hence locally integrable, so $\log(p)$ is no longer uniformly, but still Stepanov almost periodic; see [11, pp. 356–359] as well as [23, Sec. VI.4].

Now, we have to deal with the small local minima of p. By assumption, there is a $\delta > 0$ such that the points k with $\nabla p(k) = 0$ and $p(k) \in [0, \delta]$ are isolated. As p is a quasiperiodic function, the set of critical points of this type, Z say, must then be uniformly discrete.

Now, with a Borel–Cantelli argument, compare [11, Thm. 6.3.5] and [10, Prop. 5.1], one can derive that, for a.e. $x \in \mathbb{R}^d$, the sequence $(\alpha^n x)_{n \in \mathbb{N}_0}$ stays sufficiently far away from Z so that the average via the Birkhoff sum ultimately is not distorted by the singularities or almost singularities of $\log(p)$, and Sobol's theorem can be applied. This gives

$$\lim_{N \to \infty} \frac{1}{N} \sum_{n=0}^{N-1} \log(p(\alpha^n x)) = \mathbb{M}(\log(p))$$

for a.e. $x \in \mathbb{R}^d$ as claimed. $\quad\square$

At this point, we are set to start the spectral analysis of inflation systems via their pair correlations, where we begin with the theory in one dimension.

7.4   Results in One Dimension

Let us recall the situation in one dimension from [5, 6]. Consider a *primitive substitution* ϱ on an L-letter alphabet $\mathcal{A} = \{a_1, \ldots, a_L\}$. It defines a unique *symbolic hull* $\mathbb{X}_\varrho$, which is compact and consists of a single local indistinguishability (LI) class. This hull can be constructed as the closure of the shift orbit of a two-sided fixed point of a suitable power of ϱ. This shift space gives rise to a uniquely (in fact, strictly) ergodic dynamical system under the $\mathbb{Z}$-action of the shift, denoted as $(\mathbb{X}_\varrho, \mathbb{Z})$.

The corresponding *substitution matrix* M is the primitive $L \times L$-matrix with elements $M_{ij} = \mathrm{card}_{a_i}\left(\varrho(a_j)\right) \geqslant 0$ and Perron–Frobenius (PF) eigenvalue $\lambda > 1$. The matching (properly normalised) right eigenvector of M encodes the letter frequencies, while the left eigenvector determines the ratios of natural tile lengths for a consistent geometric *inflation rule*. The latter acts on L intervals (which are our prototiles), one for each letter, of lengths corresponding to the entries of the left eigenvector. If the L intervals do not have distinct lengths, we distinguish congruent ones by labels (or colours). The inflation map induced by ϱ then consists of a scaling of the intervals by the inflation multiplier λ and their subsequent dissection into original prototiles, according to the order determined by the substitution rule ϱ. In this setting, the inflation again defines a strictly ergodic dynamical system, now (in general) under the continuous translation action of $\mathbb{R}$, denoted as $(\mathbb{Y}, \mathbb{R})$, with $\mathbb{Y}$ the new *tiling hull*.

To capture the geometric information, let us collect the relative positions of the tiles in the inflation map in a set-valued *displacement matrix* T. Each element T_{ij} thus is a set, viewed as a list of length M_{ij} that contains the relative positions of the interval (or tile) of type i in the inflated interval (or supertile) of type j (and is the empty set if $M_{ij} = 0$). To define the distance between tiles, we assign a reference point to each tile, which we usually choose to be the left endpoint of the interval. Clearly, since the reference point determines the tile and its position, the set of (labelled or coloured) reference points is *mutually locally derivable* (MLD) with the tiling by intervals. For a given tiling, define Λ_i as the set of all reference points of tiles of type i, and $\Lambda = \dot{\bigcup}_{i=1}^{L} \Lambda_i$ as the set of all such reference points.

Let $\nu_{ij}(z)$ with $z \geqslant 0$ be the relative frequency of the occurrence of a tile of type i (left) and one of type j (right) at distance z, with the understanding that

$$\nu_{ij}(-z) = \nu_{ji}(z).$$

These are the *pair correlation coefficients* of the inflation rule, which exist for all elements of the hull and are independent of the choice of the element. Given Λ, decomposed as $\Lambda = \dot{\bigcup}_i \Lambda_i$, one can represent each coefficient as a limit,

$$\nu_{ij}(z) = \lim_{r \to \infty} \frac{\mathrm{card}\big(B_r(0) \cap \Lambda_i \cap (\Lambda_j - z)\big)}{\mathrm{card}(B_r(0) \cap \Lambda)} = \frac{\mathrm{dens}\big(\Lambda_i \cap (\Lambda_j - z)\big)}{\mathrm{dens}(\Lambda)} \geqslant 0.$$

Due to the strict ergodicity, one has $\nu_{ij}(z) > 0$ if and only if $z \in S_{ij} := \Lambda_j - \Lambda_i$, where the sets S_{ij} are independent of the choice of Λ from the hull, because the latter is minimal and thus consists of a single LI class [7].

Let us now recall the general renormalisation relations for the ν_{ij} from [4, 5, 10], which are proved in full generality in [6], also for higher dimensions; see Eq. (7.18) below.

Lemma 7.4.1 *Let ν_{ij} be the pair correlation coefficients of the geometric inflation rule induced by the primitive L-letter substitution ϱ with inflation multiplier λ, and let T be the corresponding set-valued displacement matrix. Then, they satisfy the identities*

$$\nu_{ij}(z) = \frac{1}{\lambda} \sum_{m,n=1}^{L} \sum_{r \in T_{im}} \sum_{s \in T_{jn}} \nu_{mn}\left(\frac{z + r - s}{\lambda}\right)$$

for arbitrary $z \in \mathbb{R}$. □

Remark 7.4.2 The identities of Lemma 7.4.1 have a special structure, which we call an *exact renormalisation* for the following reason. First, there is a finite subset of identities that close, and give what is known as the *self-consistency* part of the identities. Then, all remaining relations are purely *recursive*, which also implies that the solution space of the renormalisation identities is finite-dimensional. This is further discussed and explored in [5, 6]. ◇

Now, define $\Upsilon_{ij} = \sum_{z \in S_{ij}} \nu_{ij}(z)\,\delta_z$, which is a pure point measure for each $1 \leqslant i, j \leqslant L$. For the measure vector $\Upsilon = (\Upsilon_{11}, \Upsilon_{12}, \ldots, \Upsilon_{LL})$, we use $f.\Upsilon$ for the componentwise pushforward, where $f(x) = \lambda x$ as before. With this, Lemma 7.4.1 implies the matching relation for the *pair correlation measures* to be

$$\Upsilon = \frac{1}{\lambda}\big(\tilde{\delta}_T \overset{*}{\otimes} \delta_T\big) * (f.\Upsilon),$$

where δ_T is the measure-valued matrix with elements $\delta_{T_{ij}}$ and $\overset{*}{\otimes}$ denotes the Kronecker product of two measure-valued matrices with convolution as multiplication.

All elements of Υ are Fourier-transformable as measures, which follows from [5, Lemma 1]. Thus, we define the *Fourier matrix* of our inflation system as

$$B(k) := \overline{\widehat{\delta_T}(k)} = \widehat{\delta_T}(-k),$$

which is an $L \times L$ matrix function with trigonometric polynomials as entries, and thus analytic in k. Now, by Fourier transform in conjunction with the convolution theorem, one finds

$$\widehat{\Upsilon} = \frac{1}{\lambda^2} \left(B(.) \otimes \overline{B(.)} \right) \left(f^{-1}.\widehat{\Upsilon} \right), \tag{7.9}$$

to be read as a relation between measure vectors. The main advantage of this formulation is that we now actually obtain *three* equations from (7.9) as follows.

Each $\widehat{\Upsilon}_{ij}$ is a measure that has a unique decomposition into a pure point (pp) and a continuous part, with a countable supporting set for the pure point part. Taking the union of the latter over all i, j allows us to define the decomposition

$$\widehat{\Upsilon} = \widehat{\Upsilon}_{\mathsf{pp}} + \widehat{\Upsilon}_{\mathsf{cont}}$$

with a matching decomposition $\mathbb{R} = \mathcal{E}_{\mathsf{pp}} \,\dot\cup\, \mathcal{E}_{\mathsf{cont}}$. Here, $\mathcal{E}_{\mathsf{pp}}$ is a countable set, and we may assume without loss of generality that it is also invariant under f and f^{-1}, for instance by replacing $\mathcal{E}_{\mathsf{pp}}$ with $\bigcup_{n \in \mathbb{Z}} f^n(\mathcal{E}_{\mathsf{pp}})$, which is still countable. The complement then still is a valid supporting set for the continuous part, and also invariant under f and f^{-1}.

Repeating this type of argument, we can further split $\widehat{\Upsilon}_{\mathsf{cont}}$ into its singular continuous (sc) and absolutely continuous (ac) component, which goes along with a decomposition $\mathbb{R} = \mathcal{E}_{\mathsf{pp}} \,\dot\cup\, \mathcal{E}_{\mathsf{sc}} \,\dot\cup\, \mathcal{E}_{\mathsf{ac}}$, where each supporting set is invariant under f and f^{-1}; see [6] for a more detailed discussion of this point. This decomposition leads to the following result.

Lemma 7.4.3 *The measure vector $\widehat{\Upsilon}$ satisfies the three separate equations*

$$\widehat{\Upsilon}_\alpha = \frac{1}{\lambda^2} \left(B(.) \otimes \overline{B(.)} \right) \left(f^{-1}.\widehat{\Upsilon}_\alpha \right),$$

for $\alpha \in \{\mathsf{pp}, \mathsf{sc}, \mathsf{ac}\}$.

Proof This is a consequence of the fact that $B(k) \otimes \overline{B(k)}$ is analytic in k, hence cannot change the spectral type, together with $\left(f^{-1}.\widehat{\Upsilon} \right)_\alpha = f^{-1}.\widehat{\Upsilon}_\alpha$ due to f being a simple dilation, which cannot change the spectral type either. The claim now follows from restricting Eq. (7.9) to the supporting sets $\mathcal{E}_\alpha$ constructed above. $\qquad\square$

All three equations have interesting implications, as discussed in [4–6]. Here, we concentrate on the ac part. To obtain some insight into the latter, we denote the Radon–Nikodym density vector of $\widehat{\Upsilon}_{\mathsf{ac}}$ by h. Then, Lemma 7.4.3 results in the relation

$$h(k) = \frac{1}{\lambda} \left(B(k) \otimes \overline{B(k)} \right) h(\lambda k),$$

which has to hold for a.e. $k \in \mathbb{R}$ and can be iterated. Note that the different power of λ in the denominator in comparison to Lemma 7.4.3 results from a change of variable transformation. For values of k with $\det(B(k)) \neq 0$, it can also be inverted to get an iteration in the opposite direction. It is a crucial observation from [4, 6, 10] that the asymptotic behaviour can be analysed from the simpler iterations

$$v(k) = \frac{1}{\sqrt{\lambda}} B(k) v(\lambda k) \quad \text{and} \quad v(\lambda k) = \sqrt{\lambda} B^{-1}(k) v(k), \tag{7.10}$$

where the components of $v(k)$ are locally square-integrable functions. Using Fact 7.2.1, this emerges from a decomposition of $(h_{ij}(k))$, viewed as a positive semi-definite Hermitian matrix, as a sum of rank-1 matrices of the form $v_i(k) v_j^\dagger(k)$ and the observation that the overall growth rate is dictated by the maximal growth rate of these summands; see [4, 6] for details.

To capture the asymptotic behaviour, one defines the *Lyapunov exponents*, compare [57], for the iterations that emerge from Eq. (7.10), which is possible when $B(k)$ is invertible for a.e. $k \in \mathbb{R}$. It turns out that the required values can all be related to the extremal Lyapunov exponents of the matrix cocycle defined by

$$B^{(n)}(k) := B(k) B(\lambda k) \cdots B(\lambda^{n-1} k), \tag{7.11}$$

which happens to be the Fourier matrix of ϱ^n. The quantities of interest to us here are controlled by the *maximal* Lyapunov exponent of this cocycle, defined as

$$\chi^B(k) := \limsup_{n \to \infty} \frac{1}{n} \log \left\| B^{(n)}(k) \right\|, \tag{7.12}$$

where $\|.\|$ refers to any sub-multiplicative matrix norm, such as the spectral norm or the Frobenius norm. In favourable cases, $\chi^B(k)$ will exist as a limit for a.e. $k \in \mathbb{R}$, as we shall see later in several examples. The main criterion can now be formulated as follows.

Theorem 7.4.4 *Let ϱ be a primitive substitution on a finite alphabet, and consider the corresponding inflation rule with inflation multiplier $\lambda = \lambda_{\mathrm{PF}}$ for intervals of natural length. Let $\chi^B(k)$ be the maximal Lyapunov exponent of the Fourier matrix cocycle (7.11), and assume that $\det(B(k)) \neq 0$ for at least one $k \in \mathbb{R}$.*

If there is some $\varepsilon > 0$ such that $\chi^B(k) \leqslant \frac{1}{2} \log(\lambda) - \varepsilon$ holds for Lebesgue-a.e. $k \in \mathbb{R}$, one has $\widehat{\Upsilon}_{\mathrm{ac}} = 0$, and the diffraction measure of the system is singular.

Proof *(Sketch)* Under our assumptions,[2] for a.e. $k \in \mathbb{R}$ with $k \neq 0$, the sequence $(h(\lambda^n k))_{n \in \mathbb{N}}$ of Radon–Nikodym density vectors, as $n \to \infty$, displays an exponential growth of order $\mathrm{e}^{2(D-\delta)n}$, where $D = \frac{1}{2} \log(\lambda) - \chi^B(k) \geqslant \varepsilon > 0$ and $\delta > 0$ can be chosen such that $D - \delta > 0$. The implied constant will depend on k and δ. Such a behaviour is incompatible with the translation-boundedness of the components of $\widehat{\Upsilon}_{\mathrm{ac}}$, which is a contradiction unless $h(k) = 0$ for a.e. $k \in \mathbb{R}$, hence

---

[2]Since $\det(B(k))$ is a trigonometric polynomial, it is either identically 0 or has isolated zeros.

$\widehat{\Upsilon}_{ac} = 0$. For further details, we refer to [4, Sec. 6.7 and App. B] as well as to the general treatment in [6].   □

Remark 7.4.5 The statement of Theorem 7.4.4 can be strengthened and extended in various ways. First of all, one can show that $\chi^B(k) \leqslant \log \sqrt{\lambda}$ holds for a.e. $k \in \mathbb{R}$. As a consequence, a non-trivial **ac** diffraction component is only possible when $\chi^B(k) = \log \sqrt{\lambda}$ is true for k in a subset of positive measure in every interval of the form $[-\lambda a, -a]$ or $[a, \lambda a]$ with $a > 0$. When λ is a PV number without any further restriction, which thus also covers all primitive inflation rules of constant length as well as those with integer inflation factor, the relation must even hold for Lebesgue-a.e. $k \in \mathbb{R}$; see [6] for details. This poses severe restrictions on the existence of **ac** diffraction in inflation systems beyond the necessary criterion of Berlinkov and Solomyak [17].   ◇

7.5   Consequences and an Application

For the Fibonacci inflation, the exact renormalisation for the pair correlation functions was used to establish a spectral purity result and then pure point spectrum [5], thus confirming a known property in an independent way. The same line of thought works for all noble means inflations in complete analogy.

It is tempting to expect a similar result for all irreducible PV inflations, but one quickly realises that spectral purity is essentially equivalent to almost everywhere injectivity of the factor map onto the maximal equicontinuous factor (MEF). While the existence of non-trivial point spectrum in one-dimensional inflation tilings requires λ to be a PV number [54], it is the exclusion of any continuous spectral component that would settle the (still open) Pisot substitution conjecture.

A less ambitious task thus is to establish the mere absence of absolutely continuous diffraction or spectral measures. It has long been 'known' (without mathematical proof) that the presence of **ac** diffraction requires a particular scaling property of the diffraction measure as a function of the system size. This stems from the heuristic expectation that a structure has an **ac** diffraction spectrum if its fluctuations are somewhat similar to those of a disordered random structure, so fluctuations growing as $\sqrt{N}$ for a chain of length N, in line with the law of large numbers. This behaviour corresponds to a wandering exponent equal to $\frac{1}{2}$; see [2, 35, 43] for an application to aperiodic structures.

In the case of constant-length substitutions, this effectively corresponds to a condition on the spectrum of the substitution matrix M. Namely, if λ is its PF eigenvalue, M must also have an eigenvalue $\sqrt{\lambda}$ or one of that modulus. The necessity of an eigenvalue of modulus $\sqrt{\lambda}$ for the existence of an **ac** spectral measure was recently proved in [17]. That this criterion is necessary, but not sufficient, can be shown by an example, for instance using the constant-length substitution

$$a \mapsto ab, \quad b \mapsto ca, \quad c \mapsto bd, \quad d \mapsto dc \tag{7.13}$$

on the 4-letter alphabet $\{a, b, c, d\}$. The substitution matrix reads

$$M = \begin{pmatrix} 1 & 1 & 0 & 0 \\ 1 & 0 & 1 & 0 \\ 0 & 1 & 0 & 1 \\ 0 & 0 & 1 & 1 \end{pmatrix}$$

and has spectrum $\{2, \pm\sqrt{2}, 0\}$, hence clearly satisfies the $\sqrt{\lambda}$-criterion. Nevertheless, as was shown in [20] on the basis of Bartlett's algorithmic classification of spectral types [15], all spectral measures of this substitution are singular.

Let us apply Lyapunov exponents to reach this conclusion in an independent way. It is straight-forward to calculate

$$T = \begin{pmatrix} \{0\} & \{1\} & \varnothing & \varnothing \\ \{1\} & \varnothing & \{0\} & \varnothing \\ \varnothing & \{0\} & \varnothing & \{1\} \\ \varnothing & \varnothing & \{1\} & \{0\} \end{pmatrix} \quad \text{and} \quad B(k) = \begin{pmatrix} 1 & z & 0 & 0 \\ z & 0 & 1 & 0 \\ 0 & 1 & 0 & z \\ 0 & 0 & z & 1 \end{pmatrix}$$

where $z = e^{2\pi i k}$. One has $\det(B(k)) = z^4 - 1$ which vanishes only for $k \in \frac{1}{4}\mathbb{Z}$, so that $B(k)$ is invertible for a.e. $k \in \mathbb{R}$. Let us now, for $n \in \mathbb{N}$, define the matrices

$$B^{(n)}(k) := B(k)B(2k)B(4k)\cdots B(2^{n-1}k). \tag{7.14}$$

By definition, $B^{(1)} = B$ is the Fourier matrix of ϱ, while $B^{(n)}$ is the Fourier matrix of ϱ^n, and hence a natural object to study in this context.[3]

Since the substitution is of constant length, $B^{(n)}$ defines a cocycle over the compact dynamical system defined by $k \mapsto 2k$ modulo 1 on $\mathbb{T}$. We thus have Oseledec's multiplicative ergodic theorem [57] at our disposal, which implies that the Lyapunov exponents exist for a.e. $k \in \mathbb{R}$ and satisfy forward Lyapunov regularity, hence in particular sum to

$$\lim_{n \to \infty} \frac{1}{n} \sum_{\ell=0}^{n-1} \log\left|\det\left(B(2^\ell k)\right)\right| = \mathbb{M}\left(\log\left|\det(B(.))\right|\right) = \mathfrak{m}(z^4 - 1) = 0,$$

---

[3]Notice that, while the Fourier matrices $B(k)$ for different k generally do not commute, the matrices $B^{(2)}(k) = B(k)B(2k)$, which correspond to the square of the substitution rule (7.13), form a commuting family of matrices. This corresponds to the fact that the substitution is non-Abelian in the sense of [48], meaning that the column-wise letter permutations do not commute, while its square becomes Abelian.

where the first equality is a consequence of Birkhoff's ergodic theorem, as detailed in Fact 7.3.6, while the last step follows directly from Fact 7.2.2.

To continue, it is helpful to observe that $B(k)$ admits a k-independent splitting of $\mathbb{C}^4$ into a two-dimensional and two one-dimensional subspaces. Concretely, one finds

$$
U B(k) U^{-1} = \begin{pmatrix} 1+z & 0 & 0 & 0 \\ 0 & 1-z & 0 & 0 \\ 0 & 0 & -z & 1 \\ 0 & 0 & 1 & z \end{pmatrix} \quad \text{with} \quad U = \frac{1}{2} \begin{pmatrix} 1 & 1 & 1 & 1 \\ 1 & -1 & -1 & 1 \\ 1 & -1 & 1 & -1 \\ 1 & 1 & -1 & -1 \end{pmatrix},
$$

where the unitary matrix U is an involution, so $U^{-1} = U$. By standard arguments, it is clear that two of the four exponents are given by $\mathrm{m}(1+z) = 0$ and $\mathrm{m}(1-z) = 0$, which derives from the invariant one-dimensional subspaces. The remaining two exponents must still sum to 0, and can be determined from the induced cocycle $\tilde{B}^{(n)}(k) = \tilde{B}(k)\tilde{B}(2k)\cdots\tilde{B}(2^{n-1}k)$ with $\tilde{B}(k) = \begin{pmatrix} -z & 1 \\ 1 & z \end{pmatrix}$. With $p_N(k) := \|\tilde{B}^{(N)}(k)\|_{\mathrm{F}}^2$, which is a trigonometric polynomial due to the use of the Frobenius norm, we know that

$$
\chi^B(k) = \chi^{\tilde{B}}(k) \leqslant \frac{1}{N} \mathbb{M}\big(\log \|\tilde{B}^{(N)}(k)\|_{\mathrm{F}}\big) = \frac{\mathrm{m}(p_N)}{2N} =: m_N \tag{7.15}
$$

holds for a.e. $k \in \mathbb{R}$ and every $N \in \mathbb{N}$. In particular, one has the (almost sure) estimate $\chi^B(k) \leqslant \liminf_{N\to\infty} m_N$.

Now, employing Jensen's formula again, the numbers m_N can easily be calculated numerically with high precision, and are given in Table 7.1 for $N \leqslant 12$. These values clearly show that $\chi^B(k) \leqslant \frac{1}{5} < \log\sqrt{2} \approx 0.346574$, which implies the absence of ac diffraction.

Since we are in the constant-length case, this result translates into one on the spectral measures via the general results of [48, Prop. 7.2] on the maximal spectral type of a constant-length substitution; see also [15, Thm. 3.4]. The crucial point to observe here is that we do not need to consider the spectral measures of all functions that are square-integrable over the hull, but only those of the (possibly weighted) lookup functions for the type of level-m supertile at 0, for all $m \in \mathbb{N}_0$.

Our diffraction measure provides the result for the spectral measure of the lookup functions of the prototiles themselves, compare [14], while we can repeat our analysis for any supertile in noting that this will simply lead to a rescaling, as a result of Lemma 7.2.5. Concretely, the spectral measures will then be Riesz products of the

Table 7.1 Some values of the means m_N from Eq. (7.15), calculated via Eq. (7.2). The numerical error is always less than 10^{-3}

N	1	2	3	4	5	6	7	8	9	10	11	12
m_N	0.693	0.478	0.379	0.334	0.302	0.274	0.252	0.235	0.220	0.208	0.198	0.189

same type in the sense that only finitely many initial factors are missing. Since they clearly have the same Lyapunov exponents and growth rates, our result translates to their spectral measures as well. In line with [20], but by a completely different method, we have thus arrived at the following result.

Corollary 7.5.1 *Consider the dynamical system* $(\mathbb{X}_\varrho, \mathbb{Z})$ *defined by the primitive constant-length substitution* ϱ *from* (7.13)*, which has inflation multiplier 2. Although its substitution matrix also has an eigenvalue* $\sqrt{2}$*, and thus satisfies the necessary criterion for the presence of an absolutely continuous spectral measure, no such measure exists, and all spectral measures are singular.* $\qquad\square$

It follows from the full analysis in [20] that the extremal spectral measures are either pure point or singular continuous, and that both possibilities occur here. Let us briefly mention that [6] presents a method to construct infinitely many other examples of this kind, which demonstrates that the $\sqrt{\lambda}$-criterion alone is far from sufficient for the emergence of ac spectral components.

7.6 Results in Higher Dimensions

One advantage of the geometric language with tilings is its generalisability to higher dimensions. Here, a *tile* in $\mathbb{R}^d$ is a compact set t that is the closure of its interior, and we will only consider cases where t is simply connected, though this is not required for the general theory. A *prototile* is a representative of a tile and all its translates under the action of $\mathbb{R}^d$.

Given a finite set $\mathcal{T} = \{t_1, \ldots, t_L\}$ of L prototiles and an expansive linear map Q, one speaks of a *stone inflation* relative to Q (otherwise often called a self-affine inflation) if there is a rule how to exactly subdivide each level-1 supertile $Q(t_i)$ into translated copies of the original tiles. Iterating such an inflation rule, called ϱ as before, leads to tilings that cover $\mathbb{R}^d$, and via the orbit closure in the standard local rubber topology also to a compact hull $\mathbb{Y}$. If the inflation is primitive, see [7, 12, 34] for details, this hull is minimal and consists of a single LI class, which is to say that any two elements of the hull are LI. It is an interesting and important fact that this property is not restricted to the FLC situation, but still holds for more general inflation tilings [30], with the properly adjusted notions of indistinguishability and repetitivity; see also [41].

To keep track of the relative positions of the tiles under the inflation procedure, we need to equip each t_i with a reference or control point. While there are usually many ways to do so, some will be more 'natural' than others. What really counts is that the tiling and the point set contain the same information. So, it is imperative to choose the control points such that they are MLD with the tiling. When congruent tiles exist, the control points are *coloured* to distinguish them according to the tile type. This means that the space of (coloured) control point sets and the tiling hull are topologically conjugate as dynamical systems under the translation action of $\mathbb{R}^d$

in a *local* way. For this reason, we usually identify the two pictures, and speak of tilings or point sets interchangeably, always using $\mathbb{Y}$ to denote the hull.

Now, we can define the displacement sets T_{ij} essentially as before, so

$$T_{ij} = \{\text{all relative positions of } t_i \text{ in } Q(t_j)\}, \tag{7.16}$$

where the relative positions are defined via the control points. Note that all quantities are defined in complete analogy to the one-dimensional case. In particular, the corresponding *Fourier matrix* is once again given by

$$B(k) = \widehat{\delta_T}(k). \tag{7.17}$$

Note that $k \in \mathbb{R}^d$ reflects the dimension of the Euclidean space the tiling lives in, while $B(k) \in \text{Mat}(L, \mathbb{C})$ covers the combinatorial structure of the inflation rule. As before,

$$M = B(0)$$

is the non-negative inflation or *incidence* matrix, with leading eigenvalue $\lambda = |\det(Q)|$ by construction of the stone inflation. Many explicit examples are discussed in [7, Ch. 6] as well as in [26, 29, 33]; see also the Tilings Encyclopedia.[4] Quite frequently, Q will be a homothety, simply meaning $Q(x) = \lambda x$ and thus referring to the case of a self-similar inflation, but it can also contain a rotation (as in Gähler's shield tiling; see [7, Sec. 6.3.2]) or scale differently in different directions (as in general block substitutions; see Fig. 7.1 below for an example). The crucial point here is that space and combinatorial information are properly separated for the renormalisation approach.

The renormalisation equations for the pair correlation coefficients are derived [6] by the same arguments used in Lemma 7.4.1 above, where the local recognisability in the aperiodic case follows from [53]. The result reads

$$\nu_{ij}(z) = \frac{1}{|\det(Q)|} \sum_{m,n=1}^{L} \sum_{r \in T_{im}} \sum_{s \in T_{jn}} \nu_{mn}\big(Q^{-1}(z + r - s)\big), \tag{7.18}$$

which, in terms of the corresponding pair correlation measures, becomes

$$\Upsilon = \frac{1}{|\det(Q)|} \big(\widetilde{\delta_T} \overset{*}{\otimes} \delta_T\big) * (Q.\Upsilon).$$

---

[4]The Tilings Encyclopedia is maintained by Dirk Frettlöh and Franz Gähler, and is accessible online at http://tilings.math.uni-bielefeld.de.

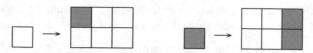

Fig. 7.1 A simple example of a primitive binary block substitution in the plane. It is a stone inflation for the linear map $Q = \mathrm{diag}(3, 2)$. The lower left corners of all blocks are used as control points (not shown), which are coloured according to the block type

Note that these relations also apply to *periodic* inflation tilings, as shown in [6]. Taking Fourier transforms, with the dual map

$$Q^* := (Q^T)^{-1},$$

we obtain the relations

$$\widehat{\Upsilon} = \frac{1}{|\det(Q)|^2} \left(B(.) \otimes \overline{B(.)} \right) (Q^* . \widehat{\Upsilon}) \tag{7.19}$$

by Lemma 7.2.5. Once again, they have to hold separately for the pure point, singular continuous and absolutely continuous components, respectively, as in Lemma 7.4.3.

Due to the appearance of Q^*, one defines the Fourier matrix cocycle as

$$B^{(n)}(k) = B(k) B(Q^T k) \cdots B((Q^T)^{n-1} k), \tag{7.20}$$

where the transpose can also be seen as a consequence of Eqs. (7.16) and (7.17) via a simple calculation with the Fourier transform. Now, as in the one-dimensional case, one has the following result [6].

Fact 7.6.1 *If $B(k)$ is the Fourier matrix of the primitive stone inflation rule ϱ, the Fourier matrix of ϱ^n is given by $B^{(n)}(k)$ from Eq. (7.20).* $\qquad\square$

With χ^B as defined in Eq. (7.12), and in complete analogy to the one-dimensional case, one can now derive [6, Thm. 5.7] the following criterion for the absence of ac diffraction components.

Theorem 7.6.2 *Consider a finite set $\mathcal{T}$ of prototiles in $\mathbb{R}^d$ and a primitive stone inflation for $\mathcal{T}$, with expansive linear map Q, and suppose that this defines an FLC tiling system. Assume further that each prototile is equipped with a control point, possibly coloured, such that the tilings and the corresponding control point sets are MLD. Define Fourier matrix and Lyapunov exponents as explained above.*

Suppose that $B(k)$ is invertible for a.e. $k \in \mathbb{R}^d$ and that there is some $\varepsilon > 0$ such that $\chi^B(k) \leqslant \frac{1}{2} \log|\det(Q)| - \varepsilon$ holds for a.e. $k \in \mathbb{R}^d$. Then, one has $\widehat{\Upsilon}_{ac} = 0$ and the diffraction measure of the tiling system is singular. $\qquad\square$

Remark 7.6.3 A closer inspection of the proof in [6] reveals that the FLC condition is actually not necessary. Indeed, if one starts form a stone inflation with finitely many prototiles up to translations, the criterion from Theorem 7.6.2 works without further modifications. We shall see and discuss several examples later on. ◇

Let us note in passing that the comments of Remark 7.4.5, with the obvious adjustments, apply to this higher-dimensional case as well. In particular, the conditions for the appearance of ac spectral components in higher dimensions are as restrictive as in one dimension.

7.7 Binary Block Substitutions of Constant Size

An interesting class is provided by primitive, binary block substitutions in d dimensions, where we have two types of unit blocks, white (0) and black (1) say, which are both substituted into a block of equal size and shape. Here, we assume the corresponding linear expansion to be $Q = \mathrm{diag}(n_1, \ldots, n_d)$ with all $n_i \geqslant 2$, so $Q = Q^T$ in this case.

Let us now place the inflated white block on top of the inflated black one (in $\mathbb{R}^{d+1}$ that is), so that one can easily identify bijective and coincident positions via the corresponding columns. We cast them into polynomials as follows. Let p be the polynomial in $z = (z_1, \ldots, z_d)$ for *all* positions, which means that

$$p(z) = \prod_{j=1}^{d} \left(1 + z_j + \ldots + z_j^{n_j - 1}\right).$$

Likewise, q and r are the polynomials for bijective columns of type $\begin{bmatrix} 0 \\ 1 \end{bmatrix}$ and $\begin{bmatrix} 1 \\ 0 \end{bmatrix}$, while s_0 and s_1 stand for the polynomials of the coincident columns of type $\begin{bmatrix} 0 \\ 0 \end{bmatrix}$ and $\begin{bmatrix} 1 \\ 1 \end{bmatrix}$, respectively. Clearly, one has $q + r + s_0 + s_1 = p$. With $k = (k_1, \ldots, k_d) \in \mathbb{R}^d$, the Fourier matrix then has the form

$$B(k) = \begin{pmatrix} q(z) + s_0(z) & r(z) + s_0(z) \\ r(z) + s_1(z) & q(z) + s_1(z) \end{pmatrix} \quad \text{with} \quad z_j = e^{2\pi i k_j}.$$

Since $\det(B(k)) = p(z)(q(z) - r(z))$, the matrix $B(k)$ is invertible for a.e. $k \in \mathbb{R}^d$; see [8] for more on bijective block substitutions and [26] for some general results.

The Fourier matrix cocycle belongs to the compact dynamical system defined by $k \mapsto Qk$ modulo 1 on $\mathbb{T}^d$. In this situation, we may use Oseledec's multiplicative ergodic theorem [57], which tells us that the two Lyapunov exponents exist for a.e. $k \in \mathbb{R}^d$ and add up to

$$\lim_{N \to \infty} \frac{1}{N} \sum_{\ell=0}^{N-1} \log \left| \det(B(Q^\ell k)) \right|$$

whenever this limit exists, which is true for a.e. $k \in \mathbb{R}^d$ by Lemma 7.3.7. The limit is

$$\int_{\mathbb{T}^d} \log|\det(B(k))| \, dk = \mathfrak{m}(p) + \mathfrak{m}(q - r) = \mathfrak{m}(q - r),$$

because $\mathfrak{m}(p) = 0$ by Fact 7.2.2.

Since $(1, 1)$ is a left eigenvector of $B(k)$ for all $k \in \mathbb{R}^d$, with eigenvalue $p(z_1, \ldots, z_d)$ and the z_j from above, one obtains

$$\frac{1}{N} \log \left\| (1, 1) B^{(N)}(k) \right\| \xrightarrow{N \to \infty} \mathfrak{m}(p) = \sum_{j=1}^{d} \mathfrak{m}\left(1 + z_j + \ldots + z_j^{n_j - 1}\right) = 0$$

for a.e. $k \in \mathbb{R}^d$. Since we thus know one exponent together with the sum, we get that

$$\chi^B(k) = \chi^B_{\max}(k) = \mathfrak{m}(q - r)$$

holds for a.e. $k \in \mathbb{R}^d$. By standard estimates [6, 10, 44], one now finds

$$\exp\left(\mathfrak{m}(q - r)\right) < \|q - r\|_1 \leqslant \|q - r\|_2 = \sqrt{\det(Q) - n_c} \leqslant \sqrt{\det(Q)},$$

where the first step follows from Jensen's inequality; see [42] for a suitable formulation. Moreover, with n_c denoting the total number of coincident columns, the equality is a result of Parseval's identity. Together, we get the inequality $\mathfrak{m}(q - r) < \log \sqrt{\det(Q)}$, which gives the required criterion for the absence of absolutely continuous components in $\widehat{\Upsilon}$.

Clearly, the corresponding property also holds in higher dimensions, and we have the following general result; see also [6].

Theorem 7.7.1 *The diffraction measure of a primitive binary block substitution of constant size in dimension $d \geqslant 2$ is always singular.* □

Example 7.7.2 For the block substitution of Fig. 7.1, $Q = \text{diag}(3, 2)$ is the linear expansion. With $(z_1, z_2) = (x, y)$, the polynomials are $p(x, y) = (1 + x + x^2)(1 + y)$ together with $q(x, y) = x^2(1 + y)$, $r(x, y) = y$, $s_0(x, y) = 1 + x(1 + y)$ and $s_1(x, y) = 0$. This gives

$$B(k) = \begin{pmatrix} 1 + (x + x^2)(1 + y) & (1 + x)(1 + y) \\ y & x^2(1 + y) \end{pmatrix} \quad \text{with } (x, y) = \left(e^{2\pi i k_1}, e^{2\pi i k_2}\right)$$

and $\det(B(k)) = p(x, y)(x^2 + x^2 y - y)$. The existence of the various limits of Birkhoff sums we need here, for a.e. $k \in \mathbb{R}^2$, follows once again from Lemma 7.3.7.

The non-zero Lyapunov exponent is given by

$$m(x^2 + x^2y - y) = m(x^2 + x^2y + y) = m(1 + x^{-2}y + y) = m(1 + x + y)$$

$$= \frac{3\sqrt{3}}{4\pi} L(2, \chi_{-3}) = 2 \int_0^{1/3} \log(2\cos(\pi t)) \, dt \approx 0.323066,$$

where $L(z, \chi_{-3})$ is the L-function for the principal Dirichlet character χ_{-3} of the imaginary quadratic field $\mathbb{Q}(\sqrt{-3})$. Here, the first equality follows from a change of variable transformation, while the third emerges by a standard formula from [25]. The connection between logarithmic Mahler measures and special values of L-functions is a famous result that first appeared in [58] as the groundstate entropy[5] of the anti-ferromagnetic Ising model on the triangular lattice; see [3, 49] and references therein for more.   ◊

Remark 7.7.3 The absence of ac diffraction immediately implies that the spectral measure for the 'one-point lookup function' must be singular. As in the one-dimensional case, this implies that the spectral measure is also singular for any function that looks up the level-n supertile at the origin; see the discussion before Corollary 7.5.1. Then, by [15], *any* spectral measure must be singular, and our system has singular dynamical spectrum. This gives another, independent proof of a result that was previously shown in [8, 26]; see also [29].   ◊

7.8   Block Substitutions with Squares

Here, we are interested in inflation rules with a single prototile of unit area and linear expansion Q, but some added complexity from the set S of relative positions of the tiles within the supertile. In particular, this will be our first class of examples where we go beyond the FLC case. The special interest in this class originates in the fact that the Fourier matrix cocycle, $B^{(n)}(k)$, simply is a sequence of multivariate trigonometric polynomials. We thus write $P^{(n)}(k)$ to indicate this. The renormalisation equation becomes an equation directly for the autocorrelation and reads

$$\gamma = \nu * (f.\gamma) \quad \text{with} \quad \nu = \frac{\delta_S * \delta_{-S}}{|\det(Q)|}. \tag{7.21}$$

Let us begin with a particularly simple case.

---

[5]It is interesting historically that this connection was overlooked for a long time because the numerical value given there (for the correct integral) was erroneous, which was corrected in an erratum 23 years later.

Example 7.8.1 The staggered block substitution defined by

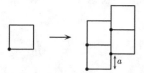

with arbitrary $a \in \mathbb{R}$ results in a tiling that is lattice-periodic, with lattice of periods $\Gamma = \langle v, e_2 \rangle_{\mathbb{Z}}$, where $v = e_1 + a e_2$. In particular, the resulting tiling is 1-periodic in e_2-direction. Since $\mathrm{dens}(\Gamma) = 1$, the autocorrelation is $\gamma = \delta_\Gamma$, where we use a reference point in the lower left corner of every unit square as indicated. The diffraction measure is $\widehat{\gamma} = \delta_{\Gamma^*}$ with the dual lattice $\Gamma^* = \langle e_1, e_2 - a e_1 \rangle_{\mathbb{Z}}$.

Let us look at this result from the renormalisation point of view, which is also applicable here because the inflation can easily be changed into a stone inflation without changing the control point positions; see Remark 7.8.2 below. As mentioned above, the approach also works for periodic cases, like the one at hand. By an iteration of the renormalisation equation (7.21) and an application of Lemma 7.2.4, the autocorrelation is

$$\gamma = \mathop{\Large *}_{m \geqslant 0} f^m . \nu \quad \text{with} \quad \nu = \left(\delta_0 + \tfrac{1}{2}(\delta_{e_2} + \delta_{-e_2})\right) * \left(\delta_0 + \tfrac{1}{2}(\delta_v + \delta_{-v})\right),$$

where $f(x) = 2x$. Now, Fourier transform leads to the two-dimensional Riesz product

$$\widehat{\gamma} = \prod_{m \geqslant 0} \left(1 + \cos(2\pi\, 2^m (k_1 + a k_2))\right)\left(1 + \cos(2\pi\, 2^m k_2)\right) = \sum_{\ell \in \mathbb{Z}} \widehat{\gamma}^{(1)}_\ell \times \delta^{(2)}_\ell$$

with $\widehat{\gamma}^{(1)}_\ell = \prod_{m \geqslant 0}\left(1 + \cos(2\pi\, 2^m (k_1 + a\ell))\right)$ and $k = (k_1, k_2)$. Here, we have adopted the standard notation for product distributions or measures, where the upper index refers to the two coordinate directions. Clearly, one has $\widehat{\gamma}^{(1)}_0 = \delta^{(1)}_{\mathbb{Z}}$ by Lemma 7.2.6. Moreover, the distribution $\widehat{\gamma}^{(1)}_\ell$ is 1-periodic for every $\ell \in \mathbb{Z}$, which means that $\widehat{\gamma}$ is 1-periodic in e_1-direction, independently of a.

Whenever $a \in \mathbb{Z}$, one finds $\widehat{\gamma}^{(1)}_\ell = \delta^{(1)}_{\mathbb{Z}}$ by Lemma 7.2.6, and hence $\widehat{\gamma} = \delta_{\mathbb{Z}^2}$ as required. More generally, given $k_2 = \ell$, the only contribution to $\widehat{\gamma}^{(1)}_\ell$ emerges for $k_1 + a\ell = r \in \mathbb{Z}$, hence for $k = r e_1 + \ell(e_2 - a e_1) \in \Gamma^*$, which means that the Riesz product representation also gives $\widehat{\gamma} = \delta_{\Gamma^*}$, as it must.

Now, looking at this result from the cocycle perspective, we find that $P(k) = (1 + y)(1 + x y^a)$ with $x = \mathrm{e}^{2\pi \mathrm{i} k_1}$ and $y = \mathrm{e}^{2\pi \mathrm{i} k_2}$, so that

$$\chi^P(k) := \lim_{n \to \infty} \frac{1}{n} \sum_{\ell=0}^{n-1} \log|P(2^\ell k)| = \mathbb{M}(\log|P|)$$

holds for a.e. $k \in \mathbb{R}^2$ by Lemma 7.3.8, where one observes that $\log|P| = \frac{1}{2}\log|P|^2$ with $|P|^2$ satisfying the required conditions. Now, the mean can be calculated as

$$\mathbb{M}\big(\log|P|\big) = \mathfrak{m}_y(1+y) + \lim_{T\to\infty} \frac{1}{T} \int_0^T \mathfrak{m}_x(1+x\,y^a)\,\mathrm{d}k_2 = 0,$$

where both Mahler measures are zero because all roots of the polynomials lie on the unit circle. This fits with the explicit calculation of $\widehat{\gamma}$ from above. $\Diamond$

Remark 7.8.2 By standard methods, which are explained in [7] and in [33], one can replace the square in Example 7.8.1 with a new prototile so that the inflation rule is turned into a stone inflation. This has no effect on the position of the control point, wherefore we continue with the simpler formulation as a block substitution.

In the same vein, one can see that our approach also works for more general inflation rules, certainly as long as they are MLD with a stone inflation. For examples of such rules and their reduction to stone inflations, see [7, Ch. 5] and [33]. $\Diamond$

Extending this initial example, we may consider a block substitution with M columns of N blocks each, where entire columns can be shifted in vertical direction by an arbitrary amount, as indicated in the next diagram,

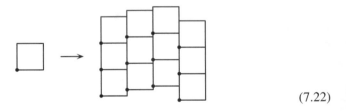

$$(7.22)$$

When using the lower left corner as reference point for each square, it is clear that a modification into a stone inflation according to Remark 7.8.2 does not change the resulting point set. For this reason, we stick to the formulation with squares for simplicity.

Let us now assume that the ith column is shifted by $a_i \in \mathbb{R}$ in vertical direction, with $i \in \{0, 1, \ldots, M-1\}$. Since a_0 only results in a global shift of the entire block, we set $a_0 = 0$ without loss of generality, and consider the remaining a_i as shifts relative to column 0.

As in the previous case, which had the FLC property, we define the hull as the orbit closure of a fixed point tiling, where the closure is now taken with respect to the *local rubber topology* [13]. This defines a compact tiling space [30], without any change in the FLC case. However, this slight modification takes care of the potential occurrence of a tiling with infinite local complexity. As before, we obtain a dynamical system, under the continuous translation action of $\mathbb{R}^2$, which is uniquely ergodic; see also [41]. Each tiling in the hull is 1-periodic in the e_2-direction.

From the set S of relative displacements, one finds

$$P(k) = \overline{\delta_S}(k) = (1 + y + \ldots + y^{N-1})(1 + xy^{a_1} + x^2 y^{a_2} + \ldots + x^{M-1} y^{a_{M-1}})$$

with $x = e^{2\pi i k_1}$ and $y = e^{2\pi i k_2}$. Note that the trigonometric polynomial P is quasiperiodic, but 1-periodic in k_1. Now, with $Q = \mathrm{diag}(M, N)$, the cocycle is defined by

$$P^{(n)}(k) = P(k) P(Qk) \cdots P(Q^{n-1}k),$$

with $P^{(1)} = P$ as usual. Our Lyapunov exponent can be calculated as follows,

$$\chi^P(k) = \lim_{n \to \infty} \frac{1}{n} \log |P^{(n)}(k)| = \lim_{n \to \infty} \frac{1}{n} \sum_{\ell=0}^{n-1} \log |P(Q^\ell k)|.$$

By an obvious variant of Lemma 7.3.8, where α is replaced by the expansion Q, the limit exists for a.e. $k \in \mathbb{R}^2$ and is given by

$$\mathbb{M}(\log|P|) = \lim_{T \to \infty} \frac{1}{T} \int_0^T \int_0^1 \log |P(k)| \, dk_1 \, dk_2$$

$$= \mathfrak{m}_y(1 + y + \ldots + y^{N-1})$$

$$+ \lim_{T \to \infty} \frac{1}{T} \int_0^T \mathfrak{m}_x(1 + xy^{a_1} + \ldots + x^{M-1} y^{a_{M-1}}) \, dk_2.$$

As $1 + y + \ldots + y^{N-1}$ is cyclotomic, the first term vanishes. The integrand in the second is the logarithmic Mahler measure of a polynomial (in x) with all coefficients on the unit circle, which is known as a *unimodular polynomial*. By an application of Jensen's inequality in conjunction with Parseval's equation, one can show that its logarithmic Mahler measure is bounded by $\log \sqrt{M}$ for every $k_2 \in \mathbb{R}$. Consequently, we have

$$\mathbb{M}(\log|P|) \leqslant \log \sqrt{M} < \log \sqrt{MN}$$

because $N > 1$ by assumption. This implies that we cannot have any absolutely continuous diffraction, and $\widehat{\gamma}$ must be singular.

The remarkable aspect of this simple class of examples is that the tilings are generally *not* FLC; compare [33] and references therein. One can say a bit more about the explicit structure of the diffraction measure. First of all, it is 1-periodic in e_1-direction, and it consists of parallel arrangements of one-dimensional layers, distinct in general, which have their own Riesz product representation. We leave further details to the interested reader.

Remark 7.8.3 The above results can be generalised to $\mathbb{R}^{d+1}$ with $d \geqslant 1$ as follows. Consider a block of $M_1 \times \cdots \times M_d \times N$ cubes, with all $M_i \geqslant 2$ and $N \geqslant 2$. Now, modify this block as an arrangement of $M_1 \times \cdots \times M_d$ columns of N cubes each, where column $(m_1, \ldots, m_d)$ is shifted by an arbitrary real number $a_{m_1,\ldots,m_d}$ in e_{d+1}-direction. We may set $a_{0,\ldots,0} = 0$ without loss of generality. With the expansion $Q = \mathrm{diag}(M_1, \ldots, M_d, N)$, one can now repeat the above analysis. Here, one obtains tilings of $\mathbb{R}^{d+1}$ that are 1-periodic in e_{d+1}-direction.

Writing

$$z = (z_1, \ldots, z_d, z_{d+1}) = \left(\mathrm{e}^{2\pi \mathrm{i} k_1}, \ldots, \mathrm{e}^{2\pi \mathrm{i} k_d}, \mathrm{e}^{2\pi \mathrm{i} k_{d+1}}\right) = (x_1, \ldots, x_d, y),$$

one finds the polynomial

$$P(k) = \left(1 + y + \ldots + y^{N-1}\right) R(k) \quad \text{with}$$

$$R(k) = \sum_{m_1=0}^{M_1-1} \cdots \sum_{m_d=0}^{M_d-1} x_1^{m_1} \cdots x_d^{m_d} \, y^{a_{m_1,\ldots,m_d}},$$

where R, and hence also P, is 1-periodic in e_i-direction for all $1 \leqslant i \leqslant d$. The maximal Lyapunov exponent, for a.e. $k \in \mathbb{R}^{d+1}$, is now given by

$$\chi^B(k) = \lim_{T \to \infty} \frac{1}{T} \int_0^T \mathrm{m}_x(R) \, \mathrm{d}k_{d+1} \leqslant \log \sqrt{M_1 \cdots M_d}$$

$$= \frac{1}{2} \sum_{i=1}^{d} \log(M_i) < \frac{1}{2} \log\big(\det(Q)\big),$$

where the last estimate is a consequence of $N \geqslant 2$, while the intermediate steps work in complete analogy to our above treatment for $d = 1$. The conclusion is, once again, the absence of absolutely continuous diffraction. $\Diamond$

Obviously, one can extend this class of examples by colouring blocks. This will lead to higher-dimensional Fourier matrices again, with an uncoloured tiling of the above type as a factor system. We leave further details to the interested reader, and turn to a perhaps more interesting non-FLC example.

7.9 The Frank–Robinson Tiling

Let us take a closer look at the tiling dynamical system defined by the stone inflation

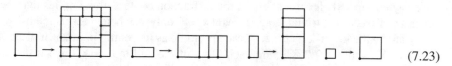

$$(7.23)$$

where the short edge has length 1 and the long one length

$$\lambda = \frac{1}{2}\big(1 + \sqrt{13}\big) \approx 2.303,$$

while the linear expansion is $Q = \lambda \mathbb{1}_2$. This inflation defines the Frank–Robinson tiling [30], see also [27], a patch of which is shown in Fig. 7.2. Note that the algebraic integer λ is neither a PV number nor a unit. By standard PF theory, the relative prototile frequencies in any Frank–Robinson tiling are given by

$$(\nu_1, \ldots, \nu_4) = \frac{1}{9}(4 - \lambda, 4\lambda - 7, 4\lambda - 7, 19 - 7\lambda);\qquad (7.24)$$

see [7, Ex. 5.8] for details. With the chosen edge lengths, the density of the control point set Λ induced by Eq. (7.23) is $\mathrm{dens}(\Lambda) = (3 + \lambda)/13 \approx 0.408$.

Fig. 7.2 A patch of the Frank–Robinson tiling defined by the stone inflation rule (7.23), obtained by three inflation steps from a single large square

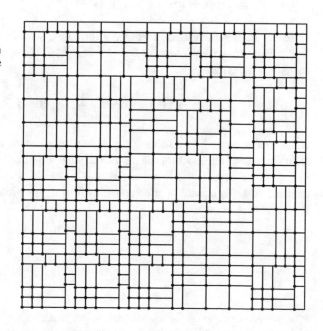

If we define the hull as the orbit closure of a fixed point tiling (under the square of the rule (7.23)) in the local rubber topology, we get a compact tiling space of infinite local complexity; see [30] and [7, Ex. 5.8] for more. As in the FLC case, it is also true here [41, Cor. 5.7 and Ex. 6.3] that the tiling dynamical system does not have any non-trivial eigenfunction. In the diffraction context, this implies that the pp part consists of the trivial Bragg peak at $k = 0$ only; see below for its intensity.

By taking the lower-left corner of each prototile as its control point, as indicated in Eq. (7.23) and Fig. 7.2, we turn each tiling of the hull into a Delone set that is MLD with the tiling. Now, the absence of non-trivial eigenfunctions translates to the diffraction of this Delone set by asserting that the trivial Bragg peak at $k = 0$ is the only contribution to the pure point part of the diffraction measure.

The Fourier matrix B is given by

$$B(x, y) = \begin{pmatrix} x^2 y^2 & 1 & 1 & 1 \\ p(x, y) & 0 & r(y) & 0 \\ p(y, x) & r(x) & 0 & 0 \\ q(x, y) & 0 & 0 & 0 \end{pmatrix}, \quad \text{with } (x, y) = \left(e^{2\pi i k_1}, e^{2\pi i k_2} \right) \quad (7.25)$$

and the (trigonometric) polynomials

$$\begin{aligned} r(x) &= x^\lambda + x^{\lambda+1} + x^{\lambda+2}, \\ p(x, y) &= x^2 + x^2 y + y^{\lambda+2}, \\ q(x, y) &= 1 + x + y + xy + x^\lambda y^\lambda \left(x^2 + y^2 + xy^2 + x^2 y + x^2 y^2 \right). \end{aligned} \quad (7.26)$$

Note that $B(0)$ is the inflation matrix of the tiling, with PF eigenvalue λ^2.

Now, the cocycle is given by $B^{(n)}(k) = B(k)B(\lambda k) \cdots B(\lambda^{n-1} k)$, with maximal Lyapunov exponent

$$\chi^B(k) = \limsup_{n \to \infty} \frac{1}{n} \log \| B^{(n)}(k) \|,$$

where the choice of the (sub-multiplicative) matrix norm is arbitrary. Absence of absolutely continuous components of the diffraction will be implied if we show that $\chi^B(k) \leqslant \log(\lambda) - \varepsilon$ for some $\varepsilon > 0$ and a.e. $k \in \mathbb{R}^2$. Since it is convenient to work with the square of the Frobenius norm,[6] we prefer to compare $2\chi^B$ with $\log(\lambda^2) \approx 1.668$ instead.

---

[6]The spectral norm gives better bounds, but is harder to calculate. Also, computing means is easier with simple trigonometric polynomials, via harvesting their quasiperiodicity.

By standard subadditive arguments along the lines used for our previous examples, one finds that, for any $N \in \mathbb{N}$ and then a.e. $k \in \mathbb{R}^2$,

$$2\chi^B(k) \leqslant \frac{1}{N} \mathbb{M}\big(\log \|B^{(N)}(.)\|_F^2\big) =: \mathfrak{m}_N, \tag{7.27}$$

where $\mathbb{M}$ again denotes the mean. Since $\log \|B^{(N)}(.)\|_F^2$ is a quasiperiodic function in two variables, with fundamental frequencies 1 and λ, the mean can be expressed as an integral over the 4-torus, $\mathbb{T}^4$. To this end, one introduces new variables u_1, u_2 and v_1, v_2 such that

$$B(k) = \tilde{B}(u_1, u_2, v_1, v_2)\big|_{u_1 = \lambda k_1,\, u_2 = k_1,\, v_1 = \lambda k_2,\, v_2 = k_2}$$

where $\tilde{B}$ is 1-periodic in each variable. Here, $\tilde{B}$ is defined in complete analogy to (7.25), with the corresponding modifications on r, p and q from Eq. (7.26); compare [4] for a related one-dimensional case analysed previously. One now finds

$$\begin{aligned}
\mathfrak{m}_N &= \frac{1}{N} \mathbb{M}\big(\log \|\tilde{B}^{(N)}(.)\|_F^2\big) \\
&= \frac{1}{N} \int_{\mathbb{T}^4} \log \|\tilde{B}^{(N)}(u_1, u_2, v_2, v_2)\|_F^2 \, du_1 \, du_2 \, dv_1 \, dv_2,
\end{aligned} \tag{7.28}$$

which can be calculated numerically with good precision. Figure 7.3 illustrates the result.

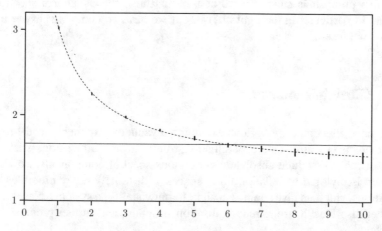

Fig. 7.3 Numerical values of the upper bounds $\mathfrak{m}_N$ of $2\chi^B$ from Eq. (7.28), for $1 \leqslant n \leqslant 10$. The horizontal line is at height $2\log(\lambda) \approx 1.668$. The estimated numerical errors are indicated by vertical bars (the dotted line is for eye guidance only)

Let us summarise this section as follows.

Theorem 7.9.1 *Let* Λ *be the set of control points of any element of the Frank–Robinson tiling hull. Then, the diffraction measure of the corresponding Dirac comb,* δ_Λ, *is of the form* $\widehat{\gamma} = \mathrm{dens}(\Lambda)^2 \, \delta_0 + \widehat{\gamma}_{\mathsf{sc}}$, *where the singular continuous part can be expressed in terms of a generalised Riesz product.*

More generally, if one assigns general complex weights $u_1, \ldots, u_4$ *to the four types of control points, not all zero, the corresponding diffraction measure is still singular, where the only point measure is the central peak at* 0 *with intensity*

$$I_0 = \mathrm{dens}(\Lambda)^2 \left| v_1 u_1 + \ldots + v_4 u_4 \right|^2,$$

where $\mathrm{dens}(\Lambda) = (3 + \lambda)/13$ *and where the* v_i *are the prototile frequencies from Eq. (7.24).* □

The natural next step consists in defining the (integrated) *distribution function* for $\widehat{\gamma}_{\mathsf{sc}}$ in the positive quadrant, as

$$F(k_1, k_2) = \widehat{\gamma}_{\mathsf{sc}}\big([0, k_1] \times [0, k_2]\big),$$

with the matching extension to the other quadrants. This leads to a continuous function (which requires an extra argument along the directions of e_1 and e_2; compare [8, Sec. 5] for a similar analysis) which behaves as $F(k_1, k_2) \sim \gamma(\{0\}) \, k_1 k_2$ for large values of k_1 and k_2. As such, it does not reveal the interesting structure of the sc measure. A better understanding of the latter requires a multi-fractal analysis, which is outside the scope of this survey.

At this point, it is suggestive to assume that also the dynamical spectrum is singular, in particular in the light of [14], but we have no complete answer to this question at present.

7.10   Closing Remarks

As we have illustrated by various examples, Lyapunov exponents lead to useful insight on the spectral nature of inflation tilings in any dimension. They are a powerful tool to exclude absolutely continuous spectral components. So far, our approach is taylored to inflation tiling spaces with finitely many prototiles up to translations, and thus gives no new insight to pinwheel-type systems. Nevertheless, the latter also have a strong renormalisation structure, and further progress seems possible.

To explore the absence of ac diffraction and spectral measures in more generality, one would need a more analytic (rather than numerical) approach to the estimates for upper bounds, or, ideally, exact expressions of Fürstenberg type for the exponents. Also, it would help to establish the almost sure existence of Lyapunov exponents as limits, which does not seem to be an easy task outside the constant-length or the Pisot case.

In this exposition, we have mainly considered the absolutely continuous part of the spectrum. It is not difficult to analyse the pure point part as well, where some results are discussed in [5, 6]. Considerably more involved seems the singular continuous part, which originates from the different scalings one encounters. Various results on the spectral measures in one dimension are derived in [18, 19] via matrix Riesz products, which are not restricted to the self-similar case. It would be interesting to establish a connection with the topological constraints on size and shape changes [21, 22], which should at least be possible in the irreducible Pisot case.

Acknowledgments It is our pleasure to thank Alan Bartlett, Michael Coons, Natalie Frank, Franz Gähler, Alan Haynes, Neil Mañibo, Robbie Robinson, Dan Rust, Lorenzo Sadun and Boris Solomyak for discussions and helpful comments on the manuscript. Various useful hints from an anonymous reviewer are gratefully acknowledged. This work was supported by the German Research Foundation (DFG), within the CRC 1283 at Bielefeld University, and by EPSRC through grant EP/S010335/1.

References

1. S. Akiyama, M. Barge, V. Berthé, J.-Y. Lee, A. Siegel, On the Pisot substitution conjecture, in *Mathematics of Aperiodic Order*, ed. by J. Kellendonk, D. Lenz, J. Savinien (Birkhäuser, Basel, 2015), pp. 33–72
2. S. Aubry, C. Godrèche, J.M. Luck, Scaling properties of a structure intermediate between quasiperiodic and random. J. Stat. Phys. **51**, 1033–1074 (1988)
3. M. Baake, M. Coons, N. Mañibo, Binary constant-length substitutions and Mahler measures of Borwein polynomials, in *From Analysis to Visualization: JBCC 2017*, ed. by D. Bailey, N.S. Borwein, R.P. Brent, R.S. Burachik, J.-A.H. Osborn, B. Sims, Q.J. Zhu (Springer, Cham, 2020), pp. 303–322; arXiv:1711.02492
4. M. Baake, N.P. Frank, U. Grimm, E.A. Robinson, Geometric properties of a binary non-Pisot inflation and absence of absolutely continuous diffraction. Stud. Math. **247**, 109–154 (2019); arXiv:1706.03976
5. M. Baake, F. Gähler, Pair correlations of aperiodic inflation rules via renormalisation: some interesting examples. Topol. Appl. **205**, 4–27 (2016); arXiv: 1511.00885
6. M. Baake, F. Gähler, N. Mañibo, Renormalisation of pair correlation measures for primitive inflation rules and absence of absolutely continuous diffraction. Commun. Math. Phys. **370**, 591–635 (2019); arXiv:1805.09650
7. M. Baake, U. Grimm, *Aperiodic Order. Vol. 1: A Mathematical Invitation* (Cambridge University Press, Cambridge, 2013)
8. M. Baake, U. Grimm, Squirals and beyond: Substitution tilings with singular continuous spectrum. Ergodic Theory Dyn. Syst. **34**, 1077–1102 (2014); arXiv:1205.1384
9. M. Baake, U. Grimm (eds.), *Aperiodic Order. Vol. 2: Crystallography and Almost Periodicity* (Cambridge University Press, Cambridge, 2017)
10. M. Baake, U. Grimm, N. Mañibo, Spectral analysis of a family of binary inflations rules. Lett. Math. Phys. **108**, 1783–1805 (2018); arXiv:1709.09083
11. M. Baake, A. Haynes, D. Lenz, Averaging almost periodic functions along exponential sequences, in *Aperiodic Order. Vol. 2: Crystallography and Almost Periodicity*, ed.

by M. Baake, U. Grimm (Cambridge University Press, Cambridge, 2017), pp. 343–362; arXiv:1704.08120

12. M. Baake, D. Lenz, Dynamical systems on translation bounded measures: pure point dynamical and diffraction spectra. Ergodic Theory Dyn. Syst. **24**, 1867–1893 (2004); arXiv:math.DS/0302061

13. M. Baake, D. Lenz, Spectral notions of aperiodic order. Discrete Contin. Dyn. Syst. S **10**, 161–190 (2017); arXiv:1601.06629

14. M. Baake, D. Lenz, A.C.D. van Enter, Dynamical versus diffraction spectrum for structures with finite local complexity. Ergodic Theory Dyn. Syst. **35**, 2017–2043 (2015); arXiv:1307.7518

15. A. Bartlett, Spectral theory of $\mathbb{Z}^d$ substitutions. Ergodic Theory Dyn. Syst. **38**, 1289–1341 (2018); arXiv:1410.8106

16. C. Berg, G. Forst, *Potential Theory on Locally Compact Abelian Groups* (Springer, Berlin, 1975)

17. A. Berlinkov, B. Solomyak, Singular substitutions of constant length. Ergodic Theory Dyn. Syst. **39**, 2384–2402 (2019); arXiv:1705.00899

18. A. Bufetov, B. Solomyak, On the modulus of continuity for spectral measures in substitution dynamics. Adv. Math. **260**, 84–129 (2014); arXiv:1305.7373

19. A. Bufetov, B. Solomyak, A spectral cocycle for substitution systems and translation flows. J. Anal. Math. (in press); arXiv:1802.04783

20. L. Chan, U. Grimm, Spectrum of a Rudin–Shapiro-like sequence. Adv. Appl. Math. **87**, 16–23 (2017); arXiv:1611.04446

21. A. Clark, L. Sadun, When size matters: Subshifts and their related tiling spaces. Ergodic Theory Dyn. Syst. **23**, 1043–1057 (2003); arXiv:math.DE/0201152

22. A. Clark, L. Sadun, When shape matters: deformation of tiling spaces. Ergodic Theory Dyn. Syst. **26**, 69–86 (2006); arXiv:math.DE/0306214

23. C. Corduneanu, *Almost Periodic Functions*, 2nd English edn. (Chelsea, New York, 1989)

24. M. Einsiedler, T. Ward, *Ergodic Theory — with a View Towards Number Theory* (Springer, London, 2011)

25. G. Everest, T. Ward, *Heights of Polynomials and Entropy in Algebraic Dynamics* (Springer, London, 1999)

26. N.P. Frank, Multi-dimensional constant-length substitution sequences. Topol. Appl. **152**, 44–69 (2005)

27. N.P. Frank, A primer of substitution tilings of the Euclidean plane. Expo. Math. **26**, 295–326 (2008); arXiv:0705.1142

28. N.P. Frank, Tilings with infinite local complexity, in *Mathematics of Aperiodic Order*, ed. by J. Kellendonk, D. Lenz, J. Savinien (Birkhäuser, Basel, 2015), pp. 223–257; arXiv: 1312.4987

29. N.P. Frank, Introduction to hierarchical tiling dynamical systems, in *Substitution and Tiling Dynamics: Introduction to Self-inducing Structures*, ed. by S. Akiyama, P. Arnoux. Lecture Notes in Mathematics, vol. 2273 (Springer, Cham, 2020); arXiv:1802.09956

30. N.P. Frank, E.A. Robinson, Generalized β-expansions, substitution tilings, and local finiteness. Trans. Am. Math. Soc. **360**, 1163–1177 (2008); arXiv: math.DS/0506098

31. N.P. Frank, L. Sadun, Topology of (some) tiling spaces without finite local complexity. Discrete Contin. Dyn. Syst. A **23**, 847–865 (2009); arXiv:math.DS/ 0701424

32. N.P. Frank, L. Sadun, Fusion: a general framework for hierarchical tilings of $\mathbb{R}^d$. Geom. Dedicata **171**, 149–186 (2014); arXiv:1101.4930

33. D. Frettlöh, More inflation tilings, in *Aperiodic Order. Vol.* 2: *Crystallography and Almost Periodicity*, ed. by M. Baake, U. Grimm (Cambridge University Press, Cambridge, 2017), pp. 1–37

34. D. Frettlöh, C. Richard, Dynamical properties of almost repetitive Delone sets. Discrete Contin. Dyn. Syst. A **34**, 531–556 (2014); arXiv:1210.2955

35. C. Godrèche, J.M. Luck, Multifractal analysis in reciprocal space and the nature of the Fourier transform of self-similar structures. J. Phys. A: Math. Gen. **23**, 3769–3797 (1990)

36. P.R. Halmos, *Measure Theory* (Springer, New York, 1974, reprint)

37. J. Hartinger, R.F. Kainhofer, R.F. Tichy, Quasi-Monte Carlo algorithms for unbounded, weighted integration problems. J. Complexity **20**, 654–668 (2004)

38. J. Kellendonk, D. Lenz, J. Savinien (eds.), *Mathematics of Aperiodic Order* (Birkhäuser, Basel, 2015)

39. L. Kuipers, H. Niederreiter, *Uniform Distribution of Sequences*, reprint (Dover, New York, 2006)

40. J.-Y. Lee, R.V. Moody, B. Solomyak, Pure point dynamical and diffraction spectra. Ann. Henri Poincaré **3**, 1003–1018 (2002); arXiv:0910.4809

41. J.-Y. Lee, B. Solomyak, On substitution tilings and Delone sets without finite local complexity. Discrete Contin. Dyn. Syst. A **39**, 3149–3177 (2019); arXiv:1804.10235

42. E.H. Lieb, M. Loss, *Analysis*, 2nd edn. (American Mathematical Society, Providence, RI, 2001)

43. J.M. Luck, A classification of critical phenomena on quasi-crystals and other aperiodic structures. Europhys. Lett. **24**, 359–364 (1993)

44. N. Mañibo, Lyapunov exponents for binary substitutions of constant length. J. Math. Phys. **58**, 113504 (9 pp) (2017); arXiv:1706.00451

45. N. Mañibo, Spectral analysis of primitive inflation rules. Oberwolfach Rep. **14**, 2830–2832 (2017)

46. R.V. Moody, N. Strungaru, Almost periodic measures and their Fourier transforms, in *Aperiodic Order. Vol. 2: Crystallography and Almost Periodicity*, ed. by M. Baake, U. Grimm (Cambridge University Press, Cambridge, 2017), pp. 173–270

47. Y. Peres, W. Schlag, B. Solomyak, Sixty years of Bernoulli convolutions, in *Fractal Geometry and Stochastics II*, ed. by C. Bandt, S. Graf, M. Zähle (Birkhäuser, Basel, 2000), pp. 39–65

48. M. Queffélec, *Substitution Dynamical Systems — Spectral Analysis*. Lecture Notes in Mathematics, vol. 1294, 2nd edn. (Springer, Berlin, 2010)

49. K. Schmidt, *Dynamical Systems of Algebraic Origin* (Birkhäuser, Basel, 1995)

50. B. Simon, *Analysis, Part I: Real Analysis* (American Mathematical Society, Providence, RI, 2014)

51. B. Sing, *Pisot Substitutions and Beyond*, PhD thesis, Bielefeld University (2007). Available electronically at urn:nbn:de:hbz:361-11555

52. I.M. Sobol, Calculation of improper integrals using uniformly distributed sequences. Soviet Math. Dokl. **14**, 734–738 (1973)

53. B. Solomyak, Nonperiodicity implies unique composition for self-similar translationally finite tilings. Discrete Comput. Geom. **20**, 265–279 (1989)

54. B. Solomyak, Dynamics of self-similar tilings. Ergodic Theory Dyn. Syst. **17**, 695–738 (1997) and **19**, 1685 (1999) (Erratum)

55. B. Solomyak, Delone sets and dynamical systems, in *Substitution and Tiling Dynamics: Introduction to Self-inducing Structures*, ed. by S. Akiyama, P. Arnoux. Lecture Notes in Mathematics, vol. 2273 (Springer, Cham, 2020)

56. N. Strungaru, Almost periodic pure point measures, in *Aperiodic Order. Vol. 2: Crystallography and Almost Periodicity*, ed. by M. Baake, U. Grimm (Cambridge University Press, Cambridge, 2017), pp. 271–342; arXiv: 1501.00945

57. M. Viana, *Lectures on Lyapunov Exponents* (Cambridge University Press, Cambridge, 2013)

58. G.H. Wannier, Antiferromagnetism. The triangular Ising net. Phys. Rev. **79**, 357–364 (1950) and Phys. Rev. B **7**, 5017 (1973) (Erratum)

Chapter 8
Yet Another Characterization of the Pisot Substitution Conjecture

Paul Mercat and Shigeki Akiyama

Abstract We give a sufficient geometric condition for a subshift to be measurably isomorphic to a domain exchange and to a translation on a torus. This gives another characterization of the Pisot substitution conjecture. For an irreducible unit Pisot substitution, we introduce a new topology on the discrete line and give a simple necessary and sufficient condition for the symbolic system to have pure discrete spectrum. This condition gives rise to an algorithm based on computation of automata. To see the power of this criterion, we provide families of substitutions that are shown, using different methods, to satisfy the Pisot substitution conjecture:

- $a \mapsto a^k bc, b \mapsto c, c \mapsto a$, for $k \in \mathbb{N}$
- $a \mapsto a^l ba^{k-l}, b \mapsto c, c \mapsto a$, for $k \in \mathbb{N}_{\geq 1}$, for $0 \leq l \leq k$.

We also provide an example of S-adic system with pure discrete spectrum.

8.1 Introduction

Sturmian systems are well-known examples of subshifts that are conjugate to translations on the torus $\mathbb{R}/\mathbb{Z}$. In 1982, Gérard Rauzy (see [18]) gave a generalization to higher dimension for the subshift generated by the infinite fixed point of the Tribonacci substitution:

$$\begin{cases} a \mapsto ab \\ b \mapsto ac \\ c \mapsto a \end{cases}.$$

P. Mercat (✉)
Aix-Marseille Université, Marseille, France
e-mail: paul.mercat@univ-amu.fr

S. Akiyama
Institute of Mathematics, University of Tsukuba, Tsukuba, Japan
e-mail: akiyama@math.tsukuba.ac.jp

© The Editor(s) (if applicable) and The Author(s), under exclusive license
to Springer Nature Switzerland AG 2020
S. Akiyama, P. Arnoux (eds.), *Substitution and Tiling Dynamics: Introduction to Self-inducing Structures*, Lecture Notes in Mathematics 2273,
https://doi.org/10.1007/978-3-030-57666-0_8

He constructed a compact subset of $\mathbb{R}^2$ (which we call now the Rauzy fractal) that tiles the plane by translation and on which we can define a domain exchange which is measurably conjugate both to the symbolic subshift, and to a translation on the two dimensional torus $\mathbb{R}^2/\mathbb{Z}^2$.

In 2001, Arnoux and Ito (see [2]) generalized the work of Rauzy to any irreducible unit Pisot substitution. They introduced a combinatorial condition which is easy to check, called the strong coincidence, that permits to get a measurable conjugacy between the subshift and a domain exchange, which is also a finite extension of a translation on a torus.

All the known examples satisfied the condition, and this led to one of the main open questions in the field, the Pisot substitution conjecture. This conjecture has also been cited in Chaps. 1, 2, 3, and 7, and we cite below the statement given in Chap. 2 as Conjecture 2.6.1, which is the simplest and most restricted form of the conjecture.

Conjecture 8.1.1 (Pisot Substitution Conjecture: Symbolic Substitutive Case) If σ is an irreducible Pisot substitution then the associated substitutive system $(\Omega_\sigma, \mathbb{Z})$ has pure discrete spectrum.

This conjecture has been proved in the case of two letters (see [3]), but not for larger alphabets. It admits several generalizations, to tilings and to higher dimensions, and it can be formulated in several equivalent ways, symbolic or geometric, see [1]. One way to attack this conjecture is to understand well the construction of [2]. Then we soon realize that the strong coincidence (which is still open for all irreducible Pist substitutions) seems a little too weak to achieve this goal. To obtain a measurable conjugacy between the subshift of an irreducible unit Pisot substitution and a translation on a torus, several equivalent conditions (super coincidence, Geometric coincidence) have been studied, see for example [4, 10]. This article gives another formulation of these various coincidences and a short proof of their equivalence. The new criterion is checked by automata computation.

We introduce a topology on $\mathbb{Z}^d$ that permits to characterize easily when the subshift of a given irreducible unit Pisot substitution over d letters is measurably isomorphic to a translation on a $(d-1)$-dimensional torus: see Theorem 8.3.3. We show that this condition is equivalent to the non-emptiness of some computable regular language: see Theorem 8.5.11.

In the last section, we use this condition to prove pure discreteness for the family of substitution

$$s_k : \begin{cases} a \mapsto a^k bc \\ b \mapsto c \\ c \mapsto a \end{cases}$$

for $k \in \mathbb{N}$, where a^k means that the letter a is repeated k times. And we also prove pure discreteness for the family of substitution

$$s_{l,k} : \begin{cases} a \mapsto a^l b a^{k-l} \\ b \mapsto c \\ c \mapsto a \end{cases}$$

for $k \in \mathbb{N}_{\geq 1}$, $0 \leq l \leq k$, by computing explicitly a automaton describing algebraic relations, and showing that pure discreteness for the substitution

$$s_k : \begin{cases} a \mapsto a^k b \\ b \mapsto c \\ c \mapsto a \end{cases}$$

implies pure discreteness for the other substitutions.

We also use the criterion to prove pure discreteness for the S-adic system with the two substitutions

$$s_1 : \begin{cases} a \mapsto aab \\ b \mapsto \quad c \\ c \mapsto \quad a \end{cases} \quad \text{and} \quad s_2 : \begin{cases} a \mapsto aba \\ b \mapsto \quad c \\ c \mapsto \quad a \end{cases} ,$$

for all words in $\{s_1, s_2\}^{\mathbb{N}}$.

8.2   A Criterion for a Subshift to Have Purely Discrete Spectrum

In this section, we describe a general geometric criterion for a subshift to be measurably isomorphic to a translation on a torus. Let us start by introducing some notations.

8.2.1   Subshift

We denote by $A^{\mathbb{N}}$ (respectively $A^{\mathbb{Z}}$) the set of infinite (respectively bi-infinite) words over an alphabet A. We denote by $|u|$ the length of a word u, and $|u|_a$ denotes the number of occurrences of the letter a in a word $u \in A^*$. And we denote by

$$\mathrm{Ab}(u) = (|u|_a)_{a \in A} \in \mathbb{N}^A$$

the *abelianisation vector* of a word $u \in A^*$. The canonical basis of $\mathbb{R}^A$ will be denoted by $(e_a)_{a \in A} = (\mathrm{Ab}(a))_{a \in A}$.

The *shift* on infinite words is the application

$$S : \begin{array}{c} A^{\mathbb{N}} \longrightarrow A^{\mathbb{N}} \\ (u_i)_{i \in \mathbb{N}} \longmapsto (u_{i+1})_{i \in \mathbb{N}} \end{array}$$

We can also define the shift on bi-infinite words in an obvious way, and it becomes invertible.

We denote $S^{\mathbb{N}u} = \{S^n u \mid n \in \mathbb{N}\}$ and $S^{\mathbb{Z}u} = \{S^n u \mid n \in \mathbb{Z}\}$. We use the usual metric on $A^{\mathbb{N}}$:

$$d(u, v) = 2^{-n} \text{ where } n \text{ is the length of the maximal common prefix.}$$

The map S is continuous for this metric. Given an infinite word u, the compact set $\overline{S^{\mathbb{N}}u}$ is S-invariant. We call *subshift* generated by u, the dynamical system $(\overline{S^{\mathbb{N}}u}, S)$. The same can be done for bi-infinite words.

8.2.2 Discrete Line Associated to a Word

Let $u \in A^{\mathbb{N}}$ be an infinite word over the alphabet A. Then, the associated *discrete line* is the following subset of $\mathbb{Z}^A$:

$$D_u := \left\{ Ab(v) \in \mathbb{Z}^A \mid v \text{ finite prefix of } u \right\}.$$

If $u \in A^{\mathbb{Z}}$ is a bi-infinite word, then the corresponding discrete line is

$$D_u := -D_v \cup D_w,$$

where $v, w \in A^{\mathbb{N}}$ are infinite words such that $u = {}^t v w$, where ${}^t v = \ldots v_n \ldots v_2 v_0$ denotes the mirror of the word $v = v_0 v_2 \ldots v_n \ldots$.

For $u \in A^{\mathbb{N}}$, we can partition this discrete line into $d = |A|$ pieces. For every $a \in A$, let

$$D_{u,a} := \left\{ Ab(v) \in \mathbb{Z}^A \mid va \text{ finite prefix of } u \right\}.$$

The sets $D_{u,a} + e_a$, $a \in A$, also gives almost a partition of D_u:

$$D_u = \{0\} \cup \bigcup_{a \in A} D_{u,a} + e_a.$$

For a bi-infinite word $u \in A^{\mathbb{Z}}$, we have the same, but we get a real partition, without the $\{0\}$. In both cases, these partitions permit to see the shift S on the word u as a domain exchange E:

$$E : \begin{array}{l} D_u \longrightarrow D_u \\ x \longmapsto x + e_a \text{ for } a \in A \text{ such that } x \in D_{u,a}. \end{array}$$

There is also a property of tiling for this discrete line: we have the following

Proposition 8.2.1 *Let Γ_0 be the subgroup of $\mathbb{Z}^A$ generated by $(e_a - e_b)_{a,b \in A}$, and let u be any bi-infinite aperiodic word over the alphabet A. Then D_u is a fundamental domain for the action of Γ_0 on $\mathbb{Z}^A$. Moreover the translation T by e_a (for any $a \in A$) on $\mathbb{Z}^A / \Gamma_0$ is conjugate to the domain exchange E on D_u by the natural quotient map $\pi_0 : \mathbb{Z}^A \to \mathbb{Z}^A / \Gamma_0$, and the shift $(S^{\mathbb{Z}} u, S)$ is conjugate to the domain exchange (D_u, E) by the map*

$$c : \begin{array}{l} S^{\mathbb{Z}} u \to D_u \\ S^n u \mapsto E^n 0 \end{array}.$$

Remark 8.2.2 We have the same for infinite non-eventually periodic words, but we get a fundamental domain for the action on the half-space

$$\left\{ (x_a)_{a \in A} \in \mathbb{Z}^A \,\middle|\, \sum_{a \in A} x_a \geq 0 \right\},$$

and a conjugacy with the shift on $S^{\mathbb{N}} u$.

Proof The vectors $(e_a)_{a \in A}$ are equivalent modulo the group Γ_0. Hence, this discrete line is equivalent to $\mathbb{Z} e_a$ for any letter $a \in A$, and this is an obvious fundamental domain of $\mathbb{Z}^A$ for the action of Γ_0. The map c is well-defined and one-to-one because the word u is aperiodic. And it gives a conjugacy between the shift $(S^{\mathbb{Z}} u, S)$ and the domain exchange (D_u, E): $c \circ S = E \circ c$. The natural quotient map $\pi_0 : \mathbb{Z}^A \to \mathbb{Z}^A / \Gamma_0$ restricted to D_u is bijective, and it gives a conjugacy between the domain exchange (D_u, E) and the translation $(\mathbb{Z}^A / \Gamma_0, T)$: $\pi_0 \circ E = T \circ \pi_0$.

If the discrete line D_u stays near a given line of $\mathbb{R}^A$ (this will be the case for example for periodic points of Pisot substitutions), then we can project onto a hyperplane $\mathcal{P}$ of $\mathbb{R}^A$ (for example the hyperplane of equation $\sum_{a \in A} x_a = 0$) along this line. The projection of $\mathbb{Z}^A$ is dense in the hyperplane for almost all lines, and the group Γ_0 becomes a lattice in the hyperplane. If the projection of the discrete line is not so bad, we can expect that the closure gives a tiling of the hyperplane, and that the closure of each piece of the partition of the discrete line doesn't intersect each other. And we can expect that the conjugacy given by the previous proposition becomes a conjugacy of the closures. Figure 8.1 shows the conjugacy given by the Proposition 8.2.1, and what we get if everything goes well.

Let us now give a general geometric criterion that permits to know that everything works well as in Fig. 8.1.

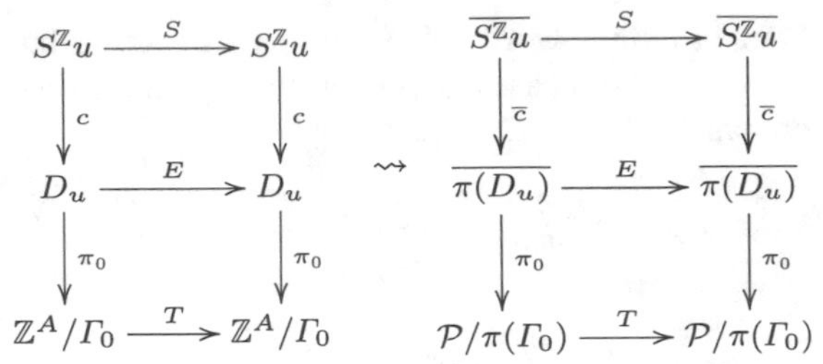

Fig. 8.1 Commutative
diagrams of the conjugacy
between the shift S, the
domain exchange E and the
translation T on the quotient
torus, before and after taking
the closure

$$
\begin{array}{ccc}
S^{\mathbb{Z}}u & \xrightarrow{\ S\ } & S^{\mathbb{Z}}u \\
\downarrow{c} & & \downarrow{c} \\
D_u & \xrightarrow{\ E\ } & D_u \\
\downarrow{\pi_0} & & \downarrow{\pi_0} \\
\mathbb{Z}^A/\Gamma_0 & \xrightarrow{\ T\ } & \mathbb{Z}^A/\Gamma_0
\end{array}
\quad\rightsquigarrow\quad
\begin{array}{ccc}
\overline{S^{\mathbb{Z}}u} & \xrightarrow{\ S\ } & \overline{S^{\mathbb{Z}}u} \\
\downarrow{\overline{c}} & & \downarrow{\overline{c}} \\
\overline{\pi(D_u)} & \xrightarrow{\ E\ } & \overline{\pi(D_u)} \\
\downarrow{\pi_0} & & \downarrow{\pi_0} \\
\mathcal{P}/\pi(\Gamma_0) & \xrightarrow{\ T\ } & \mathcal{P}/\pi(\Gamma_0)
\end{array}
$$

8.2.3 Geometrical Criterion for the Pure Discreteness of the Spectrum

Here is the main general geometric criterion for a subshift to have a pure discrete
spectrum. We use the notations defined in Sect. 8.2.2.

Theorem 8.2.3 *Let $u \in A^{\mathbb{N}}$ be an infinite word over an alphabet A, and let π
be a linear projection from $\mathbb{R}^A$ onto a hyperplane $\mathcal{P}$. We assume that we have the
following:*

- *the restriction of π to $\mathbb{Z}^A$ is injective and has a dense image,*
- *the set $\pi(D_u)$ is bounded,*
- *the subshift $(\overline{S^{\mathbb{N}}u}, S)$ is minimal,*
- *the boundaries of $\overline{\pi(D_{u,a})}$, $a \in A$, have zero Lebesgue measure,*
- *the union $\pi(D_u) = \bigcup_{a \in A} \pi(D_{u,a})$ is disjoint in measure.*

   *Then there exists a σ-algebra and a S-invariant measure μ such that the subshift
$(\overline{S^{\mathbb{N}}u}, S, \mu)$ is a finite extension of the translation of the torus $(\mathcal{P}/\pi(\Gamma_0), T, \lambda)$,
where T is the translation by $\pi(e_a)$ (for any $a \in A$) on the torus $\mathcal{P}/\pi(\Gamma_0)$, Γ_0 is
the group generated by $\{e_a - e_b \,|\, a, b \in A\}$, and λ is the Lebesgue measure. And it
is also a topological semi-conjugacy.*

   If moreover the union

$$
\bigcup_{t \in \pi(\Gamma_0)} \overline{\pi(D_u)} + t = \mathcal{P}
$$

*is disjoint in Lebesgue measure, then the subshift $(\overline{S^{\mathbb{N}}u}, S, \mu)$ is uniquely ergodic
and is isomorphic to the translation on the torus $(\mathcal{P}/\pi(\Gamma_0), T, \lambda)$ and to a domain
exchange on $\overline{\pi(D_u)}$.*

Remark 8.2.4 The disjointness in measure of the union

$$
\bigcup_{t \in \pi(\Gamma_0)} \overline{\pi(D_u)} + t = \mathcal{P},
$$

implies the disjointness in measure of the union

$$\bigcup_{a \in A} \overline{\pi(D_{u,a})} = \overline{\pi(D_u)},$$

and it also implies that the boundaries of $\overline{\pi(D_{u,a})}$, $a \in A$, have zero Lebesgue measure.

Indeed, if we have $\lambda(\overline{\pi(D_{u,a})} \cap \overline{\pi(D_{u,b})}) > 0$, then we have

$$\lambda\left((\overline{\pi(D_u)} + \pi(e_a - e_b)) \cap \overline{\pi(D_u)}\right) \geq \lambda\left((\overline{\pi(D_{u,a})} \cap \overline{\pi(D_{u,b})}) + \pi(e_a)\right) > 0,$$

so we have $a = b$.

And we obtain that the boundary of each $\overline{\pi(D_{u,a})}$, $a \in A$, has zero Lebesgue measure, since

$$\partial\overline{\pi(D_{u,a})} \subseteq \overline{\pi(D_{u,a})} \cap \left(\bigcup_{b \in A \setminus \{a\}} \overline{\pi(D_{u,b})} \cup \bigcup_{t \in \pi(\Gamma_0) \setminus \{0\}} \overline{\pi(D_u)} + t\right).$$

In order to prove this theorem, we start by showing that we can extend by continuity the map $\pi \circ c : S^{\mathbb{N}}u \to \pi(D_u)$ that gives the conjugacy between the shift $(S^{\mathbb{N}}u, S)$ and the domain exchange $(\pi(D_u), E)$.

Lemma 8.2.5 *Let $u \in A^{\mathbb{N}}$ be a non-eventually periodic infinite word over an alphabet A, and let π be a projection from $\mathbb{R}^A$ onto a hyperplane $\mathcal{P}$. We assume that $\pi(D_u)$ is bounded. Then the map*

$$\pi \circ c : \begin{array}{rcl} S^{\mathbb{N}}u & \to & \pi(D_u) \\ S^n u & \mapsto & E^n 0 \end{array}$$

can be extended by continuity at any point of the closure whose orbit is dense in $\overline{S^{\mathbb{N}}u}$.

To prove this lemma, we need the following geometric lemma, saying that we can always translate a bounded set of $\mathbb{R}^d$ in order to have a non empty but arbitrarily small intersection with the initial set.

Lemma 8.2.6 *Let Ω be a bounded subset of $\mathbb{R}^d$. Then, we have*

$$\inf_{t \in \Omega - \Omega} \operatorname{diam}(\Omega \cap (\Omega - t)) = 0.$$

The proof is left as an exercise. It can be proven for example by considering a diameter and using the parallelogram law.

Proof of Lemma 8.2.5 Let $w \in \overline{S^{\mathbb{N}}u}$ having dense orbit in $\overline{S^{\mathbb{N}}u}$ and let $\epsilon > 0$. By Lemma 8.2.6, there exists $t \in D_u - D_u$ such that $\operatorname{diam}(\pi(D_u) \cap (\pi(D_u) - \pi(t))) \leq \epsilon$.

Let n_1 and $n_2 \in \mathbb{N}$ such that $c(S^{n_2}u) - c(S^{n_1}u) = t$. We can assume that $n_1 \leq n_2$ up to replace t by $-t$. Then, there exists $n_0 \in \mathbb{N}$ such that $d(S^{n_0}w, u) \leq 2^{-n_2}$. Now, for all $v \in S^{\mathbb{N}}u$ such that $d(w, v) \leq 2^{-(n_0+n_2)}$, we have that $c(S^{n_0+n_1}v) \in D_u \cap (D_u - t)$, because $c(S^{n_0+n_2}v) - c(S^{n_0+n_1}v) = t$. Hence, if we let $\eta = 2^{-(n_0+n_2)}$, we have for all $v, v' \in D_u$,

$$\left.\begin{array}{c} d(v, w) \leq \eta \\ \text{and} \\ d(v', w) \leq \eta \end{array}\right\} \Rightarrow d(\pi \circ c(v), \pi \circ c(v')) = d(\pi \circ c(S^{n_0+n_1}v), \pi \circ c(S^{n_0+n_1}v')) \leq \epsilon.$$

This proves that we can extend $\pi \circ c$ by continuity at point w.

Lemma 8.2.7 *Let $u \in A^{\mathbb{N}}$ be an infinite non-eventually periodic word over an alphabet A, and let π be a projection from $\mathbb{R}^A$ onto a hyperplane $\mathcal{P}$. We assume that we have the following conditions:*

- *the restriction of the projection π to $\mathbb{Z}^A$ is injective and has a dense image,*
- *the set $\pi(D_u)$ is bounded,*
- *for every $a \in A$, the boundary of $\overline{\pi(D_{u,a})}$ has zero Lebesgue measure,*
- *the union $\bigcup_{a \in A} \overline{\pi(D_{u,a})} = \overline{\pi(D_u)}$, is disjoint in Lebesgue measure.*

Then the natural coding cod of $(\pi(D_u), E)$ for the partition $D_u = \bigcup_{a \in A} D_{u,a}$, can be extended by continuity to a full measure part M of the closure. And we have

$$\forall x \in M, \quad \lim_{\substack{y \to x \\ y \in \pi(D_u)}} (\pi \circ c)^{-1}(y) = \text{cod}(x).$$

Proof Let $\Omega = \overline{\pi(D_u)}$ and $\forall a \in A$, $\Omega_a = \overline{\pi(D_{u,a})}$. We can extend the domain exchange E in an obvious way:

$$E' : \begin{array}{ccc} \bigcup_{a \in A} \mathring{\Omega}_a & \longrightarrow & \Omega \\ x & \longmapsto & x + \pi(e_a) \text{ for } a \in A \text{ such that } x \in \mathring{\Omega}_a. \end{array}$$

The part of full Lebesgue measure that we consider is the E'-invariant set

$$M := \bigcap_{n \in \mathbb{N}} E'^{-n}\Omega.$$

Let $\epsilon > 0$ and let $x \in M$. Let $n_0 \in \mathbb{N}_{\geq 1}$ such that $2^{-n_0} \leq \epsilon$. The set

$$M_{n_0} := \bigcap_{n=0}^{n_0} E'^{-n}\Omega$$

is an open set containing x, because E' is continuous and $E'^{-1}\Omega = \bigcup_{a \in A} \mathring{\Omega}_a$ is open. Hence there exists $\eta > 0$ such that $B(x, \eta) \subseteq M_{n_0}$. And for every $y \in B(x, \eta) \cap M$, the natural coding of (M, E') for the partition $M = \bigcup_{a \in A} M \cap \Omega_a + \pi(e_a)$ coincides with the coding of x for the n_0 first steps. Hence, cod is continuous on M. We get also the last part of the lemma by observing that if $y \in B(x, \eta) \cap \pi(D_u)$, then the coding of y (which is equal to $(\pi \circ c)^{-1}(y)$) also coincide with the coding of x for the n_0 first steps.

Now we can prove the main theorem of this section. We start by extending the map $\pi \circ c$ by continuity, and we show that this map is almost everywhere one-to-one. It gives us an isomorphism between the subshift $(\overline{S^{\mathbb{N}}u}, S, \mu)$, for some measure μ, and a domain exchange defined Lebesgue-almost everywhere on $\overline{\pi(D_u)}$. Then, we show that the map $\pi_0 : \overline{\pi(D_u)} \to \mathcal{P}/\pi(\Gamma_0)$ is finite-to-one, and it gives us that the subshift is a finite extension of the translation on the torus $(\mathcal{P}/\pi(\Gamma_0), T, \lambda)$. Then if we assume that we have also the last hypothesis that $\overline{\pi(D_u)}$ tiles the hyperplane $\mathcal{P}$, then we deduce that we have the isomorphism with the translation on the torus, and we show that the unique ergodicity of the translation on the torus implies the unique ergodicity of the subshift.

Proof of the Theorem 8.2.3 The hypothesis on the projection π show that u cannot be eventually periodic. Indeed, if u was eventually periodic with a period $v \in A^*$, then the hypothesis that $\pi(D_u)$ is bounded implies that $\pi(Ab(v)) = 0$, but this contradict the hypothesis that the restriction of π to $\mathbb{Z}^A$ is injective.

The Lemma 8.2.5 shows that we can extend the map $\pi \circ c$ by continuity to a map $\bar{c} : \overline{S^{\mathbb{N}}u} \to \overline{\pi(D_u)}$. If we compose $\bar{c}$ with the natural projection π_0 onto the torus $\mathcal{P}/\pi(\Gamma_0)$, we get a continuous function which is onto, because of the equality $\pi(\Gamma_0) + \pi(D_u) = \mathcal{P}$ that comes from $\Gamma_0 + D_u = \mathbb{Z}^A$. And we have the equality

$$\pi_0 \circ \bar{c} \circ S = T \circ \pi_0 \circ \bar{c},$$

where T is the translation by $\pi(e_a)$ (for any $a \in A$) on the torus $\mathcal{P}/\pi(\Gamma_0)$. Indeed, this equality is true on the dense subset $S^{\mathbb{N}}u$ by the Proposition 8.2.1, and the maps π_0, S and T are continuous. This proves that the translation on the torus $(\mathcal{P}/\pi(\Gamma_0), T)$ is a topological factor of the subshift $(\overline{S^{\mathbb{N}}u}, S)$.

Let's consider the σ-algebra that we get from the Borel σ-algebra with the continuous map $\pi_0 \circ \bar{c} : \overline{S^{\mathbb{N}}u} \to \mathcal{P}/\pi(\Gamma_0)$. A measure μ on this σ-algebra can be defined by $\mu((\pi_0 \circ \bar{c})^{-1}(A)) = \lambda(A)$ for any Borel set A of $\mathcal{P}/\pi(\Gamma_0)$, where λ is the Lebesgue measure. By continuity, this measure μ that we get on $\overline{S^{\mathbb{N}}u}$ is S-invariant, and for this measure the translation of the torus $(\mathcal{P}/\pi(\Gamma_0), T, \lambda)$ is a factor of the subshift $(\overline{S^{\mathbb{N}}u}, S, \mu)$. Then, the Lemma 8.2.7 gives

$$\forall x \in \bar{c}^{-1}(M), \; x = \lim_{\substack{y \to x \\ y \in S^{\mathbb{N}}u}} (\pi \circ c)^{-1} \circ \pi \circ c(y) = \text{cod}(x) \circ \bar{c}.$$

So the map $\bar{c}$ is one-to-one on the subset of full μ-measure $\bar{c}^{-1}(M)$. Hence, the map $\bar{c} : \overline{S^{\mathbb{N}}u} \to \overline{\pi(D_u)}$ is a measurable conjugacy between the subshift $(\overline{S^{\mathbb{N}}u}, S, \mu)$ and the domain exchange $(\overline{\pi(D_u)}, E, \lambda)$.

To prove that the subshift is a finite extension of the translation on the torus, it remains to show that the number of preimages by π_0 is bounded and almost everywhere constant. The boundedness is a consequence of the hypothesis that $\pi(D_u)$ is bounded, and because $\pi(\Gamma_0)$ is a discrete subgroup of $\mathcal{P}$. But this number of preimages is invariant by the translation, so by ergodicity of the translation on the torus, it is almost everywhere constant. Hence we get that the domain exchange $(\overline{\pi(D_u)}, E, \lambda)$ (which is isomorphic to the subshift $(\overline{S^{\mathbb{N}}u}, S, \mu)$) is a finite extension of the translation on the torus $(\mathcal{P}/\pi(\Gamma_0), T, \lambda)$.

If we assume moreover that the sets $\pi(D_u)+t, t \in \pi(\Gamma_0)$, are disjoint in measure, then the map $\pi_0 : \overline{\pi(D_u)} \to \mathcal{P}/\pi(\Gamma_0)$ is invertible almost everywhere, is one-to-one on $\bar{c}^{-1}(M)$, and is a measurable conjugacy between the domain exchange $(\overline{\pi(D_u)}, E, \lambda)$ and the translation on the torus $(\mathcal{P}/\pi(\Gamma_0), T, \lambda)$. And the map $\pi_0 \circ \bar{c} : \overline{S^{\mathbb{N}}u} \to \mathcal{P}/\pi(\Gamma_0)$ is a measurable conjugacy between the subshift $(\overline{S^{\mathbb{N}}u}, S, \mu)$ and the translation on the torus $(\mathcal{P}/\pi(\Gamma_0), T, \lambda)$.

Then, the unique ergodicity of the translation on the torus implies that the subshift is also uniquely ergodic. Indeed, if μ' is an S-invariant measure of $\overline{S^{\mathbb{N}}u}$, then the pushforward $(\pi_0 \circ \bar{c})_*(\mu')$ is a T-invariant measure of the torus $\mathcal{P}/\pi(\Gamma_0)$, so it is proportional to the Lebesgue measure. And the μ'-measure of the complementary of the set $\bar{c}^{-1}(M)$ is 0, so the support of μ' is included in the support of μ. And the restriction of $\pi_0 \circ \bar{c}$ to $\bar{c}^{-1}(M)$ is injective and bi-continuous, thus we have $\mu' = \mu$ up to a scaling constant.

8.2.4 An Easy Example: Generalization of Sturmian Sequences

An easy example where all works fine is obtained by taking a random line of $\mathbb{R}^d$ with a positive direction vector. We consider the natural $\mathbb{Z}^d$-tiling by hypercubes, and we take the sequence of hyperfaces that intersect the line. Almost surely, this gives a discrete line corresponding to some word u over the alphabet of the d type of hyperfaces. It is not difficult to see that the orthogonal projection π along the line onto a hyperplane $\mathcal{P}$ behave correctly for almost every choice of line. It gives a set whose closure tile the plane, on which a domain exchange acts. This dynamics is conjugate to the subshift generated by the word u. It is also conjugate to the translation by $\pi(e_1)$ on the torus $\mathcal{P}/\pi(\Gamma_0) \simeq \mathbb{T}^{d-1}$. Figure 8.2 shows the domain exchange for a line whose a direction vector is around $(0.54973, 0.36490, 0.99501)$ in $\mathbb{R}^3$.

Remark 8.2.8 The word appearing in this last example is obtain by a simple algorithm: If the positive direction vector of the line is (v_1, v_2, v_3), and if the line

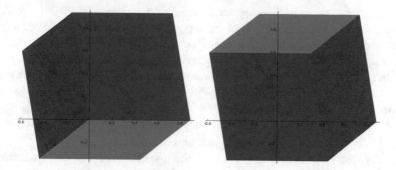

Fig. 8.2 Domain exchange conjugate to a translation on the torus $\mathbb{T}^2$, and also conjugate to the subshift generated by a word corresponding to a discrete approximation of a line of $\mathbb{R}^3$

goes through the point (c_1, c_2, c_3) then we have almost surely

$$\exists (k_1, k_2, k_3) \in \mathbb{Z}^3, \; \forall j \in \{1, 2, 3\}, \; k_j = \lfloor v_j \frac{k_{i_n} - c_{i_n}}{v_{i_n}} + c_j \rfloor$$

$$\Rightarrow \exists! i_{n+1} \in \{1, 2, 3\}, \; \forall j \in \{1, 2, 3\} \backslash \{i_{n+1}\}, \; k_j = \lfloor v_j \frac{k_{i_{n+1}} + 1 - c_{i_{n+1}}}{v_{i_{n+1}}} + c_j \rfloor.$$

The sequence $(i_n)_{n \in \mathbb{N}}$ define an infinite word over the alphabet $\{1, 2, 3\}$, and we get a bi-infinite word by invertibility of this algorithm.

8.3 Pure Discreteness for Irreducible Unit Pisot Substitution

In this section, we define a topology on $\mathbb{N}^A$ that permits to give a simple condition to get the pure discreteness of the spectrum of the subshift coming from an irreducible Pisot unit substitution, using the criterion of the previous section. And in the next section, we show that the reciprocal is true.

8.3.1 Substitutions

Let s be a substitution (i.e. a word morphism) over a finite alphabet A of cardinality d. We denote by A^* the set of finite words over the alphabet A. Let M_s (or simply M when there is no ambiguity) be the *incidence matrix* of s. It is the $d \times d$ matrix whose coefficients are

$$m_{a,b} = |s(b)|_a, \; \forall (a, b) \in A^2,$$

where $|u|_a$ denotes the number of occurrences of the letter a in a word $u \in A^*$. A *periodic point* of s is a fixed point of some power of s. It is an infinite word $u \in A^{\mathbb{N}}$ such that there exists $k \in \mathbb{N}_{\geq 1}$ such that $s^k(u) = u$.

A substitution s is *primitive* if there exists a $n \in \mathbb{N}$ such that for all a and $b \in A$, the letter b appears in $s^n(a)$. A substitution s is *irreducible* if its incidence matrix is irreducible—i.e. if the degree of the Perron eigenvalue of the matrix equals the number of letters of the substitution.

If u is a periodic point of a primitive substitution, we can check that the subshift $(\overline{S^{\mathbb{N}} u}, S)$ depends only on the substitution and is minimal.

We say that a substitution s is *Pisot* if the maximal eigenvalue of its incidence matrix is a Pisot number—i.e. an algebraic integer greater than one, and whose conjugates have modulus less than one. If a substitution is Pisot irreducible, we can verify that the projection π onto a hyperplane, along the eigenspace for the Pisot eigenvalue is bounded. We say that a Pisot number is an *unit* if its inverse is an algebraic integer. We say that a substitution is an *irreducible Pisot unit substitution* if the substitution is irreducible (i.e. the characteristic polynomial of the incidence matrix is irreducible), the highest eigenvalue of the incidence matrix is a Pisot number, and the determinant of the incidence matrix is ± 1. It is equivalent to say that the incidence matrix has only one eigenvalue of modulus greater or equal to one, and that this eigenvalue is a Pisot unit number.

8.3.2 Topology and Main Criterion

Let A be a finite set, $\mathcal{P}$ be an hyperplane of $\mathbb{R}^A$ (for example the hyperplane of equation $\sum_{a \in A} x_a = 0$), and π be an irrational projection onto this hyperplane—that is a projection such that $\pi(\mathbb{Z}^A)$ is dense in $\mathcal{P}$. We define, for any subset S of $\mathcal{P}$, the *discrete line* of points that project to S:

$$Q_S = \left\{ x \in \mathbb{N}^A \,\middle|\, \pi(x) \in S \right\}.$$

This permits to define a topology on $\mathbb{N}^A$ by taking the following set of open sets

$$\left\{ Q_U \,\middle|\, U \text{ open subset of } \mathcal{P} \right\}.$$

And we can extend this topology to $\mathbb{Z}^A$ by considering the open sets of $\mathbb{Z}^A$

$$\left\{ S \subseteq \mathbb{Z}^A \,\middle|\, S \cap \mathbb{N}^A \text{ is an open subset of } \mathbb{N}^A \right\}.$$

Remark 8.3.1 We could define the topology directly on the whole space $\mathbb{Z}^A$ and work with bi-infinite words, but this topology permits to deal with infinite words.

Properties 8.3.2 *The topology that we just defined has the following properties:*

- *If the projection π is such that $\pi(\mathbb{Z}^A)$ is dense in $\mathcal{P}$, then for any open subset U of $\mathcal{P}$, we have that $\pi(Q_U)$ is dense in U, and we have*

$$Q_U = \emptyset \Longleftrightarrow U = \emptyset.$$

- *For any bounded subset $S \subseteq \mathcal{P}$ and any $t \in \mathbb{Z}^A$, the symmetric difference*

$$(Q_S + t)\Delta Q_{S+t}$$

is finite. In particular, we have $\overset{\circ}{Q}_S = \emptyset \Longleftrightarrow \overset{\circ}{Q}_{S+t} = \emptyset \Longleftrightarrow \overset{\circ}{\overline{Q_S + t}} = \emptyset.$
- *If $\det(M) \in \{-1, 1\}$, then for any bounded subset S of $\mathcal{P}$, the symmetric difference $(MQ_S)\Delta Q_{\pi(MS)}$ is finite. In particular, we have $\overset{\circ}{\overline{MQ_S}} = \emptyset \Longleftrightarrow \overset{\circ}{Q}_{\pi(MS)} = \emptyset.$*
- *The space $\mathbb{N}^A$ is a Baire space for this topology.*

The fact that $\mathbb{N}^A$ is a Baire space follows from the fact that $\mathcal{P}$ is a Baire space, by the Baire category theorem. Indeed, if Q is a dense open set of $\mathbb{Z}^d$, then there exists a dense open set U of $\mathcal{P}$ such that $Q = Q_U$. Hence, a countable intersection of dense open subsets of $\mathbb{Z}^d$ is a dense subset of $\mathbb{Z}^d$.

This topology gives a necessary and sufficient condition for the subshift of a Pisot irreducible substitution, to have a pure discrete spectrum:

Theorem 8.3.3 *Let s be an irreducible unit Pisot substitution over an alphabet A, and let $u \in A^{\mathbb{N}}$ be a periodic point of s. Then the subshift $(\overline{S^{\mathbb{N}}u}, S)$ has pure discrete spectrum if and only if $\exists a \in A, \overset{\circ}{D}_{u,a} \neq \emptyset.$*

8.3.3 Proof That an Inner Point Implies the Pure Discreteness of the Spectrum

In this subsection, we prove the first statement and the sufficiency of the second statement. The necessity is proven in the next section.

Proof *(First Part of the Proof of the Theorem 8.3.3)* Up to replace the substitution s by a power, we can assume that the periodic point u is a fixed point. Let us show that the hypothesis of the Theorem 8.2.3 are satisfied.

- The restriction of the projection π to $\mathbb{Z}^A$ is injective, and $\pi(\mathbb{Z}^A)$ is dense in $\mathcal{P}$: this is known for every irreducible Pisot substitution.
- The set $\pi(D_u)$ is bounded: it is well known that for any Pisot irreducible substitution, the Rauzy fractal $\overline{\pi(D_u)}$ is compact.

- The subshift $(\overline{S^{\mathbb{N}}u}, S)$ is minimal: this is true for every primitive substitution, see [17] proposition 5.5.

Now if we assume that $\exists a \in A,\ \overset{\circ}{D}_{u,a} \neq \emptyset$, then we have the following

Lemma 8.3.4 *We have for all $a \in A$, $\overset{\circ}{D}_{u,a} \neq \emptyset$ and $\overline{D_{u,a}} = \overline{\overset{\circ}{D}_{u,a}}$.*

Proof We have the equality

$$D_{u,a} = \bigcup_{b \overset{t}{\to} a \in \mathcal{A}^s} M D_{u,b} + t,$$

where $b \overset{t}{\to} a \in \mathcal{A}^s$ means that it is a transition in the automaton $\mathcal{A}^s$ (i.e. there exists words $u, v \in A^*$ such that $s(b) = uav$ and $\mathrm{Ab}(u) = t$). By primitivity, up to iterate enough this equality, every set $D_{u,b}$ appears in the union, so every set $D_{u,a}$ has non-empty interior as soon as one of them has non-empty interior. This ends the proof of Lemma 8.3.4.

If we iterate $n + 1$ times the equality, we get

$$D_{u,a} = \bigcup_{b \overset{t_n}{\to} \ldots \overset{t_0}{\to} a \in \mathcal{A}^s} M^n D_{u,b} + \sum_{k=0}^{n} M^k t_k,$$

where $b \overset{t_n}{\to} \ldots \overset{t_0}{\to} a \in \mathcal{A}^s$ means that there exists states $(q_i)_{i=0}^{n+1}$ of $\mathcal{A}^s$ with $q_0 = a$ and $q_{n+1} = b$, such that for every i, $q_{i+1} \overset{t_i}{\to} q_i$ is a transition in $\mathcal{A}^s$. Each term of this union has non-empty interior, because each $D_{u,b}$ has non-empty interior and $\det(M) \in \{-1, 1\}$. And the diameter of each $\pi(M^n D_{u,b})$ tends to zero as n tends to infinity, so it proves that the interior of $D_{u,a}$ is dense in $D_{u,a}$.

Hence, $\overset{\circ}{D}_u$ is a dense open subset of $\overline{\pi(D_u)}$. By Baire's theorem, for all $t \in \Gamma_0 \backslash \{0\}$, the empty set $\overset{\circ}{D}_u \cap (\overset{\circ}{D}_u + t)$ is a dense subset of $\overline{\overset{\circ}{D}_u} \cap (\overline{\overset{\circ}{D}_u} + t)$, therefore the sets $\overline{\overset{\circ}{D}_u}$ and $(\overline{\overset{\circ}{D}_u} + t)$ are disjoint. Moreover the boundary of $\overline{\pi(D_u)}$ has zero Lebesgue measure since the substitution s is Pisot unimodular. This gives the wanted disjointness in measure.

The hypothesis of the Theorem 8.2.3 are satisfied, thus the subshift $(\overline{S^{\mathbb{N}}u}, S)$ is uniquely ergodic and measurably conjugate to the rotation by $\pi(e_a)$ on the torus $\mathcal{P}/\pi(\Gamma_0)$ with respect to the Lebesgue measure. In particular, it has pure discrete spectrum.

8.4   Algebraic Coincidence Ensures an Inner Point

In this section, we prove that pure discreteness of the subshift $(\overline{S^{\mathbb{N}}u}, S)$ ensures the non emptiness of $\overset{\circ}{D}_{u,a}$ for some a.

8.4.1   Algebraic Coincidence of Substitutive Delone Set

A Delone set is a relatively dense and uniformly discrete subset of $\mathbb{R}^d$. We say that $\Lambda = (\Lambda_a)_{a \in A}$ is a *Delone multi-color set* in $\mathbb{R}^d$ (also called *Delone m-set* in chapter 1) if each Λ_a is a Delone set and $\cup_{a \in A}\Lambda_a \subset \mathbb{R}^d$ is Delone. Here a 'multi-set' Λ is simply a vector whose entries are Delone sets. We introduce this concept instead of taking their union, only because $\Lambda_a \cap \Lambda_b$ may not be empty for $a \neq b$. We think that each element of Λ_a has color a. A set $\Lambda \subset \mathbb{R}^d$ is a *Meyer set* if it is a Delone set and there exists a finite set F that $\Lambda - \Lambda \subset \Lambda + F$. A Delone set is a Meyer set if and only if $\Lambda - \Lambda$ is uniformly discrete in $\mathbb{R}^d$ [13]. Note that a Meyer set has finite local complexity (FLC), i.e., for any $r > 0$ there are only finitely many transitionally inequivalent clusters (configurations of points) in a ball of radius r. $\Lambda = (\Lambda_a)_{a \in A}$ is called a *substitution Delone multi-color set* if Λ is a Delone multi-color set and there exist an expansive matrix B and finite sets $\mathcal{D}_{ab}$ for $a, b \in A$ such that

$$\Lambda_a = \bigcup_{b \in A}(B\Lambda_b + \mathcal{D}_{ab}), \quad a \in \Lambda, \tag{8.1}$$

where the union on the right side is disjoint. The translation closure of the multi-color Delone set gives a topological dynamical system, which is minimal and uniquely ergodic if the substitution matrix $(^{\#}\mathcal{D}_{ab})$ is primitive. Lagarias and Wang [12] proved that $|\det B|$ must be equal to the Perron Frobenius root of the substitution matrix.

Self-affine tiling dynamical system is the minimal and uniquely ergodic topological dynamical system given by a self-affine tiling with translation action (Solomyak [21]). One can restates its translation dynamics by the translation action on the corresponding multi-colored Delone set Λ (see [15]) Here a point in Λ_a represents a tile colored by a. The points are located in relatively the same position in the same colored tile. Lee [14] introduced *algebraic coincidence* of substitutive multi-color Meyer set in $\mathbb{R}^d$ which is equivalent to pure discreteness of the corresponding dynamical system.

In this section, we prove that if 1-dimensional substitutive Meyer set associated to an irreducible Pisot unit substitution satisfy the algebraic coincidence, then there exists $a \in A$ such that $\overset{\circ}{D}_{u,a} \neq \emptyset$, which completes the proof of the main theorem.

8.4.2   Substitutive Meyer Set from D_u

Let s be a primitive substitution over an alphabet A whose substitution matrix is M. Assume that $v = \ldots v_n \ldots v_1$ and $u = u_1 \ldots u_n \ldots$ are one-sided infinite words such that $vu \in A^{\mathbb{Z}}$ is a 2-sided fixed point of s. The u (resp. v) is a right (resp. left) infinite fixed point of s and $v_1 u_1$ is the subword of $\sigma^n(a)$ for some $n \in \mathbb{N}$ and $a \in A$. The left abelianisation $D_{v,a}$ is defined by

$$\left\{ -\sum_{i=1}^{n} e_{v_i} \,\middle|\, v_n = a \right\}.$$

and $D_{vu,a} = D_{v,a} \cup D_{u,a}$. As s is a substitution, $D_{u,a}$ (resp. $D_{v,a}$) has one to one correspondence to the words $\{u_1 \ldots u_n \mid n \in \mathbb{N}\}$ (resp. $\{v_n \ldots v_1 \mid n \in \mathbb{N}\}$). We also define $D_{vu} = \bigcup_{a \in A} D_{vu,a}$. Then D_{vu} is a geometric realization of the fixed point $vu \in A^{\mathbb{Z}}$, that is, the set of vertices of a broken line naturally generated by corresponding fundamental unit vectors e_a ($a \in A$).

We project this broken line to make a self-similar tiling of the real line by tiles (intervals) corresponding to each letter. This is done by associating intervals whose lengths are given by the entry of a left eigenvector $\ell = (\ell_a)_{a \in A}$. The corresponding expanding matrix is of size 1 and equal to the Perron Frobenius root of M. Define $\psi : D_{u,v} \to \mathbb{R}$ by $\psi(\sum_i^n e_{u_i}) = \sum_{i=1}^{n} \ell_{u_i}$ and $\psi(-\sum_i^n e_{v_i}) = -\sum_{i=1}^{n} \ell_{v_i}$ according to the domain D_u or D_v. Put $\Lambda_a = \{\psi(v) \mid v \in D_{u,a}\}$ for $a \in A$. We normalize the eigenvector ℓ so that ψ becomes the orthogonal projection to the 1-dimensional subspace $\pi^{-1}(0)$ generated by the expanding vector of M. Then this is exactly the set of left end points of intervals which consists the tiling. It is clear that $\psi : D_{u,a} \to \Lambda_a$ is bijective and preserves addition structure, i.e., if $x \pm y \in D_{u,a}$ for $x, y \in D_{u,a}$ then $\psi(x \pm y) = \psi(x) \pm \psi(y)$ holds in Λ_a and vice versa. By this choice of the length, $\Lambda = (\Lambda_a)_{a \in A}$ forms a substitution multi-colored Delone set. When s is a Pisot substitution, Λ is a substitution multi-colored Meyer set. The closure of the set of translations $\{\Lambda - t \mid t \in \mathbb{R}\}$ by local topology forms a compact set X and $(X, \mathbb{R})$ is a topological dynamical system. By primitivity of s, this system is minimal and uniquely ergodic. Moreover the system $(X, \mathbb{R})$ is not weakly mixing if and only if s is a Pisot substitution [21]. Clark and Sadun [6] showed that if s is an irreducible Pisot substitution, then $(X, \mathbb{R})$ shows pure discrete spectrum if and only if $(\overline{S^{\mathbb{N}}u}, S)$ does. Therefore we can use techniques developed in the tiling dynamical system to our problem.

8.4.3   Algebraic Coincidence for D_u

In this setting the algebraic coincidence in [14] reads

$$\exists a \,\exists n \in N \,\exists \eta' \in \mathbb{R} \quad \beta^n \bigcup_{a \in A} (\Lambda_a - \Lambda_a) \in \Lambda_a - \eta' \tag{8.2}$$

and the projection ψ is bijective, (8.2) is equivalent to

$$\exists a \; \exists n \in N \; \exists \eta \in \mathbb{R}^d \quad M^n(\cup_{a \in A}(D_{u,a} - D_{u,a})) \in D_{u,a} - \eta. \tag{8.3}$$

Clearly we see $\eta \in D_{u,a}$. By primitivity of s, we easily see that for any $a, b \in A$, there exists $k \in \mathbb{N}$ that

$$\beta^k(\Lambda_a - \Lambda_a) \subset (\Lambda_b - \Lambda_b) \tag{8.4}$$

Yet we need another result depending heavily on irreducibility of substitution:

Lemma 8.4.1 ([4, 20]) *Let s be a primitive irreducible substitution and Λ be an associated substitution Delone multi-color set in $\mathbb{R}$. Then we have*

$$\left\langle \bigcup_{a \in A}(\Lambda_a - \Lambda_a) \right\rangle = \left\langle (\bigcup_{a \in A} \Lambda_a) - (\bigcup_{a \in A} \Lambda_a) \right\rangle. \tag{8.5}$$

Here $\langle X \rangle$ stands for the additive subgroup of $\mathbb{R}^d$ generated by the set described in X. Since

$$\psi^{-1}\left(\left\langle (\bigcup_{a \in A} \Lambda_a) - (\bigcup_{a \in A} \Lambda_a) \right\rangle\right)$$

contains all fundamental unit vector e_a ($a \in A$), it clearly coincides with $\mathbb{Z}^d$. Therefore Lemma 8.4.1 implies

$$\left\langle \bigcup_{a \in A}(D_{u,a} - D_{u,a}) \right\rangle = \mathbb{Z}^d \tag{8.6}$$

8.4.4 Proof of the Existence of an Inner Point

Note that the substitution matrix M is contained in $GL(d, \mathbb{Z})$, because s is a Pisot unit substitution.

Without loss of generality, we assume that u begins with $a \in A$, which implies $0 \in D_{u.a}$. We will prove that there exists $N \in \mathbb{N}$ that $\pi(\eta)$ is an inner point of $D_{u,a}$ where $\eta \in D_{u,a}$ appeared in (8.3).

Let

$$\varphi : \begin{array}{c} \mathcal{P}(\mathbb{Z}^A) \to \mathcal{P}(\mathbb{Z}^A) \\ S \mapsto M^n(S - S) \end{array} \qquad \mathcal{D} : \begin{array}{c} \mathcal{P}(\mathbb{Z}^A) \to \mathcal{P}(\mathbb{Z}^A) \\ S \mapsto S - S \end{array}'$$

where $n \in \mathbb{N}$ is such that

$$\forall b \in A, \ M^n(D_{u,b} - D_{u,b}) \subseteq D_{u,a} - \eta.$$

Lemma 8.4.2 *For all $k \in \mathbb{N}_{\geq 1}$, we have*

$$\varphi^k \left(\bigcup_{b \in A} D_{u,b} - D_{u,b} \right) \subseteq D_{u,a} - \eta.$$

Proof Easy, by induction.

Lemma 8.4.3 *Let $P \subseteq \mathbb{Z}^A$ such that $\psi(P)$ is relatively dense in $\mathbb{R}_+$ and $\pi(P)$ is bounded. Then, there exists $R > 0$ such that*

$$Q_{B(0,1)} \subseteq \bigcup_{x \in P} B(x, R),$$

where $B(x, R)$ is the ball of $\mathbb{Z}^A$ of center x and radius R.

Proof Let $M > 0$ such that $\forall x \in \mathbb{R}_+, \ d(x, \psi(P)) \leq M$. There exists $C_1 > 0$ and $C_2 > 0$ such that

$$\forall (x, y) \in (\mathbb{R}^A)^2, \ d(x, y) \leq C_1 d(\psi(x), \psi(y)) + C_2 d(\pi(x), \pi(y))$$

depending on the choice of the left eigenvector for the map ψ, and the choice of the linear projection π. We choose $R = C_1 M + C_2(\text{diam}(\pi(P \cup \{0\})) + 1$. Let $x \in Q_{B(0,1)}$, then we have $\psi(x) \in \mathbb{R}_+$, and for $y \in P$ such that $d(x, P) = d(x, y)$ we have

$$\begin{aligned} d(x, y) \ &\leq C_1 d(\psi(x), \psi(P)) + C_2 d(\pi(x), \pi(y)) \\ &\leq C_1 M + C_2(\text{diam}(\pi(P \cup \{0\})) + 1) \leq R. \end{aligned}$$

Therefore, we have $x \in \bigcup_{y \in P} B(y, R)$.

Lemma 8.4.4 *There exists $N \in \mathbb{N}$ such that*

$$B(0, 1) \subseteq \mathcal{D}^N \left(\bigcup_{b \in A} D_{u,b} - D_{u,b} \right),$$

where $B(0, 1)$ is the unit ball of $\mathcal{P}$.

Proof The set $\psi(D_{u,a})$ is relatively dense in $\mathbb{R}_+$ and $\pi(D_{u,a})$ is bounded. So, by Lemma 8.4.3, there exists a $R > 0$ such that

$$Q_{B(0,1)} \subseteq \bigcup_{x \in D_{u,a}} B(x, R).$$

And we have, by (8.6),

$$\bigcup_{N \in \mathbb{N}} \mathcal{D}^N \left(\bigcup_{b \in A} D_{u,b} - D_{u,b} \right) = \left\langle \bigcup_{b \in A} (D_{u,b} - D_{u,b}) \right\rangle = \mathbb{Z}^A.$$

Thus there exists $N \in \mathbb{N}_{\geq 3}$ large enough such that

$$B(0, R) \subseteq \mathcal{D}^{N-1} \left(\bigcup_{b \in A} D_{u,b} - D_{u,b} \right),$$

where $B(0, R)$ is the ball of $\mathbb{Z}^A$ of center 0 and radius R. Then, we have

$$Q_{B(0,1)} \subseteq \bigcup_{x \in D_{u,a}} B(x, R) \subseteq B(0, R) - \mathcal{D}^2(D_{u,a}) \subseteq \mathcal{D}^N \left(\bigcup_{b \in A} D_{u,b} - D_{u,b} \right). \quad \square$$

Using these lemmas, we have the inclusion

$$M^{nN} Q_{B(0,1)} \subseteq M^{nN} \mathcal{D}^N \left(\bigcup_{b \in A} D_{u,b} - D_{u,b} \right)$$
$$= \varphi^N \left(\bigcup_{b \in A} D_{u,b} - D_{u,b} \right) \subseteq D_{u,a} - \eta.$$

And this implies that $D_{u,a}$ contains $M^{nN} Q_{B(0,1)} + \eta$, therefore it has non-empty interior.

8.5 Computation of the Interior

In this section, we show that the interior of some subsets of $\mathbb{Z}^d$, for the topology defined in the Sect. 8.3.2, can be described by a computable regular language. This gives a way to decide the Pisot substitution conjecture for any given irreducible Pisot unit substitution.

8.5.1 Regular Languages

Let Σ be a finite set, and let $\Sigma^* = \bigcup_{n\in\mathbb{N}} \Sigma^n$ be the set of finite words over the alphabet Σ. A subset of $\mathcal{P}(\Sigma^*)$, is called a *language* over the alphabet Σ. We say that a language L over an alphabet Σ is *regular* if the set

$$\left\{u^{-1}L \mid u \in \Sigma^*\right\}$$

is finite, where $u^{-1}L := \left\{v \in \Sigma^* \mid uv \in L\right\}$.

An *automaton* is a quintuplet $\mathcal{A} = (\Sigma, Q, I, F, T)$, where Σ is a finite set called *alphabet*, Q is a finite set called *states*, $I \subseteq Q$ is the set of *initial states*, $F \subseteq Q$ is the set of *final states*, and $T \subseteq Q \times \Sigma \times Q$ is the set of *transitions*. We denote by $p \xrightarrow{t} q$ a transition $(p, t, q) \in T$, and we will write

$$q_0 \xrightarrow{t_1} q_1 \xrightarrow{t_2} \dots \xrightarrow{t_n} q_n \in T$$

when for all $i = 1, 2, \ldots, n$ we have $(q_{i-1}, t_i, q_i) \in T$. We call *language recognized* by $\mathcal{A}$ the language $L_{\mathcal{A}}$ over the alphabet Σ defined by

$$L_{\mathcal{A}} = \{u \in \Sigma^* \mid \exists(q_i)_i \in Q^{|u|+1},\ q_0 \in I, q_{|u|} \in F,$$

$$\text{and } q_0 \xrightarrow{u_1} q_1 \xrightarrow{u_2} \dots \xrightarrow{u_{|u|}} q_{|u|} \in T\}.$$

The following proposition is a classical result about regular languages (see [5, 9, 11, 19]).

Proposition 8.5.1 *A language is regular if and only if it is the language recognized by some automaton.*

We say that an automaton is deterministic if I has cardinality one, and if for every state $q \in Q$ and every letter $t \in \Sigma$, there exists at most one state $q' \in Q$ such that $(q, t, q') \in T$ is a transition.

The *minimal automaton* of a regular language L, is a deterministic automaton recognizing L and having the minimal number of states. Such automaton exists, is unique, and the number of states is equal to the cardinal of the set $\left\{u^{-1}L \mid u \in \Sigma^*\right\} \setminus \{\emptyset\}$. To an automaton, we can associate the *adjacency matrix* in $M_Q(\mathbb{Z})$ whose (s', s) coefficient is the number of transitions from state s to state s'. We denote by ^tL the mirror of a language L.

$$^tL := \left\{u_n u_{n-1} \dots u_1 u_0 \mid u_0 u_1 \dots u_{n-1} u_n \in L\right\}.$$

8.5.2 Discrete Line Associated to a Regular Language

Given a word u over an alphabet $\Sigma \subseteq \mathbb{Z}^d$, and a matrix $M \in M_d(\mathbb{Z})$, we define

$$Q_{u,M} = \sum_{k=0}^{|u|-1} M^i u_i.$$

Given a language L over an alphabet $\Sigma \subseteq \mathbb{Z}^d$, and a matrix $M \in M_d(\mathbb{Z})$, we define the following subset of $\mathbb{Z}^d$.

$$Q_{L,M} = \left\{ Q_{u,M} \,\middle|\, u \in L \right\} = \left\{ \sum_{k=0}^{|u|-1} M^i u_i \,\middle|\, u \in L \right\}.$$

We will also call this set a discrete line, because when M has a Pisot number as eigenvalue and no other eigenvalue of modulus greater than one, then this set stays at bounded distance of a line of $\mathbb{R}^d$—line which is the eigenspace of the matrix M for the Pisot eigenvalue. And we show now that every discrete line coming from a substitution is also the discrete line of some regular language. When it will be clear from the context what is the matrix, we will simply write Q_u and Q_L.

Remark 8.5.2 The notation Q_S was also defined for a part $S \subseteq \mathcal{P}$, but there is no ambiguity, because parts of $\mathcal{P}$ and languages are always different objects, and we use the same notation because in both cases it represents a discrete line.

To a substitution s over the alphabet A, and $a, b \in A$, we associated the following deterministic automaton $\mathcal{A}_{a,b}^s$ with

- set of states A,
- initial state a,
- set of final states $\{b\}$,
- alphabet $\Sigma = \{t \in \mathbb{Z}^d | \exists (c, u, v) \in A \times A^* \times A^*, \ s(c) = uv \text{ with } \mathrm{Ab}(u) = t \text{ and } |v| > 0\}$,
- set of transitions $T = \{(c, t, d) \in A \times \Sigma \times A | \exists u, v \in A^*, \ s(c) = udv \text{ and } \mathrm{Ab}(u) = t\}$.

We denote by $L_{a,b}^s$ the language of this automaton. We denotes by $\mathcal{A}^s$ the automaton $\mathcal{A}_{a,b}^s$ where we forget the data of the initial state and the set of final states.

Remark 8.5.3 This automaton is the abelianisation of what we usually call the prefix automaton.

Remark 8.5.4 For a substitution s and two letters a and b, the language $L_{a,b}^s$ has little to do with what we usually call the language of the substitution s (i.e. the set of finite factors of periodic points of s). The alphabet of $L_{a,b}^s$ is not even the same as the alphabet of the substitution s.

Proposition 8.5.5 *If u is a fixed point of a substitution s whose first letter is a, then we have for every letter b,*

$$D_{u,b} = Q_{^tL_{a,b}^s, M_s}.$$

Remark 8.5.6 This proposition corresponds to write elements of the discrete line $D_{u,b}$ using the Dumont-Thomas numeration.

Remark 8.5.7 If we want to describe the left infinite part of the discrete line associated to a bi-infinite fixed point of the substitution s, we have to consider the automata $A_{a,b}^{^ts}$ where ^ts is the reverse substitution of s—that is $\forall a \in A$, $^ts(a) = {}^t(s(a))$. We can also describe a bi-infinite discrete line with only one automaton over the bigger alphabet $\Sigma_s \cup -\Sigma_{^ts}$.

Remark 8.5.8 The automaton A^s permits to compute easily the map $E_1(s)$ defined in [2]:

$$E_1(s)(x, e_a) = \sum_{a \xrightarrow{t} b \in T} (Mx + t, e_b),$$

where T is the set of transitions of A^s.

We can also compute easily the map $E_1^*(s)$ when $\det(M) \in \{-1, 1\}$:

$$E_1^*(s)(x, e_b^*) = \sum_{a \xrightarrow{t} b \in T} (M^{-1}(x - t), e_a^*).$$

And we have

$$(y, e_b) \in E_1(s)(x, e_a) \iff a \xrightarrow{y - Mx} b \in T \iff (x, e_a^*) \in E_1^*(s)(y, e_b^*).$$

Proof of the Proposition 8.5.5 The first idea is to describe the set of prefixes of $s^n(a)$ followed by a letter b, by words of length n in the regular language $^tL_{a,b}^s$, like in the Fig. 8.3 but with a different alphabet.

Lemma 8.5.9 *For every $n \in \mathbb{N}$, there exists a natural map*

$$\varphi_n : \bigcup_{b \in A} \{v \in A^* \mid vb \text{ prefix of } s^n(a)\} \to \bigcup_{b \in A} \{w \in L_{a,b}^s \mid |w| = n\}$$

such that

$$\forall b \in A, \ \varphi_n(\{v \in A^* \mid vb \text{ prefix of } s^n(a)\}) = \{w \in L_{a,b}^s \mid |w| = n\}.$$

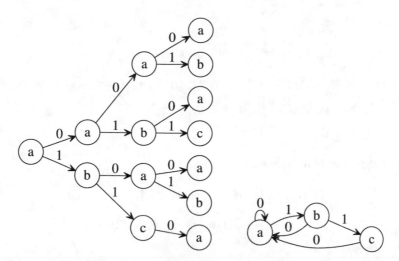

Fig. 8.3 Different ways of describing the same thing. Equivalence between the choice of a prefix of $s^n(a)$ followed by a letter b, a path of length n in a tree to a letter b, a path of length n in the prefix automaton with initial state a and final state b, for the substitution $s : a \mapsto ab, b \mapsto ac, c \mapsto a$. For example, the prefix $abaca$ of $s^3(a)$ corresponds to the word 101, and the prefix a corresponds to the word 001

Proof By induction on n. For $n = 0$, the map φ is uniquely defined. If $n \geq 1$, for $v \in A^*$ such that vb is a prefix of $s^n(a)$, there exists a unique uplet of words $(v', v'', v''') \in (A^*)^3$, and an unique letter $c \in A$ such that $s(v'c) = vbv'''$, where $s(c) = v''bv'''$. We define $\varphi_n(v) = \varphi_{n-1}(v') \text{Ab}(v'')$. This is a word of length n in the regular language $L^s_{a,b}$, because by induction we have $\varphi_{n-1}(v') \in L^s_{a,c}$, with $\varphi_{n-1}(v')$ of length $n - 1$, and the equality $s(c) = v''bv'''$ implies that there exists a transition for state c to state b labeled by $\text{Ab}(v'')$ in the automaton $\mathcal{A}^s$.

We check that the formulae linking the abelianisation of the prefix and the corresponding word in $^tL^s_{a,b}$ is the one expected.

Lemma 8.5.10 *For every $v \in A^*$ such that vb is a prefix of $s^n(a)$, we have*

$$\text{Ab}(v) = Q\,^t\varphi_n(v),$$

where φ_n is the map defined by Lemma 8.5.9 above.

Proof For every such word v, the map φ_n gives a unique sequence (v_k, c_k, w_k) in $A^* \times A \times A^*$ such that we have $\forall\, 0 \leq k \leq n - 1$, $s(c_k) = v_{k+1}c_{k+1}w_{k+1}$, with $c_0 = a$, $v_0 = w_0 = \epsilon$, and $c_n = b$. And, we have

$$s^n(a) = vbw = s^{n-1}(v_1)s^{n-2}(v_2)\ldots s(v_{n-1})v_n c_n w_n s(w_n)\ldots s^{n-2}(w_2)s^{n-1}(w_1).$$

Therefore we have

$$\mathrm{Ab}(v) = \mathrm{Ab}\left(s^{n-1}(v_1)s^{n-2}(v_2)\ldots s(v_{n-1})v_n\right)$$

$$= M^{n-1}\,\mathrm{Ab}(v_1) + M^{n-2}\,\mathrm{Ab}(v_2) + \ldots + M\,\mathrm{Ab}(v_{n-1}) + \mathrm{Ab}(v_n)$$

$$= Q_{\mathrm{Ab}(v_n)\,\mathrm{Ab}(v_{n-1})\ldots\,\mathrm{Ab}(v_2)\,\mathrm{Ab}(v_1)}$$

$$= Q_{{}^t\varphi_n(v)}.$$

With these two lemmas, we get

$$D_{u,b} = \left\{\mathrm{Ab}(v)\,\big|\, vb \text{ prefix of } u\right\}$$

$$= \bigcup_{n \subset \mathbb{N}} \left\{\mathrm{Ab}(v)\,\big|\, vb \text{ prefix of } s^n(a)\right\}$$

$$= \bigcup_{n \in \mathbb{N}} \left\{Q_{{}^tw}\,\big|\, w \in L^s_{a,b},\ |w| = n\right\}$$

$$= Q_{{}^tL^s_{a,b}}.$$

This ends the proof of the Proposition 8.5.5.

8.5.3 Computation of the Interior

We have seen in the previous section that the subshift associated to an irreducible Pisot substitution has pure discrete spectrum as soon as the interior of a piece of the discrete line is non-empty (see Theorem 8.3.3), for the topology defined in Sect. 8.3.2. In this section, we give a way to compute the interior (and hence to test the Pisot substitution conjecture) with the following

Theorem 8.5.11 *Let L be a regular language over an alphabet $\Sigma \subseteq \mathbb{Z}^A$, M be an irreducible Pisot unimodular matrix, and π be the projection on a hyperplane $\mathcal{P}$ along the eigenspace of M for its maximal eigenvalue β. Then, there exists a regular language $\overset{\circ}{L} \subseteq L$ such that $Q_{\overset{\circ}{L}} = \overset{\circ}{Q_L}$. Moreover, this language $\overset{\circ}{L}$ is computable from L.*

Remark 8.5.12 The language $\overset{\circ}{L}$ doesn't depend on the choice of the hyperplane $\mathcal{P}$.

With this theorem, the criterion given by the Theorem 8.3.3 gives the following result:

Corollary 8.5.13 *Let s be an irreducible Pisot unit substitution over an alphabet A. If there exist letters $a, b \in A$ such that a is the first letter of a fixed point of s,* *and the regular language* ${}^t\overset{\circ}{L^s}_{a,b}$ *is non-empty, then the subshift* $(\overline{S^{\mathbb{N}}u}, S)$ *has pure discrete spectrum.*

And, the Pisot substitution conjecture is equivalent to

Conjecture 8.5.14 For any irreducible Pisot unit substitution s over an alphabet A and for any letters $a, b \in A$, the regular language ${}^t\overset{\circ}{L^s}_{a,b}$ is non-empty.

8.5.4   Proof of the Theorem 8.5.11

In order to compute the interior, we need a big enough alphabet.

Lemma 8.5.15 *For any Pisot unit primitive matrix* $M \in M_d(\mathbb{N})$, *there exists* $\Sigma' \subseteq \mathbb{Z}^A$ *such that* $0 \in \overset{\circ}{Q}_{\Sigma'^*, M}$.

Proof Let's consider any substitution s whose incidence matrix is the irreducible unit Pisot matrix M. Let u be a periodic point for this substitution. We know that $\pi(D_u)$ is bounded and is a fundamental domain for the action of the lattice $\pi(\Gamma_0)$ on $\pi(\mathbb{Z}^A)$, where Γ_0 is the subgroup of $\mathbb{Z}^A$ spanned by $e_a - e_b, a, b \in A$. Hence, there exists a finite subset $S \subseteq \Gamma_0$ such that $D_u + S = \{x + y \mid (x, y) \in D_u \times S\}$ contains zero in its interior. Then, the alphabet $\Sigma' = \Sigma_s + S$ satisfy that $0 \in \overset{\circ}{Q}_{\Sigma'\Sigma_s^*} \subseteq \overset{\circ}{Q}_{\Sigma'^*}$.

The alphabet given by this lemma is not optimal. Here are two conjectures that gives natural choices of alphabet. The first one gives an alphabet of minimal size, and the second one gives the alphabet Σ that naturally comes from the substitution.

Conjecture 8.5.16 For all irreducible unit Pisot matrix M with spectral radius β, we have $0 \in \overset{\circ}{Q}_{\Sigma'^*}$, for $\Sigma' = \{-1, 0, 1, 2, \ldots, \lceil\beta\rceil - 2\}$.

Conjecture 8.5.17 For all irreducible unit Pisot substitution s, we have $\overset{\circ}{Q}_{\Sigma_s^*} \neq \emptyset$.

Remark 8.5.18 This last conjecture is a consequence of the Pisot substitution conjecture. But it should be easier to solve.

Remark 8.5.19 We cannot assume in this last conjectures that the interior always contains 0, since we can only get the positive part of the hyperplane $\mathcal{P}$ with Pisot numbers whose conjugates are positive reals numbers. Nevertheless, if we have only $\overset{\circ}{Q}_{\Sigma'^*} \neq \emptyset$, then the set L_{int} computed in the proof of the Theorem 8.5.11 satisfy

$$\overset{\circ}{Q}_L \subseteq Q_{L_{int}} \subseteq \overline{\overset{\circ}{Q}_L},$$

so we have $\overset{\circ}{Q}_L = \emptyset \Longleftrightarrow L_{int} = \emptyset$. Hence we can decide if Q_L has empty interior or not by computing L_{int} with this alphabet Σ'.

The following theorem is also useful to compute the interior. It is a variant of the main theorem of [16].

Theorem 8.5.20 *Consider two alphabets Σ and Σ' in $\mathbb{Z}^A$, and a matrix $M \in M_A(\mathbb{Z})$ without eigenvalue of modulus one. Then the language*

$$L^{\mathrm{rel}} := \left\{ (u, v) \in (\Sigma' \times \Sigma)^* \,\middle|\, Q_u = Q_v \right\}$$

is regular.

Remark 8.5.21 This language L^{rel} is related to what is usually called the zero automaton. See [7] and [8] for more details.

Proof of the Theorem 8.5.11 Consider the language

$$L_{int} := Z(S(Z(p_1(\Sigma'^* \times L0^* \cap L^{\mathrm{rel}})))),$$

where

- Σ' is an alphabet given by the Lemma 8.5.15 and containing 0,
- $p_1 : (\Sigma' \times \Sigma)^* \to \Sigma'^*$ is the word morphism such that $\forall (x, y) \in \Sigma' \times \Sigma$, $p_1((x, y)) = x$,
- L^{rel} is the language defined in Theorem 8.5.20,
- for any language L over the alphabet Σ', $S(L) := \left\{ u \in \Sigma'^* \,\middle|\, u\Sigma'^* \subseteq L \right\}$,
- for any language $L \subseteq \Sigma'^*$, $Z(L) := \left\{ u \in \Sigma'^* \,\middle|\, \exists n \in \mathbb{N}, \ u0^n \in L \right\}$.

Then, we have

$$Q_{L_{int}} = \overset{\circ}{Q}_L,$$

and we have that the language L_{int} is "complete", that is:

$$L_{int} = \left\{ u \in \Sigma'^* \,\middle|\, Q_u \in Q_{L_{int}} \right\}.$$

Indeed, for all $u \in \Sigma'^*$ we have

$$u \in L_{int} \iff \exists n \in \mathbb{N}, \ u0^n \in S(Z(p_1(\Sigma'^* \times L0^* \cap L^{\mathrm{rel}})))$$

$$\iff \exists n \in \mathbb{N}, \ u0^n \Sigma'^* \subseteq Z(p_1(\Sigma'^* \times L0^* \cap L^{\mathrm{rel}}))$$

$$\iff \exists n \in \mathbb{N}, \ \forall v \in \Sigma'^*, \ \exists k \in \mathbb{N}, \ u0^n v0^k \in p_1(\Sigma'^* \times L0^* \cap L^{\mathrm{rel}})$$

$$\iff \exists n \in \mathbb{N}, \ \forall v \in \Sigma'^*, \ \exists k \in \mathbb{N}, \ \exists w \in L0^*, \ (u0^n v0^k, w) \in L^{\mathrm{rel}}$$

$$\iff \exists n \in \mathbb{N}, \ \forall v \in \Sigma'^*, \ Q_{u0^n v} \in Q_L,$$

$$\iff \exists n \in \mathbb{N}, \ Q_u + M^{n+|u|} Q_{\Sigma'^*} \subseteq Q_L,$$

$$\iff Q_u \in \overset{\circ}{Q}_L.$$

This last equivalence is true, if we assume that the alphabet Σ' is positive (which is not always possible). This problem disappears if we consider the bi-infinite topology, where open subsets of $\mathbb{Z}^A$ are $\pi^{-1}(U)$ for U open subset of $\mathcal{P}$. Or one could replace Σ'^* by $\Sigma'\Sigma_s^*$, and it computes the interior up to a finite number of points.

And we can assume that $\Sigma \subseteq \Sigma'$ up to replace Σ' by $\Sigma \cup \Sigma'$. Then, we get the language $\overset{\circ}{L}$ by taking the intersection with L:

$$\overset{\circ}{L} = L \cap L_{int}.$$

This language verify what we want because $Q_{L_{int}} \subseteq Q_L$ and because L_{int} is complete.

Remark 8.5.22 If we just want to test the non-emptiness of the language $\overset{\circ}{L}$, it is not necessary to compute all what is done in this proof. For example, the computation of the language L_{int} is enough (and we do not need that $\Sigma \subseteq \Sigma'$). And we don't even need to compute completely L_{int} if we only want to test if it is non-empty. And it is enough to have Σ' such that $Q_{\Sigma'^*}$ has non-empty interior.

8.5.5 Examples

Example 8.5.23 For the Fibonacci and for the Tribonacci substitutions, we get $\overset{\circ}{{}^lL^s_{a,b}} = {}^lL^s_{a,b}$, for a the first letter of the fixed point u, and any letter b. Therefore the sets $D_{u,b}$ are open: $\overset{\circ}{D}_{u,b} = D_{u,b}$ (and we can check that they are also closed).

Example 8.5.24 For the "flipped" Tribonacci substitution:

$$a \mapsto ab$$
$$b \mapsto ca$$
$$c \mapsto a$$

the minimal automaton of the language $\overset{\circ}{{}^lL^s_{a,a}}$ has 79 states (80 states for $\overset{\circ}{{}^lL^s_{a,b}}$, 81 for $\overset{\circ}{{}^lL^s_{a,c}}$). This automaton is plotted in Fig. 8.4, and the sets $\pi(D_{u,a})$ and $\pi(\overset{\circ}{D}_{u,a})$ for the fixed point u are drawn in Fig. 8.5.

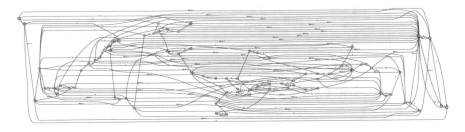

Fig. 8.4 Minimal automaton of the language $^tL^s_{a,a}$ of the Example 8.5.24. The labels 0 correspond to the null vector, the labels 1 correspond to the vector e_a, and the labels $b^2 - b - 1$ correspond to the vector e_c. Final states are the double circles, and the initial state is the bold circle

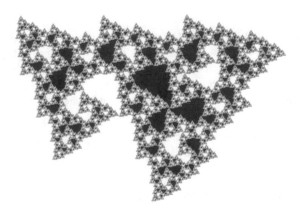

Fig. 8.5 The sets $\pi(D_{u,a})$ (in gray and blue) and $\pi(\mathring{D}_{u,a})$ (in blue) for the Example 8.5.24

Example 8.5.25 For the following substitution associated to the smallest Pisot number:

$$a \mapsto b$$
$$b \mapsto c$$
$$c \mapsto ab$$

the minimal automaton of the language $^tL^s_{a,a}$ has 1578 states (1576 states for $^tL^s_{a,b}$, 1577 for $^tL^s_{a,c}$). The sets $\pi(D_{u,a})$ and $\pi(\mathring{D}_{u,a})$ are plotted on Fig. 8.6, where u is the periodic point starting by letter a.

Remark 8.5.26 The first author have implemented the computing of the interior in the Sage mathematical software. The above examples has been computed using this implementation which is partially available here: https://pypi.org/project/badic or https://gitlab.com/mercatp/badic.

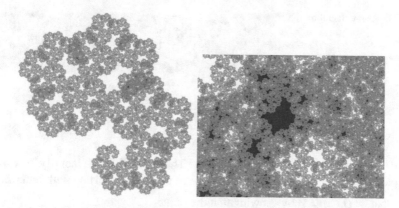

Fig. 8.6 The sets $\pi(D_{u,a})$ (in gray and blue) and $\pi(\overset{\circ}{D}_{u,a})$ (in blue) for the Example 8.5.25. Whole set at the left, and a zoom on it at the right

Remark 8.5.27 To prove the Pisot substitution conjecture, it is enough for each irreducible Pisot substitution s and for any letter a, to find one particular "canonical" word in the language $^t\overset{\circ}{L}{}^s_{a,a}$ in order to prove it is non-empty.

8.6 Pure Discreteness for Various Infinite Family of Substitutions

8.6.1 Proof of Pure Discreteness Using a Geometrical Argument

Using the Theorem 8.3.3, we can prove the Pisot substitution conjecture for a new infinite family of substitutions:

Theorem 8.6.1 *Let $k \in \mathbb{N}$, and let*

$$s_k : \begin{cases} a \mapsto a^k bc \\ b \mapsto c \\ c \mapsto a \end{cases}$$

where a^k means that the letter a is repeated k times. The subshift generated by the substitution s_k is measurably conjugate to a translation on the torus $\mathbb{T}^2$ (Fig. 8.7).

Fig. 8.7 Rauzy fractal of s_{20}

Proof The strategy of the proof is to use the Theorem 8.3.3. We identify $\mathcal{P}$ with the complex plane $\mathbb{C}$. For $k \geq 1$, let u be the fixed point of s_k starting with letter a. We show that $\pi(\overset{\circ}{D}_{u,a}) \neq \emptyset$ by showing that the point

$$t_k := \frac{k}{2} - \frac{\sqrt{k}}{2} I$$

is not in the closure of $\pi(\mathbb{Z}^A \setminus D_a)$:

$$t_k \notin \bigcup_{l \in A \setminus \{a\}} \overline{\pi(D_{u,l})} \cup \bigcup_{t \in \Gamma_0 \setminus \{0\}} \overline{\pi(D_u + t)},$$

where $A = \{a, b, c\}$ is the alphabet of the substitution s_k, Γ_0 is the group generated by $(e_i - e_j)_{i,j \in A}$, where $(e_i)_{i \in A}$ is the canonical basis of $\mathbb{R}^A$, I denotes a complex number such that $I^2 = -1$, and π is the projection along the eigenspace for the maximal eigenvalue of the incidence matrix

$$M = \begin{pmatrix} k & 0 & 1 \\ 1 & 0 & 0 \\ 1 & 1 & 0 \end{pmatrix}$$

of the substitution s_k, such that $\pi(e_a) = 1$, $\pi(e_b) = -\beta^2 + (k+1)\beta - (k-1)$ and $\pi(e_c) = \beta^2 - k\beta - 1$, where β is the complex eigenvalue of M such that $\text{Im}(\beta) < 0$.

In order to do that, we approximate the sets $D_{u,l}$ by union of balls (Fig. 8.8):

Lemma 8.6.2 *For all $k \geq 3$ and for every $l \in A$, we have the inclusion*

$$\overline{\pi(D_{u,l})} \subseteq \bigcup_{t \in S_l} B(t, \frac{1}{1 - \frac{1}{\sqrt{k}}}),$$

Fig. 8.8 Strategy to prove that $t_k \notin \overline{\pi(\mathbb{Z}^d \setminus D_{u,a})}$. Approximation of the sets $\pi(D_{u,l})$ and their translated copies, by disks, for $k = 20$

where

$$S_a = \{\gamma\beta\} \cup \left\{i + \beta j \,\middle|\, (i, j) \in \{0, 1, \ldots, k-1\}^2\right\},$$
$$S_b = \left\{k + \beta i \,\middle|\, i \in \{0, 1, \ldots, k-1\}\right\},$$
$$S_c = \{k\beta\} \cup \left\{\gamma + \beta i \,\middle|\, i \in \{0, 1, \ldots, k-1\}\right\},$$

where $\gamma = -\beta^2 + (k+1)\beta + 1 = \beta - \frac{1}{\beta}$.

Proof For every $l \in A$, we have the equality

$$\pi(D_{u,l}) = \left\{\textstyle\sum_{k=0}^{|u|-1} u_i \beta^i \,\middle|\, u \in {}^t L_l\right\}$$

where L_l is the language of the automaton of Fig. 8.9 where we replace the set of final states by $\{l\}$.

We get the proof of the lemma by considering words of length two, and by the inequality

$$\left|\sum_{k=2}^{|u|-1} u_i \beta^i\right| \leq \sum_{k=2}^{|u|-1} \max\left\{|t| \,\middle|\, t \in \Sigma\right\} |\beta|^i \leq \frac{1}{1 - \frac{1}{\sqrt{k}}}$$

for any word u over the alphabet Σ, where $\Sigma = \{0, 1, \ldots, k-1, k, \gamma\}$ is the alphabet of the languages L_l. Indeed, we have $\max\left\{|t| \,\middle|\, t \in \Sigma\right\} = k$ and $|\beta| \leq \frac{1}{\sqrt{k}}$ for $k \geq 3$.

Lemma 8.6.3 *For every $k \geq 1$, we have the inequalities*

$$\frac{1}{\sqrt{k + \frac{2}{k}}} < |\beta| < \frac{1}{\sqrt{k}}$$

$$\sqrt{k} - \frac{1}{\sqrt{k}} < |\gamma| < \sqrt{k + \frac{2}{k}} + \frac{1}{\sqrt{k}}$$

Fig. 8.9 Automaton describing $\pi(D_u)$

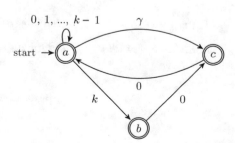

$$-\frac{1}{k} < \mathrm{Re}(\beta) < -\frac{1}{k+\frac{2}{k}}$$

$$-\frac{1}{\sqrt{k}} < \mathrm{Im}(\beta) < -\frac{1}{\sqrt{k+\frac{2}{k}}} + \frac{1}{k}$$

where $\mathrm{Re}(\beta)$ *is the real part of* β, *and* $\mathrm{Im}(\beta)$ *is the imaginary part.*

Proof Let β_+ be the real conjugate of β. We have $\beta_+ = k + \frac{1}{\beta_+} + \frac{1}{\beta_+^2} > 0$, so

$$k \le \beta_+ \le k + \frac{2}{k}.$$

And we have $|\beta|^2 = \frac{1}{\beta_+}$, hence we get the wanted inequalities for $|\beta|$. The inequalities for $\gamma = \beta - \frac{1}{\beta}$ follow. To get the real part, remarks that we have $k = \beta_+ + \beta + \overline{\beta} = \beta_+ + 2\mathrm{Re}(\beta)$, and this gives $\mathrm{Re}(\beta) = -\frac{1}{2\beta_+} - \frac{1}{2\beta_+^2}$. The inequalities for the imaginary part follow.

Lemma 8.6.4 *For all* $k \ge 14$, *we have* $t_k \notin \overline{\pi(D_{u,b})}$

Proof For all $i \in \{0, 1, \ldots, k-1\}$, we have $|k + i\beta - t_k| = \left|\frac{k}{2} + i\beta + \frac{\sqrt{k}}{2}I\right| \ge \frac{k}{2} - |i\beta| - \frac{\sqrt{k}}{2} \ge \frac{k}{2} - \frac{3\sqrt{k}}{2}$. This is greater than $\frac{1}{1-\frac{1}{\sqrt{k}}}$ for $k \ge 14$.

Lemma 8.6.5 *For all* $k \ge 31$, *we have* $t_k \notin \overline{\pi(D_{u,c})}$

Proof We have $|\gamma + i\beta - t_k| = \left|-\frac{k}{2} + \gamma + i\beta + \frac{\sqrt{k}}{2}I\right| \ge \frac{k}{2} - |i\beta| - |\gamma| - \frac{\sqrt{k}}{2} \ge \frac{k}{2} - \frac{3\sqrt{k}}{2} - \sqrt{k+\frac{2}{k}} - \frac{1}{\sqrt{k}}$. This is greater than $\frac{1}{1-\frac{1}{\sqrt{k}}}$ for $k \ge 31$.

We have $|k\beta - t_k| \ge \frac{k}{2} - \frac{3\sqrt{k}}{2}$. This is greater than $\frac{1}{1-\frac{1}{\sqrt{k}}}$ for $k \ge 14$.

Let us show now that the point t_k is not in the translated copies of $\overline{\pi(D_u)}$ by the group $\pi(\Gamma_0)$. The group $\pi(\Gamma_0)$ is

$$\pi(\Gamma_0) = \left\{c(\beta - k - 2) + d(\beta^2 - k\beta - 2) \,\middle|\, (c, d) \in \mathbb{Z}^2\right\}.$$

Let $t_{c,d} := c(\beta - k - 2) + d(\beta^2 - k\beta - 2)$.

Lemma 8.6.6 *For all* $k \ge 8$ *and for all* $(c, d) \in \mathbb{Z}^2$ *such that* $|c| \ge 1$ *and* $2|c| \ge |d|$, *we have*

$$\left|t_{c,d}\right| \ge k - 3\sqrt{k}.$$

Proof We have

$$\left|t_{c,d}\right| \geq |c|\,(k+2) - \frac{|c|}{\sqrt{k}} - |d|\,(\frac{1}{k} + \sqrt{k} + 2) \geq |c|\,(k - \frac{1}{\sqrt{k}} - \frac{2}{k} - 2\sqrt{k} - 2).$$

This is greater than $k - 3\sqrt{k}$ for $k \geq 8$.

Lemma 8.6.7 *For all $k \geq 9$ and for all $(c, d) \in \mathbb{Z}^2$ such that $|c| \leq |d|$ and $2 \leq |d|$, we have*

$$\left|\mathrm{Im}(t_{c,d})\right| \geq 2\sqrt{k} - 3.$$

And if moreover $|d| \geq 3$, then we have $\left|\mathrm{Im}(t_{c,d})\right| \geq 3\sqrt{k} - 5$.

Proof We have $\left|\mathrm{Im}(\beta^2 - k\beta - 2)\right| \geq k\,|\mathrm{Im}(\beta)| - |\beta|^2 \geq \frac{k}{\sqrt{k+\frac{2}{k}}} - 1 - \frac{1}{k}$. And we have $|\mathrm{Im}(\beta - k - 2)| = |\mathrm{Im}(\beta)| \leq \frac{1}{\sqrt{k}}$. Hence,

$$\left|\mathrm{Im}(t_{c,d})\right| \geq |d|\left(\frac{k}{\sqrt{k+\frac{2}{k}}} - 1 - \frac{1}{k}\right) - \frac{|c|}{\sqrt{k}} > 2\left(\frac{k}{\sqrt{k+\frac{2}{k}}} - 1 - \frac{1}{k} - \frac{1}{\sqrt{k}}\right).$$

This is greater than $2\sqrt{k} - 3$ for $k \geq 9$. If moreover $|d| \geq 3$, then we have

$$\left|\mathrm{Im}(t_{c,d})\right| \geq 3\left(\frac{k}{\sqrt{k+\frac{2}{k}}} - 1 - \frac{1}{k} - \frac{1}{\sqrt{k}}\right),$$

and this is greater than $3\sqrt{k} - 5$ for $k \geq 6$ (Fig. 8.10).

Fig. 8.10 Zone covered by the Lemma 8.6.6 (in green), and by the Lemma 8.6.7 (in blue), and remaining points (the red points are remaining only for $l = c$)

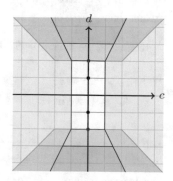

Lemma 8.6.8 *For all $k \geq 8$, $l \in A$ and $t \in S_l$, we have*

$$|t - t_k| \leq \frac{k}{2} + 2\sqrt{k} \qquad and$$

$$|\mathrm{Im}(t - t_k)| \leq \begin{cases} \frac{3}{2}\sqrt{k} & if\ t \notin \{\gamma + \beta i \mid i \in \{0, 1, \ldots, k-1\}\} \\ 2\sqrt{k} & otherwise. \end{cases}$$

Proof For every $(i, j) \in \{0, 1, \ldots, k-1\}^2$, we have $|i + \beta j - t_k| \leq \frac{k}{2} + \frac{3}{2}\sqrt{k}$, $|k + \beta i - t_k| \leq \frac{k}{2} + \frac{3}{2}\sqrt{k}$, $|\gamma + \beta i - t_k| \leq \sqrt{k + \frac{2}{k}} + \frac{1}{\sqrt{k}} + 1 + \frac{k}{2} + \frac{\sqrt{k}}{2}$ (because the imaginary part of β is negative), $|k\beta - t_k| \leq \frac{k}{2} + \frac{3}{2}\sqrt{k}$ and $|\gamma\beta - t_k| = \left|\beta^2 - 1 - \frac{k}{2} + \frac{\sqrt{k}}{2}I\right| \leq \frac{1}{k} + 1 + \frac{k}{2} + \frac{\sqrt{k}}{2}$. Hence, the first inequality is true for $k \geq 8$.

We have $|\mathrm{Im}(i + \beta j - t_k)| = |\mathrm{Im}(k + \beta j - t_k)| = \left|j\,\mathrm{Im}(\beta) - \frac{\sqrt{k}}{2}\right| \leq \frac{3}{2}\sqrt{k}$, $|\mathrm{Im}(\gamma + \beta i - t_k)| \leq \left|\mathrm{Im}(\gamma) + \frac{\sqrt{k}}{2}\right| + i\,|\beta| \leq \sqrt{k + \frac{2}{k}} + \frac{1}{\sqrt{k}} - \frac{\sqrt{k}}{2} + \sqrt{k}$, $|\mathrm{Im}(k\beta - t_k)| \leq k\,|\beta| + \frac{\sqrt{k}}{2} \leq \frac{3}{2}\sqrt{k}$, $|\mathrm{Im}(\gamma\beta - t_k)| \leq |\gamma\beta| + \frac{\sqrt{k}}{2} \leq \frac{\sqrt{k}}{2} + 1 + \frac{1}{k}$. Hence, we get the wanted inequality for $k \geq 3$.

Lemma 8.6.9 *For all $k \geq 69$, we have $t_k \notin (\overline{\pi(D_u)} + t_{0,1}) \cup (\overline{\pi(D_u)} + t_{0,-1})$, and we have $t_k \notin \bigcup_{d \in \{-2,-1,0,1,2\}} (\overline{\pi(D_{u,c})} + t_{0,d})$.*

Proof For all $(i, j) \in \{0, 1, \ldots, k\}$, we have

$$\left|i + \beta j + t_{0,\pm 1} - t_k\right| \geq \left|\mathrm{Im}(\beta j + t_{0,\pm 1} - t_k)\right|$$

$$= \left|\mathrm{Im}(\beta)(j \mp k) \pm \mathrm{Im}(\beta^2) + \frac{\sqrt{k}}{2}\right|.$$

If $\pm = +$, we have $\left|i + \beta j + t_{0,1} - t_k\right| \geq \frac{\sqrt{k}}{2} - \frac{1}{k}$ because $\mathrm{Im}(\beta) < 0$. This is greater than $\frac{1}{1 - \frac{1}{\sqrt{k}}}$ for $k \geq 10$.

If $\pm = -$, we have $\left|i + \beta j + t_{0,-1} - t_k\right| \geq \frac{k}{\sqrt{k + \frac{2}{k}}} - \frac{\sqrt{k}}{2} - 1 - \frac{1}{k}$. This is greater than $\frac{1}{1 - \frac{1}{\sqrt{k}}}$ for $k \geq 22$.

For $|d| \leq 1$, we have

$$\left|\gamma\beta + t_{0,\pm 1} - t_k\right| \geq \frac{k}{2} - \frac{\sqrt{k}}{2} - \left|\beta^2 - 1\right| - \left(\frac{1}{k} + \sqrt{k} + 2\right) \geq \frac{k}{2} - 3\frac{\sqrt{k}}{2} - 3 - \frac{2}{k}.$$

This is greater than $\frac{1}{1 - \frac{1}{\sqrt{k}}}$ for $k \geq 24$.

For all $i \in \{0, 1, \ldots, k-1\}$, and $|d| \leq 2$, we have

$$\left| \gamma + \beta i + t_{0,d} - t_k \right| \geq \frac{k}{2} - \sqrt{k + \frac{2}{k} - \frac{1}{\sqrt{k}}} - \sqrt{k} - \left| t_{0,d} \right| - \frac{\sqrt{k}}{2}$$

$$\geq \frac{k}{2} - \frac{1}{\sqrt{k}} - \frac{3}{2}\sqrt{k} - |d|\frac{1}{k} - |d|\sqrt{k} - 2|d|$$

$$= \frac{k}{2} - \frac{1}{\sqrt{k}} - \frac{7}{2}\sqrt{k} - \frac{2}{k} - 4.$$

This is greater than $\dfrac{1}{1 - \frac{1}{\sqrt{k}}}$ for $k \geq 69$.

For $|d| \leq 2$, we have $\left| k\beta + t_{0,d} - t_k \right| \geq \frac{k}{2} - \frac{3}{2}\sqrt{k} - |d|\left(\frac{1}{k} + \sqrt{k} + 2\right) \geq$
$\frac{k}{2} - \frac{7}{2}\sqrt{k} - \frac{2}{k} - 4$. This is greater than $\dfrac{1}{1 - \frac{1}{\sqrt{k}}}$ for $k \geq 69$.

Using the Lemmas 8.6.6 and 8.6.7, we have that for all the cases not covered by the Lemma 8.6.9

$$\left| t_{c,d} \right| \geq k - 3\sqrt{k} \quad \text{or} \quad \left| \operatorname{Im}(t_{c,d}) \right| \geq 3\sqrt{k} - 5 \quad \text{or} \quad \left| \operatorname{Im}(t_{c,d}) \right| \geq 2\sqrt{k} - 3.$$

Hence, for all $l \in A$ and all $t \in S_l$, we have

$$\left| t + t_{c,d} - t_k \right| \geq k - 3\sqrt{k} - \frac{\sqrt{k}}{2} - \left| t - \frac{k}{2} \right| \geq \frac{k}{2} - \frac{9}{2}\sqrt{k} - \sqrt{k + \frac{2}{k} - \frac{1}{k}} \quad \text{or}$$

$$\left| t + t_{c,d} - t_k \right| \geq 3\sqrt{k} - 5 - \frac{\sqrt{k}}{2} - |\operatorname{Im}(t)| \geq \frac{3}{2}\sqrt{k} - 5 - \sqrt{k + \frac{2}{k} - \frac{1}{k}} \quad \text{or}$$

$$\left| t + t_{c,d} - t_k \right| \geq 2\sqrt{k} - 3 - \frac{\sqrt{k}}{2} - |\operatorname{Im}(t)| \geq \frac{1}{2}\sqrt{k} - 3,$$

if $t \notin \left\{ \gamma + i\beta \,\middle|\, i \in \{0, 1, \ldots, k-1\} \right\}$.

This is greater than $\dfrac{1}{1 - \frac{1}{\sqrt{k}}}$ for $k \geq 126$ in the first case, for $k \geq 149$ in the second case and for $k \geq 69$ in the third case.

Consequently, we have proven that for every $k \geq 149$, we have $\pi(\mathring{D}_{u,a}) \neq \emptyset$ because t_k is not in the closure of $\pi(\mathbb{Z}^A \setminus D_{u,a})$. By the Theorem 8.3.3, we obtain the conclusion.

For $0 \leq k < 149$, we can check by computer, using what is done in the Sect. 8.5, that the interior of $D_{u,a}$ is non-empty, by computing explicitly a regular language describing this interior and checking that this language is non-empty.

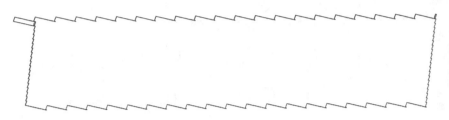

Fig. 8.11 $\pi(D_{u,a}\backslash\mathring{D}_{u,a})$ for s_{20}

When we compute the interior of $D_{u,a}$ for these substitutions s_k, it appears that we get automata of the same shape for k large enough.

Conjecture 8.6.10 For all $k \geq 4$, the minimal automaton of the regular language $\mathring{L} := \left\{ u \in {}^t L_{a,a}^{s_k} \,\middle|\, Q_u \in \mathring{D}_{u,a} \right\}$ has 45 states (Fig. 8.11).

8.6.2 Proof of Pure Discreteness Using Automata

In this subsection, we prove the pure discreteness using completely different technics but still as a corollary of Theorem 8.3.3, for an another infinite family of substitutions:

$$s_{l,k} : \begin{cases} a \mapsto a^l b a^{k-l} \\ b \mapsto c \qquad\qquad \forall\, 0 \leq l \leq k, \\ c \mapsto a \end{cases}$$

using the pure discreteness for the substitution

$$s_k : \begin{cases} a \mapsto a^k b \\ b \mapsto c \qquad\qquad \forall\, k \in \mathbb{N}_{\geq 1}. \\ c \mapsto a \end{cases}$$

This last substitution is a β-substitution, and the associated symbolic system is pure discrete after [4] (but a similar argument as the one of the previous subsection can also be used to prove the pure discreteness for this family). We use this fact to prove the pure discreteness for the other family of substitutions.

The idea is to show that the part corresponding to letter a of the discrete line associated to $s_{l,k}$ contains a homothetic copy of the one for s_k. More precisely, we prove that $M^2 D_{u,a} \subseteq D_{v,a}$, where M is the incidence matrix of $s_{l,k}$ (it doesn't depends on l and k), v is the infinite fixed point of $s_{l,k}$ and u is the infinite fixed point of s_k. Hence, we have $\mathring{D}_{u,a} \neq \emptyset \Longrightarrow \mathring{D}_{v,a} \neq \emptyset$, and we can use the Theorem 8.3.3.

Remark 8.6.11 The symbolic system associated to $s_{l,k}$ is conjugate to the one associated to $s_{k-l,k}$ by word-reversal. Therefore, we can assume without loss of generality that $1 \leq l \leq \left\lceil \dfrac{k}{2} \right\rceil$.

8.6.2.1   Description of $D_{u,a}$ and $D_{v,a}$

In all the following, u is the infinite fixed point of s_k, and v is the infinite fixed point of $s_{l,k}$, for integers $l, k \in \mathbb{N}$, with $k \geq l \geq 1$. We consider the following map

$$\varphi : \begin{cases} \mathbb{Z}^A & \to & \mathbb{Q}(\beta) \\ (v_l)_{l \in A} & \mapsto & v_a + (\beta - k)v_b + (\beta^2 - \beta k)v_c. \end{cases}$$

where $A = \{a, b, c\}$. This linear map is one-to-one and has the property that the multiplication by the incidence matrix M in $\mathbb{R}^A$ becomes a multiplication by β in $\mathbb{Q}(\beta)$, because $(1, \beta - k, \beta^2 - \beta k)$ is a left eigenvector of M for the eigenvalue β. For a language over an alphabet $\Sigma \subseteq \mathbb{Q}(\beta)$, we denote

$$Q_L := \varphi(Q_{\varphi^{-1}(L)}) = \left\{ \sum_{i=0}^{|u|} u_i \beta^i \,\middle|\, u \in L \right\}.$$

Using the Proposition 8.5.5, we have $\varphi(D_{u,a}) = Q_{L_k}$ and $\varphi(D_{v,a}) = Q_{L_{l,k}}$ where L_k is defined in Fig. 8.12 and $L_{l,k}$ is the regular language defined on Fig. 8.13, for β root of the polynomial $X^3 - kX^2 - 1$.

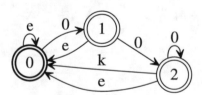

Fig. 8.12   Automaton defining a language L_k such that $\varphi(D_{u,a}) = Q_{L_k}$, where a transition labeled by e means that there are $k - 1$ transitions labeled by $1, 2, \ldots, k - 1$

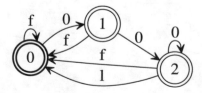

Fig. 8.13   Automaton defining a language $L_{l,k}$ such that $\varphi(D_{v,a}) = Q_{L_{l,k}}$, where a transition labeled by f means that there are $l - 1$ transitions labeled by $1, 2, \ldots, l - 1$ and $k - l$ transitions labeled by $\beta - k + l, \beta - k + l + 1, \ldots, \beta - 2, \beta - 1$

8.6.2.2 Zero-Automaton

In order to show that $M^2 D_{u,a} \subseteq D_{v,a}$, we need a way to go from the language $L_{l,k}$ to the language L_k. The following proposition permits to do it by describing algebraic relations between a word over the alphabet of L_k and a word over the alphabet of $L_{l,k}$. It works for $1 \leq l \leq k - 2$, but by the Remark 8.6.11 we can assume it without loss of generality as soon as $k \geq 4$.

Proposition 8.6.12 *Let L_0 be the language defined in Fig. 8.15. We have*

$$L_0 \subseteq \left\{ u \in (\Sigma_k - \Sigma_{l,k})^* \,\middle|\, \sum_{i=0}^{|u|} u_i \beta^i = 0 \right\},$$

if $1 \leq l \leq k - 2$, where

- $\Sigma_k = \{0, 1, \ldots, k\}$ *is the alphabet of the language L_k,*
- $\Sigma_{l,k} = \{0, 1, \ldots, l, \beta - k + l, \beta - k + l + 1, \ldots, \beta - 1\}$ *is the alphabet of the language $L_{l,k}$, and*

Proof Let L_0' be the language of the automaton depicted in the Fig. 8.14. We verify easily that L_0' is the transposed (i.e. the word reversal) of the language L_0. And we easily check that the transitions of the automaton of Fig. 8.14, satisfy the following.

$$x \xrightarrow{t} y \qquad \Longrightarrow \qquad y = \beta x + t, \quad t \in \Sigma_k - \Sigma_{l,k}.$$

Hence, if we have a word $u_0 u_1 u_2 \ldots u_n \in L_0'$, it corresponds to a path from 0 to 0, so we have

$$0 \xrightarrow{u_0} u_0 \xrightarrow{u_1} \beta u_0 + u_1 \xrightarrow{u_2} \ldots \xrightarrow{u_{n-1}} \sum_{i=0}^{n-1} \beta^{n-1-i} u_i \xrightarrow{u_n} \sum_{i=0}^{n} \beta^{n-i} u_i = 0. \qquad \square$$

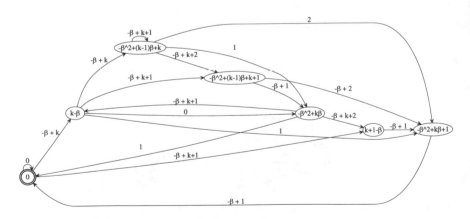

Fig. 8.14 Automaton recognizing the language L_0'

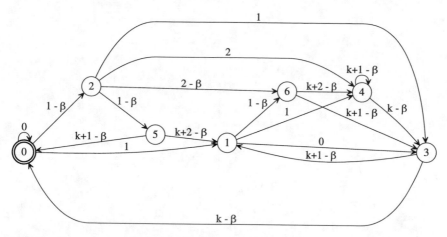

Fig. 8.15 Automaton recognizing the language L_0

8.6.2.3 Proof That $M^2 D_{u,a} \subseteq D_{v,a}$

We define a language L by the automaton $\mathcal{A}$ of Fig. 8.16.

Lemma 8.6.13 *The transitions of the automaton $\mathcal{A}$ of Fig. 8.16 satisfy*

$$X \xrightarrow{b} Y \implies Y \subseteq \left\{ y \in Q_0 \mid \exists x \in X, \ \exists c \in \Sigma_{l,k}, \ x \xrightarrow{b-c} y \in \mathcal{A}_0 \right\}$$

where $Q_0 = \{0, 1, 2, 3, 4, 5, 6\}$ is the set of states of the automaton $\mathcal{A}_0$ of Fig. 8.15.

Proof We can check that using the following array: a star or a letter l means that we have $a \in b - \Sigma_{l,k}$ for a given $(a, b) \in \Sigma_0 \times \Sigma_k$ (the converse is false, but we don't need it), where $\Sigma_0 = \{0, 1, 2, 1 - \beta, 2 - \beta, k - \beta, k + 1 - \beta, k + 2 - \beta\}$ is the alphabet of $\mathcal{A}_0$.

Σ_0 \ Σ_k	0	1	2...l−1	l	l+1	l+2...k−1	k
0	*	*	*	l			
1		*	*	*	l		
2			*	*	*		
1 − β	*						
2 − β	*	*					
k − β				*	*	*	
k + 1 − β					*	*	*
k + 2 − β						*	*

□

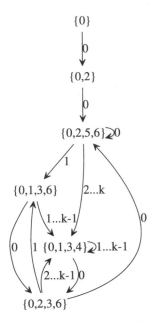

Fig. 8.16 The automaton $\mathcal{A}$, recognizing a language L. $\{0\}$ is the initial state and every state is final

We remark that every state of the automaton $\mathcal{A}$ of Fig. 8.16 contains 0. Hence, if we have a path $\{0\} \overset{b_1}{\to} X_1 \overset{b_2}{\to} \ldots \overset{b_n}{\to} X_n$ in this automaton, we have $0 \in X_n$, and by the Lemma 8.6.13, we can find $(c_i)_{i=1}^n \in \Sigma_{l,k}^n$ such that we have the following path in $\mathcal{A}_0$:

$$0 \overset{b_1-c_1}{\longrightarrow} x_1 \overset{b_2-c_2}{\longrightarrow} \ldots \overset{b_n-c_n}{\longrightarrow} 0.$$

Then, by definition of the automaton $\mathcal{A}_0$, we have

$$Q_b = \sum_{i=1}^n b_i \beta^i = \sum_{i=1}^n c_i \beta^i = Q_c.$$

And we have the following.

Lemma 8.6.14 *The sequence* $(c_i)_{i=1}^n \in \Sigma_{l,k}^*$ *can be chosen such that the word* $c_1 c_2 \ldots c_n$ *is in* $L_{l,k}$.

Proof The language $L_{l,k}$ is the set of words over the alphabet $\Sigma_{l,k}^*$ such that every letter l is preceded by two letters 0. And we can check that in the proof of the Lemma 8.6.13, the only place where we need to take $c_i = l$ is when we follow a transition of $\mathcal{A}$ labeled by l or by $l + 1$. And this occurs only when we follow an transition labeled by 0 or 1 in the automaton $\mathcal{A}_0$.

The only transitions of $\mathcal{A}$ that needs to take $c_i = l$ are $\{0, 2, 5, 6\} \xrightarrow{l} \{0, 1, 3, 4\}$, $\{0, 2, 5, 6\} \xrightarrow{l+1} \{0, 1, 3, 4\}$ and $\{0, 1, 3, 6\} \xrightarrow{l+1} \{0, 1, 3, 4\}$. But we can check that when we reach the state $\{0, 2, 5, 6\}$, we have read at least two zeroes, so we can assume that the 0 of the state $\{0, 2, 5, 6\}$ has been reached by following the path

$$0 \xrightarrow{0-0} 0 \xrightarrow{0-0} 0$$

in $\mathcal{A}_0$. This allows us to consider the transitions $0 \xrightarrow{l-l} 0$ and $0 \xrightarrow{(l+1)-l} 1$ of $\mathcal{A}_0$ and getting a word $c_1 c_2 \ldots c_n$ that stays in the language $L_{l,k}$. In the same way, we reach the state $\{0, 1, 3, 6\}$ after reading a 0 and then a 1, so we can assume that the 1 in the state $\{0, 1, 3, 6\}$ has been reached by following the path

$$0 \xrightarrow{0-0} 0 \xrightarrow{1-0} 1$$

in $\mathcal{A}_0$. This allows us to consider the transition $1 \xrightarrow{(l+1)-l} 4$ of $\mathcal{A}_0$ and getting a word $c_1 c_2 \ldots c_n$ that stays in the language $L_{l,k}$.

We deduce from this lemma and from the equality $Q_h = Q_c$ that we have $Q_b \in Q_{L_{s,k}}$ for every word b in the language L. Hence, we have the inclusion $\beta^2 Q_{L_k} = Q_{0^2 L_k} \subseteq Q_L \subseteq Q_{L_{s,k}}$. So we have $M^2 D_{u,a} \subseteq D_{v,a}$.

By the Theorem 8.3.3, we have for every $1 \leq l \leq k - 2$,

$$s_k \text{ satisfy the Pisot substitution conjecture}$$

$$\Longrightarrow \mathring{D}_{u,a} \neq \emptyset$$

$$\Longrightarrow \mathring{D}_{v,a} \neq \emptyset \quad (\text{because} M^2 D_{u,a} \subseteq D_{v,a})$$

$$\Longrightarrow s_{l,k} \text{ satisfy the Pisot substitution conjecture.}$$

And it implies that $s_{l,k}$ satisfy the Pisot conjecture for every $0 \leq l \leq k, k \geq 4$, up to take the mirror. For $1 \leq k < 4$, there is a finite number of possibilities, and we can check that it also works, for example by computing the interior of the discrete line.

8.7 Pure Discreteness for a $\mathcal{S}$-adic System

Let

$$\sigma : \begin{cases} a \mapsto aab \\ b \mapsto c \\ c \mapsto a \end{cases} \quad \text{and} \quad \tau : \begin{cases} a \mapsto aba \\ b \mapsto c \\ c \mapsto a \end{cases}$$

be two substitutions over the alphabet $A = \{a, b, c\}$ having the same incidence

matrix $M = \begin{pmatrix} 2 & 0 & 1 \\ 1 & 0 & 0 \\ 0 & 1 & 0 \end{pmatrix}$.

Given an infinite word $s_0 s_1 \ldots \in \mathcal{S}^{\mathbb{N}}$, where $\mathcal{S} = \{\sigma, \tau\}$, we define a word $u \in A^{\mathbb{N}}$ by

$$u = \lim_{n \to \infty} s_0 s_1 \ldots s_n(a).$$

Remark that $s_0 s_1 \ldots s_n(a)$ is a strict prefix of $s_0 s_1 \ldots s_n s_{n+1}(a)$, so the limit exists.

We have the following

Theorem 8.7.1 *For every word $s_0 s_1 \ldots \in \mathcal{S}^{\mathbb{N}}$, the subshift $(\overline{\mathcal{S}^{\mathbb{N}} u}, S)$ is measurably isomorphic to a translation on a torus.*

The idea of the proof is similar to the one of the previous subsection: we prove that for every sequence $s_0 s_1 \ldots \in \mathcal{S}^{\mathbb{N}}$ we have an inclusion of the form

$$t + M^k D_{u_\sigma, a} \subseteq D_{u, a},$$

for some $t \in \mathbb{Z}^3$ and $k \in \mathbb{N}$, where u_σ is the infinite fixed point of the substitution σ, and we use the Theorem 8.2.3.

8.7.1 Representation of $D_{u,a}$ by an Automaton

Like for fixed points of substitutions, we can represent $D_{u,a}$ by a finite automaton. For simplicity, we will consider rather $\psi(D_{u,a}) \subseteq \mathbb{Q}(\beta)$, where β is the highest eigenvalue of M, and $\psi : \mathbb{R}^A \to \mathbb{R}$ is the linear map such that $\psi(e_a) = 1$, $\psi(e_b) = \beta - 2$ and $\psi(e_c) = \beta^2 - 2\beta$. This map is such that $\psi(MX) = \beta\psi(X)$ for every $X \in \mathbb{R}^3$.

Proposition 8.7.2 *We have*

$$\psi(D_{u,a}) = \left\{ \sum_{i=0}^{n} u_i \beta^i \,\middle|\, n \in \mathbb{N}, (u_0, s_0)(u_1, s_1) \ldots (u_n, s_n) \in L \right\},$$

$$\psi(D_{u_\sigma, a}) = \left\{ \sum_{i=0}^{n} u_i \beta^i \,\middle|\, n \in \mathbb{N}, (u_0, \sigma)(u_1, \sigma) \ldots (u_n, \sigma) \in L \right\},$$

where L is the regular language recognized by the automaton of Fig. 8.17.

Proof The proof is very similar to the proof of the Proposition 8.5.5. We start by constructing a natural map between the prefixes of the word $s_0 s_1 \ldots s_n(a) \in A^*$ that are followed by a letter d, and the words $v_0 v_1 \ldots v_n \in \Sigma^*$ of length $n + 1$ such that $(v_n, s_n)(v_{n-1}, s_{n-1}) \ldots (v_1, s_1)(v_0, s_0)$ is in a regular language L_d coming from prefix automata, where $\Sigma = \{0, 1, 2, \beta - 1\}$.

Fig. 8.17 Automaton $\mathcal{A}$, recognizing a language L

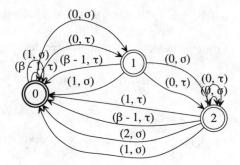

Fig. 8.18 Automaton $\mathcal{A}'$

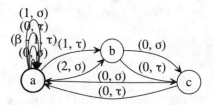

By combining prefix automata of the substitution σ and τ, we define the automaton $\mathcal{A}'$ of the Fig. 8.18. The following lemma give a direct definition of $\mathcal{A}'$.

Lemma 8.7.3 *The automaton $\mathcal{A}'$ of the Fig. 8.18 has set of states A, and has a transition $d \xrightarrow{(x,t)} e$ if and only if $t \in \mathcal{S} = \{\sigma, \tau\}$, and there exists words $v', v'' \in A^*$ such that $t(d) = v'ev''$ with $\psi(\mathrm{Ab}(v')) = x$.*

Proof Easy verification.

Lemma 8.7.4 *For every $n \in \mathbb{N}$, there exists a natural map*

$$\varphi_{s_0 s_1 \ldots s_n} : \bigcup_{d \in A} \left\{ v \in A^* \mid vd \text{ prefix of } s_0 s_1 \ldots s_n(a) \right\} \to \bigcup_{d \in A} p_1(L_d^{n+1}),$$

such that

$$\forall d \in A, \ \varphi_{s_0 s_1 \ldots s_n} \left(\left\{ v \in A^* \mid vd \text{ prefix of } s_0 s_1 \ldots s_n(a) \right\} \right) = p_1(L_d^{n+1})$$

where

$$p_1(L_d^{n+1}) = \left\{ w_n \ldots w_1 w_0 \in \Sigma^* \mid (w_n, s_n)(w_{n-1}, s_{n-1}) \ldots (w_1, s_1)(w_0, s_0) \in L_d \right\},$$

where L_d is the regular language of the automaton $\mathcal{A}'$ of the Fig. 8.18, with initial state a and final state $d \in A$.

Proof By induction on the length of the word $s_0 s_1 \ldots s_n \in \mathcal{S}^*$. The map φ_ϵ is uniquely defined. Let $s_0 s_1 \ldots s_n \in \mathcal{S}^*$ be a word of length at least one, $d \in A$, and $v \in A^*$ such that vd is a prefix of $s_0 s_1 \ldots s_n(a)$. Then, there exists an unique uplet $(v', v'', v''') \in A^*$ and an unique letter $e \in A$ such that $vdv''' = s_0(v'e)$, and hence we have $s_0(e) = v''dv'''$. We define $\varphi_{s_0s_1...s_n}(v) := \varphi_{s_1s_2...s_n}(v')\psi(\mathrm{Ab}(v''))$. From the induction hypothesis, we deduce that $(\varphi_{s_1s_2...s_n}(v'), s_ns_{n-1}...s_2s_1) \in L_e$, and by the Lemma 8.7.3 there exists a transition from state e to state d labeled by $(\psi(\mathrm{Ab}(v'')), s_0)$ in the automaton $\mathcal{A}'$, so we get that

$$(\varphi_{s_0s_1...s_n}(v), s_ns_{n-1}...s_1s_0) \in L_d$$

(we identify $\Sigma^* \times \mathcal{S}^*$ with $(\Sigma \times \mathcal{S})^*$).

Lemma 8.7.5 *For every word $v \in A^*$ and every letter $d \in A$ such that vd is a prefix of $s_0s_1...s_n(a)$, we have*

$$\psi(\mathrm{Ab}(v)) = \sum_{i=0}^{n} w_i \beta^i,$$

where $w_nw_{n-1}...w_1w_0 = \varphi_{s_0s_1...s_n}(v)$.

Proof By construction of $\varphi_{s_0s_1...s_n}(v)$, there exists a sequence of letters $d_0d_1...d_nd_{n+1} \in A^{n+2}$ and two sequences of words $v_0v_1...v_n$ and $w_0w_1...w_n \in (A^*)^{n+1}$ such that $\forall\, 0 \le k \le n$, $s_k(d_{k+1}) = v_kd_kw_k$, with $d_{n+1} = a$ and $d_0 = d$. Then, we have

$$s_0s_1...s_n(a) = (s_0s_1...s_{n-1})(v_n)(s_0s_1...s_{n-2})(v_{n-1})...s_0(v_1)v_0d_0w_0s_0(w_1)...$$

$$...(s_0s_1...s_{n-2})(w_{n-1})(s_0s_1...s_{n-1})(w_n),$$

and

$$v = (s_0s_1...s_{n-1})(v_n)(s_0s_1...s_{n-2})(v_{n-1})...s_0(v_1)v_0.$$

Hence, we get

$$\mathrm{Ab}(v) = (M_0M_1...M_{n-1})\,\mathrm{Ab}(v_n) + (M_0M_1...M_{n-2})\,\mathrm{Ab}(v_{n-1}) + ... + \mathrm{Ab}(v_0),$$

where $M_i = M_{s_i}$ is the matrix of the substitution s_i. But, here we have $M_i = M$ for all $i \in \mathbb{N}$, because $M_\sigma = M_\tau = M$, so we get

$$\psi(\mathrm{Ab}(v)) = \beta^n w_n + \beta^{n-1} w_{n-1} + ... + \beta w_1 + w_0,$$

where $w_i = \psi(\mathrm{Ab}(v_i))$. And we have $\varphi_{s_0s_1...s_n}(v) = w_nw_{n-1}...w_1w_0$ by definition of $\varphi_{s_0s_1...s_n}$.

Lemma 8.7.6 *The language L is the mirror of the language L_a.*

Proof Easy verification. We get the automaton $\mathcal{A}$ from the automaton $\mathcal{A}'$ with initial state a and final state a, by reversing the transitions, and then using an usual algorithm, called power set construction, to compute a deterministic automaton from it. With this construction, the state 0 of $\mathcal{A}$ corresponds to $\{a\}$, the state 1 corresponds to $\{a, b\}$, and the state 2 corresponds to $\{a, b, c\}$.

These lemma give a proof of the Proposition 8.7.2, because we have

$$\psi(D_{u,a}) = \psi(\mathrm{Ab}(\bigcup_{n \in \mathbb{N}} \{v \in A^* \mid va \text{ prefix of } s_0 s_1 \dots s_n(a)\}))$$

$$= \bigcup_{n \in \mathbb{N}} \left\{ \sum_{i=0}^{n} w_i \beta^i \,\middle|\, \begin{array}{l} \exists v \in \Sigma^*, w_0 w_1 \dots w_n = \varphi_{s_0 s_1 \dots s_n}(v), \\ \text{with } va \text{ prefix of } s_0 s_1 \dots s_n(a) \end{array} \right\}$$

$$= \left\{ \sum_{i=0}^{n} w_i \beta^i \,\middle|\, (w_n, s_n)(w_{n-1}, s_{n-1}) \dots (w_1, s_1)(w_0, s_0) \in L_a \right\}$$

$$= \left\{ \sum_{i=0}^{n} w_i \beta^i \,\middle|\, (w_0, s_0)(w_1, s_1) \dots (w_{n-1}, s_{n-1})(w_n, s_n) \in L \right\}.$$

And the second equality of the proposition is a particular case of the first one, where we take for all $i \in \mathbb{N}$, $s_i = \sigma$.

Now that we have a description of the discrete lines $D_{u,a}$ and $D_{u_\sigma,a}$, we use it to show that for every word $s_0 s_1 \dots \in \mathcal{S}^{\mathbb{N}}$ we have an inclusion of the form

$$M^k D_{u_\sigma,a} + t \subseteq D_{u,a},$$

for some $t \in \mathbb{Z}^3$.

8.7.2 Proof of the Inclusion

Let $\Sigma_\sigma = \{0, 1, 2\}$ and $\Sigma_\tau = \{0, 1, \beta - 1\}$. We define a regular language L_* over the alphabet $\Sigma_\sigma \times \mathcal{S}$ by

$$L_* = (\Sigma_\sigma \times \mathcal{S})^* \cap m(L_0 \times L_\sigma \times L),$$

where

$$L_\sigma = \left\{ u_0 u_1 \dots u_n \in \Sigma_\sigma^* \,\middle|\, n \in \mathbb{N}, (u_0, \sigma)(u_1, \sigma) \dots (u_n, \sigma) \in L \right\}$$

$$L_0 = \left\{ u_0 u_1 \dots u_n \in \Sigma'^* \,\middle|\, n \in \mathbb{N}, \sum_{i=0}^{n} u_i \beta^i = 0 \right\}$$

$$\Sigma' = \Sigma_\sigma - \Sigma_\tau = \{-1, 0, 1, 2, 1 - \beta, 2 - \beta, 3 - \beta\}$$

and m is the word morphism defined by

$$
m : \begin{array}{ccc}
\Sigma' \times \Sigma_\sigma \times \Sigma_L & \to & \Sigma_\sigma \times \mathcal{S} \cup \{*\} \\
(t, x, (y, i)) & \mapsto & \begin{cases} (x, i) & \text{if } x - y = t \\ * & \text{otherwise} \end{cases}
\end{array}
$$

where $\Sigma_L = (\Sigma_\sigma \cup \Sigma_\tau) \times \mathcal{S}$ is the alphabet of the language L.

Lemma 8.7.7 *We have*

$$
\psi(D_{u,a}) \supseteq \left\{ \sum_{i=0}^{n} u_i \beta^i \,\middle|\, n \in \mathbb{N}, (u_0, s_0)(u_1, s_1) \ldots (u_n, s_n) \in L_* \right\}.
$$

Proof For all $n \in \mathbb{N}$, we have

$$
(x_0, s_0)(x_1, s_1) \ldots (x_n, s_n) \in L_*
$$

$$
\implies \exists (y_0, s_0)(y_1, s_1) \ldots (y_n, s_n) \in L, \; \sum_{i=0}^{n} (x_i - y_i) \beta^i = 0
$$

$$
\implies \sum_{i=0}^{n} x_i \beta^i \in \psi(D_{u,a}). \qquad \square
$$

Lemma 8.7.8 *We have*

$$
L_* \supseteq L_u,
$$

where L_u is the language defined in Fig. 8.19.

Proof Computation done by computer. The language L_0 is regular thanks to [16], and its minimal automaton has 62 states. The minimal automaton of the language L_* has 210 states.

We deduce from these two lemma that for every sequence $s_0 s_1 \ldots \in \mathcal{S}^{\mathbb{N}}$, we have an inclusion of the form $\psi(D_{u,a}) \supseteq t + \beta^k \psi(D_{u_\sigma,a})$, for some $t \in \mathbb{Z}[\beta]$ and some $k \in \mathbb{N}$. Thus we have

$$
t' + M^k D_{u_\sigma,a} \subseteq D_{u,a}
$$

for some $t' \in \mathbb{Z}^3$.

For example, if $s_0 = \sigma$, $s_1 = \sigma$, $s_2 = \tau$ and $s_3 = \tau$, then we have the inclusion

$$
e_a + M e_a + M^3 e_a + M^4 D_{u_\sigma,a} \subseteq D_{u,a}.
$$

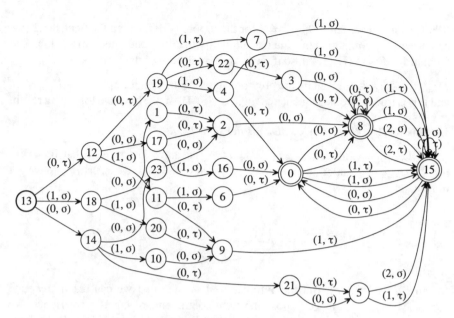

Fig. 8.19 Automaton recognizing a language L_u

8.7.3   Pure Discreteness of the Spectrum

In order to use the Theorem 8.2.3, we need the following property.

Lemma 8.7.9 *The subshift $(\overline{S^{\mathbb{N}}u}, S)$ is minimal for every sequence $s_0 s_1 \ldots \in \mathcal{S}^{\mathbb{N}}$.*

Proof In the word u, there are at most five letters between two consecutive letters a. Indeed, $u = s_0 s_1(v)$ for an infinite word $v \in A^{\mathbb{N}}$, and we have that $s_0 s_1(a)$ is a word of length 7 with four letters a, $s_0 s_1(b) = a$ and $s_0 s_1(c)$ is a word of length 3 with two letters a.

Thus, there exists a constant C such that every factor of u of length $\geq C\beta^n$ contains $s_0 s_1 \ldots s_n(a)$. Indeed, it suffices to see that $u = s_0 s_1 \ldots s_n(w)$ with a word w that satisfies the above property.

If a word is in $\overline{S^{\mathbb{N}}u}$, then it contains arbitrarily large factors of u, so it contains $s_0 s_1 \ldots s_n(a)$ for every $n \in \mathbb{N}$. Therefore this word is dense in $S^{\mathbb{N}}u$.

A projection π along the eigenspace for the eigenvalue β of M onto some plane $\mathcal{P}$ is such that the restriction of π to $\mathbb{Z}^3$ is injective and has a dense image in $\mathcal{P}$. It is not difficult to see that the set $\pi(D_u)$ is bounded, by using for example the description of $\psi(D_u)$ given in the Proposition 8.7.2.

In order to prove the disjointness in measure of the translated copies of $\overline{\pi(D_u)}$ by $\pi(\Gamma_0)$, we use the same strategy than in the proof of the Theorem 8.3.3: we show that the interior of $D_{u,a}$ is non-empty, and we show that the interior of D_u is dense.

It is known that we have $\overset{\circ}{D}_{u_{s_1},a} \neq \emptyset$ for the topology defined on the Sect. 8.3.2 (we can prove it by computing this interior explicitly thanks to the Theorem 8.5.11). And the matrix M is in $GL(3, \mathbb{Z})$, so if we have an inclusion of the form $t + M^k D_{u_{s_1},a} \subseteq D_{u,a}, t \in \mathbb{Z}^3, k \in \mathbb{N}$ given by the previous subsection, then it implies that $\overset{\circ}{D}_{u,a} \neq \emptyset$, and this is true for every sequence $s_0 s_1 \ldots \in \mathcal{S}^{\mathbb{N}}$. Then, we have the following result.

Lemma 8.7.10 $\forall l \in \{a, b, c\}$, $\overset{\circ}{D}_{u,l}$ is dense in $D_{u,l}$.

Proof Let $u_n = \lim_{k \to \infty} s_n s_{n+1} \ldots s_{n+k}(a)$. We have just proven that for every $n \in \mathbb{N}$, $D_{u_n,a}$ has non-empty interior. And we have $u = s_0 s_1 \ldots s_{n-1}(u_n)$, so we get the equality

$$\psi(D_{u,i}) = \bigcup_{i \xrightarrow{(t_0,s_0)} \ldots \xrightarrow{(t_{n-1},s_{n-1})} j \in \mathcal{A}} \beta^n \psi(D_{u_n,j}) + \sum_{k=0}^{n-1} t_k \beta^k$$

for all $i, j \in \{a, b, c\}$. But the automaton $\mathcal{A}$ is such that we can reach any state from any state, even if we impose the right coefficients of labels read. Hence, we can approach (for our topology) any point of $D_{u,i}$ by subsets of $D_{u,i}$ of the form $M^k D_{u_k,a} + t$, $t \in \mathbb{Z}^3$, $k \in \mathbb{N}$. Such subsets have non-empty interior since $M \in GL(3, \mathbb{Z})$. This ends the proof.

Lemma 8.7.11 *The boundary of $\pi(D_u)$ has zero Lebesgue measure.*

In order to prove this lemma, let introduce some notations. For all $n \in \mathbb{N}$, let $u_n = \lim_{k \to \infty} s_n s_{n+1} \ldots s_{n+k}(a)$, and for all $a \in A$, $R_a^n = \overline{\pi(D_{u_n,a})}$. We have the following

Lemma 8.7.12 *For every $a \in A$, the sequence $(\lambda(R_a^n))_{n \in \mathbb{N}}$ is increasing and bounded.*

Proof By the Proposition 8.7.2, we have the following equality

$$\psi(D_{u_n,a}) = \bigcup_{b \xrightarrow{(t,s_{n+1})} u \in \mathcal{A}} \beta \psi(D_{u_{n+1},b}) + t,$$

where $\mathcal{A}$ is the automaton of the Fig. 8.17. Without loss of generality, we can assume that $\pi = \sigma_- \circ \psi : \mathbb{Z}^A \to \mathbb{C}$, where σ_- is the Galois morphism

$$\sigma_- : \begin{matrix} \mathbb{Q}(\beta) \to \mathbb{Q}(\gamma), \\ \beta \mapsto \gamma \end{matrix},$$

where γ is a complex conjugate of β. Thus, we have the equality

$$R_a^n = \bigcup_{b \xrightarrow{(t, s_{n+1})} a \in \mathcal{A}''} \gamma R_b^{n+1} + t,$$

where $\mathcal{A}''$ is the automaton $\mathcal{A}$ where we apply the Galois morphism σ_-.
   Then, we have

$$\lambda(R_a^n) \leq \sum_{b \xrightarrow{(t, s_{n+1})} a \in \mathcal{A}''} \frac{1}{\beta} \lambda(R_b^{n+1}).$$

If we take the vector $X_n = (\lambda(R_a^n))_{a \in A} \in \mathbb{R}^A$, the previous inequality becomes

$$X_n \leq \frac{1}{\beta} M X_{n+1}.$$

But by the Perron-Frobenius theorem, we have the inequality $MX \leq \beta X$ for every $X \in \mathbb{R}_+^A$, so we get that X_n is increasing. The coefficient of X_n are also bounded by $\dfrac{\max_{t \in \Sigma''} |t|}{1 - |\gamma|}$, where Σ'' is the alphabet of the automaton $\mathcal{A}''$.

   This lemma give the existence of the limit $\lambda_a^\infty = \lim_{n \to \infty} \lambda(R_a^n)$. We have the following lemma.

Lemma 8.7.13 *There exists $\epsilon > 0$ and $\eta > 0$ such that for every $n \in \mathbb{N}$ and every $a \in A$, there exists $t \in \mathbb{C}$ and $r > 0$ such that the ball $B(t, r + \epsilon)$ is included in R_a, and such that $\lambda(B(t, r)) \geq \eta \lambda_a^\infty$.*

Proof It is an immediate consequence of the inclusions proven in the Sect. 8.7.2.

Lemma 8.7.14 *There exists $n_0 \in \mathbb{N}$ such that for every $n \geq n_0$ and every $a \in A$, we have $\lambda(\partial R_a^n) = 0$.*

Proof Let $n_0 \in \mathbb{N}$ such that

$$\forall a \in A, \ \lambda_a^\infty \leq (1 + \eta) \lambda(R_a^{n_0}),$$

and let $k \in \mathbb{N}$ such that for every $a \in A$, every $(t, t') \in \mathbb{C}^2$, every $r > 0$, and every $n \in \mathbb{N}$,

$$\gamma^k R_a^n + t \cap B(t', r) \neq \emptyset \implies \gamma^k R_a^n + t \subseteq B(t', r + \epsilon).$$

   Let us show that for every $n \geq n_0$ we have

$$\forall a \in A, \ \lambda(\partial R_a^{n+k}) \leq c \lambda(R_a^{n+k}) \implies \forall a \in A, \ \lambda(\partial R_a^n) \leq c(1 - \eta^2) \lambda(R_a^n).$$

Let $a \in A$ and $n \geq n_0$. Let $t \in \mathbb{C}$ and $r > 0$ such that $B(t, r + \epsilon) \subseteq R_a^n$ and $\lambda(B(t, r)) \geq \eta \lambda_a^\infty$. Let

$$T_b = \left\{ \sum_{j=n}^{n+k-1} \gamma^{n+k-j-1} t_j \,\middle|\, b \xrightarrow{(t_n, s_n)} \cdots \xrightarrow{(t_{n+k-1}, s_{n+k-1})} a \in \mathcal{A}' \right\},$$

$$T_b' = \left\{ \sum_{j=n}^{n+k-1} \gamma^{n+k-j-1} t_j \,\middle|\, \begin{array}{c} b \xrightarrow{(t_n, s_n)} \cdots \xrightarrow{(t_{n+k-1}, s_{n+k-1})} a \in \mathcal{A}' \\ \text{and } (\gamma^k \partial R_b^{n+k} + t) \cap B(t, r) = \emptyset \end{array} \right\}.$$

Then we have

$$\lambda(\partial R_a^n) \leq \lambda \left(\bigcup_{b \in A} \bigcup_{t \in T_b'} (\gamma^k \partial R_b^{n+k} + t) \right)$$

$$\leq \sum_{b \in A,\, t \in T_b'} \frac{1}{\beta^k} \lambda(\partial R_b^{n+k})$$

$$\leq \frac{c}{\beta^k} \sum_{b \in A,\, t \in T_b'} \lambda(R_b^{n+k})$$

$$\leq \frac{c}{\beta^k} \left[\left(\sum_{b \in A,\, t \in T_b} \lambda(R_b^{n+k}) \right) - \beta^k \lambda(B(t, r)) \right]$$

$$\leq \frac{c}{\beta^k} \left(\beta^k \lambda(R_a^{n+k}) - \beta^k \eta \lambda_a^\infty \right)$$

$$\leq c(1 - \eta) \lambda_a^\infty$$

$$\leq c(1 - \eta^2) \lambda(R_a^n).$$

We deduce from these equalities that we have

$$\lambda(\partial R_a^n) \leq (1 - \eta^2)^k \lambda(R_a^n) \xrightarrow[k \to \infty]{} 0. \qquad \square$$

Proof of the Lemma 8.7.11 We have the inclusion

$$\partial R_a \subseteq \bigcup_{b \xrightarrow{(t_0, s_0)} \cdots \xrightarrow{(t_{n-1}, s_{n-1})} a \in \mathcal{A}'} \gamma^n \partial R_b^n + \sum_{j=0}^{n-1} \gamma^{n-1-j} t_j.$$

And by the Lemma 8.7.14, we have $\lambda(\partial R_b^n) = 0$ for $n \geq n_0$ and $b \in A$. Thus the boundary of R_a has zero Lebesgue measure.

Thanks to the Lemma 8.7.10, for every $t \in \Gamma_0 \backslash \{0\}$, the empty intersection $\overset{\circ}{D_u} \cap$ $\overset{\circ}{D_u} + t$ is a dense open subset of $\overline{D_u} \cap \overline{D_u} + t$. Hence, the interior of $\overline{\pi(D_u)}$ and $\pi(D_u + t)$ are disjoint. By the Lemma 8.7.11, it proves that the Lebesgue measure of the intersection is zero.

Every hypothesis of the Theorem 8.2.3 is satisfied, thus the subshift $(\overline{S^{\mathbb{N}}u}, S, \mu)$ is uniquely ergodic and measurably conjugate to the translation on the torus $(\mathcal{P}/\pi(\Gamma_0), T, \lambda)$. This ends the proof of the Theorem 8.7.1.

References

1. S. Akiyama, M. Barge, V. Berthé, J.-Y. Lee, A. Siegel, On the Pisot substitution conjecture. *Mathematics of Aperiodic Order.* Progr. Math., vol. 309 (Birkhäuser/Springer, Basel, 2015), pp. 33–72. MR 3381478

2. P. Arnoux, S. Ito, Pisot substitutions and Rauzy fractals. Bull. Belg. Math. Soc. **8**, 181–207 (2001), http://iml.univ-mrs.fr/~arnoux/ArnouxIto.pdf

3. M. Barge, B. Diamond, Coincidence for substitutions of Pisot type. Bull. Soc. Math. France **130**(4), 619–626 (2002). MR 1947456

4. M. Barge, J. Kwapisz, Geometric theory of unimodular Pisot substitutions. Am. J. Math. **128**(5), 1219–1282 (2006)

5. O. Carton, *Langages formels, Calculabilité et Complexité* (Vuibert, Paris, 2014), https://gaati. org/bisson/tea/lfcc.pdf. ISBN 978-2-311-01400-6

6. A. Clark, L. Sadun, When size matters: subshifts and their related tiling spaces. Ergodic Theory Dyn. Syst. **23**(4), 1043–1057 (2003)

7. Ch. Frougny, E. Pelantová, Beta-representations of 0 and Pisot numbers, JTNB, 2017, https:// arxiv.org/pdf/1512.04234.pdf

8. Ch. Frougny, J. Sakarovitch, Number representation and finite automata, Chapter 2, in *Combinatorics, Automata and Number Theory*, ed. by V. Berthé, M. Rigo. Encyclopedia of Mathematics and its Applications, vol. 135 (Cambridge University Press, Cambridge, 2010), https://www.irif.fr/~cf/publications/cant-ch1.pdf

9. J.E. Hopfcroft, J.D.Ullman, *Introduction to Automata Theory, Languages and Computation* (Addison-Wesley, Boston, 1979)

10. Sh. Ito, H. Rao, Atomic surfaces, tilings and coincidence. I. Irreducible case. Israel J. Math. **153**, 129–155 (2006)

11. B. Khoussainov, A. Nerode, *Automata Theory and its Applications* (Springer, Berlin, 2012), ISBN 978-1-4612-0171-7

12. J.C. Lagarias, Y. Wang, Substitution Delone sets. Discrete Comput. Geom. **29**, 175–209 (2003)

13. J.C. Lagarias, Meyer's concept of quasicrystal and quasiregular sets. Commun. Math. Phys. **179**(2), 365–376 (1996)

14. J.-Y. Lee, Substitution Delone sets with pure point spectrum are inter-model sets. J. Geom. Phys. **57**(11), 2263–2285 (2007)

15. J.-Y. Lee, R.V. Moody, B. Solomyak, Consequences of pure point diffraction spectra for multiset substitution systems. Discrete Comput. Geom. **29**(4), 525–560 (2003)

16. P. Mercat, Semi-groupes fortement automatiques. Bull. Soc. Math. France **141**, fascicule 3, Paris (2013)

17. M. Queffélec, *Substitution Dynamical Systems - Spectral Analysis*. Lecture Notes in Maths (Springer, Cham, 2010), ISBN 978-3-642-11212-6

18. G. Rauzy, Nombres algébriques et substitutions, Bull. Soc. Math. France **110**, 147–178 (1982), http://www.numdam.org/article/BSMF_1982__110__147_0.pdf
19. J. Sakarovitch, *Elements of Automata Theory* (Cambridge University Press, Cambridge, 2009)
20. B. Sing, Pisot substitutions and beyond, PhD Thesis, 2006
21. B. Solomyak, Dynamics of self-similar tilings. Ergodic Theory Dyn. Syst. **17**(3), 695–738 (1997)

Index

LECTURE NOTES IN MATHEMATICS ⟪🐴⟫ Springer

Editors in Chief: J.-M. Morel, B. Teissier;

Editorial Policy

1. Lecture Notes aim to report new developments in all areas of mathematics and their applications – quickly, informally and at a high level. Mathematical texts analysing new developments in modelling and numerical simulation are welcome.

   Manuscripts should be reasonably self-contained and rounded off. Thus they may, and often will, present not only results of the author but also related work by other people. They may be based on specialised lecture courses. Furthermore, the manuscripts should provide sufficient motivation, examples and applications. This clearly distinguishes Lecture Notes from journal articles or technical reports which normally are very concise. Articles intended for a journal but too long to be accepted by most journals, usually do not have this "lecture notes" character. For similar reasons it is unusual for doctoral theses to be accepted for the Lecture Notes series, though habilitation theses may be appropriate.

2. Besides monographs, multi-author manuscripts resulting from SUMMER SCHOOLS or similar INTENSIVE COURSES are welcome, provided their objective was held to present an active mathematical topic to an audience at the beginning or intermediate graduate level (a list of participants should be provided).

   The resulting manuscript should not be just a collection of course notes, but should require advance planning and coordination among the main lecturers. The subject matter should dictate the structure of the book. This structure should be motivated and explained in a scientific introduction, and the notation, references, index and formulation of results should be, if possible, unified by the editors. Each contribution should have an abstract and an introduction referring to the other contributions. In other words, more preparatory work must go into a multi-authored volume than simply assembling a disparate collection of papers, communicated at the event.

3. Manuscripts should be submitted either online at www.editorialmanager.com/lnm to Springer's mathematics editorial in Heidelberg, or electronically to one of the series editors. Authors should be aware that incomplete or insufficiently close-to-final manuscripts almost always result in longer refereeing times and nevertheless unclear referees' recommendations, making further refereeing of a final draft necessary. The strict minimum amount of material that will be considered should include a detailed outline describing the planned contents of each chapter, a bibliography and several sample chapters. Parallel submission of a manuscript to another publisher while under consideration for LNM is not acceptable and can lead to rejection.

4. In general, **monographs** will be sent out to at least 2 external referees for evaluation.

   A final decision to publish can be made only on the basis of the complete manuscript, however a refereeing process leading to a preliminary decision can be based on a pre-final or incomplete manuscript.

   Volume Editors of **multi-author works** are expected to arrange for the refereeing, to the usual scientific standards, of the individual contributions. If the resulting reports can be forwarded to the LNM Editorial Board, this is very helpful. If no reports are forwarded or if other questions remain unclear in respect of homogeneity etc, the series editors may wish to consult external referees for an overall evaluation of the volume.

5. Manuscripts should in general be submitted in English. Final manuscripts should contain at least 100 pages of mathematical text and should always include

   – a table of contents;
   – an informative introduction, with adequate motivation and perhaps some historical remarks: it should be accessible to a reader not intimately familiar with the topic treated;
   – a subject index: as a rule this is genuinely helpful for the reader.
   – For evaluation purposes, manuscripts should be submitted as pdf files.

6. Careful preparation of the manuscripts will help keep production time short besides ensuring satisfactory appearance of the finished book in print and online. After acceptance of the manuscript authors will be asked to prepare the final LaTeX source files (see LaTeX templates online: https://www.springer.com/gb/authors-editors/book-authors-editors/manuscriptpreparation/5636) plus the corresponding pdf- or zipped ps-file. The LaTeX source files are essential for producing the full-text online version of the book, see http://link.springer.com/bookseries/304 for the existing online volumes of LNM). The technical production of a Lecture Notes volume takes approximately 12 weeks. Additional instructions, if necessary, are available on request from lnm@springer.com.

7. Authors receive a total of 30 free copies of their volume and free access to their book on SpringerLink, but no royalties. They are entitled to a discount of 33.3 % on the price of Springer books purchased for their personal use, if ordering directly from Springer.

8. Commitment to publish is made by a *Publishing Agreement*; contributing authors of multiauthor books are requested to sign a *Consent to Publish form*. Springer-Verlag registers the copyright for each volume. Authors are free to reuse material contained in their LNM volumes in later publications: a brief written (or e-mail) request for formal permission is sufficient.

Addresses:
Professor Jean-Michel Morel, CMLA, École Normale Supérieure de Cachan, France
E-mail: moreljeanmichel@gmail.com

Professor Bernard Teissier, Equipe Géométrie et Dynamique,
Institut de Mathématiques de Jussieu – Paris Rive Gauche, Paris, France
E-mail: bernard.teissier@imj-prg.fr

Springer: Ute McCrory, Mathematics, Heidelberg, Germany,
E-mail: lnm@springer.com